SYSTEM RELIABILITY THEORY
Second Edition

WILEY SERIES IN PROBABILITY AND STATISTICS

Established by WALTER A. SHEWHART and SAMUEL S. WILKS

Editors: *David J. Balding, Noel A. C. Cressie, Nicholas I. Fisher,
Iain M. Johnstone, J. B. Kadane, Geert Molenberghs, Louise M. Ryan,
David W. Scott, Adrian F. M. Smith, Jozef L. Teugels*
Editors Emeriti: *Vic Barnett, J. Stuart Hunter, David G. Kendall*

A complete list of the titles in this series appears at the end of this volume.

6.5 Associated Variables, 223
     Problems, 228

7. Counting Processes   231

7.1 Introduction, 231
7.2 Homogeneous Poisson Processes, 240
7.3 Renewal Processes, 246
7.4 Nonhomogeneous Poisson Processes, 277
7.5 Imperfect Repair Processes, 287
7.6 Model Selection, 295
     Problems, 298

8. Markov Processes   301

8.1 Introduction, 301
8.2 Markov Processes, 303
8.3 Asymptotic Solution, 315
8.4 Parallel and Series Structures, 322
8.5 Mean Time to First System Failure, 328
8.6 Systems with Dependent Components, 334
8.7 Standby Systems, 339
8.8 Complex Systems, 346
8.9 Time-Dependent Solution, 351
8.10 Semi-Markov Processes, 353
     Problems, 355

9. Reliability of Maintained Systems   361

9.1 Introduction, 361
9.2 Types of Maintenance, 363
9.3 Downtime and Downtime Distributions, 364
9.4 Availability, 367
9.5 System Availability Assessment, 373
9.6 Preventive Maintenance Policies, 380
9.7 Maintenance Optimization, 400
     Problems, 416

10. Reliability of Safety Systems   419

10.1 Introduction, 419
10.2 Safety Instrumented Systems, 420
10.3 Probability of Failure on Demand, 426
10.4 Safety Unavailability, 436
10.5 Common Cause Failures, 442
10.6 IEC 61508, 446
10.7 The PDS Approach, 452

10.7 The PDS Approach, 452
10.8 Markov Approach, 453
Problems, 459

11. Life Data Analysis   465

11.1 Introduction, 465
11.2 Complete and Censored Data Sets, 466
11.3 Nonparametric Methods, 469
11.4 Parametric Methods, 500
11.5 Model Selection, 515
Problems, 518

12. Accelerated Life Testing   525

12.1 Introduction, 525
12.2 Experimental Designs for ALT, 526
12.3 Parametric Models Used in ALT, 527
12.4 Nonparametric Models Used in ALT, 535
Problems, 537

13. Bayesian Reliability Analysis   539

13.1 Introduction, 539
13.2 Basic Concepts, 541
13.3 Bayesian Point Estimation, 544
13.4 Credibility Interval, 546
13.5 Choice of Prior Distribution, 547
13.6 Bayesian Life Test Sampling Plans, 553
13.7 Interpretation of the Prior Distribution, 555
13.8 The Predictive Density, 557
Problems, 558

14. Reliability Data Sources   561

14.1 Introduction, 561
14.2 Types of Reliability Databases, 562
14.3 Generic Reliability Databases, 564
14.4 Data Analysis and Data Quality, 569

Appendix A. The Gamma and Beta Functions   573
A.1 The Gamma Function, 573
A.2 The Beta Function, 574

Appendix B. Laplace Transforms   577

Appendix C. Kronecker Products   581

SYSTEM RELIABILITY THEORY
Models, Statistical Methods, and Applications

SECOND EDITION

Marvin Rausand
École des Mines de Nantes
Departement Productique et Automatique
Nantes Cedex 3 France

Arnljot Høyland

A JOHN WILEY & SONS, INC., PUBLICATION

Copyright © 2004 by John Wiley & Sons, Inc. All rights reserved.

Published by John Wiley & Sons, Inc., Hoboken, New Jersey.
Published simultaneously in Canada.

No part of this publication may be reproduced, stored in a retrieval system or transmitted in any form or by any means, electronic, mechanical, photocopying, recording, scanning or otherwise, except as permitted under Section 107 or 108 of the 1976 United States Copyright Act, without either the prior written permission of the Publisher, or authorization through payment of the appropriate per-copy fee to the Copyright Clearance Center, Inc., 222 Rosewood Drive, Danvers, MA 01923, (978) 750-8400, fax (978) 646-8600, or on the web at www.copyright.com. Requests to the Publisher for permission should be addressed to the Permissions Department, John Wiley & Sons, Inc., 111 River Street, Hoboken, NJ 07030, (201) 748-6011, fax (201) 748-6008.

Limit of Liability/Disclaimer of Warranty: While the publisher and author have used their best efforts in preparing this book, they make no representation or warranties with respect to the accuracy or completeness of the contents of this book and specifically disclaim any implied warranties of merchantability or fitness for a particular purpose. No warranty may be created or extended by sales representatives or written sales materials. The advice and strategies contained herein may not be suitable for your situation. You should consult with a professional where appropriate. Neither the publisher nor author shall be liable for any loss of profit or any other commercial damages, including but not limited to special, incidental, consequential, or other damages.

For general information on our other products and services please contact our Customer Care Department within the U.S. at 877-762-2974, outside the U.S. at 317-572-3993 or fax 317-572-4002.

Wiley also publishes its books in a variety of electronic formats. Some content that appears in print, however, may not be available in electronic format.

Library of Congress Cataloging-in-Publication Data:

Rausand, Marvin.
   System reliability theory : models, statistical methods, and applications / Marvin Rausand, Arnljot Høyland. — 2nd ed.
     p. cm. — (Wiley series in probability and mathematics. Applied probability and statistics)
   Høyland's name appears first on the earlier edition.
   Includes bibliographical references and index.
   ISBN 0-471-47133-X (acid-free paper)
   1. Reliability (Engineering)—Statistical methods.   I. Høyland, Arnljot, 1924–   II. Title.   III. Series.

TA169.H68 2004
620'.00452—dc22   2003057631

Printed in the United States of America.

10 9 8 7 6 5 4 3

*The second edition is dedicated to the memory of
Professor Arnljot Høyland (1924–2002)*

Contents

Preface to the Second Edition   *xiii*

Preface to the First Edition   *xvii*

Acknowledgments   *xix*

1. Introduction   **1**

     1.1   A Brief History, 1
     1.2   Different Approaches to Reliability Analysis, 2
     1.3   Scope of the Text, 4
     1.4   Basic Concepts, 5
     1.5   Application Areas, 8
     1.6   Models and Uncertainties, 11
     1.7   Standards and Guidelines, 14

2. Failure Models   **15**

     2.1   Introduction, 15
     2.2   State Variable, 16
     2.3   Time to Failure, 16
     2.4   Reliability Function, 17
     2.5   Failure Rate Function, 18
     2.6   Mean Time to Failure, 22
     2.7   Mean Residual Life, 23
     2.8   The Binomial and Geometric Distributions, 25
     2.9   The Exponential Distribution, 26
     2.10   The Homogeneous Poisson Process, 31
     2.11   The Gamma Distribution, 33
     2.12   The Weibull Distribution, 37
     2.13   The Normal Distribution, 41
     2.14   The Lognormal Distribution, 43
     2.15   The Birnbaum-Saunders Distribution, 47
     2.16   The Inverse Gaussian Distribution, 50
     2.17   The Extreme Value Distributions, 54

2.18 Stressor-Dependent Modeling, 58
2.19 Some Families of Distributions, 59
2.20 Summary of Failure Models, 63
Problems, 65

3. Qualitative System Analysis   73

3.1 Introduction, 73
3.2 Systems and Interfaces, 74
3.3 Functional Analysis, 77
3.4 Failures and Failure Classification, 83
3.5 Failure Modes, Effects, and Criticality Analysis, 88
3.6 Fault Tree Analysis, 96
3.7 Cause and Effect Diagrams, 106
3.8 Bayesian Belief Networks, 107
3.9 Event Tree Analysis, 108
3.10 Reliability Block Diagrams, 118
3.11 System Structure Analysis, 125
Problems, 139

4. Systems of Independent Components   147

4.1 Introduction, 147
4.2 System Reliability, 148
4.3 Nonrepairable Systems, 153
4.4 Quantitative Fault Tree Analysis, 160
4.5 Exact System Reliability, 166
4.6 Redundancy, 173
Problems, 178

5. Component Importance   183

5.1 Introduction, 183
5.2 Birnbaum's Measure, 185
5.3 Improvement Potential, 189
5.4 Risk Achievement Worth, 190
5.5 Risk Reduction Worth, 191
5.6 Criticality Importance, 192
5.7 Fussell-Vesely's Measure, 193
5.8 Examples, 197
Problems, 204

6. Dependent Failures   207

6.1 Introduction, 207
6.2 How to Obtain Reliable Systems, 210
6.3 Modeling of Dependent Failures, 214

Appendix E. Maximum Likelihood Estimation	**587**
Appendix F. Statistical Tables	**591**
Acronyms	**595**
Glossary	**599**
References	**605**
Author Index	**625**
Subject Index	**629**

Preface to the Second Edition

The second edition of *System Reliability Theory* is a major upgrade compared to the first edition. Two new chapters have been added, and most of the original chapters have been significantly revised. Most of the text has been rewritten, and all the figures have been redrawn. The new chapters are:

Chapter 9, Reliability of Maintained Systems, where reliability assessment of repairable systems is discussed, together with models and methods for optimization of age-based and condition-based replacement policies. A description of reliability centered maintenance (RCM) and total productive maintenance (TPM) is also given.

Chapter 10, Reliability of Safety Systems, where reliability assessment of periodically tested safety-critical systems is discussed. The terminology from the international standard IEC61508 is used, and an approach to document compliance with this standard is outlined.

New material has been included in all the original chapters, with the greatest number of additions in Chapters 3, 5, and 7. Various approaches to functional modelling and analysis are included in Chapter 3, Qualitative System Analysis, as a basis for failure analysis. Chapter 3 is very fundamental and it may be beneficial to read this chapter before reading Chapter 2.

The second edition has more focus on practical application of reliability theory than the first edition. This is mainly shown by the two new chapters and by the high number of new worked examples that are based on real industry problems and real data.

A glossary of the main terms used in the book has been included at the end of the book, together with a list of acronyms and abbreviations.

The revision of the book is based on experience from using the book in various courses in reliability and life data analysis at the Norwegian University of Science and Technology (NTNU) in Trondheim, continuing education courses arranged for industry both in Norway and abroad. Many instructors who have used the first edition have sent very useful comments and suggestions. Feedback has also been received from people working in industry and consulting companies who have used the book as a reference in practical reliability studies. These comments have led to improvements in the second edition.

Audience and Assumed Knowledge. The book has primarily been written as a textbook for university courses at senior undergraduate and graduate level. The book is also intended as a reference book for practicing engineers in industry and consulting companies, and for engineers who which to do self-study.

The reader should have some knowledge of calculus and of elementary probability theory and statistics. We have tried to avoid heavy mathematical formalism, especially in the first six chapters of the book. Several worked examples are included to illustrate the use of the various methods.

A number of problems are included at the end of almost all the chapters. The problems give the readers a chance to test their knowledge and to verify that they have understood the material. We have tried to arrange the problems such that the easiest problems come first. Some problems are rather complex and cover extensions of the theory presented in the chapter.

Use as a Textbook. The second edition should be applicable as a text for several types of courses, both at senior undergraduate as well as graduate level. Some suggested courses are listed in Table P.1. Each course in Table P.1 is a one-semester course with two to three lectures per week. Several alternatives to these courses may be defined based on the desired focus of the course and the background of the students. It should be possible to cover the whole book in a two-semester course with three to four lectures per week. Examples of detailed course programs at NTNU may be found on the associated web site.

Solutions to the problems are not provided as part of the book. A solutions manual, which contains full worked-out solutions to selected problems is, however, available to instructors and self-learning practicing engineers. A free copy can be obtained by contacting the author (marvin.rausand@ipk.ntnu.no).

Associated web site. The first edition contained detailed references to computer programs for the various methods and approaches. These have been removed from the second edition and included in a web site that is associated to the book (see end

Table P.1. Suggested Courses Based on the Book

Course	Chapters													
	1	2	3	4	5	6	7	8	9	10	11	12	13	14
System reliability theory (undergraduate course)	x	(x)	x	x	(x)			(x)	(x)				(x)	x
System reliability theory (graduate course)	x	x	(x)	x	x	x		x		(x)			x	x
Reliability of safety systems	x	(x)	x	x	(x)	x		(x)		x				x
Reliability and maintenance modelling	x	x	x	x			x	(x)	x	(x)				x
Analysis of life data	x	x	(x)				x				x	x	x	x

(x) means that this chapter may be an option or that only part of the chapter is required.

of Chapter 1). The reason for this is that such references will be outdated rather fast and are easier maintained on a web site. The intention is to keep this web site as up-to-date as possible, including additional information and links to other sites that are potentially useful to instructors, students, and other users of the book.

MARVIN RAUSAND

Preface to the First Edition

The main purpose of this book is to present a comprehensive introduction to system reliability theory. We have structured our presentation such that the book may be used as a text in introductory as well as graduate level courses. For this purpose we treat simple situations first. Then we proceed to more complicated situations requiringadvanced analytical tools.

At the same time the book has been developed as a reference and handbook for industrial statisticians and reliability engineers.

The reader ought to have some knowledge of calculus and of elementary probability theory and statistics.

In the first five chapters we confine ourselves to situations where the state variables of components and systems are binary and independent. Failure models, qualitative system analysis, and reliability importance are discussed. These chapters constitute an elementary, though comprehensive introduction to reliability theory. They may be covered in a one-semester course with three weekly lectures over fourteen weeks.

The remaining part of the book is somewhat more advanced and may serve as a text for a graduate course. In Chapter 6 situations where the components and systems may be in two or more states are discussed. This situation is modeled by Markov processes. Renewal theory is treated in Chapter 7, and dependent failures in Chapter 8. A rather broad introduction to life data analysis is given in Chapter 9, accelerated life testing in Chapter 10, and Bayesian reliability analysis in Chapter 11. The book concludes with information about reliability data sources in Chapter 12.

The book contains a large number of worked examples, and each chapter ends with a selection of problems, providing exercises and additional applications.

A forerunner of this book, written in Norwegian by professor Arne T. Holen and the present authors, appeared in 1983 as an elementary introduction to reliability analysis. It was published by TAPIR and reprinted in 1988. However, we have rewritten all the chapters of the earlier book and added new material as well as several new chapters. The present book contains approximately twice as many pages as its forerunner and can be considered as a completely new book.

We have already tried much of the material in the present book in courses on reliability and risk analysis at the university level in Norway and Sweden, including continuing education courses for engineers working in industry. The feedback from participants in these courses has significantly improved the quality of the book.

We are grateful to Bjarne Stolpnessæter for drawing most of the figures, and to Anne Kajander for typing a first draft of the manuscript. We are further grateful for economic support by Conoco Norway. Permission from various publishers to reproduce tables and figures is also appreciated.

ARNLJOT HØYLAND AND MARVIN RAUSAND

Trondheim, 1993

Acknowledgments

First of all I would like to express my deepest thanks to Professor Arnljot Høyland. Professor Høyland died in December 2002, 78 years old, and could not participate in writing this second edition. I hope that he would have approved and appreciated the changes and additions I have made.

The second edition was written during my sabbatical year at École des Mines de Nantes (EMN) in France. I am very grateful to Pierre Dejax, Philippe Castagliola, and their colleagues at EMN, who helped me in various ways and made my stay a very positive experience. Special thanks go to Bruno Castanier, EMN, who helped me in all possible ways, and also co-authored a section in Chapter 9 of this second edition. Per Hokstad at SINTEF read drafts to several chapters and gave a lot of constructive comments. Also thanks to Bo Lindqvist, Jørn Vatn, and Knut Øien at NTNU, Tørris Digernes at Aker Kværner, Leif T. Sunde at FMC Kongsberg Subsea, Enrico Zio at Politecnico di Milano, and several anonymous referees for many helpful comments.

Special thanks go to my family for putting up with me during the preparation of this edition.

M.R.

1
Introduction

1.1 A BRIEF HISTORY

Reliability, as a human attribute, has been praised for a very long time. For technical systems, however, the reliability concept has not been applied for more than some 60 years. It emerged with a technological meaning just after World War I and was then used in connection with comparing operational safety of one-, two-, and four-engine airplanes. The reliability was measured as the number of accidents per hour of flight time.

At the beginning of the 1930s, Walter Shewhart, Harold F. Dodge, and Harry G. Romig laid down the theoretical basis for utilizing statistical methods in quality control of industrial products. Such methods were, however, not brought into use to any great extent until the beginning of World War II. Products that were composed of a large number of parts often did not function, despite the fact that they were made up of individual high-quality components.

During World War II a group in Germany was working under Wernher von Braun developing the V-1 missile. After the war, it was reported that the first 10 V-1 missiles were all fiascos. In spite of attempts to provide high-quality parts and careful attention to details, all the first missiles either exploded on the launching pad or landed "too soon" (in the English Channel). Robert Lusser, a mathematician, was called in as a consultant. His task was to analyze the missile system, and he quickly derived the *product probability law of series components*. This theorem concerns systems functioning only if all the components are functioning and is valid under special assumptions. It says that the reliability of such a system is equal to the product of the reliabilities of the individual components which make up the system. If the system comprises a large number of components, the system reliability may therefore be rather low, even though the individual components have high reliabilities.

In the United States, attempts were made to compensate a low system reliability by improving the quality of the individual components. Better raw materials and better designs for the products were demanded. A higher system reliability was obtained, but extensive systematic analysis of the problem was probably not carried out at that time.

After World War II, the development continued throughout the world as increasingly more complicated products were produced, composed of an ever-increasing number of components (television sets, electronic computers, etc.). With automation, the need for complicated control and safety systems also became steadily more pressing.

Toward the end of the 1950s and the beginning of the 1960s, interest in the United States was concentrated on intercontinental ballistic missiles and space research, especially connected to the Mercury and Gemini programs. In the race with the Russians to be the first nation to put men on the moon, it was very important that the launching of a manned spacecraft be a success. An association for engineers working with reliability questions was soon established. The first journal on the subject, IEEE Transactions on reliability came out in 1963, and a number of textbooks on the subject were published in the 1960s.

In the 1970s interest increased, in the United States as well as in other parts of the world, in risk and safety aspects connected to the building and operation of nuclear power plants. In the United States, a large research commission, led by Professor Norman Rasmussen was set up to analyze the problem. The multimillion dollar project resulted in the so-called Rasmussen report, WASH-1400 (NUREG-75/014). Despite its weaknesses, this report represents the first serious safety analysis of so complicated a system as a nuclear power plant.

Similar work has also been carried out in Europe and Asia. In the majority of industries a lot of effort is presently put on the analysis of risk and reliability problems. The same is true in Norway, particularly within the offshore oil industry. The offshore oil and gas development in the North Sea is presently progressing into deeper and more hostile waters, and an increasing number of remotely operated subsea production systems are put into operation. The importance of the reliability of subsea systems is in many respects parallel to the reliability of spacecrafts. A low reliability cannot be compensated by extensive maintenance.

A more detailed history of reliability technology is presented, for example, by Knight (1991), and Villemeur (1988).

1.2 DIFFERENT APPROACHES TO RELIABILITY ANALYSIS

We can distinguish between three main branches of reliability:

- Hardware reliability
- Software reliability

Fig. 1.1 Load and the strength distributions.

- Human reliability

The present textbook is concerned with the first of these branches: the reliability of technical components and systems. Many technical systems will also involve software and humans in many different roles, like designers, operators, and maintenance personnel. The interactions between the technical system, software, and humans are very important, but not a focused topic in this book. Within hardware reliability we may use two different approaches:

- The physical approach

- The actuarial approach

In the *physical approach* the strength of a technical item is modeled as a random variable S. The item is exposed to a load L that is also modeled as a random variable. The distributions of the strength and the load at a specific time t are illustrated in Fig. 1.1. A failure will occur as soon as the load is higher than the strength. The reliability R of the item is defined as the probability that the strength is greater than the load,

$$R = \Pr(S > L)$$

where $\Pr(A)$ denotes the probability of event A.

The load will usually vary with time and may be modeled as a time-dependent variable $L(t)$. The item will deteriorate with time, due to failure mechanisms like corrosion, erosion, and fatigue. The strength of the item will therefore also be a function of time, $S(t)$. A possible realization of $S(t)$ and $L(t)$ is illustrated in Fig. 1.2. The time to failure T of the item is the (shortest) time until $S(t) < L(t)$,

$$T = \min\{t;\, S(t) < L(t)\}$$

and the reliability $R(t)$ of the item may be defined as

$$R(t) = \Pr(T > t)$$

The physical approach is mainly used for reliability analyses of structural elements, like beams and bridges. The approach is therefore often called *structural reliability*

4 INTRODUCTION

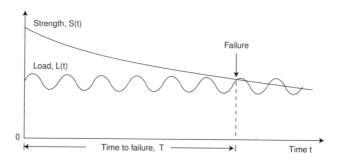

Fig. 1.2 Possible realization of the load and the strength of an item.

analysis (Melchers 1999). A structural element, like a leg on an offshore platform, may be exposed to loads from waves, current, and wind. The loads may come from different directions, and the load must therefore be modeled as a vector $\mathbf{L}(t)$. In the same way, the strength will also depend on the direction and has to be modeled as a vector $\mathbf{S}(t)$. The models and the analysis may therefore become rather complex.

In the *actuarial approach*, we describe all our information about the operating loads and the strength of the component in the probability distribution function $F(t)$ of the time to failure T. No explicit modeling of the loads and the strength is carried out. Reliability characteristics like *failure rate* and *mean time to failure* are deduced directly from the probability distribution function $F(t)$. Various approaches can be used to model the reliability of systems of several components and to include maintenance and replacement of components. When several components are combined into a system, the analysis is called a *system reliability analysis*.

1.3 SCOPE OF THE TEXT

This book provides a thorough introduction to component and system reliability analysis by the actuarial approach. When we talk about reliability and reliability studies, it is tacitly understood that we follow the actuarial approach.

The main objectives of the book are:

1. To present and discuss the terminology and the main models used in reliability studies.

2. To present the analytical methods that are fundamental within reliability engineering and analysis of reliability data.

The methods described in the book are applicable during any phase of a system's lifetime. They have, however, their greatest value during the design phase. During this phase reliability engineering can have the greatest effect for enhancing the system's safety, quality, and operational availability.

Some of the methods described in the book may also be applied during the operational phase of the system. During this phase, the methods will aid in the evaluation of the system and in improving the maintenance and the operating procedures.

The book does not specifically deal with how to build a reliable system. The main topics of the book are connected to how to evaluate, measure, and predict the reliability of a system.

1.4 BASIC CONCEPTS

The main concept of this book is *reliability*. During the preceding sections the concept of reliability has been used without a precise definition. It is, however, very important that all main concepts are defined in an unambiguous way. We fully agree with Kaplan (1990) who states: "When the words are used sloppily, concepts become fuzzy, thinking is muddled, communication is ambiguous, and decisions and actions are suboptimal, to say the least."

A precise definition of reliability and some associated concepts like quality, availability, safety, security, and dependability are given below. All of these concepts are more or less interconnected, and there is a considerable controversy concerning which is the broadest and most general concept. Further concepts are defined in the Glossary at the end of the book.

Until the 1960s reliability was defined as "the probability that an item will perform a required function under stated conditions for a stated period of time." Some authors still prefer this definition, for example, Smith (1997) and Lakner and Anderson (1985). We will, however, in this book use the more general definition of reliability given in standards like ISO 8402 and British Standard BS 4778:

Reliability
The ability of an item to perform a required function, under given environmental and operational conditions and for a stated period of time (ISO 8402).

- The term "item" is used here to denote any component, subsystem, or system that can be considered as an entity.

- A required function may be a single function or a combination of functions that is necessary to provide a specified service.

- All technical items (components, subsystems, systems) are designed to perform one or more (required) functions. Some of these functions are active and some functions are passive. Containment of fluid in a pipeline is an example of a passive function. Complex systems (e.g., an automobile) usually have a wide range of required functions. To assess the reliability (e.g., of an automobile), we must first specify the required function(s) we are considering.

- For a hardware item to be reliable, it must do more than meet an initial factory performance or quality specification—it must operate satisfactorily for a specified period of time in the actual application for which it is intended.

Remark: The North American Electric Reliability Council (NERC) has introduced a more comprehensive definition of the reliability of an electric system. NERC defines the reliability of an electric systems in terms of two basic functional aspects:

1. *Adequacy*. The ability of the electric system to supply the aggregate electrical demand and energy requirements of customers at all times, taking into account scheduled and reasonably expected unscheduled outages of system elements.

2. *Security*. The ability of the electric system to withstand sudden disturbances such as electric short circuits or unanticipated loss of system elements. □

Quality
The totality of features and characteristics of a product or service that bear on its ability to satisfy stated or implied needs (ISO 8402).

- Quality is also sometimes defined as conformance to specifications (e.g., see Smith 1997).

- The quality of a product is characterized not only by its conformity to specifications at the time it is supplied to the user, but also by its ability to meet these specifications over its entire lifetime.

However, according to common usage, quality denotes the conformity of the product to its specification as manufactured, while reliability denotes its ability to continue to comply with its specification over its useful life. *Reliability is therefore an extension of quality into the time domain.*

Remark: In common language we often talk about the *reliability and quality* of a product. Some automobile journals publish regular surveys of reliability and quality problems of the various cars. Under reliability problems they list problems related to the essential functions of the car. A reliability problem is present when the car cannot be used for transport. Quality problems are secondary problems that may be considered a nuisance. □

Availability
The ability of an item (under combined aspects of its reliability, maintainability and maintenance support) to perform its required function at a stated instant of time or over a stated period of time (BS 4778).

- We may distinguish between the availability $A(t)$ at time t and the average availability A_{av}. The availability at time t is

$$A(t) = \Pr(\text{item is functioning at time } t)$$

The term "functioning" means here that the item is either in active operation or that it is able to operate if required.

The average availability A_{av} denotes the mean proportion of time the item is functioning. If we have an item that is repaired to an "as good as new" condition every time it fails, the average availability is

$$A_{av} = \frac{\text{MTTF}}{\text{MTTF} + \text{MTTR}} \qquad (1.1)$$

where MTTF (mean time to failure) denotes the mean functioning time of the item, and MTTR (mean time to repair) denotes the mean downtime after a failure. Sometimes MDT (mean downtime) is used instead of MTTR to make it clear that it is the total mean downtime that should be used in (1.1) and not only the mean active repair time.

- When considering a production system, the average availability of the production (i.e., the mean proportion of time the system is producing) is sometimes called the *production regularity*.

Maintainability
The ability of an item, under stated conditions of use, to be retained in, or restored to, a state in which it can perform its required functions, when maintenance is performed under stated conditions and using prescribed procedures and resources (BS 4778).

- "Maintainability" is a main factor determining the availability of the item.
- RAM is often used as an acronym for reliability, availability, and maintainability. We also use the notions RAM studies and RAM engineering.

Safety
Freedom from those conditions that can cause death, injury, occupational illness, or damage to or loss of equipment or property (MIL-STD-882D).

- This definition has caused considerable controversy. A number of alternative definitions have therefore been proposed. The main controversy is connected to the term "freedom from." Most activities involve some sort of risk and are never totally *free* from risk. In most of the alternative definitions safety is defined as an *acceptable level of risk*.
- The concept *safety* is mainly used related to random hazards, while the concept *security* is used related to deliberate actions.

Security
Dependability with respect to prevention of deliberate hostile actions.

- Security is often used in relation to information and computer systems. In this context, security may be defined as "dependability with respect to prevention of unauthorized access to and/or handling of information" (Laprie 1992).
- The security of critical infrastructures is thoroughly discussed in CCIP (1997)

Dependability
The collective term used to describe the availability performance and its influencing factors: reliability performance, maintainability performance and maintenance support performance (IEC 60300).

8 INTRODUCTION

- A slightly different definition is given by Laprie (1992). He defines dependability to be: "Trustworthiness of a system such that reliance can justifiably be placed on the service it delivers." In comments to this definition, Laprie (1992) claims that dependability is a global concept which subsumes the attributes of reliability, availability, safety, and security. This is also in accordance with the definition used by Villemeur (1988).

- If safety and security are included in the definition of dependability as influencing factors, dependability will be identical to the RAMS concept (RAMS is an acronym for reliability, availability, maintainability, and safety).

- According to Laprie (1992) the definition of dependability is synonymous to the definition of reliability. Some authors, however, prefer to use the concept of dependability instead of reliability. This is also reflected in the important series of standards IEC 60300 "Dependability Management."

In this book we will use reliability as a global, or general, concept with the same main attributes as listed under the definition of dependability.

The reliability may be measured in different ways depending on the particular situation, for example as:

1. Mean time to failure (MTTF)

2. Number of failures per time unit (*failure rate*)

3. The probability that the item does not fail in a time interval $(0, t]$ (*survival probability*)

4. The probability that the item is able to function at time t (*availability at time t*)

If the item is not repaired after failure, 3 and 4 coincide. All these measures are given a mathematically precise definition in Chapter 2 with concepts from probability theory.

1.5 APPLICATION AREAS

The main objective of a reliability study should always be to provide information as a basis for decisions. Before a reliability study is initiated, the decision maker should clarify the decision problem, and then the objectives and the boundary conditions and limitations for the study should be specified such that the relevant information needed as input to the decision is at hand, in the right format, and on time.

Reliability technology has a potentially wide range of application areas. Some of these areas are listed below to illustrate the wide scope of application of reliability technology.

1. *Risk analysis.* The main steps of a quantitative risk analysis (QRA) are, as illustrated in Fig. 1.3:

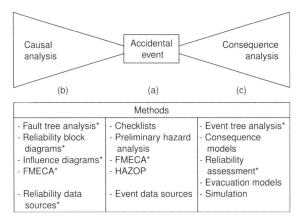

Fig. 1.3 Main steps of a risk analysis, with main methods.

(a) Identification and description of potential *accidental events* in the system. An accidental event is usually defined as a significant deviation from normal operating conditions that may lead to unwanted consequences. In an oil/gas processing plant a gas leak may, for example, be defined as an accidental event.

(b) The potential causes of each accidental event are identified by a *causal analysis*. The causes are usually identified in a hierarchical structure starting with the main causes. Main causes, and subcauses may be described by a tree structure called a *fault tree*. If probability estimates are available, these may be input into the fault tree, and the probability/frequency of the accidental event may be calculated.

(c) Most well-designed systems include various barriers and safety functions that have been installed to stop the development of accidental events or to reduce the consequences of accidental events. In the gas leak example in step (a), the barriers and safety functions may comprise gas detection systems, emergency shutdown systems, fire-fighting systems, fire containment systems (e.g., fire walls), and evacuation systems/procedures. The final consequences of an accidental event will depend on whether or not these systems are functioning adequately. The *consequence analysis* is usually carried out by an *event tree analysis*. The event tree analysis is often supplemented by calculations of fire and explosion loads, simulations of the escalation of fires, reliability assessment of emergency shutdown systems, and so on. Specific methods may be required to analyze consequences to:

- Humans
- The environment
- Material assets
- Production regularity (if relevant)

The methods that are most commonly used during the three steps of a risk analysis are listed beneath the relevant step in Fig. 1.3. Methods that are described in this book are marked with (*). Reliability analysis is a main part of any QRA, and several methods are common for risk and reliability analyses.

2. *Environmental protection.* Reliability studies may be used to improve the design and operational regularity of antipollution systems like gas/water cleaning systems.

Many industries have realized that the majority of the pollution from their plants is caused by production irregularities and that consequently the production regularity of the plant is the most important factor in order to reduce pollution. Reliability and regularity studies are among the most important tools to optimize production regularity.

An environmental risk analysis is carried out according to the same procedure as a standard risk analysis and has the same interfaces with reliability analysis.

3. *Quality.* Quality management and assurance is increasingly focused, stimulated by the almost compulsory application of the ISO 9000 series of standards.

The concepts of quality and reliability are closely connected. Reliability may in some respects be considered to be a quality characteristic (perhaps the most important characteristic). Complementary systems are therefore being developed and implemented for reliability management and assurance as part of a total quality management (TQM) system. Note the relation between the ISO 9000 and the IEC 60300 series of standards as discussed by Strandberg (1992).

4. *Optimization of maintenance and operation.* Maintenance is carried out to prevent system failures and to restore the system function when a failure has occurred. The prime objective of maintenance is thus to maintain or improve the system reliability and production/operation regularity.

Many industries (e.g., the nuclear power, aviation, defense, and the offshore and shipping industry) have fully realized the important connection between maintenance and reliability and have implemented the reliability centered maintenance (RCM) approach. The RCM approach is a main tool to improve the cost-effectiveness and control of maintenance in all types of industries, and hence to improve availability and safety. Reliability assessment is also an important element of the following applications: life cycle cost (LCC), life cycle profit (LCP), logistic support, spare part allocation, and manning level analysis.

5. *Engineering design.* Reliability is considered to be one of the most important quality characteristics of technical products. Reliability assurance should therefore be an important topic during the engineering design process.

Many industries have realized this and integrated a reliability program in the design process. This is especially the case within the nuclear power, the aviation, the aerospace, the automobile, and the offshore industries. Such integration may be accomplished through concepts like concurrent engineering (Kusiak

1993) and design for X (Huang 1996) that focus on the total product perspective from inception through product delivery.

6. *Verification of quality/reliability.* A number of official bodies require that the producer and/or the user of technical systems are able to verify that their equipment satisfies specified requirements. Such requirements usually have a basis in safety and/or environmental protection. Some industries also meet strict requirements with respect to production regularity. This is especially the case within the power generation and petroleum industries.

As part of the formation of the European Union (EU) a number of new EU directives have been issued. Among these are the machinery safety directive, the product safety directive, and the product liability directive. The producers of equipment must, according to these directives, verify that their equipment comply with the requirements. Reliability analyses and reliability demonstration testing are necessary tools in the verification process.

During the last few years it has become more and more common that buyers of technical equipment require a quantitative assessment of the quality and reliability as part of the total system documentation. The documentation required varies a lot, from filled-in failure modes, effects, and criticality analysis (FMECA) forms to detailed results from life testing of the equipment (e.g., see DNV-RP-A203). Documented quality/reliability has been required by some industries for many years (e.g., aircraft, aerospace, automobile, nuclear, defense).

1.6 MODELS AND UNCERTAINTIES

In practical situations the analyst will have to derive (stochastic) models of the system at hand, or at least have to choose from several possible models before an analysis can be performed. To be "realistic" the model must describe the essential features of the system, but do not necessarily have to be exact in all details. One of the pioneers in mathematical statistics, Jerzy Neyman (1945), expresses this in the following way:

> Every attempt to use mathematics to study some real phenomena must begin with building a mathematical model of these phenomena. Of necessity, the model simplifies the matters to a greater or lesser extent and a number of details are ignored. The success depends on whether or not the details ignored are really unimportant in the development of the phenomena studied. The solution of the mathematical problem may be correct and you may be in violent conflict with realities, simply because the original assumptions of the mathematical model diverge essentially from the conditions of the practical problem considered. Beforehand, it is impossible to predict with certainty whether or not a given mathematical model is adequate. To find this out, it is necessary to deduce a number of consequences of the model and to compare them with observation.

Another pioneer in statistics, George E. P. Box, repeatedly points out that "no model is absolutely correct. In particular situations, however, some models are more useful than others."

12   INTRODUCTION

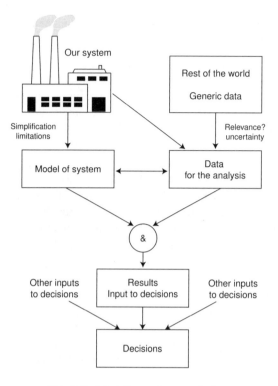

Fig. 1.4 Modeling and uncertainties.

In reliability and safety studies of technical systems, one will always have to work with models of the systems. These models may be graphical (networks of different types) or mathematical. A mathematical model is necessary in order to be able to bring in data and use mathematical and statistical methods to estimate reliability, safety, or risk parameters. For such models, two conflicting interests always apply:

- The model should be sufficiently simple to be handled by available mathematical and statistical methods.

- The model should be sufficiently "realistic" such that the deducted results are of practical relevance.

We should, however, always bear in mind that we are working with an idealized, simplified model of the system. Furthermore, the results we derive are, strictly speaking, valid only for the model, and are accordingly only "correct" to the extent that the model is realistic.

The modeling situation is illustrated in Fig. 1.4. Before we start developing a model, we should clearly understand what type of decision the results from our analysis should provide input to, and also the required format of the input to the decision. This was also mentioned on page 8. To estimate the system reliability from a model, we need input data. The data will usually come from generic data sources, as discussed in Chapter 14. The generic data may not be fully relevant for our system and may have to be adjusted by expert judgment. This is especially the case when we are introducing new technology. Some data may also come from the specific system. When establishing the system model, we have to consider the type, amount, and quality of the available input data. It has limited value to establish a very detailed model of the system if we cannot find the required input data.

Now and then there is the contention that use of "probabilistic reliability analysis" has only bounded validity and is of little practical use. When something goes wrong, it is usually attributed to human error. That is to say, someone has failed to do what he should have done in a certain situation, or has done something that should *not* have been done. In principle, however, there is nothing to prevent key persons from counting as "components" of the system in the same way as the technical components do. It would obviously be difficult to derive numerical estimates for the probability of human errors in a given situation, but that is another issue.

From what has been said, we understand that many subject areas are involved in a reliability analysis of technical systems.

- Detailed knowledge is needed of the technical aspects of the system and of the physical mechanisms that may lead to failure.

- Knowledge of mathematical/statistical concepts and statistical methods is a necessary (but far from sufficient) condition to be able to carry out such analyses.

- If humans are treated as components in the system, medical, psychological, and sociological insight into their behavior patterns is needed, and last but not least, knowledge of how humans react under stress.

- Data must be available for estimation of parameters and checking of models.

- Analysis of complicated systems must be accompanied by appropriate computer programs.

The above list is not complete but illustrates that a reliability analysis requires many different areas of knowledge and has to be a multidiscipline task.

Boundary Conditions for the Analysis A reliability analysis of a system will always be based on a wide range of assumptions and boundary conditions. Here we briefly mention a few such considerations:

- Precisely which parts of the system are going to be included in the analysis and which parts are not?

- Precisely what are the objectives of the analysis? Different objectives may necessitate different approaches.

- What system interfaces will be used? Operator and software interfaces have to be identified and defined.

14   INTRODUCTION

- What level of detail is required?

- Which operational phases are to be included in the analysis (e.g., start-up, steady state, maintenance, disposal)?

- What are the environmental conditions for the system?

- Which external stresses should be considered (e.g., sabotage, earthquakes, lightning strikes)?

1.7 STANDARDS AND GUIDELINES

A wide range of standards and guidelines containing requirements with respect to reliability and safety have been issued. Any reliability engineer should be familiar with the standards and guidelines that are applicable within his or her subject areas. A survey of relevant standards and guidelines may be found on the the book's web page.

Contents of associated web page	
- Supplementary notes - Overhead presentations - Additional problems - Control questions - Misprints - Laws and regulations - Standards - Guidelines - Reliability data sources - Computer programs	- Scientific journals - Other books - Conferences Links to: - Universities offering educational programs - Organizations - Consulting companies - Other resources

The web page may be found by following the link from the book's presentation page at www.wiley.com, or by sending an email to marvin.rausand@ipk.ntnu.no.

2
Failure Models

2.1 INTRODUCTION

We will now introduce several quantitative measures for the reliability of a *nonrepairable* item. This item can be anything from a small component to a large system. When we classify an item as nonrepairable, we are only interested in studying the item until the first failure occurs. In some cases the item may be literally nonrepairable, meaning that it will be discarded by the first failure. In other cases, the item may be repaired, but we are not interested in what is happening with the item after the first failure.

First we will introduce four important measures for the reliability of a nonrepairable item. These are:

- The reliability (survivor) function $R(t)$
- The failure rate function $z(t)$
- The mean time to failure (MTTF)
- The mean residual life (MRL)

Thereafter, we introduce a number of probability distributions that may be used to model the lifetime of a nonrepairable item. The following life distributions are discussed:

- The exponential distribution
- The gamma distribution

16 FAILURE MODELS

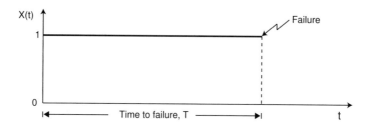

Fig. 2.1 The state variable and the time to failure of an item.

- The Weibull distribution
- The normal distribution
- The lognormal distribution
- The Birnbaum-Saunders distribution
- The inverse Gaussian distribution

We also introduce three discrete distributions: the binomial, the geometric, and the Poisson distributions. Finally, we discuss some extreme value distributions and how we can model the lifetime of an item as a function of various stress levels (stressors). The chapter is concluded by a survey of some broader classes of life distributions.

2.2 STATE VARIABLE

The state of the item at time t may be described by the state variable $X(t)$:

$$X(t) = \begin{cases} 1 & \text{if the item is functioning at time } t \\ 0 & \text{if the item is in a failed state at time } t \end{cases}$$

The state variable of a nonrepairable item is illustrated in Fig. 2.1 and will generally be a random variable.

2.3 TIME TO FAILURE

By the *time to failure* of an item we mean the time elapsing from when the item is put into operation until it fails for the first time. We set $t = 0$ as the starting point. At least to some extent the time to failure is subject to chance variations. It is therefore natural to interpret the time to failure as a random variable, T. The connection between the state variable $X(t)$ and the time to failure T is illustrated in Fig. 2.1.

Note that the time to failure T is not always measured in calendar time. It may also be measured by more indirect time concepts, such as:

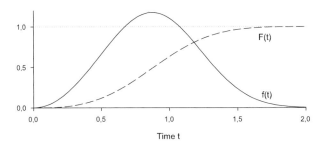

Fig. 2.2 Distribution function $F(t)$ and probability density function $f(t)$.

- Number of times a switch is operated
- Number of kilometers driven by a car
- Number of rotations of a bearing
- Number of cycles for a periodically working item

From these examples, we notice that time to failure may often be a discrete variable. A discrete variable can, however, be approximated by a continuous variable. Here, unless stated otherwise, we will assume that the time to failure T is continuously distributed with probability density function $f(t)$ and distribution function

$$F(t) = \Pr(T \leq t) = \int_0^t f(u)\,du \quad \text{for } t > 0 \tag{2.1}$$

$F(t)$ thus denotes the probability that the item fails within the time interval $(0, t]$.

The probability density function $f(t)$ is defined as

$$f(t) = \frac{d}{dt} F(t) = \lim_{\Delta t \to 0} \frac{F(t + \Delta t) - F(t)}{\Delta t} = \lim_{\Delta t \to 0} \frac{\Pr(t < T \leq t + \Delta t)}{\Delta t}$$

This implies that when Δt is small,

$$\Pr(t < T \leq t + \Delta t) \approx f(t) \cdot \Delta t$$

The distribution function $F(t)$ and the probability density function $f(t)$ are illustrated is Fig. 2.2.

2.4 RELIABILITY FUNCTION

The reliability function of an item is defined by

$$R(t) = 1 - F(t) = \Pr(T > t) \quad \text{for } t > 0 \tag{2.2}$$

18 FAILURE MODELS

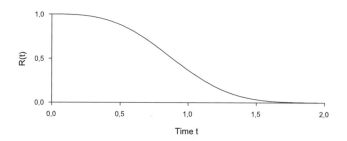

Fig. 2.3 The reliability (survivor) function $R(t)$.

or equivalently

$$R(t) = 1 - \int_0^t f(u)\,du = \int_t^\infty f(u)\,du \qquad (2.3)$$

Hence $R(t)$ is the probability that the item does not fail in the time interval $(0, t]$, or, in other words, the probability that the item survives the time interval $(0, t]$ and is still functioning at time t. The reliability function $R(t)$ is also called the *survivor function* and is illustrated in Fig. 2.3.

2.5 FAILURE RATE FUNCTION

The probability that an item will fail in the time interval $(t, t + \Delta t]$ when we know that the item is functioning at time t is

$$\Pr(t < T \le t + \Delta t \mid T > t) = \frac{\Pr(t < T \le t + \Delta t)}{\Pr(T > t)} = \frac{F(t + \Delta t) - F(t)}{R(t)}$$

By dividing this probability by the length of the time interval, Δt, and letting $\Delta t \to 0$, we get the *failure rate function* $z(t)$ of the item

$$\begin{aligned} z(t) &= \lim_{\Delta t \to 0} \frac{\Pr(t < T \le t + \Delta t \mid T > t)}{\Delta t} \\ &= \lim_{\Delta t \to 0} \frac{F(t + \Delta t) - F(t)}{\Delta t} \frac{1}{R(t)} = \frac{f(t)}{R(t)} \end{aligned} \qquad (2.4)$$

This implies that when Δt is small,

$$\Pr(t < T \le t + \Delta t \mid T > t) \approx z(t) \cdot \Delta t$$

Remark: Note the similarity and the difference between the probability density function $f(t)$ and the failure rate function $z(t)$.

$$\Pr(t < T \le t + \Delta t) \approx f(t) \cdot \Delta t \qquad (2.5)$$
$$\Pr(t < T \le t + \Delta t \mid T > t) \approx z(t) \cdot \Delta t \qquad (2.6)$$

Say that we start out with a new item at time $t = 0$ and at time $t = 0$ ask: "What is the probability that this item will fail in the interval $(t, t + \Delta t]$?" According to (2.5) this probability is approximately equal to the probability density function $f(t)$ at time t multiplied by the length of the interval Δt. Next consider an item that has survived until time t, and ask: "What is the probability that this item will fail in the next interval $(t, t + \Delta t]$?" This (conditional) probability is according to (2.6) approximately equal to the failure rate function $z(t)$ at time t multiplied by the length of the interval, Δt. □

If we put a large number of identical items into operation at time $t = 0$, then $z(t) \cdot \Delta t$ will roughly represent the relative proportion of the items still functioning at time t, failing in $(t, t + \Delta t]$. Since

$$f(t) = \frac{d}{dt} F(t) = \frac{d}{dt}(1 - R(t)) = -R'(t)$$

then

$$z(t) = -\frac{R'(t)}{R(t)} = -\frac{d}{dt} \ln R(t) \qquad (2.7)$$

Since $R(0) = 1$, then

$$\int_0^t z(t)\, dt = -\ln R(t) \qquad (2.8)$$

and

$$R(t) = \exp\left(-\int_0^t z(u)\, du\right) \qquad (2.9)$$

The reliability (survivor) function $R(t)$ and the distribution function $F(t) = 1 - R(t)$ are therefore uniquely determined by the failure rate function $z(t)$. From (2.4) and (2.9) we see that the probability density function $f(t)$ can be expressed by

$$f(t) = z(t) \cdot \exp\left(-\int_0^t z(u)\, du\right) \quad \text{for } t > 0 \qquad (2.10)$$

In actuarial statistics the failure rate function is called the *force of mortality* (FOM). This term has also been adopted by several authors of reliability textbooks to avoid the confusion between the failure rate function and the *rate of occurrence of failures* (ROCOF) of a repairable item. The failure rate function (FOM) is a function of the life distribution of a single item and an indication of the "proneness to failure" of the item after time t has elapsed, while ROCOF is the occurrence rate of failures for a stochastic process; see Chapter 7. A thorough discussion of these concepts is given by Ascher and Feingold (1984). Some authors (e.g., Thompson, 1988) prefer the term *hazard rate* instead of failure rate. The term failure rate is, however, now well established in applied reliability. We have therefore decided to use this term instead of FOM in this textbook, although we realize that the use of this term may lead to some confusion.

Table 2.1 Relationship between the Functions $F(t)$, $f(t)$, $R(t)$, and $z(t)$

Expressed by	$F(t)$	$f(t)$	$R(t)$	$z(t)$
$F(t) =$	–	$\int_0^t f(u)\,du$	$1 - R(t)$	$1 - \exp\left(-\int_0^t z(u)\,du\right)$
$f(t) =$	$\dfrac{d}{dt} F(t)$	–	$-\dfrac{d}{dt} R(t)$	$z(t) \cdot \exp\left(-\int_0^t z(u)\,du\right)$
$R(t) =$	$1 - F(t)$	$\int_t^\infty f(u)\,du$	–	$\exp\left(-\int_0^t z(u)\,du\right)$
$z(t) =$	$\dfrac{dF(t)/dt}{1 - F(t)}$	$\dfrac{f(t)}{\int_t^\infty f(u)\,du}$	$-\dfrac{d}{dt} \ln R(t)$	–

The relationships between the functions $F(t)$, $f(t)$, $R(t)$, and $z(t)$ are presented in Table 2.1.

From (2.9) we see that the reliability (survivor) function $R(t)$ is uniquely determined by the failure rate function $z(t)$. To determine the form of $z(t)$ for a given type of items, the following experiment may be carried out:

Split the time interval $(0, t)$ into disjoint intervals of equal length Δt. Then put n identical items into operation at time $t = 0$. When an item fails, note the time and leave that item out. For each interval record:

- The number of items $n(i)$ that fail in interval i.

- The functioning times for the individual items $(T_{1i}, T_{2i}, \ldots, T_{ni})$ in interval i. Hence T_{ji} is the time item j has been functioning in time interval i. T_{ji} is therefore equal to 0 if item j has failed before interval i, where $j = 1, 2, \ldots, n$.

Thus $\sum_{j=1}^n T_{ji}$ is the total functioning time for the items in interval i. Now

$$z(i) = \frac{n(i)}{\sum_{j=1}^n T_{ji}}$$

which shows the number of failures per unit functioning time in interval i is a natural estimate of the "failure rate" in interval i for the items that are functioning at the start of this interval.

Let $m(i)$ denote the number of items that are functioning at the start of interval i:

$$z(i) \approx \frac{n(i)}{m(i)\Delta t}$$

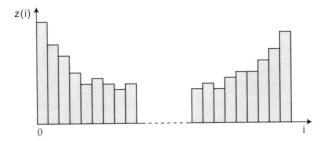

Fig. 2.4 Empirical bathtub curve.

and hence

$$z(i)\Delta t \approx \frac{n(i)}{m(i)}$$

A histogram depicting $z(i)$ as a function of i typically is of the form given in Fig. 2.4. If n is very large, we may use very small time intervals. If we let $\Delta t \to 0$, is it expected that the step function $z(i)$ will tend toward a "smooth" curve, as illustrated in Fig. 2.5, which may be interpreted as an estimate for the failure rate function $z(t)$.

This curve is usually called a *bathtub curve* after its characteristic shape. The failure rate is often high in the initial phase. This can be explained by the fact that there may be undiscovered defects (known as "infant mortality") in the items; these soon show up when the items are activated. When the item has survived the infant mortality period, the failure rate often stabilizes at a level where it remains for a certain amount of time until it starts to increase as the items begin to wear out. From the shape of the bathtub curve, the lifetime of an item may be divided into three typical intervals: the *burn-in period*, the *useful life period* and the *wear-out period*. The useful life period is also called the *chance failure period*. Often the items are tested at the factory before they are distributed to the users, and thus much of the infant mortality will be removed before the items are delivered for use. For the majority of mechanical items the failure rate function will usually show a slightly increasing tendency in the useful life period.

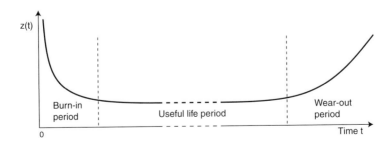

Fig. 2.5 The bathtub curve.

2.6 MEAN TIME TO FAILURE

The mean time to failure (MTTF) of an item is defined by

$$\text{MTTF} = E(T) = \int_0^\infty t f(t)\, dt \tag{2.11}$$

When the time required to repair or replace a failed item is very short compared to MTTF, MTTF also represents the mean time between failures (MTBF). If the repair time cannot be neglected, MTBF also includes the mean time to repair (MTTR).

Since $f(t) = -R'(t)$,

$$\text{MTTF} = -\int_0^\infty t R'(t)\, dt$$

By partial integration

$$\text{MTTF} = -[t R(t)]_0^\infty + \int_0^\infty R(t)\, dt$$

If MTTF $< \infty$, it can be shown that $[t R(t)]_0^\infty = 0$. In that case

$$\text{MTTF} = \int_0^\infty R(t)\, dt \tag{2.12}$$

It is often easier to determine MTTF by (2.12) than by (2.11).

The mean time to failure of an item may also be derived by using Laplace transforms. The Laplace transform of the survivor function $R(t)$ is (see Appendix B)

$$R^*(s) = \int_0^\infty R(t)\, e^{-st}\, dt \tag{2.13}$$

When $s = 0$, we get

$$R^*(0) = \int_0^\infty R(t)\, dt = \text{MTTF} \tag{2.14}$$

The MTTF may thus be derived from the Laplace transform $R^*(s)$ of the survivor function $R(t)$, by setting $s = 0$.

Median Life The MTTF is only one of several measures of the "center" of a life distribution. An alternative measure is the median life t_m, defined by

$$R(t_m) = 0.50 \tag{2.15}$$

The median divides the distribution in two halves. The item will fail before time t_m with 50% probability, and will fail after time t_m with 50% probability.

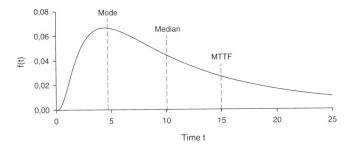

Fig. 2.6 Location of the MTTF, the median life, and the mode of a distribution.

Mode The mode of a life distribution is the most likely failure time, that is, the time t_{mode} where the probability density function $f(t)$ attains its maximum:

$$f(t_{\text{mode}}) = \max_{0 \leq t < \infty} f(t) \qquad (2.16)$$

Fig. 2.6 shows the location of the MTTF, the median life t_m, and the mode t_{mode} for a distribution that is skewed to the right.

Example 2.1
Consider an item with reliability (survivor) function

$$R(t) = \frac{1}{(0.2\,t + 1)^2} \quad \text{for } t \geq 0$$

where the time t is measured in months. The probability density function is

$$f(t) = -R'(t) = \frac{0.4}{(0.2\,t + 1)^3}$$

and the failure rate function is from (2.4):

$$z(t) = \frac{f(t)}{R(t)} = \frac{0.4}{0.2\,t + 1}$$

The mean time to failure is from (2.12):

$$\text{MTTF} = \int_0^\infty R(t)\,dt = 5 \text{ months}$$

The functions $R(t)$, $f(t)$, and $z(t)$ are illustrated in Fig. 2.7. □

2.7 MEAN RESIDUAL LIFE

Consider an item with time to failure T that is put into operation at time $t = 0$ and is still functioning at time t. The probability that the item of age t survives an additional

24 FAILURE MODELS

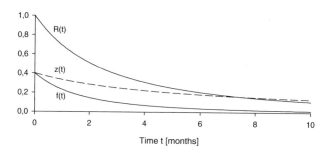

Fig. 2.7 The survivor function $R(t)$, the probability density function $f(t)$, and the failure rate function $z(t)$ (dashed line) in Example 2.1.

interval of length x is

$$R(x \mid t) = \Pr(T > x + t \mid T > t) = \frac{\Pr(T > x + t)}{\Pr(T > t)} = \frac{R(x + t)}{R(t)} \qquad (2.17)$$

$R(x \mid t)$ is called the *conditional survivor function* of the item at age t. The mean residual (or, remaining) life, MRL(t), of the item at age t is

$$\text{MRL}(t) = \mu(t) = \int_0^\infty R(x \mid t)\, dx = \frac{1}{R(t)} \int_t^\infty R(x)\, dx \qquad (2.18)$$

When $t = 0$, the item is new, and we have $\mu(0) = \mu = $ MTTF. It is sometimes of interest to study the function

$$g(t) = \frac{\text{MRL}(t)}{\text{MTTF}} = \frac{\mu(t)}{\mu} \qquad (2.19)$$

When an item has survived up to time t, then $g(t)$ gives the MRL(t) as a percentage of the initial MTTF. If, for example, $g(t) = 0.60$, then the mean residual lifetime, MRL(t) at time t, is 60% of mean residual lifetime at time 0.

By differentiating $\mu(t)$ with respect to t it is straightforward to verify that the failure rate function $z(t)$ can be expressed as

$$z(t) = \frac{1 + \mu'(t)}{\mu(t)} \qquad (2.20)$$

Example 2.2
Consider an item with failure rate function $z(t) = t/(t+1)$. The failure rate function is increasing and approaches 1 when $t \to \infty$. The corresponding survivor function is

$$R(t) = \exp\left(-\int_0^t \frac{u}{u+1}\, du\right) = (t+1)\, e^{-t}$$

and
$$\text{MTTF} = \int_0^\infty (t+1)\, e^{-t}\, dt = 2$$

The conditional survival function is
$$R(x \mid t) = \Pr(T > x + t \mid T > t) = \frac{(t + x + 1)\, e^{-(t+x)}}{(t+1)\, e^{-t}} = \frac{t + x + 1}{t + 1}\, e^{-x}$$

The mean residual life is
$$\text{MRL}(t) = \int_0^\infty R(x \mid t)\, dx = 1 + \frac{1}{t+1}$$

We see that MRL(t) is equal to 2 (= MTTF) when $t = 0$, that MRL(t) is a decreasing function in t, and that MRL$(t) \to 1$ when $t \to \infty$. □

2.8 THE BINOMIAL AND GEOMETRIC DISTRIBUTIONS

The binomial distribution is one of the most widely used discrete distributions in reliability engineering. The distribution is used in the following situation:

1. We have n independent trials.
2. Each trial has two possible outcomes A and A^*.
3. The probability $\Pr(A) = p$ is the same in all the n trials.

This situation is called a *binomial situation*, and the trials are sometimes referred to as *Bernoulli trials*. Let X denote the number of the n trials that have outcome A. Then X is a discrete random variable with distribution

$$\Pr(X = x) = \binom{n}{x} p^x (1-p)^{n-x} \quad \text{for } x = 0, 1, \ldots, n \tag{2.21}$$

where $\binom{n}{x}$ is the binomial coefficient
$$\binom{n}{x} = \frac{n!}{x!(n-x)!}$$

The distribution (2.21) is called the *binomial distribution* (n, p), and we sometimes write $X \sim \text{bin}(n, p)$. The mean value and the variance of X are

$$E(X) = np \tag{2.22}$$
$$\text{var}(X) = np(1-p) \tag{2.23}$$

Assume that we carry out a sequence of Bernoulli trials, and want to find the number Z of trials until the first trial with outcome A. If $Z = z$, this means that the first $(z-1)$

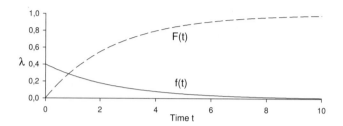

Fig. 2.8 Exponential distribution ($\lambda = 1$).

trials have outcome A^*, and that the first A will occur in trial z. The distribution of Z is

$$\Pr(Z = z) = (1-p)^{z-1} p \quad \text{for } z = 1, 2, \ldots \tag{2.24}$$

The distribution (2.24) is called the *geometric distribution*. We have that

$$\Pr(Z > z) = (1-p)^z$$

The mean value and the variance of Z are

$$E(Z) = \frac{1}{p} \tag{2.25}$$

$$\operatorname{var}(X) = \frac{1-p}{p^2} \tag{2.26}$$

2.9 THE EXPONENTIAL DISTRIBUTION

Consider an item that is put into operation at time $t = 0$. The time to failure T of the item has probability density function

$$f(t) = \begin{cases} \lambda e^{-\lambda t} & \text{for } t > 0, \lambda > 0 \\ 0 & \text{otherwise} \end{cases} \tag{2.27}$$

This distribution is called the *exponential distribution* with parameter λ, and we sometimes write $T \sim \exp(\lambda)$.

The reliability (survivor) function of the item is

$$R(t) = \Pr(T > t) = \int_t^\infty f(u)\, du = e^{-\lambda t} \quad \text{for } t > 0 \tag{2.28}$$

The probability density function $f(t)$ and the survivor function $R(t)$ for the exponential distribution are illustrated in Fig. 2.8. The mean time to failure is

$$\text{MTTF} = \int_0^\infty R(t)\,dt = \int_0^\infty e^{-\lambda t}\,dt = \frac{1}{\lambda} \qquad (2.29)$$

and the variance of T is

$$\text{var}(T) = \frac{1}{\lambda^2}$$

The probability that an item will survive its mean time to failure is

$$R(\text{MTTF}) = R\left(\frac{1}{\lambda}\right) = e^{-1} \approx 0.3679$$

The failure rate function is

$$z(t) = \frac{f(t)}{R(t)} = \frac{\lambda e^{-\lambda t}}{e^{-\lambda t}} = \lambda \qquad (2.30)$$

Accordingly, the failure rate function of an item with exponential life distribution is constant (i.e., independent of time). By comparing with Fig. 2.5, we see that this indicates that the exponential distribution may be a realistic life distribution for an item during its useful life period, at least for certain types of items.

The results (2.29) and (2.30) compare well with the use of the concepts in everyday language. If an item on the average has $\lambda = 4$ failures/year, the MTTF of the item is 1/4 year.

Consider the conditional survivor function (2.17)

$$\begin{aligned} R(x \mid t) &= \Pr(T > t + x \mid T > t) = \frac{\Pr(T > t + x)}{\Pr(T > t)} \\ &= \frac{e^{-\lambda(t+x)}}{e^{-\lambda t}} = e^{-\lambda x} = \Pr(T > x) = R(x) \end{aligned} \qquad (2.31)$$

The survivor function of an item that has been functioning for t time units is therefore equal to the survivor function of a new item. A new item, and a used item (that is still functioning), will therefore have the same probability of surviving a time interval of length t. The MRL for the exponential distribution is

$$\text{MRL}(t) = \int_0^\infty R(x \mid t)\,dx = \int_0^\infty R(x)\,dx = \text{MTTF}$$

The MRL(t) of an item with exponential life distribution is hence equal to its MTTF irrespective of the age t of the item. The item is therefore *as good as new* as long as it is functioning, and we often say that the exponential distribution has *no memory*.

Therefore, an assumption of exponentially distributed lifetime implies that

- A used item is stochastically *as good as new*, so there is no reason to replace a functioning item.

- For the estimation of the reliability function, the mean time to failure, and so on, it is sufficient to collect data on the number of hours of observed time in operation and the number of failures. The age of the items is of no interest in this connection.

The exponential distribution is the most commonly used life distribution in applied reliability analysis. The reason for this is its mathematical simplicity and that it leads to realistic lifetime models for certain types of items.

Example 2.3
A rotary pump has a constant failure rate $\lambda = 4.28 \cdot 10^{-4}$ hours^{-1} (data from OREDA 2002). The probability that the pump survives one month ($t = 730$ hours) in continuous operation is

$$R(t) = e^{-\lambda t} = e^{-4.28 \cdot 10^{-4} \cdot 730} \approx 0.732$$

The mean time to failure is

$$\text{MTTF} = \frac{1}{\lambda} = \frac{1}{4.28 \cdot 10^{-4}} \text{ hours} \approx 2336 \text{ hours} \approx 3.2 \text{ months}$$

Suppose that the pump has been functioning without failure during its first 2 months ($t_1 = 1\,460$ hours) in operation. The probability that the pump will fail during the next month ($t_2 = 730$ hours) is

$$\Pr(T \leq t_1 + t_2 \mid T > t_1) = \Pr(T \leq t_2) = 1 - e^{-4.28 \cdot 10^{-4} \cdot 730} \approx 0.268$$

since the pump is as good as new when it is still functioning at time t_1. ☐

Example 2.4
Consider a system of two independent components with failure rates λ_1 and λ_2, respectively. The probability that component 1 fails before component 2 is

$$\begin{aligned}
\Pr(T_2 > T_1) &= \int_0^\infty \Pr(T_2 > t \mid T_1 = t) f_{T_1}(t)\, dt \\
&= \int_0^\infty e^{-\lambda_2 t} \lambda_1 e^{-\lambda_1 t}\, dt \\
&= \lambda_1 \int_0^\infty e^{-(\lambda_1 + \lambda_2)t}\, dt = \frac{\lambda_1}{\lambda_1 + \lambda_2}
\end{aligned}$$

This result can easily be generalized to a system of n independent components with failure rates $\lambda_1, \lambda_2, \ldots, \lambda_n$. The probability that component j is the first component to fail is

$$\Pr(\text{component } j \text{ fails first}) = \frac{\lambda_j}{\sum_{i=1}^n \lambda_i}$$

☐

Example 2.5 Mixture of Exponential Distributions
Assume that the same type of items are produced at two different plants. The items are

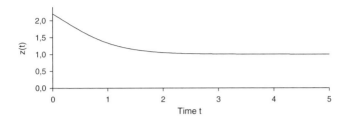

Fig. 2.9 The failure rate function of the mixture of two exponential distributions in Example 2.5 ($\lambda_1 = 1$, $\lambda_2 = 3$, and $p = 0.4$).

assumed to be independent and have constant failure rates. The production process is slightly different at the two plants, and the items will therefore have different failure rates. Let λ_i denote the failure rate of the items coming from plant i, for $i = 1, 2$. The items are mixed up before they are sold. A fraction p is coming from plant 1, and the rest $(1 - p)$ is coming from plant 2. If we pick one item at random, the survival function of this item is

$$R(t) = p \cdot R_1(t) + (1 - p) \cdot R_2(t) = p\, e^{-\lambda_1 t} + (1 - p)\, e^{-\lambda_2 t}$$

The mean time to failure is

$$\text{MTTF} = \frac{p}{\lambda_1} + \frac{1 - p}{\lambda_2}$$

and the failure rate function is

$$z(t) = \frac{p\lambda_1\, e^{-\lambda_1 t} + (1 - p)\lambda_2\, e^{-\lambda_2 t}}{p\, e^{-\lambda_1 t} + (1 - p)\, e^{-\lambda_2 t}}$$

The failure rate function, which is illustrated in Fig. 2.9, is seen to be decreasing. If we assume that $\lambda_1 > \lambda_2$, early failures should have a failure rate close to λ_1. After a while all the "weak" components have failed, and we are left with components with a lower failure rate λ_2.

The example can easily be extended to a mixture of more than two exponential distributions. □

Example 2.6 Phase-Type Distribution
Consider an item that is exposed to three failure mechanisms. A random overstress may occur that will cause a critical (C) failure. The component is further exposed to wear that may cause a well-defined degraded (D) failure. In degraded mode, a new failure mechanism may cause a degraded critical (DC) failure. The random overstress occurs independent of the state of the component. The failure transitions are illustrated in Fig. 2.10 where O denotes full operating state. (The diagram in Fig. 2.10 is an example of a *state transition diagram* that is further discussed in Chapter 8).

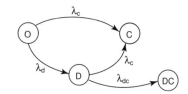

Fig. 2.10 Failure transitions for the component in Example 2.6.

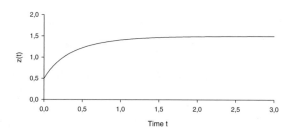

Fig. 2.11 The failure rate function of the component in Example 2.6 for $\lambda_c = 0.5$, $\lambda_d = 1$, and $\lambda_{dc} = 3$.

The various times to failure are assumed to be independent and exponentially distributed. The time T_c to a C-failure has failure rate λ_c, and the time T_d to a D-failure has failure rate λ_d. The time T_{dc} for a component in degraded mode to fail to a DC-failure has failure rate λ_{dc}. The time T to critical (overstress or degraded) failure is therefore $T = \min\{T_c, T_d + T_{dc}\}$. The survival function is

$$\begin{aligned} R(t) &= \Pr(\min\{T_c, T_d + T_{dc}\} > t) = \Pr(T_c > t \cap T_d + T_{dc} > t) \\ &= \Pr(T_c > t) \cdot \Pr(T_d + T_{dc} > t) \end{aligned}$$

By using Problem 2.5, $R(t)$ may be written

$$R(t) = e^{-\lambda_c t} \cdot \frac{1}{\lambda_{dc} - \lambda_d} \left(\lambda_{dc} e^{-\lambda_d t} - \lambda_d e^{-\lambda_{dc} t} \right) \tag{2.32}$$

The failure rate function is

$$z(t) = \frac{-R'(t)}{R(t)} = \lambda_c + \frac{\lambda_d \lambda_{dc} \left(e^{-\lambda_d t} - e^{-\lambda_{dc} t} \right)}{\lambda_{dc} e^{-\lambda_d t} - \lambda_d e^{-\lambda_{dc} t}} \tag{2.33}$$

The failure rate function $z(t)$ is illustrated in Fig. 2.11 for selected values of λ_c, λ_d, and λ_{dc}. The mean time to failure is

$$\text{MTTF} = \int_0^\infty R(t)\, dt = \frac{\lambda_{dc} + \lambda_d + \lambda_c}{(\lambda_d + \lambda_c)(\lambda_{dc} + \lambda_c)} \tag{2.34}$$

This example is further discussed by Hokstad and Frøvig (1996). □

2.10 THE HOMOGENEOUS POISSON PROCESS

We will now briefly introduce the homogeneous Poisson process (HPP). The HPP is discussed in more detail in Chapter 7. The HPP is used to model occurrences of a specific event $\mathcal{A}$ in the course of a given time interval. The event $\mathcal{A}$ may, for example, be a failure or an accident. The following conditions are assumed to be fulfilled:

1. The event $\mathcal{A}$ may occur at any time in the interval, and the probability of $\mathcal{A}$ occurring in the interval $(t, t + \Delta t]$ is independent of t and may be written as $\lambda \cdot \Delta t + o(\Delta t)$,[1] where λ is a positive constant.

2. The probability of more that one event $\mathcal{A}$ in the interval $(t, t + \Delta t]$ is $o(\Delta t)$.

3. Let $(t_{11}, t_{12}], (t_{21}, t_{22}], \ldots$ be any sequence of disjoint intervals in the time period in question. Then the events "$\mathcal{A}$ occurs in $(t_{j1}, t_{j2}]$," $j = 1, 2, \ldots$, are independent.

Without loss of generality we let $t = 0$ be the starting point of the process.

Let $N(t)$ denote the number of times the event $\mathcal{A}$ occurs during the interval $(0, t]$. The stochastic process $\{N(t), t \geq 0\}$ is then an HPP with rate λ. The rate λ is sometimes called the *intensity* of the process or the *frequency* of events $\mathcal{A}$. A consequence of assumption 1 is that the rate of events $\mathcal{A}$ is constant, and does not change with time. The HPP can therefore not be used to model processes where the rate of events changes with time, for example, processes that have a long-term trend or are exposed to seasonal variations.

The time t may be measured as calendar time or operational time. In many cases several subprocesses are running in parallel and the time t must then be measured as total, or accumulated, time in service. This is, for example, the case when we observe failures in a population of repairable items.

The probability that $\mathcal{A}$ occurs exactly n times in the time interval $(0, t]$ is

$$\Pr(N(t) = n) = \frac{(\lambda t)^n}{n!} e^{-\lambda t} \quad \text{for } n = 0, 1, 2, \ldots \tag{2.35}$$

The distribution (2.35) is called the Poisson distribution. When we observe the occurrence of events $\mathcal{A}$ in an interval $(s, s + t]$, the probability that $\mathcal{A}$ occurs exactly n times in $(s, s + t]$ is

$$\Pr(N(s + t) - N(s) = n) = \frac{(\lambda t)^n}{n!} e^{-\lambda t} \quad \text{for } n = 0, 1, 2, \ldots$$

that is, the same probability as we found in (2.35). The important quantity is therefore the length t of the time interval we are observing the process, not when this interval starts.

---

[1] $o(\Delta t)$ denotes a function of Δt with the property that $\lim_{\Delta t \to 0} \frac{o(\Delta t)}{\Delta t} = 0$.

The mean number of events in $(0, t]$ is

$$E(N(t)) = \sum_{n=0}^{\infty} n \cdot \Pr(N(t) = n) = \lambda t \qquad (2.36)$$

and the variance is

$$\mathrm{var}(N(t)) = \lambda t \qquad (2.37)$$

From equation (2.36), the parameter λ may be written as $\lambda = N(t)/t$, that is, the mean number of events per time unit. This is why λ is called the *rate* of the HPP. When the event $\mathcal{A}$ is a failure, λ is the ROCOF of the HPP.

A natural unbiased estimator of λ is

$$\hat{\lambda} = \frac{N(t)}{t} = \frac{\text{No. of events observed in an interval of length } t}{\text{Length } t \text{ of the interval}} \qquad (2.38)$$

Let T_1 denote the time when $\mathcal{A}$ occurs for the first time, and let $F_{T_1}(t)$ denote the distribution function of T_1. Since the event $(T_1 > t)$ means that no event has occurred in the interval $(0, t]$, we get

$$\begin{aligned} F_{T_1}(t) &= \Pr(T_1 \le t) = 1 - \Pr(T_1 > t) \\ &= 1 - \Pr(N(t) = 0) = 1 - e^{-\lambda t} \quad \text{for } t \ge 0 \end{aligned} \qquad (2.39)$$

The time T_1 to the first $\mathcal{A}$ is seen to be *exponentially* distributed with parameter λ. It may be shown, see Chapter 7, that the times between events, $T_1, T_2, \ldots$ are independent, and exponentially distributed with parameter λ. The times between events $T_1, T_2, \ldots$ are called the *interoccurrence times* of the process.

Example 2.7
Consider a repairable item that is put into operation at time $t = 0$. The first failure (event $\mathcal{A}$) occurs at time T_1. When the item has failed, it is replaced with a new item of the same type. The replacement time is so short that it can be neglected. The second failure occurs at time T_2, and so on. We thus get a sequence of times $T_1, T_2, \ldots$. The number of failures $N(t)$ in the time interval $(0, t]$ is assumed to be Poisson distributed with ROCOF λ. The interoccurrence times $T_1, T_2, \ldots$ are then independent and exponentially distributed with failure rate λ. Note the important difference in meaning between the two concepts "failure rate" and ROCOF. □

Let us consider an HPP with rate λ, and assume that we are interested in determining the distribution of the time S_k where $\mathcal{A}$ occurs for the kth time (k is accordingly an integer). We let t be an arbitrarily chosen point of time on the positive real axis. The event $(T_k > t)$ is then obviously synonymous with the event that $\mathcal{A}$ is occurring at most $(k-1)$ times in the time interval $(0, t]$. Therefore

$$\Pr(S_k > t) = \Pr(N(t) \le k - 1) = \sum_{j=0}^{k-1} \frac{(\lambda t)^j}{j!} e^{-\lambda t}$$

Hence

$$F_{S_k}(t) = 1 - \sum_{j=0}^{k-1} \frac{(\lambda t)^j}{j!} e^{-\lambda t} \qquad (2.40)$$

where $F_{S_k}(t)$ is the distribution function for S_k. The probability density function $f_{S_k}(t)$ is obtained by differentiating $F_{S_k}(t)$ with respect to t:

$$\begin{aligned}
f_{S_k}(t) &= -\sum_{j=1}^{k-1} \frac{j\lambda(\lambda t)^{j-1}}{j!} e^{-\lambda t} + \lambda \sum_{j=0}^{k-1} \frac{(\lambda t)^j}{j!} e^{-\lambda t} \\
&= \lambda e^{-\lambda t} \left(\sum_{j=0}^{k-1} \frac{(\lambda t)^j}{j!} - \sum_{j=1}^{k-1} \frac{(\lambda t)^{j-1}}{(j-1)!} \right) \\
&= \lambda e^{-\lambda t} \left(\sum_{j=0}^{k-1} \frac{(\lambda t)^j}{j!} - \sum_{j=0}^{k-2} \frac{(\lambda t)^j}{j!} \right) \\
&= \frac{\lambda}{(k-1)!} (\lambda t)^{k-1} e^{-\lambda t} \quad \text{for } t \geq 0 \text{ and } \lambda > 0 \qquad (2.41)
\end{aligned}$$

where k is a positive integer. This distribution is called the *gamma* distribution with parameters k and λ. The gamma distribution is further discussed in Section 2.11. We can therefore conclude that the waiting time until the kth occurrence of $\mathcal{A}$ in an HPP with rate λ is gamma distributed (k, λ).

The HPP is further discussed in Section 8.2.

2.11 THE GAMMA DISTRIBUTION

Consider an item that is exposed to a series of shocks that occur according to an HPP with rate λ. The time intervals $T_1, T_2, \ldots$, between consecutive shocks are then independent and exponentially distributed with parameter λ (see Section 2.10). Assume that the item fails exactly at shock k, and not earlier. The time to failure of the item is then

$$T = T_1 + T_2 + \cdots + T_k$$

and according to (2.41) T is gamma distributed (k, λ), and we sometimes write $T \sim \text{gamma}(k, \lambda)$. The probability density function is

$$f(t) = \frac{\lambda}{\Gamma(k)} (\lambda t)^{k-1} e^{-\lambda t} \qquad (2.42)$$

where $\Gamma(\cdot)$ denotes the gamma function (see Appendix A), $t > 0$, $\lambda > 0$, and k is a positive integer. The probability density function $f(t)$ is sketched in Fig. 2.12 for selected values of k.

34 FAILURE MODELS

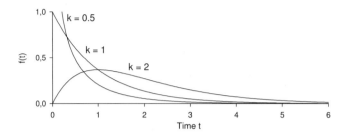

Fig. 2.12 The gamma probability density, $\lambda = 1.0$.

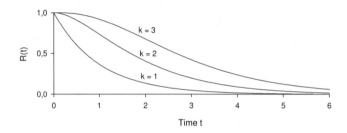

Fig. 2.13 Reliability function for the gamma distribution, $\lambda = 1.0$.

The parameter λ denotes the rate (frequency) of shocks and is an external parameter for the item. The integer k may be interpreted as a measure of the ability to resist the shocks and will from now on generally not be restricted to integer values but be a positive constant. Equation (2.42) will still be a probability density function.

From (2.42) we find that

$$\text{MTTF} = \frac{k}{\lambda} \qquad (2.43)$$

$$\text{var}(T) = \frac{k}{\lambda^2} \qquad (2.44)$$

For integer values of k the reliability function [see (2.40)] is given by

$$R(t) = 1 - F(t) = \sum_{n=0}^{k-1} \frac{(\lambda t)^n}{n!} e^{-\lambda t} \qquad (2.45)$$

A sketch of $R(t)$ is given in Fig. 2.13 for some values of k.

The corresponding failure rate function is

$$z(t) = \frac{f(t)}{R(t)} = \frac{\lambda (\lambda t)^{k-1} e^{-\lambda t} / \Gamma(k)}{\sum_{n=0}^{k-1} (\lambda t)^n e^{-\lambda t}/n!} \qquad (2.46)$$

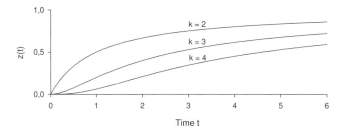

Fig. 2.14 Failure rate function of the gamma distribution, $\lambda = 1$.

For $k = 2$ the failure rate function is

$$z(t) = \frac{\lambda^2 t}{1 + \lambda t}$$

and the distribution in Example 2.2 is therefore a gamma distribution with $k = 2$ and $\lambda = 1$.

When k is not an integer, we have to use the general formulas (2.3) and (2.7) to find the reliability function $R(t)$ and the failure rate function $z(t)$, respectively. It may be shown (e.g., see Cocozza-Thivent, 1997, p. 10) that

$$\lim_{t \to 0} z(t) = \infty \quad \text{and} \quad \lim_{t \to \infty} z(t) = \lambda \quad \text{when } 0 < k < 1$$
$$\lim_{t \to 0} z(t) = 0 \quad \text{and} \quad \lim_{t \to \infty} z(t) = \lambda \quad \text{when } k > 1$$

The failure rate function $z(t)$ is illustrated in Fig. 2.14 for some integer values of k.

Let $T_1 \sim$ gamma(k_1, λ) and $T_2 \sim$ gamma(k_2, λ) be independent. It is then easy to show (see Problem 2.16) that $T_1 + T_2 \sim$ gamma$(k_1 + k_2, \lambda)$. Gamma distributions with a common λ are therefore *closed under addition*.

2.11.1 Special Cases

For special values of the parameters k and λ, the gamma distribution is know under other names:

1. When $k = 1$, we have the *exponential distribution* with failure rate λ.

2. When $k = n/2$ and $\lambda = 1/2$, the gamma distribution is called a *chi-square* (χ^2) *distribution* with n degrees of freedom (n is an integer). Percentile values for the χ^2 distribution are given in Appendix E.

3. When k is an integer, the gamma distribution is called an *Erlangian distribution* with parameters k and λ.

Example 2.8 Mixture of Exponential Distributions
Assume that items of a specific type are produced in a plant where the production

process is unstable such that failure rate λ of the items varies with time. If we pick an item at random, the conditional probability density function of the time to failure T, given λ, is

$$f(t \mid \lambda) = \lambda e^{-\lambda t} \quad \text{for } t > 0$$

Assume that the variation in λ can be modeled by a gamma distribution with parameters k and α:

$$\pi(\lambda) = \frac{\alpha^k}{\Gamma(k)} \lambda^{k-1} e^{-\alpha \lambda} \quad \text{for } \lambda > 0, \ \alpha > 0, \ k > 0$$

The unconditional density of T is thus

$$f(t) = \int_0^\infty f(t \mid \lambda) \pi(\lambda) \, d\lambda = \frac{k\alpha^k}{(\alpha + t)^{k+1}} \tag{2.47}$$

The survivor function is

$$R(t) = \Pr(T > t) = \int_t^\infty f(u) \, du = \frac{\alpha^k}{(\alpha + t)^k} = \left(1 + \frac{t}{\alpha}\right)^{-k} \tag{2.48}$$

The mean time to failure is

$$\text{MTTF} = \int_0^\infty R(t) \, dt = \frac{\alpha}{k-1} \quad \text{for } k > 1$$

Note that the MTTF does not exist for $0 < k \leq 1$. The failure rate function is

$$z(t) = \frac{f(t)}{R(t)} = \frac{k}{\alpha + t} \tag{2.49}$$

and hence is monotonically *decreasing* as a function of t.

A factory is producing a specific type of gas detectors. Experience has shown that the *mean* failure rate of the detectors is $\lambda_m = 1.15 \cdot 10^{-5}$ hours^{-1} (data from OREDA 2002). The corresponding mean MTTF is $1/\lambda_m \approx 9.93$ years. The production is, however, unstable and the standard deviation of the failure rate is estimated to be $4 \cdot 10^{-6}$ hours^{-1}. As above, we assume that the failure rate is a random variable Λ with a gamma (k, α) distribution. From (2.44) we have $E(\Lambda) = k/\alpha = 1.15 \cdot 10^{-5}$, and $\text{var}(\Lambda) = k/\alpha^2 = [4 \cdot 10^{-6}]^2$. We can now solve for k and α and get

$$k \approx 8.27 \quad \text{and} \quad \alpha \approx 7.19 \cdot 10^6$$

The mean time to failure is then

$$\text{MTTF} = \frac{\alpha}{k-1} \approx 9.9 \cdot 10^5 \text{ hours} \approx 11.3 \text{ years}$$

The corresponding failure rate function $z(t)$ may be found from (2.49). □

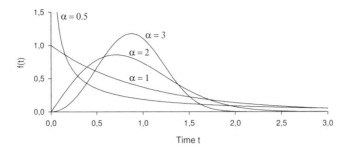

Fig. 2.15 The probability density function of the Weibull distribution for selected values of the shape parameter α ($\lambda = 1$).

Remark: Example 2.8 is similar to Example 2.5 where we mixed two different exponential distributions and got a decreasing failure rate function. The results from these examples are very important for collection and analysis of field data. Suppose that the (true) failure rate of a specific type of items is constant and equal to λ. When we collect data from different installations and from different operational and environmental conditions, the failure rate λ will vary. If we pool all the data into one single sample and analyze the data, we will conclude that the failure rate function is decreasing. □

2.12 THE WEIBULL DISTRIBUTION

The Weibull distribution is one of the most widely used life distributions in reliability analysis. The distribution is named after the Swedish professor Waloddi Weibull (1887–1979) who developed the distribution for modeling the strength of materials. The Weibull distribution is very flexible, and can, through an appropriate choice of parameters, model many types of failure rate behaviors.

The time to failure T of an item is said to be Weibull distributed with parameters $\alpha (> 0)$ and $\lambda (> 0)$ [$T \sim \text{Weibull}(\alpha, \lambda)$] if the distribution function is given by

$$F(t) = \Pr(T \leq t) = \begin{cases} 1 - e^{-(\lambda t)^\alpha} & \text{for } t > 0 \\ 0 & \text{otherwise} \end{cases} \qquad (2.50)$$

The corresponding probability density is

$$f(t) = \frac{d}{dt} F(t) = \begin{cases} \alpha \lambda^\alpha t^{\alpha-1} e^{-(\lambda t)^\alpha} & \text{for } t > 0 \\ 0 & \text{otherwise} \end{cases} \qquad (2.51)$$

where λ is a *scale* parameter, and α is referred to as the *shape* parameter. Note that when $\alpha = 1$, the Weibull distribution is equal to the exponential distribution. The probability density function $f(t)$ is illustrated in Fig. 2.15 for selected values of α.

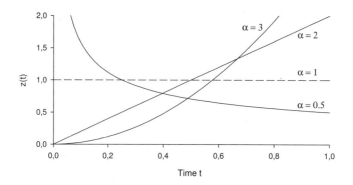

Fig. 2.16 Failure rate function of the Weibull distribution, $\lambda = 1$.

The survivor function is

$$R(t) = \Pr(T > 0) = e^{-(\lambda t)^\alpha} \quad \text{for } t > 0 \tag{2.52}$$

and the failure rate function is

$$z(t) = \frac{f(t)}{R(t)} = \alpha \lambda^\alpha t^{\alpha-1} \quad \text{for } t > 0 \tag{2.53}$$

When $\alpha = 1$, the failure rate is constant; when $\alpha > 1$, the failure rate function is increasing; and when $0 < \alpha < 1$, $z(t)$ is decreasing. When $\alpha = 2$, the resulting distribution is known as the *Rayleigh distribution*. The failure rate function $z(t)$ of the Weibull distribution is illustrated in Fig. 2.16 for some selected values of α. The Weibull distribution is seen to be flexible and may be used to model life distributions, where the failure rate function is decreasing, constant, or increasing.

Remark: Notice that the failure rate function is discontinuous as a function of the shape parameter α at $\alpha = 1$. It is important to be aware of this discontinuity in numerical calculations, since, for example, $\alpha = 0.999$, $\alpha = 1.000$, and $\alpha = 1.001$ will give significantly different failure rate functions for small values of t. □

From (2.52) it follows that

$$R\left(\frac{1}{\lambda}\right) = \frac{1}{e} \approx 0.3679 \quad \text{for all } \alpha > 0$$

Hence $\Pr(T > 1/\lambda) = 1/e$, independent of α. The quantity $1/\lambda$ is sometimes called the *characteristic* lifetime. The mean time to failure is

$$\text{MTTF} = \int_0^\infty R(t)\,dt = \frac{1}{\lambda}\Gamma\left(\frac{1}{\alpha}+1\right) \tag{2.54}$$

The median life t_m is

$$R(t_m) = 0.50 \quad \Rightarrow \quad t_m = \frac{1}{\lambda}(\ln 2)^{1/\alpha} \tag{2.55}$$

The variance of T is

$$\text{var}(T) = \frac{1}{\lambda^2}\left(\Gamma\left(\frac{2}{\alpha}+1\right) - \Gamma^2\left(\frac{1}{\alpha}+1\right)\right) \tag{2.56}$$

Note that MTTF/$\sqrt{\text{var}(T)}$ is independent of λ.

The Weibull distribution also arises as a limit distribution for the smallest of a large number of independent, identically distributed, nonnegative random variables. The Weibull distribution is therefore often called the *weakest link* distribution. This is further discussed in Section 2.17.

The Weibull distribution has been widely used in reliability analysis of semiconductors, ball bearings, engines, spot weldings, biological organisms, and so on.

Example 2.9
The time to failure T of a variable choke valve is assumed to have a Weibull distribution with shape parameter $\alpha = 2.25$ and scale parameter $\lambda = 1.15 \cdot 10^{-4}$ hours^{-1}. The valve will survive 6 months ($t = 4380$ hours) in continuous operation with probability:

$$R(t) = e^{-(\lambda t)^\alpha} = e^{-(1.15 \cdot 10^{-4} \cdot 8760)^{2.25}} \approx 0.808$$

The mean time to failure is

$$\text{MTTF} = \frac{1}{\lambda}\Gamma\left(\frac{1}{\alpha}+1\right) = \frac{\Gamma(1.44)}{1.15 \cdot 10^{-4}} \text{ hours} \approx 7706 \text{ hours}$$

and the median life is

$$t_m = \frac{1}{\lambda}(\ln 2)^{1/\alpha} \approx 7389 \text{ hours}$$

A valve that has survived the first 6 months ($t_1 = 4380$ hours), will survive the next 6 months ($t_2 = 4380$ hours) with probability

$$R(t_1 + t_2 \mid t_1) = \frac{R(t_1 + t_2)}{R(t_1)} = \frac{e^{-(\lambda(t_1+t_2))^\alpha}}{e^{-(\lambda t_1)^\alpha}} \approx 0.448$$

that is, significantly less than the probability that a new valve will survive 6 months.

The mean residual life when the valve has been functioning for 6 months ($t = 4380$ hours) is

$$\text{MRL}(t) = \frac{1}{R(t)}\int_0^\infty R(t+x)\,dx \approx 4730 \text{ hours}$$

The MRL(t) cannot be given a closed form in this case and was therefore found by using a computer. The function $g(t) = \text{MRL}(t)/\text{MTTF}$ is illustrated in Fig. 2.17. □

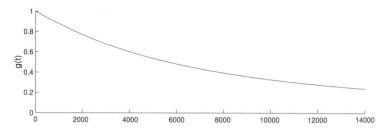

Fig. 2.17 The scaled mean residual lifetime function $g(t) = \mathrm{MRL}(t)/\mathrm{MTTF}$ for the Weibull distribution with parameters $\alpha = 2.25$ and $\lambda = 1.15 \cdot 10^{-4}$ hours^{-1}.

Example 2.10
Consider a series system of n components. The times to failure $T_1, T_2, \ldots, T_n$ of the n components are assumed to be independent and Weibull distributed:

$$T_i \sim \mathrm{Weibull}\,(\alpha, \lambda_i) \quad \text{for } i = 1, 2, \ldots, n$$

A series system fails as soon as the first component fails. The time to failure of the system, T_s is thus

$$T_s = \min\{T_1, T_2, \ldots, T_n\}$$

The survivor function of this system becomes

$$\begin{aligned} R_s(t) &= \Pr(T_s > t) = \Pr(\min_{1 \leq i \leq n} T_i > t) = \prod_{i=1}^{n} \Pr(T_i > t) \\ &= \prod_{i=1}^{n} \exp\left(-(\lambda_i t)^\alpha\right) = \exp\left(-\sum_{i=1}^{n}(\lambda_i\, t)^\alpha\right) = \exp\left(-\left[\sum_{i=1}^{n} \lambda_i^\alpha\right] t^\alpha\right) \end{aligned}$$

Hence a series system of independent components with Weibull life distribution with the same shape parameter α again has a Weibull life distribution, with scale parameter $\lambda_s = \left(\sum_{i=1}^{n} \lambda_i^\alpha\right)^{1/\alpha}$ and with the shape parameter being unchanged.

When all the n components have the same distribution, such that $\lambda_i = \lambda$ for $i = 1, 2, \ldots, n$, then the series system has a Weibull life distribution with scale parameter $\lambda \cdot n^{1/\alpha}$ and shape parameter α. □

The Weibull distribution we have discussed so far is a two-parameter distribution with shape parameter $\alpha > 0$ and scale parameter $\lambda > 0$. A natural extension of this distribution is the three-parameter Weibull distribution (α, λ, ξ) with distribution function

$$F(t) = \Pr(T \leq t) = \begin{cases} 1 - e^{-[\lambda(t-\xi)]^\alpha} & \text{for } t > \xi \\ 0 & \text{otherwise} \end{cases}$$

The corresponding density is

$$f(t) = \frac{d}{dt} F(t) = \alpha\lambda[\lambda(t-\xi)]^{\alpha-1}\, e^{-[\lambda(t-\xi)]^\alpha} \quad \text{for } t > \xi$$

The third parameter ξ is sometimes called the *guarantee* or *threshold* parameter, since a failure occurs before time ξ with probability 0 (e.g., see Mann et al. 1974, p. 185).

Since $(T - \xi)$ obviously has a two-parameter Weibull distribution (α, λ), the mean and variance of the three-parameter Weibull distribution (α, λ, ξ) follows from (2.54): and (2.56).

$$\text{MTTF} = \xi + \frac{1}{\lambda}\Gamma\left(\frac{1}{\alpha} + 1\right)$$

$$\text{var}(T) = \frac{1}{\lambda^2}\left(\Gamma\left(\frac{2}{\alpha} + 1\right) - \Gamma^2\left(\frac{1}{\alpha} + 1\right)\right)$$

In statistical literature, reference to the Weibull distribution usually means the two-parameter family, unless otherwise specified.

2.13 THE NORMAL DISTRIBUTION

The most commonly used distribution in statistics is the normal (Gaussian[2]) distribution. A random variable T is said to be normally distributed with mean ν and variance τ^2, $T \sim \mathcal{N}(\nu, \tau^2)$, when the probability density of T is

$$f(t) = \frac{1}{\sqrt{2\pi} \cdot \tau} e^{-(t-\nu)^2/2\tau^2} \quad \text{for } -\infty < t < \infty \tag{2.57}$$

The $\mathcal{N}(0, 1)$ distribution is called the *standard normal distribution*. The distribution function of the standard normal distribution is usually denoted by $\Phi(\cdot)$. The probability density of the standard normal distribution is

$$\phi(t) = \frac{1}{\sqrt{2\pi}} e^{-t^2/2} \tag{2.58}$$

The distribution function of $T \sim \mathcal{N}(\nu, \tau^2)$ may be written as

$$F(t) = \Pr(T \leq t) = \Phi\left(\frac{t - \nu}{\tau}\right) \tag{2.59}$$

The normal distribution is sometimes used as a lifetime distribution, even though it allows negative values with positive probability.

The survivor function is

$$R(t) = 1 - \Phi\left(\frac{t - \nu}{\tau}\right) \tag{2.60}$$

The failure rate function of the normal distribution is

$$z(t) = -\frac{R'(t)}{R(t)} = \frac{1}{\tau} \cdot \frac{\phi((t - \nu)/\tau)}{1 - \Phi((t - \nu)/\tau)} \tag{2.61}$$

---

[2] Named after the German mathematician Johann Carl Friedrich Gauss (1777–1855).

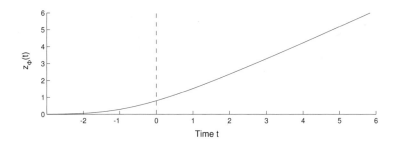

Fig. 2.18 Failure rate function of the standard normal distribution $\mathcal{N}(0,1)$.

If $z_\Phi(t)$ denotes the failure rate function of the standard normal distribution, the failure rate function of $\mathcal{N}(\nu, \tau^2)$ is seen to be

$$z(t) = \frac{1}{\tau} \cdot z_\Phi\left(\frac{t-\nu}{\tau}\right)$$

The failure rate function of the standard normal distribution, $\mathcal{N}(0,1)$, is illustrated in Fig. 2.18. The failure rate function is *increasing* for all t and approaches $z(t) = t$ when $t \to \infty$.

When a random variable has a normal distribution but with an upper bound and/or a lower bound for the values of the random variable, the resulting distribution is called a *truncated normal distribution*. When there is only a lower bound, the distribution is said to be left truncated. When there is only an upper bound, the distribution is said to be right truncated. Should there be an upper as well as a lower bound, it is said to be doubly truncated.

A normal distribution, left truncated at 0, is sometimes used as a life distribution. This left truncated normal distribution has survivor function

$$R(t) = \Pr(T > t \mid T > 0) = \frac{\Phi((\nu-t)/\tau)}{\Phi(\nu/\tau)} \quad \text{for } t \geq 0 \qquad (2.62)$$

The corresponding failure rate function becomes

$$z(t) = \frac{-R'(t)}{R(t)} = \frac{1}{\tau} \cdot \frac{\phi((t-\nu)/\tau)}{1 - \Phi((t-\nu)/\tau)} \quad \text{for } t \geq 0$$

Note that the failure rate function of the left truncated normal distribution is identical to the failure rate function of the (untruncated) normal distribution when $t \geq 0$.

Example 2.11
A specific type of car tires has an average wear-out "time" T of 50000 km, and 5% of the tires last for at least 70000 km. We will assume that T is normally distributed with mean $\nu = 50000$ km, and that $\Pr(T > 70000) = 0.05$. Let τ denote the standard deviation of T. The variable $(T - 50000)/\tau$ then has a standard normal distribution. Standardizing, we get

$$\Pr(T > 70000) = 1 - P\left(\frac{T-50000}{\tau} \leq \frac{70000-50000}{\tau}\right) = 0.05$$

Therefore
$$\Phi\left(\frac{20000}{\tau}\right) = 0.95 \approx \Phi(1.645)$$
and
$$\frac{20000}{\tau} \approx 1.645 \quad \Rightarrow \quad \tau \approx 12158$$

The probability that a tire will last more than 60 000 km is now
$$\begin{aligned} \Pr(T > 60000) &= 1 - P\left(\frac{T - 50000}{12\,158} \le \frac{60000 - 50000}{12158}\right) \\ &\approx 1 - \Phi(0.883) \approx 0.188 \end{aligned}$$

The probability of a "negative" life length is in this case
$$\Pr(T < 0) = P\left(\frac{T - 50000}{12158} < \frac{-50000}{12158}\right) \approx \Phi(-4.11) \approx 0$$

The effect of using a truncated normal distribution instead of a normal distribution is therefore negligible. □

2.14 THE LOGNORMAL DISTRIBUTION

The time to failure T of an item is said to be lognormally distributed with parameters ν and τ^2, $T \sim \text{lognormal}(\nu, \tau^2)$, if $Y = \ln T$ is normally (Gaussian) distributed with mean ν and variance τ^2 [i.e., $Y \sim \mathcal{N}(\nu, \tau^2)$]. The probability density function of T is

$$f(t) = \begin{cases} \dfrac{1}{\sqrt{2\pi}\,\tau\,t} e^{-(\ln t - \nu)^2 / 2\tau^2} & \text{for } t > 0 \\ 0 & \text{otherwise} \end{cases} \quad (2.63)$$

The lognormal probability density is sketched in Fig. 2.19 for selected values of ν and τ.

The mean time to failure is
$$\text{MTTF} = e^{\nu + \tau^2/2} \quad (2.64)$$

the median time to failure [satisfying $R(t_m) = 0.5$] is
$$t_m = e^{\nu} \quad (2.65)$$

and the mode of the distribution is
$$t_{\text{mode}} = e^{\nu - \tau^2}$$

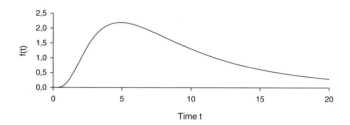

Fig. 2.19 Probability density of the lognormal distribution.

Notice that the MTTF may be written

$$\text{MTTF} = t_m \cdot e^{\tau^2/2}$$

and that the mode may be written

$$t_{\text{mode}} = t_m \cdot e^{-\tau^2}$$

It is therefore easy to see that

$$t_{\text{mode}} < t_m < \text{MTTF} \quad \text{for } \tau > 0$$

The variance of T is

$$\text{var}(T) = e^{2\nu}\left(e^{2\tau^2} - e^{\tau^2}\right) \tag{2.66}$$

The reliability (survivor) function becomes

$$\begin{aligned} R(t) &= \Pr(T > t) = \Pr(\ln T > \ln t) \\ &= P\left(\frac{\ln T - \nu}{\tau} > \frac{\ln t - \nu}{\tau}\right) = \Phi\left(\frac{\nu - \ln t}{\tau}\right) \end{aligned} \tag{2.67}$$

where $\Phi(\cdot)$ is the distribution function of the standard normal distribution.

The failure rate function of the lognormal distribution is

$$z(t) = -\frac{d}{dt}\left(\ln \Phi\left(\frac{\nu - \ln t}{\tau}\right)\right) = \frac{\phi((\nu - \ln t)/\tau)/\tau t}{\Phi((\nu - \ln t)/\tau)/\tau} \tag{2.68}$$

where $\phi(t)$ denotes the probability density of the standard normal distribution.

The shape of $z(t)$ which is illustrated in Fig. 2.20 is discussed in detail by Sweet (1990) who describes an iterative procedure to compute the time t for which the failure rate function attains its maximum value. He also proves that $z(t) \to 0$ when $t \to \infty$.

Let $T_1, T_2, \ldots, T_n$ be independent and lognormally distributed with parameters ν_i and τ_i^2 for $i = 1, 2, \ldots, n$. The product $T = \prod_{i=1}^{n} T_i$ is then lognormally distributed with parameters $\sum_{i=1}^{n} \nu_i$ and $\sum_{i=1}^{n} \tau_i^2$.

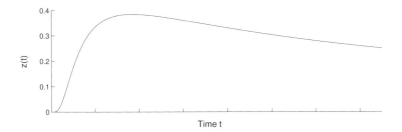

Fig. 2.20 Failure rate function of the lognormal distribution.

2.14.1 Repair Time Distribution

The lognormal distribution is commonly used as a distribution for repair time. The *repair rate* is defined analogous to the failure rate. When modeling the repair time, it is natural to assume that the repair rate is increasing, at least in a first phase. This means that the probability of completing the repair action within a short interval increases with the elapsed repair time. When the repair has been going on for a rather long time, this indicates serious problems, for example, that there are no spare parts available on the site. It is therefore natural to believe that the repair rate is decreasing after a certain period of time, namely, that the repair rate function has the same shape as the failure rate function of the lognormal distribution illustrated in Fig. 2.20.

2.14.2 Median and Error Factor

In some cases we may be interested to find an interval (t_L, t_U) such that $\Pr(t_L < T \leq t_U) = 1 - 2\alpha$, for example. If the interval is symmetric in the sense that $\Pr(T \leq t_L) = \alpha$ and $\Pr(T > t_U) = \alpha$, it is easy to verify that $t_L = e^{-u_\alpha \tau}$ and $t_U = e^{u_\alpha \tau}$, where u_α is the upper $\alpha\%$ percentile of the standard normal distribution [i.e., $\Phi(u_\alpha) = 1 - \alpha$]. By introducing the median $t_m = e^\nu$ and $k = e^{u_\alpha \tau}$, the lower limit t_L and the upper limit t_U may be written

$$t_L = \frac{t_m}{k} \quad \text{and} \quad t_U = k \cdot t_m \tag{2.69}$$

The factor k is often called the $(1 - 2\alpha)$ *error factor*, and α is usually chosen to be 0.05.

Example 2.12
In many situations the (constant) failure rate λ may vary from one item to another. In the Reactor safety study (NUREG-75/014), the variation (uncertainty) in λ was modeled by a lognormal distribution; that is, the failure rate λ is regarded as a random variable Λ with a lognormal distribution.

In the Reactor safety study the lognormal distribution was determined by the median λ_m and a 90% error factor k such that

$$P\left(\frac{\lambda_m}{k} < \Lambda < k \cdot \lambda_m\right) = 0.90$$

If we, as an example, choose the median to be $\lambda_m = 6.0 \cdot 10^{-5}$ failures per hour, and an error factor $k = 3$, then the 90% interval is equal to $(2.0 \cdot 10^{-5}, 1.8 \cdot 10^{-4})$. The parameters ν and τ of the lognormal distribution can now be determined from (2.65) and (2.69):

$$\nu = \ln(\lambda_m) = \ln 6.0 \cdot 10^{-5} \approx -9.721$$
$$\tau = \frac{1}{1.645} \ln k = \frac{1}{1.645} \ln 3 \approx 0.668$$

With these parameter values the mean failure rate λ is equal to

$$e^{\nu+\tau^2/2} \approx 7.50 \cdot 10^{-5} \text{ hours}^{-1}$$

□

Example 2.13 Fatigue Analysis
The lognormal distribution is also commonly used in the analysis of fatigue failures. Considering the following simple situation: A smooth, polished test rod of steel is exposed to sinusoidal stress cycles with a given stress range (double amplitude) s. We want to estimate the time to failure of the test rod (i.e., the number of stress cycles N, until fracture occurs). In this situation it is usually assumed that N is lognormally distributed. The justification for this is partly physical and partly mathematical convenience. A fatigue crack will always start in an area with local yield, normally caused by an impurity in the material. It seems reasonable that in the beginning the failure rate function increases with the number of stress cycles. If the test rod survives a large number of stress cycles, this indicates that there are very few impurities in the material. It is therefore to be expected that the failure rate function will decrease when the possibility for impurities in the material is reduced.

It is known that within a limited area of the stress range s, the number N of cycles to failure will roughly satisfy the equation

$$Ns^b = c \tag{2.70}$$

where b and c are constants depending on the material and the geometry of the test rod. They may also depend on the surface treatment and the environment in which the rod is used.

By taking the logarithms of both sides of (2.70) we get

$$\ln N = \ln c - b \ln s \tag{2.71}$$

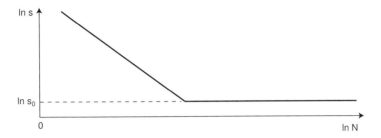

Fig. 2.21 Wöhler or s-N diagram.

If we introduce $Y = \ln N$, $\alpha = \ln c$, $\beta = -b$, and $x = \ln s$, it follows from (2.71) that Y roughly can be expressed by the relation

$$Y = \alpha + \beta x + \text{random error}$$

If N is assumed to be lognormally distributed, then $Y = \ln N$ will be normally distributed, and the usual theory for linear regression models (e.g., see Dudewicz and Mishra 1988) applies when estimating the expected number of cycles to failure for a given stress range s. Equation (2.70) represents the Wöhler or s-N diagram for the test rod. Such a diagram is illustrated in Fig. 2.21. When the stress range is below a certain value s_0, the test rod will not fracture, irrespective of how many stress cycles it is exposed to. Formula (2.71) will therefore be valid only for stress values above s_0.

The stress range s_0 is called the fatigue limit. For certain materials such as aluminium, the Wöhler curve has no horizontal asymptote. Such materials therefore have no fatigue limit. In a corrosive environment, such as salt water, neither does steel have any fatigue limit. □

2.15 THE BIRNBAUM-SAUNDERS DISTRIBUTION

If the fatigue cannot be attributed to sinusoidal stress cycles, formula (2.70) cannot be applied directly. The oldest, simplest, and most used procedure in this situation is Miner's rule (Miner 1945): The various applied stress ranges are divided into a certain number m of discrete stress ranges $s_1, s_2, \ldots, s_m$. Let n_j denote the numbers of stress cycles occurring under stress range s_j, $j = 1, 2, \ldots, m$. Then, according to Miner's rule, fatigue fracture will occur when

$$\frac{n_1}{N_1} + \frac{n_2}{N_2} + \cdots + \frac{n_m}{N_m} = 1 \tag{2.72}$$

where N_j is the hypothetic number of stress cycles until fracture if the item were exposed to pure sinusoidal stress cycles with stress range s_j, $j = 1, 2, \ldots, m$. Experience shows that by using Miner's rule the average time to failure will often be overestimated.

48 FAILURE MODELS

Birnbaum and Saunders (1969) introduced a life distribution based on a stochastic interpretation of Miner's rule. They argue as follows: Consider an item which is exposed to a sequence of work cycles each of which is composed of m individual stresses. Each of the work cycles is the source of a partial damage which is stochastic and which may depend on factors such as the material and the number of earlier stresses. Assume that the increase Z_j in partial damage in work cycle j is a random variable with mean value μ and variance σ^2 for all $j = 1, 2, \ldots$. Furthermore, assume that increases in partial damage due to different work cycles are independent of each other.

Let $W_n = Z_1 + Z_2 + \cdots + Z_n$ denote the total partial damage after n work cycles. Further let N be the smallest number of work cycles for which W_n exceeds the critical value which causes a fatigue fracture to occur. Then we have

$$\Pr(N \leq n) \approx \Pr(W_n > \omega) = 1 - P\left(\sum_{i=1}^{n} Z_i \leq \omega\right)$$

$$= 1 - P\left(\sum_{i=1}^{n} \frac{Z_i - \mu}{\sigma\sqrt{n}} \leq \frac{\omega - n\mu}{\sigma\sqrt{n}}\right) \quad (2.73)$$

Since the Z_i's are assumed to be independent and identically distributed, then W_n for large n is approximately normally distributed $\mathcal{N}(n\mu, n\sigma^2)$ by the central limit theorem (e.g., see Dudewicz and Mishra 1988, p. 315). Thus

$$\Pr(N \leq n) \approx \Phi\left(\frac{\omega - n\mu}{\sigma\sqrt{n}}\right) \quad (2.74)$$

This discrete model is now extended to a continuous model by replacing "life length" N by T, where T is a continuous variable.

The life distribution for the item is then

$$F(t) = \Pr(T \leq t) \approx 1 - \Phi\left(\frac{\omega - t\mu}{\sigma\sqrt{t}}\right) \quad (2.75)$$

By introducing $\alpha = \sigma/\sqrt{\mu\omega}$ and $\lambda = \mu/\omega$, we obtain

$$F(t) \approx \Phi\left(\frac{1}{\alpha}\left(\sqrt{\lambda t} - \frac{1}{\sqrt{\lambda t}}\right)\right) \quad (2.76)$$

The distribution on the right-hand side is called the *Birnbaum-Saunders* distribution with shape parameter α and scale parameter λ.

The probability density of the Birnbaum-Saunders distribution is

$$f(t) = \frac{\sqrt{\lambda t} + 1/\sqrt{\lambda t}}{2\alpha t} \frac{1}{\sqrt{2\pi}} e^{-(\sqrt{\lambda t} - (1/\sqrt{\lambda t}))^2/2\alpha^2} \quad \text{for } t > 0 \quad (2.77)$$

This probability density is plotted for selected values of α in Fig. 2.22.

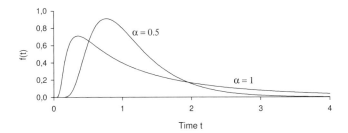

Fig. 2.22 Probability density of the Birnbaum-Saunders distribution ($\lambda = 1$).

The survivor function of the Birnbaum-Saunders distribution is

$$R(t) = \Phi\left(\frac{1}{\alpha}\left(\frac{1}{\sqrt{\lambda t}} - \sqrt{\lambda t}\right)\right)$$

The failure rate function $z(t) = f(t)/R(t)$ is illustrated in Fig. 2.23 for $\alpha = 0.5$ and $\alpha = 1$. It is relatively easy to show (see Problem 2.39) that $\lim_{t \to \infty} = \lambda/2\alpha^2$.

It can furthermore be shown that

$$U = \frac{1}{\alpha}\left(\sqrt{\lambda T} - \frac{1}{\sqrt{\lambda T}}\right)$$

has a standard normal distribution, and that

$$\text{MTTF} = \frac{1}{\lambda}\left(1 + \frac{\alpha^2}{2}\right) \qquad (2.78)$$

$$\text{var}(T) = \frac{\alpha^2}{\lambda^2}\left(1 + \frac{5\alpha^2}{4}\right) \qquad (2.79)$$

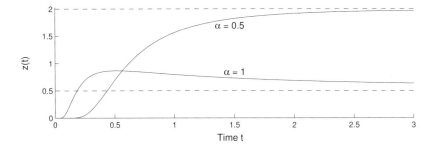

Fig. 2.23 Failure rate function of the Birnbaum-Saunders distribution ($\lambda = 1$).

50 FAILURE MODELS

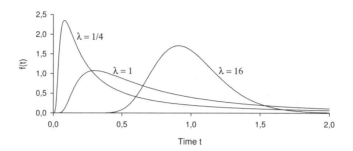

Fig. 2.24 The probability density of the inverse Gaussian distribution for the parameters $\mu = 1$ and $\lambda = 1/4, 1, 16$.

2.16 THE INVERSE GAUSSIAN DISTRIBUTION

In some situations the failure rate function $z(t)$ appears to be an increasing function of time t from $t = 0$ to $t = t_0$ (unknown), and from t_0 on monotonically decreasing. Among the life distributions having the mentioned property is the lognormal distribution. This distribution is in such situations often "shown to be the appropriate one" by use of some goodness of fit criterion, and then used without further consideration. However, if one studies the lognormal distribution a little more closely, it can be shown (Sweet 1990) that the corresponding failure rate function, after having reached its maximum, decreases monotonically to *zero*, as $t \to \infty$. A natural question is then: Is this a reasonable assumption? In many practical situations, this will *not* be the case, and hence the lognormal life distribution should *not* be used as model, even if the goodness of fit test does not reject it.

We now turn to another life distribution, namely the *inverse Gaussian* distribution. The inverse Gaussian, like the lognormal distribution, has a failure rate function, increasing with time t, from $t = 0$ to $t = t_0$ (unknown), and thereafter monotonically decreasing, however, *not toward zero*, but toward a limit, depending on the parameters of the distribution function. (As we will see, the name inverse Gaussian is rather misleading).

For the reason mentioned above, we claim that in many situations where the lognormal model has been used, it would have been more appropriate to apply the inverse Gaussian distribution. The probability density of the inverse Gaussian distribution can be expressed in many different ways, depending on how the distribution is parameterized. We are going to use the following form:

$$f_T(t; \mu, \lambda) = \sqrt{\frac{\lambda}{2\pi t^3}} e^{-(\lambda/2\mu^2)[(t-\mu)^2/t]} \text{ for } t > 0, \mu > 0, \lambda > 0$$
$$= 0 \text{ otherwise} \qquad (2.80)$$

This distribution will be referred to as the inverse Gaussian distribution with parameters μ and λ, and we sometimes write $T \sim \text{IG}(\mu, \lambda)$. The shape of the probability

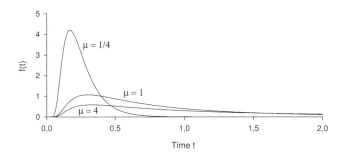

Fig. 2.25 The probability density of the inverse Gaussian distribution for the parameters $\mu = 1/4, \ 1, \ 4$ and $\lambda = 1$.

density can be varied considerably by varying μ and λ. This is illustrated in Fig. 2.24 and 2.25.

In Section 2.12 we presented a heuristic argument leading to the Birnbaum-Saunders distribution. Their derivation has been criticized by Bhattacharayya and Fries (1982), who instead propose that one, in this situation, rather should regard the accumulated fatigue in the time interval $(0, t]$, $W(t)$, to be governed by a Wiener process $\{W(t); 0 < t < \infty\}$ with positive drift η and diffusion constant δ^2; that is, for any $(0 \leq s < t)$ the distribution of $(W(t) - W(s))$ is $N(\eta(t-s), \delta^2(t-s))$ (e.g., see Cox and Miller 1965). Furthermore the item in question is supposed to fail when $W(t)$ for the first time exceeds ω (the critical level of failure). Accordingly, the time T to failure is defined as

$$T = \inf_t \{t; W(t) > \omega\}$$

Under these assumptions it follows (Cox and Miller 1965, p. 221), that the distribution of T is given by

$$\begin{aligned} F_T(t) &= \Phi\left(\frac{\eta}{\delta}\sqrt{t} - \frac{\omega}{\delta}\frac{1}{\sqrt{t}}\right) \\ &+ \Phi\left(-\frac{\eta}{\delta}\sqrt{t} - \frac{\omega}{\delta}\frac{1}{\sqrt{t}}\right) \cdot e^{2\eta\omega/\delta^2} \quad \text{for } t > 0 \end{aligned} \quad (2.81)$$

where $\Phi(\cdot)$ denotes the distribution function of the standard normal distribution.

By introducing new parameters

$$\mu = \frac{\omega}{\eta}, \quad \lambda = \frac{\omega^2}{\delta^2} \quad \text{for } \mu > 0, \ \lambda > 0 \quad (2.82)$$

52   FAILURE MODELS

$F_T(t)$ can be written

$$F_T(t) = \Phi\left(\frac{\sqrt{\lambda}}{\mu}\sqrt{t} - \sqrt{\lambda}\frac{1}{\sqrt{t}}\right)$$
$$+\Phi\left(-\frac{\sqrt{\lambda}}{\mu}\sqrt{t} - \sqrt{\lambda}\frac{1}{\sqrt{t}}\right) \cdot e^{2\lambda/\mu}$$

for $t > 0$, $\mu > 0$, $\lambda > 0$. (2.83)

The corresponding probability density becomes

$$f_T(t) = \sqrt{\frac{\lambda}{2\pi t^3}} \cdot e^{-(\lambda/2\mu^2)\,[(t-\mu)^2/t]} \quad \text{for } t > 0\; \mu > 0,\; \lambda > 0 \qquad (2.84)$$

which we recognize as the probability density of IG(μ, λ). Without loss of generality, ω can be chosen to be $= 1$. Then μ corresponds to $1/\eta$ and λ to $1/\delta^2$.

This distribution was first derived by Schrödinger (1915) in connection with his studies of Brownian motion. In an attempt to extend Schrödinger's results, Tweedie (1957) noticed the inverse relationship between the cumulant-generating function of the time to cover a unit distance and the cumulant-generating function of the distance covered in unit time. In 1956 he used for the first time "inverse Gaussian" for the first passage time of the Brownian motion. Thereby the distribution got the rather misleading name "inverse Gaussian."

Wald (1947) derived the same distribution as a limiting distribution for the sample size in connection with a certain sequential probability ratio test. A heuristic derivation is given by Whitmore and Seshadri (1987).

The moment generating function (see Dudewicz and Mishra 1988, p. 255) of the IG(μ, λ) turns out to be

$$M_T(t; \mu, \lambda) = e^{\lambda[1-(1-2\mu^2 t/\lambda)^{1/2}]/\mu} \qquad (2.85)$$

and the corresponding cumulant-generating function, defined by:

$$L_T(t; \mu, \lambda) = \ln M_T(t; \mu, \lambda) \qquad (2.86)$$

becomes

$$L_T(t; \mu, \lambda) = \frac{\lambda}{\mu}\left(1 - \left(1 - \frac{2\mu^2 t}{\lambda}\right)^{1/2}\right) \qquad (2.87)$$

The mean and variance of IG(μ, λ) are now easily determined

$$\text{MTTF} = \kappa_1 = \frac{d}{dt}L_T(t; \mu, \lambda)|_{t=0} = \mu \qquad (2.88)$$

$$\text{var}(T) = \kappa_2 = \frac{d^2}{dt^2}L_T(t; \mu, \lambda)|_{t=0} = \frac{\mu^3}{\lambda} \qquad (2.89)$$

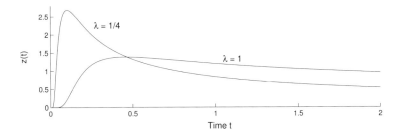

Fig. 2.26 Failure rate function of the inverse Gaussian distribution ($\mu = 1$, and $\lambda = 1/4,\ 1$).

Notice that μ enters in MTTF $= \mu$, as well as into var(T). Hence (μ, λ) are *not* location/scale parameters in the usual sense. Also notice that

$$\lambda = \frac{\kappa_1^3}{\kappa_2} \tag{2.90}$$

From (2.85) it follows that moments of arbitrarily high (positive) order exist for IG(μ, λ). From the result stated in Problem 2.41, it follows that moments of arbitrarily high "negative" order, also exist.

The failure rate function of IG(μ, λ) is

$$z(t) = z(t; \mu, \lambda) = \frac{f_T(t; \mu, \lambda)}{1 - F_T(t; \mu, \lambda)}$$

$$= \frac{\sqrt{\frac{\lambda}{2\pi t^3}}\ e^{-(\lambda/2\mu^2)[(t-\mu)^2/t]}}{\Phi(\sqrt{\lambda}/\sqrt{t} - \sqrt{\lambda}\sqrt{t}/\mu) - \Phi(-\sqrt{\lambda}/\sqrt{t} - \sqrt{\lambda}\sqrt{t}/\mu) \cdot e^{2\lambda/\mu}} \tag{2.91}$$

for $t > 0$, $\mu > 0$, and $\lambda > 0$. In Fig. 2.26 graphs of the failure rate function (2.91) are given for $\mu = 1$ and selected values of λ. Chhikara and Folks (1977, p. 155) have shown that

$$\lim_{t \to \infty} z(t) = \frac{\lambda}{2\mu^2} \tag{2.92}$$

Chhikara and Folks (1989, Section 1.3) point out a surprising analogy between sampling results for the inverse Gaussian distribution IG(μ, λ) and sampling results known for the normal distribution $\mathcal{N}(\mu, \tau^2)$. If one considers a random sample $T_1, T_2, \ldots, T_n$ from the inverse Gaussian distribution IG(μ, λ):

1. $\overline{T} = 1/n \sum_{j=1}^{n} T_j$ is inverse Gaussian IG($\mu, n\lambda$).

2. $n\lambda \sum_{j=1}^{n}(1/T_j - 1/\overline{T})$ is χ^2 distributed with $(n-1)$ degrees of freedom.

3. $\overline{T}$ and $\sum_{j=1}^{n}(1/T_j - 1/\overline{T})$ are independently distributed.

4. The term in the exponent of the probability density function of $\text{IG}(\mu, \lambda)$ is $-1/2$ times a χ^2 distributed variable.

The explanation of these and other analogies with the normal distribution has yet not been revealed.

2.17 THE EXTREME VALUE DISTRIBUTIONS

Extreme value distributions play an important role in reliability analysis. They arise in a natural way, for example, in the analysis of engineering systems, made up of n identical items with a series structure, and also in the study of corrosion of metals, of material strength, and of breakdown of dielectrics.

Let $T_1, T_2, \ldots, T_n$ be independent, identically distributed random variables (not necessarily life lengths) with a continuous distribution function $F_T(t)$, for the sake of simplicity assumed to be strictly increasing for $F_T^{-1}(0) < t < F_T^{-1}(1)$. Then

$$T_{(1)} = \min\{T_1, T_2, \ldots, T_n\} = U_n \tag{2.93}$$
$$T_{(n)} = \max\{T_1, T_2, \ldots, T_n\} = V_n \tag{2.94}$$

are called the *extreme values*.

The distribution functions of U_n and V_n are easily expressed by $F_T(\cdot)$ in the following way (e.g., see Cramér 1946, p. 370; Mann et al. 1974, p. 102).

$$F_{U_n}(u) = 1 - (1 - F_T(u))^n = L_n(u) \tag{2.95}$$
$$F_{V_n}(v) = F_T(v)^n = H_n(v) \tag{2.96}$$

Despite the simplicity of (2.95) and (2.96), these formulas are usually not easy to work with. If $F_T(t)$, say, represents a normal distribution, one is led to work with powers of $F_T(t)$, which may be cumbersome.

However, in many practical applications to reliability problems, n is very large. Hence, one is lead to look for asymptotic techniques, which under general conditions on $F_T(t)$ may lead to simple representations of $F_{U_n}(u)$ and $F_{V_n}(v)$.

Cramér (1946) suggested the following approach: Introduce

$$Y_n = n F_T(U_n) \tag{2.97}$$

where U_n is defined as in (2.93). Then for $y \geq 0$,

$$\begin{aligned}
\Pr(Y_n \leq y) &= P\left[F_T(U_n) \leq \frac{y}{n}\right] \\
&= P\left[U_n \leq F_T^{-1}\left(\frac{y}{n}\right)\right] \\
&= F_{U_n}\left[F_T^{-1}\left(\frac{y}{n}\right)\right] \\
&= 1 - \left[1 - F_T\left(F_T^{-1}\left(\frac{y}{n}\right)\right)\right] \\
&= 1 - \left(1 - \frac{y}{n}\right)^n \tag{2.98}
\end{aligned}$$

As $n \to 0$

$$\Pr(Y_n \le y) \to 1 - e^{-y} \quad \text{for } y > 0 \qquad (2.99)$$

Since the right-hand side of (2.99) expresses a distribution function,[3] continuous for $y > 0$, this implies that Y_n converges in distribution to a random variable Y, with distribution function:

$$F_Y(y) = 1 - e^{-y} \quad \text{for } y > 0 \qquad (2.100)$$

Hence it follows from (2.95) that the sequence of random variables U_n converges in distribution to the random variable $F_T^{-1}(Y/n)$.

$$U_n \xrightarrow{\mathcal{L}} F_T^{-1}\left(\frac{Y}{n}\right) \qquad (2.101)$$

Similarly, let

$$Z_n = n(1 - F_T(V_n)) \qquad (2.102)$$

where V_n is defined in (2.94). By an analogous argument it can be shown that for $z > 0$.

$$\Pr(Z_n \le z) = 1 - \left(1 - \frac{z}{n}\right)^n \qquad (2.103)$$

which implies that

$$V_n \xrightarrow{\mathcal{L}} F_T^{-1}\left(1 - \frac{Z}{n}\right) \qquad (2.104)$$

where Z has distribution function

$$\Pr(Z \le z) = 1 - e^{-z} \quad \text{for } z > 0 \qquad (2.105)$$

It is to be expected that the limit distribution of U_n and V_n will depend on the type of distribution $F_T(\cdot)$. However, it turns out that there are only three possible types of limiting distributions for the minimum extreme U_n, and only three possible types of limiting distributions for the maximum extreme V_n.

For a comprehensive discussion of the application of extreme value theory to reliability analysis, the reader is referred to Mann et al. (1974, p. 106), Nelson (1982, p. 39), Lawless (1982, p. 169), and Johnson and Kotz (1970, Vol. I, Section 2.1). Here we will content ourselves with mentioning three of the possible types of limiting distributions and indicate areas where they are applied.

---

[3] The exponential distribution with parameter $\lambda = 1$.

2.17.1 The Gumbel Distribution of the Smallest Extreme

If the probability density $f_T(t)$ of the T_i's approaches zero exponentially as $t \to \infty$, then the limiting distribution of $U_n = T_{(1)} = \min\{T_1, T_2, \ldots, T_n\}$ is of the form

$$F_{T_{(1)}}(t) = 1 - e^{-e^{(t-\vartheta)/\alpha}} \quad -\infty < t < \infty \quad (2.106)$$

where $\alpha > 0$ and ϑ are constants. α is the mode and ϑ is a scale parameter.

The corresponding "survivor" function is

$$R_{T_{(1)}}(t) = 1 - F_{T_{(1)}}(t) = e^{-e^{(t-\vartheta)/\alpha}} \quad -\infty < t < \infty \quad (2.107)$$

Gumbel (1958) denotes this distribution the type I asymptotic distribution of the smallest extreme. It is now called the *Gumbel distribution of the smallest extreme*. If standardized variables

$$Y = \frac{T - \vartheta}{\alpha} \quad (2.108)$$

are introduced, the distribution function takes the form

$$F_{Y_{(1)}}(y) = 1 - e^{-e^y} \quad \text{for } -\infty < y < \infty$$

with probability density

$$f_{Y_{(1)}}(y) = e^y \cdot e^{-e^y} \quad \text{for } -\infty < y < \infty \quad (2.109)$$

The corresponding "failure rate" is

$$z_{Y_{(1)}}(y) = \frac{f_{Y_{(1)}}(y)}{1 - F_{Y_{(1)}}(y)} = e^y \quad \text{for } -\infty < y < \infty \quad (2.110)$$

The mean value of $T_{(1)}$ is (see Lawless, 1974, p. 19)

$$E(T_{(1)}) = \vartheta - \alpha\gamma$$

where $\gamma = 0.5772\ldots$ is known as Euler's constant.

Since $T_{(1)}$ can take negative values, (2.106) is not a valid life distribution. A valid life distribution is, however, obtained by left truncating (2.106) at $t = 0$. In this way we get the *truncated* Gumbel distribution of the smallest extreme, which is given by the survivor function

$$\begin{aligned} R^0_{T_{(1)}}(t) &= \Pr(T_{(1)} > t \mid T > 0) = \frac{\Pr(T_{(1)} > t)}{\Pr(T_{(1)} > 0)} \\ &= \frac{e^{-e^{(t-\vartheta)/\alpha}}}{e^{-e^{\vartheta/\alpha}}} = e^{-e^{-(\vartheta/\alpha)(e^{t/\alpha}-1)}} \quad \text{for } t > 0 \end{aligned} \quad (2.111)$$

By introducing new parameters $\beta = e^{-\vartheta/\alpha}$ and $\varrho = 1/\alpha$, the truncated Gumbel distribution of the smallest extreme is given by the survivor function

$$R^0_{T_{(1)}}(t) = e^{-\beta(e^{\varrho t}-1)} \quad \text{for } t > 0 \quad (2.112)$$

The failure rate function of the truncated distribution is

$$z^0_{T_{(1)}}(t) = -\frac{d}{dt}\ln R^0_{T_{(1)}}(t) = \frac{d}{dt}\beta(e^{\varrho t} - 1) = \beta\varrho e^{\varrho t} \quad \text{for } t \geq 0 \quad (2.113)$$

2.17.2 The Gumbel Distribution of the Largest Extreme

If the probability density $f_T(t)$ approaches zero exponentially as $t \to \infty$, then the limiting distribution of $V_n = T_{(n)} = \max\{T_1, T_2, \ldots, T_n\}$ is of the form

$$F_{T_{(n)}}(t) = e^{-e^{-(t-\vartheta)/\alpha}} \quad \text{for } -\infty < t < \infty$$

where $\alpha > 0$ and ϑ are constants. Gumbel (1958) denotes this distribution the type I asymptotic distribution of the largest extreme. It is now called the *Gumbel distribution of the largest extreme*.

If standardized variables are introduced, see (2.89), the distribution takes the form

$$F_{Y_{(n)}}(y) = e^{-e^{-y}} \quad \text{for } -\infty < y < \infty \tag{2.114}$$

with probability density

$$f_{Y_{(n)}}(y) = e^{-y} \cdot e^{-e^{-y}} \quad \text{for } -\infty < y < \infty \tag{2.115}$$

2.17.3 The Weibull Distribution of the Smallest Extreme

Another type of such limiting distributions for the smallest extreme is the Weibull distribution

$$F_{T_{(1)}}(t) = 1 - e^{((t-\vartheta)/\eta)^\beta} \quad \text{for } t \geq \vartheta \tag{2.116}$$

where $\beta > 0$, $\eta > 0$, and $\vartheta > 0$ are constants.

Introducing standardized variables (see 2.108),

$$F_{Y_{(1)}}(y) = 1 - e^{-y^\beta} \quad \text{for } y > 0 \text{ and } \beta > 0 \tag{2.117}$$

This distribution is also denoted the type III asymptotic distribution of the smallest extreme.

Example 2.14 Pitting Corrosion

Consider a steel pipe with wall thickness θ which is exposed to corrosion. Initially the surface has a certain number n of microscopic pits. Pit i has a depth D_i, for $i = 1, 2, \ldots, n$. Due to corrosion, the depth of each pit will increase with time. Failure occurs when the first pit penetrates the surface, that is, when $\max\{D_1, D_2, \ldots, D_n\} = \theta$.

Let T_i denote the time pit i will need to penetrate the surface, for $i = 1, 2, \ldots, n$. Then the time to failure T of the item is

$$T = \min\{T_1, T_2, \ldots, T_n\}$$

We will assume that the time to penetration T_i is proportional to the remaining wall thickness, that is, $T_i = k \cdot (\theta - D_i)$. We will further assume that k is independent of time, which implies that the corrosion rate is constant.

58   FAILURE MODELS

Assume next that the random initial depths of the pits $D_1, \ldots, D_n$ are independent and identically distributed with a right truncated exponential distribution. Then the distribution function of D_i is

$$F_{D_i}(d) = \Pr(D_i \leq d \mid D_i \leq \theta) = \frac{\Pr(D_i \leq d)}{\Pr(D_i \leq \theta)}$$

$$= \frac{1 - e^{-\eta d}}{1 - e^{-\eta \theta}} \quad \text{for } 0 \leq d \leq \theta$$

The distribution function of the time to penetration, T_i, is thus

$$F_{T_i}(t) = \Pr(T_i \leq t) = \Pr(k \cdot (\theta - D_i) \leq t) = P\left(D_i \geq \theta - \frac{t}{k}\right)$$

$$= 1 - F_{D_i}\left(\theta - \frac{t}{k}\right) = \frac{e^{\eta t/k} - 1}{e^{\eta \theta} - 1} \quad \text{for } 0 \leq t \leq k\theta \quad (2.118)$$

and the survivor function $R(t)$ of the item becomes

$$R(t) = \Pr(T > t) = [1 - F_{T_i}(t)]^n \quad \text{for } t \geq 0$$

If we assume that the number n of pits is very large, then as $n \to \infty$, we get

$$R(t) = [1 - F_{T_i}(t)]^n \approx e^{-n F_{T_i}(t)} \quad \text{for } t \geq 0$$

By using (2.118)

$$R(t) \approx e^{-n \frac{e^{\eta t/k} - 1}{e^{\eta \vartheta}}} \quad \text{for } t \geq 0$$

By introducing new parameters $\beta = n/(e^{\eta \vartheta} - 1)$ and $\varrho = \eta/k$, we get

$$R(t) \approx e^{-\beta(e^{\varrho t} - 1)} \quad \text{for } t \geq 0$$

which is equal to (2.112), namely the time to failure caused by pitting corrosion has approximately a truncated Gumbel distribution of the smallest extreme.

The same example is also discussed by Lloyd and Lipow (1962, p. 140), Mann et al. (1974, p. 131), and Kapur and Lamberson (1977, p. 44). □

2.18   STRESSOR-DEPENDENT MODELING

So far we have mainly considered parametric families of time to failure distributions, tacitly assuming that the items tested are exposed to constant stress (normal stress).

If the items tested are exposed to planned variations in operational and environmental stresses (temperature, pressure, etc.), like in accelerated testing (see Chapter 12), a stressor-dependent model of the life distribution is needed.

A typical approach, which if successful, leads to a parametric stressor-dependent family of time to failure distributions in the following:

1. Establish an appropriate parametric family of time to failure distributions valid under normal stress as wells as under overstress.

2. Establish the functional relationship between the stressors and the parameters of the time to failure distribution in question.

3. Combine the result of these two steps into a stressor dependent family of time to failure distributions.

Example 2.15
Let there be only one stressor s, say temperature. Suppose that the time to failure distribution under normal stress as well as under overstress, is a Weibull distribution (see Section 2.12). Furthermore, suppose that the functional relationship between the stressor s and the parameters α and λ may be expressed as

$$\lambda(s) = cs^b, \text{ where } c \text{ and } b \text{ are unknown constants (power rule model)}$$
$$\alpha(s) = \alpha, \text{ independent of s}$$

When these two assumptions are combined, we obtain the following parametric stressor-dependent time to failure distribution:

$$\Pr(T \leq t; s) = 1 - e^{-(cs^b t)^\alpha}$$

Such models are discussed in more detail in Chapter 12. (In Chapter 12 semiparametric and nonparametric stressor-dependent models are also briefly discussed.) □

2.19 SOME FAMILIES OF DISTRIBUTIONS

2.19.1 IFR and DFR Distributions

In Section 2.5 we showed the following one-to-one correspondence between the distribution function $F(t)$ of a continuous life distribution and the corresponding failure rate function $z(t)$:

$$F(t) = 1 - e^{-\int_0^t z(u)\,du} \quad \text{for } t > 0 \tag{2.119}$$

A special family of such life distributions are those which have increasing[1] failure rate function. These are called increasing failure rate (IFR) distributions. Similarly, life distributions that have decreasing failure rate function are called decreasing failure rate (DFR) distributions. Definitions of IFR and DFR distributions that are not restricted to continuous distributions only are given below:

---

[1] In this section the concepts "increasing" and "decreasing" are used in place of "nondecreasing" and "nonincreasing," respectively.

Definition 2.1 A life distribution F is said to be IFR if $-\ln(1 - F(t))$ is convex for $0 < t < F^{-1}(1)$. A life distribution F is said to be DFR if $-\ln(1 - F(t))$ is concave for $0 < t < F^{-1}(1)$. □

Remember that when the life distribution is continuous, the failure rate function $z(t)$ can be written

$$z(t) = \frac{f(t)}{1 - F(t)} = \frac{d}{dt}(-\ln(1 - F(t))) \quad \text{for } t > 0 \qquad (2.120)$$

Since a differentiable convex function always has an increasing derivative (for continuous life distributions), the definition given above of the IFR property corresponds to an increasing $z(t)$ (for continuous life distributions). Since a differentiable concave function always has a derivative that is decreasing, by analogy the definition of the DFR property given above corresponds to a decreasing $z(t)$ (for continuous life distributions). Let us consider some commonly used life distributions and see whether they are IFR, DFR, or neither of these.

Example 2.16 The Uniform Distribution over $(0, a]$
Let T be uniformly distributed over $(0, a]$. Then

$$F(t) = \frac{t}{a} \quad \text{for } 0 < t \leq a$$

$$f(t) = \frac{1}{a} \quad \text{for } 0 < t \leq a$$

Hence

$$z(t) = \frac{1/a}{1 - (1/a)} = \frac{1}{a - t} \quad \text{for } 0 < t \leq a \qquad (2.121)$$

is strictly increasing in t for $0 < t \leq a$. The uniform distribution is accordingly IFR.
The same conclusion follows by considering $-\ln(1 - F(t))$ which in this case becomes $-\ln(1 - \frac{t}{a})$ and hence is convex for $0 < t \leq a$. □

Example 2.17 The Exponential Distribution
Let T be exponentially distributed with probability density

$$f(t) = \lambda e^{-\lambda t} \quad \text{for } t > 0$$

Then

$$z(t) = \lambda \quad \text{for } t > 0$$

$z(t)$ is thus constant, that is, both nonincreasing and nondecreasing.
The exponential distribution therefore belongs to the IFR family as well as the DFR family. Alternatively, one could argue that $-\ln(1 - F(t)) = \lambda t$, that is, convex and concave as well. □

Hence the families of IFR distributions and DRF distributions are not disjoint. The exponential distribution can be shown to be the only continuous distribution that belongs to both families (see Barlow and Proschan, 1975, p. 73).

Example 2.18 The Weibull Distribution

The distribution function of the Weibull distribution with parameters $\alpha(>0)$ and $\lambda(>0)$ is given by

$$F(t) = 1 - e^{-(\lambda t)^\alpha} \quad \text{for } t \geq 0$$

It follows that

$$-\ln(1 - F(t)) = -\ln(e^{-(\lambda t)^\alpha}) = (\lambda t)^\alpha \quad (2.122)$$

Since $(\lambda t)^\alpha$ is convex in t when $\alpha \geq 1$ and concave in t when $\alpha \leq 1$, the Weibull distribution is IFR for $\alpha \geq 1$ and DFR for $\alpha \leq 1$. For $\alpha = 1$, the distribution is "reduced" to an exponential distribution with parameter λ, and hence is IFR as well as DFR. □

Example 2.19 The Gamma Distribution

The gamma distribution is defined by the probability density

$$f(t) = \frac{\lambda}{\Gamma(\alpha)} (\lambda t)^{\alpha-1} e^{-\lambda t} \quad \text{for } t > 0$$

where $\alpha > 0$ and $\lambda > 0$. To determine whether the gamma distribution (α, λ) is IFR, DFR or neither of these, we consider the failure rate function:

$$z(t) = \frac{[\lambda(\lambda t)^{\alpha-1} e^{-\lambda t}]/\Gamma(\alpha)}{\int_t^\infty [\lambda(\lambda u)^{\alpha-1} e^{-\lambda u}]/\Gamma(\alpha)\, du}$$

By dividing the denominator by the numerator we get

$$z(t)^{-1} = \int_t^\infty \left(\frac{u}{t}\right)^{\alpha-1} e^{-\lambda(u-t)}\, du$$

By introducing $v = (u - t)$ as a new variable of integration we get

$$z(t)^{-1} = \int_0^\infty \left(1 + \frac{v}{t}\right)^{\alpha-1} e^{-\lambda v}\, dv \quad (2.123)$$

First suppose that $\alpha \geq 1$. Then $(1 + (v/t))^{\alpha-1}$ is nonincreasing in t. Accordingly the integrand is a decreasing function of t. Thus $z(t)^{-1}$ is decreasing in t. When $\alpha \geq 1$, $z(t)$ is in other words increasing in t, and the gamma distribution (α, λ) is IFR. This will in particular be the case when α is an integer (the Erlangian distribution).

Next suppose $\alpha \leq 1$. Then by an analogous argument $z(t)$ will be decreasing in t, which means that the gamma distribution (α, λ) is DRF.

For $\alpha = 1$, the gamma distribution (α, λ) is reduced to an exponential distribution with parameter λ. □

The plot of the failure rate function given in Fig. 2.20 for a lognormal distribution indicates that this distribution is neither IFR nor DFR. The following result may be useful when deciding whether a continuous distribution is DFR or not.

Theorem 2.1 If a continuous life distribution is to be DFR, its probability density $f(t)$ must be nonincreasing.

Proof
If the time to failure distribution is DFR and continuous, $z(t) = f(t)/(1 - F(t))$ must be decreasing. Knowing that $1 - F(t)$ is decreasing in t, then $f(t)$ must decrease by at least as much as $1 - F(t)$ in order for $z(t)$ to be decreasing. □

2.19.2 IFRA and DFRA Distributions

Consider a life distribution $F(t)$ with failure rate function $z(t)$. The cumulative failure rate function is according to Section 2.5:

$$-\int_0^t z(u)\,du = \ln R(t) = \ln(1 - F(t))$$

Definition 2.2 A distribution $F(t)$ with failure rate function $z(t)$ is said to have an increasing failure rate average (F is IFRA) if

$$-\frac{1}{t}\ln(1 - F(t)) \quad \text{increases with } t \geq 0$$

A distribution $F(t)$ with failure rate function $z(t)$ is said to have a decreasing failure rate average (F is DFRA) if

$$-\frac{1}{t}\ln(1 - F(t)) \quad \text{decreases with } t \geq 0$$

□

The IFRA (DFRA) property demands less of the failure rate function $z(t)$ than the IFR (DFR) property does. It is straightforward to verify that if F is IFR (DFR), then it is also IFRA (DFRA).

2.19.3 NBU and NWU Distributions

The conditional survivor function $R(t \mid x)$ was introduced in Section 2.7.

$$R(t \mid x) = \Pr(T > t + x \mid T > x) = \frac{\Pr(T > t + x)}{\Pr(T > x)} = \frac{R(t + x)}{R(x)}$$

Definition 2.3 A distribution $F(t)$ is said to be "new better than used" (F is NBU) if

$$R(t \mid x) \leq R(t) \quad \text{for } t \geq 0,\ x \geq 0$$

A distribution $F(t)$ is said to be "new worse than used" (F is NWU) if

$$R(t \mid x) \geq R(t) \quad \text{for } t \geq 0, \quad x \geq 0$$

□

A distribution $F(t)$ is thus NBU (NWU) if the conditional survivor function $R(t \mid x)$ of an item of age x is less (greater) than the corresponding survivor function $R(t)$ of a new item.

2.19.4 NBUE and NWUE Distributions

The mean residual life of an item at age x was defined in Section 2.7 as

$$\text{MRL}(x) = \int_0^\infty R(t \mid x) \, dt \qquad (2.124)$$

When $x = 0$, we start out with a new item and consequently $\text{MRL}(0) = \text{MTTF}$.

Definition 2.4 A life distribution $F(t)$ is said to be "new better than used in expectation" (F is NBUE) if

1. F has a finite mean μ.
2. $\text{MRL}(x) \leq \mu$ for $x \geq 0$.

A life distribution $F(t)$ is said to be " new worse than used in expectation" (F is NWUE) if

1. F has a finite mean μ.
2. $\text{MTTF}(x) \geq \mu$ for $x \geq 0$. □

2.19.5 A Brief Comparison

The families of life distributions presented above are further discussed, for example, by Barlow and Proschan (1975) and Gerthsbakh (1989) who prove the following chain of implications:

$$\begin{array}{ccccccc}
\text{IFR} & \Longrightarrow & \text{IFRA} & \Longrightarrow & \text{NBU} & \Longrightarrow & \text{NBUE} \\
\text{DFR} & \Longrightarrow & \text{DFRA} & \Longrightarrow & \text{NWU} & \Longrightarrow & \text{NWUE}
\end{array}$$

2.20 SUMMARY OF FAILURE MODELS

In the previous sections of this chapter we have discussed a number of different failure models or life distributions. Some main characteristics of these models are presented in Table 2.2 to provide a brief reference.

Table 2.2 Summary of Life Distributions and Parameters

Distribution	Probability density $f(t)$	Survivor function $R(t)$	Failure rate $z(t)$	MTTF
Exponential	$\lambda e^{-\lambda t}$	$e^{-\lambda t}$	λ	$1/\lambda$
Gamma	$\dfrac{\lambda}{\Gamma(k)}(\lambda t)^{k-1}e^{-\lambda t}$	$\displaystyle\sum_{x=0}^{k-1}\dfrac{(\lambda t)^x}{x!}e^{-\lambda t}$	$\dfrac{f(t)}{R(t)}$	k/λ
Weibull	$\alpha\lambda(\lambda t)^{\alpha-1}e^{-(\lambda t)^\alpha}$	$e^{-(\lambda t)^\alpha}$	$\alpha\lambda(\lambda t)^{\alpha-1}$	$\dfrac{1}{\lambda}\Gamma\left(\dfrac{1}{\alpha}+1\right)$
Lognormal	$\dfrac{1}{\sqrt{2\pi}}\dfrac{1}{\tau}\dfrac{1}{t}e^{-(\ln t - \nu)^2/2\tau^2}$	$\Phi\left(\dfrac{\nu-\ln t}{\tau}\right)$	$\dfrac{f(t)}{R(t)}$	$e^{\nu+\tau^2/2}$
Birnbaum–Saunders	$\dfrac{\sqrt{\lambda t}+\dfrac{1}{\sqrt{\lambda t}}}{2\alpha t}\dfrac{1}{\sqrt{2\pi}}\exp\left(-\left(\sqrt{\lambda t}-\dfrac{1}{\sqrt{\lambda t}}\right)^2/2\alpha^2\right)$	$\Phi\left(\dfrac{1}{\alpha}\left(\dfrac{1}{\sqrt{\lambda t}}-\sqrt{\lambda t}\right)\right)$	$\dfrac{f(t)}{R(t)}$	$\dfrac{1}{\lambda}\left(1+\dfrac{\alpha^2}{2}\right)$
Gumbel – smallest extreme	$e^t e^{-e^t}$	e^{-e^t}	e^t	–
Inverse Gaussian	$\sqrt{\dfrac{\lambda}{2\pi t^3}}e^{-(\lambda/2\mu^2)[(t-\mu)^2]/t]}$	$\Phi\left(\sqrt{\lambda}\dfrac{1}{\sqrt{t}}-\dfrac{\sqrt{\lambda}}{\mu}\sqrt{t}\right)$ $-\Phi\left(-\sqrt{\lambda}\dfrac{1}{\sqrt{t}}-\dfrac{\sqrt{\lambda}}{\mu}\sqrt{t}\right)e^{2\lambda/\mu}$	$\dfrac{f(t)}{R(t)}$	μ

PROBLEMS

2.1 A component with time to failure T has constant failure rate

$$z(t) = \lambda = 2.5 \cdot 10^{-5} \text{ (hours)}^{-1}$$

(a) Determine the probability that the component survives a period of 2 months without failure.

(b) Find the MTTF of the component.

(c) Find the probability that the component survives its MTTF.

2.2 A machine with constant failure rate λ will survive a period of 100 hours without failure, with probability 0.50.

(a) Determine the failure rate λ.

(b) Find the probability that the machine will survive 500 hours without failure.

(c) Determine the probability that the machine will fail within 1000 hours, when you know that the machine was functioning at 500 hours.

2.3 A safety valve is assumed to have constant failure rate with respect to all failure modes. A study has shown that the total MTTF of the valve is 2450 days. The safety valve is in continuous operation, and the failure modes are assumed to occur independent of each other.

(a) Determine the total failure rate of the safety valve.

(b) Determine the probability that the safety valve will survive a period of 3 months without any failure.

(c) Of all failures 45% are assumed to be *critical* failure modes. Determine the mean time to a critical failure, $\text{MTTF}_{\text{crit}}$.

2.4 The time to failure T of an item is assumed to have an exponential distribution with failure rate λ. Show that the rth moment of T is

$$E(T^r) = \frac{\Gamma(r+1)}{\lambda^r} \qquad (2.125)$$

2.5 Let T_1 and T_2 be two independent times to failure with constant failure rates λ_1 and λ_2, respectively, and assume that $\lambda_1 \neq \lambda_2$. Let $T = T_1 + T_2$.

(a) Show that the survivor function of T is

$$R(t) = \Pr(T > t) = \frac{1}{\lambda_2 - \lambda_1} \left(\lambda_2 \, e^{-\lambda_1 t} - \lambda_1 \, e^{-\lambda_2 t} \right)$$

(b) Find the corresponding failure rate function $z(t)$, and make a sketch of $z(t)$ as a function of t for selected values of λ_1 and λ_2.

2.6 Let N be a random variable with a binomial distribution with parameters (n, p). Find $E(N)$ and $\text{var}(N)$.

2.7 Let N be a random variable with value set $0, 1, \ldots$. Show that

$$E(N) = \sum_{n=1}^{\infty} \Pr(N \geq n)$$

2.8 A component with time to failure T has failure rate function

$$z(t) = kt \quad \text{for } t > 0 \text{ and } k > 0$$

(a) Determine the probability that the component survives 200 hours, when $k = 2.0 \cdot 10^{-6}$ (hours)$^{-2}$.

(b) Determine the MTTF of the component when $k = 2.0 \cdot 10^{-6}$ (hours)$^{-2}$.

(c) Determine the probability that a component which is functioning after 200 hours is still functioning after 400 hours, when $k = 2.0 \cdot 10^{-6}$ (hours)$^{-2}$.

(d) Does this distribution belong to any of the distribution classes described in this chapter?

2.9 A component with time to failure T has failure rate function

$$z(t) = \lambda_0 + \alpha t \quad \text{for } t > 0, \ \lambda_0 > 0, \text{ and } \alpha > 0$$

(a) Determine the survivor function $R(t)$ of the component.

(b) Determine the MTTF of the component.

(c) Try to give a physical interpretation of this model.

2.10 A component with time to failure T has failure rate function

$$z(t) = \frac{t}{1+t} \quad \text{for } t > 0$$

(a) Make a sketch of the failure rate function.

(b) Determine the corresponding probability density function $f(t)$.

(c) Determine the MTTF of the component.

(d) Does this distribution belong to any of the distribution classes described in this chapter?

2.11 Let Z have a geometric distribution with probability p and determine

(a) The mean value, $E(Z)$.

(b) The variance, $\text{var}(Z)$.

(c) The conditional probability, $\Pr(Z > z + x \mid Z > x)$. Describe the result you get by words.

2.12 Reconsider the item in Example 2.1 with survivor function

$$R(t) = \frac{1}{(0.2t + 1)^2} \quad \text{for } t \geq 0$$

where the time t is measured in months.

(a) Find the MRL of the item at age $t = 3$ months.

(b) Make a sketch of MRL(t) as a function of the age t.

2.13 Prove that equation (2.20) is correct. Use this equation to find the failure rate function $z(t)$ of the item in Problem 2.12.

2.14 Consider an item with survivor function $R(t)$. Show that the MTTF of the item can be written as

$$\text{MTTF} = \int_0^t R(u)\,du + R(t) \cdot \text{MRL}(t)$$

Explain the meaning of this formula.

2.15 Let N_1 and N_2 be independent Poisson random variables with $E(N_1) = \lambda_1$ and $E(N_2) = \lambda_2$.

(a) Determine the distribution of $N_1 + N_2$.

(b) Determine the conditional distribution of N_1 given that $N_1 + N_2 = n$.

2.16 Let T_1 and T_2 be independent and gamma distributed with parameters (k_1, λ) and (k_2, λ), respectively. Show that $T_1 + T_2$ has a gamma distribution with parameters $(k_1 + k_2, \lambda)$. Explain why we sometimes say that the gamma distribution is "closed under addition."

2.17 The time to failure T of an item is assumed to have a Weibull distribution with scale parameter $\lambda = 5.0 \cdot 10^{-5}$ (hours)$^{-1}$ and shape parameter $\alpha = 1.5$. Compute MTTF and $\text{var}(T)$.

2.18 Let $T \sim$ Weibull(α, λ). Show that the variable $(\lambda T)^\alpha$ has an exponential distribution with failure rate 1.

FAILURE MODELS

2.19 Let $T \sim$ Weibull(α, λ). Show that the rth moment of T is

$$E(T^r) = \frac{1}{\lambda^r} \Gamma\left(\frac{r}{\alpha} + 1\right)$$

2.20 Let T have a three-parameter Weibull distribution (α, λ, ξ) with probability density

$$f(t) = \frac{d}{dt} F(t) = \alpha \lambda [\lambda(t - \xi)]^{\alpha-1} e^{-[\lambda(t-\xi)]^\alpha} \quad \text{for } t > \xi$$

(a) Show that the density is unimodal if $\alpha > 1$. Also show that the density decreases monotonically with t if $\alpha < 1$.

(b) Show that the failure rate function is $\alpha\lambda[\lambda(t-\xi)]^{\alpha-1}$ for $t > \xi$, and hence is increasing, constant, and decreasing with t, respectively, as $\alpha > 1$, $\alpha = 1$, and $\alpha < 1$.

2.21 The failure rate function of an item is $z(t) = t^{-1/2}$. Derive:

(a) The probability density function, $f(t)$.

(b) The survivor function, $R(t)$.

(c) The mean time to failure.

(d) The variance of the time to failure, T, var(T).

2.22 The time to failure, T, has survivor function $R(t)$. Show that if $E(T^r) < \infty$, then

$$E(T^r) = \int_0^\infty rt^{r-1} R(t)\, dt \quad \text{for } r = 1, 2, \ldots$$

2.23 Consider an item with time to failure T and failure rate function $z(t)$. Show that

$$\Pr(T > t_2 \mid T > t_1) = e^{-\int_{t_1}^{t_2} z(u)\, du} \quad \text{for } t_2 > t_1$$

2.24 Assume the time to failure T to be lognormally distributed such that $Y = \ln T$ is $\mathcal{N}(\nu, \tau^2)$. Show that

$$\begin{aligned} E(T) &= e^{\nu + \tau^2/2} \\ \text{var}(T) &= e^{2\nu}\left(e^{2\tau^2} - e^{\tau^2}\right) \end{aligned}$$

and that the variance may be written as

$$\text{var}(T) = [E(T)]^2 \cdot \left(e^{\tau^2} - 1\right)$$

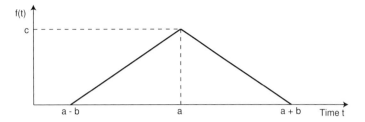

Fig. 2.27 Probability density (Problem 2.28).

2.25 Let $z(t)$ denote the failure rate function of the lognormal distribution. Show that $z(0) = 0$, that $z(t)$ increases to a maximum, and then decreases with $z(t) \to 0$ as $t \to \infty$.

2.26 Let $T \sim \text{lognormal}(\nu, \tau^2)$. Show that $1/T \sim \text{lognormal}(-\nu, \tau^2)$.

2.27 The time to failure T of a component is assumed to be uniformly distributed over $(a, b]$. The probability density is thus

$$f(t) = \frac{1}{b-a} \quad \text{for } a < t \leq b$$

Derive the corresponding survivor function $R(t)$ and failure rate function $z(t)$. Draw a sketch of $z(t)$.

2.28 The time to failure T of a component has probability density $f(t)$ as shown in Fig. 2.27.

(a) Determine c such that $f(t)$ is a valid probability density.

(b) Derive the corresponding survivor function $R(t)$.

(c) Derive the corresponding failure rate function $z(t)$, and make a sketch of $z(t)$.

2.29 Consider a system of n independent components with constant failure rates $\lambda_1, \lambda_2, \ldots, \lambda_n$, respectively. Show that the probability that component i fails first is

$$\frac{\lambda_i}{\sum_{j=1}^{n} \lambda_j}$$

2.30 A component may fail due to two different causes, excessive stresses and aging. A large number of this type of component have been tested. It has been shown that the time to failure T_1 caused by excessive stresses is exponentially distributed with density function

$$f_1(t) = \lambda_1 e^{-\lambda_1 t} \quad \text{for } t \geq 0$$

while the time to failure T_2 caused by aging has density function

$$f_2(t) = \frac{1}{\Gamma(k)} \lambda_2 (\lambda_2 t)^{k-1} e^{-\lambda_2 t} \quad \text{for } t \geq 0$$

(a) Describe the rationale behind using

$$f(t) = p \cdot f_1(t) + (1-p) \cdot f_2(t) \quad \text{for } t \geq 0$$

as the probability density function for the time to failure T of the component.

(b) Explain the meaning of p in this model.

(c) Let $p = 0.1$, $\lambda_1 = \lambda_2$, and $k = 5$, and determine the failure rate function $z(t)$ corresponding to T. Calculate $z(t)$ for some selected values of t, for example, $t = 0, \frac{1}{2}, 1, 2, \ldots$, and make a sketch of $z(t)$.

2.31 A component may fail due to two different causes, A and B. It has been shown that the time to failure T_A caused by A is exponentially distributed with density function

$$f_A(t) = \lambda_A e^{-\lambda_A t} \quad \text{for } t \geq 0$$

while the time to failure T_B caused by B has density function

$$f_B(t) = \lambda_B e^{-\lambda_B t} \quad \text{for } t \geq 0$$

(a) Describe the rationale behind using

$$f(t) = p \cdot f_A(t) + (1-p) \cdot f_B(t) \quad \text{for } t \geq 0$$

as the probability density function for the time to failure T of the component.

(b) Explain the meaning of p in this model.

(c) Show that a component with probability density $f(t)$ has a DFR function.

2.32 Consider a component with time to failure T, with IFR distribution, and MTTF $= \mu$. Show that

$$R(t) \geq e^{-t/\mu} \quad \text{for } 0 < t < \mu$$

2.33 Let $F(t)$ denote the distribution of the time to failure T. Assume $F(t)$ to be strictly increasing. Show that:

(a) $F(T)$ is uniformly distributed over $[0, 1]$.

(b) If U is a uniform $[0, 1]$ random variable, then $F^{-1}(U)$ has distribution F, where $F^{-1}(y)$ is that value of x such that $F(x) = y$.

2.34 Let T_1 and T_2 be independent lifetimes with failure rate functions $z_1(t)$ and $z_2(t)$, respectively. Show that

$$\Pr(T_1 < T_2 \mid \min\{T_1, T_2\} = t) = \frac{z_1(t)}{z_1(t) + z_2(t)}$$

2.35 Derive the Laplace transform of the survivor function $R(t)$ of the exponential distribution with failure rate λ, and use the Laplace transform to determine the MTTF of this distribution.

2.36 Let $T \sim \text{Weibull}(\alpha, \lambda)$. Show that $Y = \ln T$ has a type I asymptotic distribution of the smallest extreme. Find the mode and the scale parameter of this distribution.

2.37 Show that the median t_m of the lognormal distribution is e^ν. Compute k such that $\Pr(t_m/k \leq T \leq kt_m) = 0.90$.

2.38 Show that the failure rate function $z(t)$ of the Birnbaum-Saunders distribution is not monotonic.

2.39 The time to failure T is assumed to have a Birnbaum-Saunders distribution with scale parameter λ and "shape" parameter α. Show that the failure rate function $z(t)$ has the limit

$$\lim_{t \to \infty} z(t) = \frac{\lambda}{2\alpha^2}$$

2.40 Show that the inverse Gaussian distribution $\text{IG}(\mu, \lambda)$ is unimodal with mode

$$t_{\text{mode}} = -\frac{3\mu^2}{2\lambda} + \mu\sqrt{1 + \frac{9\mu^2}{4\lambda^2}}$$

2.41 Show that when T has an inverse Gaussian distribution $\text{IG}(\mu, \lambda)$, then

$$E(T^{-r}; \mu, \lambda) = \frac{1}{\mu^{2r+1}} E(T^{r+1}; \mu, \lambda)$$

Show that this implies that

$$E(T^{-r}; 1, \lambda) = E(T^{r+1}; 1, \lambda)$$

and furthermore that

$$E(T^{-1}; 1, \lambda) = E(T^2; 1, \lambda)$$

2.42 Prove that

$$\int_0^{t_0} z(t)\, dt \to \infty \quad \text{when} \quad t_0 \to \infty$$

2.43 Let X and Y be a pair of random variables with cumulant-generating function (cgf) $L_X(t)$ and $L_Y(t)$, respectively. According to Tweedie's definition (Tweedie 1946) the distributions of X and Y are said to constitute a pair of "inverse distributions" if there exists a function $L(t)$ such that

$$L_X(t) = \alpha L(t)$$
$$L_Y(t) = \beta L^{-1}(t) \quad \text{i.e.,} \quad L_Y(t) = \gamma L_X^{-1}(t)$$

for all values of t belonging to the domain of both cfg's, where $L(L^{-1}(t)) = t$, and α and β are appropriate constants.

Show that with Tweedie's definition:

(a) the binomial and the negative binomial distributions,

(b) the Poisson and the gamma distributions, and

(c) the normal and the inverse Gaussian distributions

constitute pairs of "inverse distributions."

3
Qualitative System Analysis

3.1 INTRODUCTION

A technical system will normally comprise a number of subsystems and components that are interconnected in such a way that the system is able to perform a set of required functions. We will use the term *functional block* to denote an element of the system, whether it is a component or a large subsystem.

The main concern of a reliability engineer is to identify potential failures and to prevent these failures from occurring. A failure of a functional block is defined as "the termination of its ability to perform a required function" (BS 4778). It is therefore necessary for the reliability engineer to identify all relevant functions and the performance criteria related to each function.

In this chapter we start by defining a technical system and its interfaces. The overall structure of the system is illustrated by a functional block diagram. We then present a classification system for the various functions of a functional block and illustrate the functions by various types of function diagrams. Failures, failure modes, and failure effects are defined and discussed and various failure classification structures are presented. We then present a number of methods for system reliability analysis:

1. *Failure modes, effects, and criticality analysis (FMEA/FMECA):* This method is used to identify the potential failure modes of each of the functional blocks of a system and to study the effects these failures might have on the system. FMEA/FMECA is primarily a tool for designers but is frequently used as a basis for more detailed reliability analyses and for maintenance planning.

2. *Fault tree analysis:* The fault tree illustrates all possible combinations of potential failures and events that may cause a specified system failure. Fault tree

construction is a deductive approach where we start with the specified system failure and then consider what caused this failure. Failures and events are combined through logic gates in a binary approach. The fault tree may be evaluated quantitatively if we have access to probability estimates for the basic events. Quantitative evaluation of fault trees is discussed in Chapter 4.

3. *Cause and effect diagrams:* Cause and effect diagrams are frequently used within quality engineering to identify and illustrate possible causes of quality problems. The same approach may also be used in reliability engineering to find the potential causes for system failures. Cause and effect diagrams are qualitative and cannot be used as a basis for quantitative analyses.

4. *Bayesian belief networks:* Bayesian belief networks may be used to identify and illustrate potential causes for system failures. Probability distributions may be allocated to the various causal factors, and the network may be evaluated quantitatively by a Bayesian approach. Bayesian belief networks are more flexible than fault trees since we do not have to use a binary representation. Quantitative evaluation of Bayesian belief networks is not covered in this book.

5. *Event tree analysis:* An event tree analysis is an inductive technique. We start with a system deviation and identify how this deviation may develop. The possible events following the deviation will usually be a function of various barriers and safety functions that are designed into the system. When we have access to probability estimates for the various barriers, we may carry out a quantitative analysis of the event tree. A brief presentation of such a quantitative evaluation is given on page 112.

6. *Reliability block diagrams:* A reliability block diagram is a success-oriented network illustrating how the functioning of the various functional blocks may secure that the system function is fulfilled. The structure of the reliability block diagram is described mathematically by structure functions. The structure functions will be used in the following chapters to calculate system reliability indices.

3.2 SYSTEMS AND INTERFACES

A technical system may be defined as: A composite, at any level of complexity, of personnel, procedures, materials, tools, equipment, facilities, and software. The elements of this composite entity are used together in the intended operational or support environment to perform a given task or achieve a specific purpose, support, or mission requirement (MIL-STD 882D).

Only technical systems are considered in this book. Any technical systems will, however, have interfaces with humans. Humans may be operators controlling or performing specific functions, they may support the system by cleaning, lubricating, testing, and repairing the system, or they may be users of the systems. The reliability

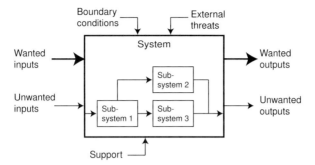

Fig. 3.1 A technical system and its interfaces.

of the system will depend on its interfaces with the rest of the world. It is therefore necessary to study how these interfaces influence the system. It is, however, not an objective of this book to study the impacts the system may have on the rest of the world.

An illustration of a technical system and its interfaces is shown in Fig. 3.1. The following elements are illustrated in Fig. 3.1:

1. *System:* The technological system that is subject to analysis. The system will usually comprise several functional blocks.

2. *System boundary:* The system boundary defines which elements that are considered as part of the system and which elements that are outside.

3. *Outputs:* The outputs from the system may be classified in two groups:

    (a) *Wanted outputs:* These are the (wanted) results of the required functions (like artifacts, materials, energy, and information).

    (b) *Unwanted outputs:* Almost all systems will produce outputs that are not wanted. Such outputs may be pollution to air, water, or ground and injuries and negative health effects to people in, or in the neighborhood of, the system.

4. *Inputs:* The inputs to the system may be classified in two groups:

    (a) *Wanted inputs:* These are the materials and the energy the system is using to perform its required functions. The quality and amount of the wanted inputs may be subject to variations.

    (b) *Unwanted inputs:* These are inputs associated to the wanted inputs that may not be considered as normal variations of the wanted inputs. An example of unwanted input is particles in the input fluid to a pump.

5. *Boundary conditions:* The operation of the system may be subject to a number of boundary conditions, like risk acceptance and environmental criteria set by authorities or by the company.

6. *Support:* The system usually needs support functions, like cleaning, lubrication, maintenance, and repair.

7. *External threats:* The system may be exposed to a wide range of external threats. Some of these threats may have direct impact on the system, others may have impact on the system inputs. External threats may be classified in four groups:

   (a) *Natural environmental threats:* Threats to the system from the external environment, like flooding, storm, lightning, and earthquake.

   (b) *Infrastructure threats:* Threats caused by deficiencies and breakdown of infrastructure, like energy supply and communication.

   (c) *Societal threats:* Threats from individuals and organizations, like arson, sabotage, hacking, and computer virus attacks.

   (d) *Threats from other technical systems:* Impact from other systems close to the system or with interfaces to the system.

   The distinction between an unwanted input and an external threat may not always be clear. It is not important how the various inputs are classified. What is important is that all inputs and threats are identified and considered in the analysis.

How we consider a technical system depends on the role we have, the phase of the system's life cycle, and on the objectives of our study. The system may generally be considered from two different points of view:

1. *Structural focus:* Here we are interested in the physical structure of the various subsystems and components of the system. The serviceman, for example, is mainly interested in the subsystems and components of a television set and how these transform and transmit electromagnetic waves.

2. *Functional focus:* Here we are interested in the various functions of the system and how these functions are fulfilled. A user of a television set, for example, is primarily interested in the information (pictures and sound) he gets from the television set.

In the early design process of a new system, we usually start with a set of desired functions. We want to develop a system that is able to fulfill these functions. No physical realization has yet been decided. In this phase we have a functional focus.

Several types of diagrams are used to illustrate the structural and the functional interrelationships in a system. Many of these diagrams are called *functional block diagrams* but may be rather different both regarding symbols and layout. A mixture between functions and physical elements is also often seen. An example of a functional block diagram of a diesel engine is shown in Fig. 3.2. Functional block diagrams are recommended by IEC 60812 and MIL-STD 1629A as a basis for failure modes, effects, and criticality analysis (FMECA), and by Smith (1993) as a basis for reliability centered maintenance (RCM). In the process industry, the systems are often illustrated by process and instrumentation diagrams (P&ID).

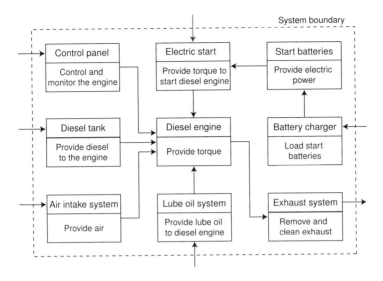

Fig. 3.2 Functional block diagram of a diesel engine.

For some systems a numbering system has been developed, where each functional block is given a unique number according to a hierarchical numbering system. This is, for example, the case in the Norwegian offshore industry where this system is called the *tag number system*. The lowest level in the numbering system is usually the smallest item that is separately maintained. Such items are called *maintainable items* or *least replaceable items*. It is usually recommended to use the same structure also in reliability analyses, since the reliability data will be available for these items.

3.3 FUNCTIONAL ANALYSIS

To be able to identify all potential failures, the reliability engineer has to have a thorough understanding of the various functions of each functional block, and the performance criteria related to the various functions. A functional analysis is therefore an important step in a system reliability analysis. The objectives of a functional analysis are to:

1. Identify all the functions of the system

2. Identify the functions required in the various operational modes of the system

3. Provide a hierarchical decomposition of the system functions

4. Describe how each function is realized

5. Identify the interrelationships between the functions

6. Identify interfaces with other systems and with the environment

A function is an intended *effect* of a functional block and should be defined such that each function has a single definite purpose. It is recommended to give the functions names that have a declarative structure, and say "what" is to be done rather than "how." The functions should preferably be expressed as a statement comprising a verb plus a noun; for example, close flow, contain fluid, pump fluid, and transmit signal. In practice, however, it is often difficult to specify a function with only two words, and we may have to add one or two extra words.

A *functional requirement* is a specification of the performance criteria related to a function. If, for example, the function is "pump water," a functional requirement may be that the output of water must be between 100 and 110 liters per minute. For some functions we may have several functional requirements.

3.3.1 Classification of Functions

A complex system may have a high number of required functions. All functions are, however, not equally important, and a classification may therefore be an aid for identification and analysis purposes. One way of classifying functions are:

1. *Essential functions:* These are the functions required to fulfill the intended purpose of the functional block. The essential functions are simply the reasons for installing the functional block. Often an essential function is reflected in the name of the functional block. An essential function of a pump is, for example, to "pump fluid."

2. *Auxiliary functions:* These are the functions that are required to support the essential functions. The auxiliary functions are usually less obvious than the essential functions but may in many cases be as important as the essential functions. Failure of an auxiliary function may in many cases be more safety critical than a failure of an essential function. An auxiliary function of a pump is, for example, to "contain fluid."

3. *Protective functions:* These functions are intended to protect people, equipment and the environment from damage and injury. The protective functions may be classified as:

    (a) Safety functions (i.e., to prevent accidental events and/or to reduce consequences to people, material assets, and the environment)

    (b) Environment functions (e.g., antipollution functions during normal operation)

    (c) Hygiene functions

4. *Information functions:* These functions comprise condition monitoring, various gauges and alarms, and so forth.

5. *Interface functions:* These functions apply to the interfaces between the functional block in question and other functional blocks. The interfaces may be

active or passive. A passive interface is, for example, present when the functional block is a support or a base for another functional block.

6. *Superfluous functions:* In some cases the functional blocks may have functions that are never used. This is sometimes the case with electronic equipment that have a wide range of "nice to have" functions that are not really necessary. Superfluous function may further be found in systems that have been modified several times. Superfluous functions may also be present when the functional block has been designed for an operational context that is different from the actual operational context. In some cases failure of a superfluous function may cause failure of other functions.

These classes are not necessarily disjoint. Some functions may be classified in more that one class.

In many applications it is important to distinguish between evident and hidden (dormant) failures. The following classification of functions may therefore prove necessary:

1. *On-line functions:* These are functions operated either continuously or so often that the user has current knowledge about their state. The termination of an on-line function is called an *evident* failure.

2. *Off-line functions:* These are functions that are used intermittently or so infrequently that their availability is not known by the user without some special check or test. An example of an off-line function is the essential function of an emergency shutdown (ESD) system. Many protective functions are off-line functions. The termination of the ability to perform an off-line function is called a *hidden* failure.

3.3.2 Operational Modes

A system and its functional blocks may in general have several operational modes and several functions for each operational mode. Operational modes should include normal operating modes, test modes, transition between modes, and contingency modes induced by failures, faults, or operator errors. The establishment of the different operational modes is recommended for two reasons:

1. It reveals other functions that might be overlooked when focusing too much on the essential function.

2. It provides a structured basis for the identification of failure modes that are completely connected to, and dependent on, the given operational mode.

Operational modes are therefore an aid in identifying both functions and failure modes.

3.3.3 Function Tree

For complex systems it is sometimes beneficial to illustrate the various functions as a tree structure, called a *function tree*. A function tree is a hierarchical functional

80   QUALITATIVE SYSTEM ANALYSIS

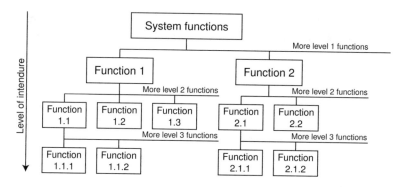

Fig. 3.3   Function tree.

breakdown structure starting with a system function or a system mission and illustrating the corresponding necessary functions on lower levels of indenture. The function tree is created by asking *how* an already established function is accomplished. This is repeated until functions on the lowest level are reached. The diagram may also be developed in the opposite direction by asking *why* a function is necessary. This is repeated until functions on the system level are reached. Function trees may be represented in many different ways. An example is shown in Fig. 3.3. A lower level function may be required by a number of main functions and may therefore appear several places in the function tree.

An alternative to the function tree is the function analysis system technique (FAST) that was introduced in 1965 by the Society of American Value Engineers (Fox 1993; Lambert et al. 1999). The FAST diagram is drawn from left to right. We start with a system function on the left side and ask *how* this function is (or may be) accomplished. The functions on the first level are then identified and entered into the diagram. We continue asking *how* until we reach the intended level of detail. The lower level functions can be connected by AND and OR relations as illustrated in Fig. 3.4. Functions that have to be performed at the same time may be indicated by vertical arrows. An illustration of a FAST diagram is given in Fig. 3.5.

When we are analyzing an existing system, it is often more obvious to use a *physical breakdown* of the system instead of a functional breakdown. A physical

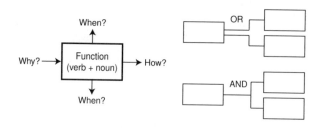

Fig. 3.4   Symbols used in FAST diagrams.

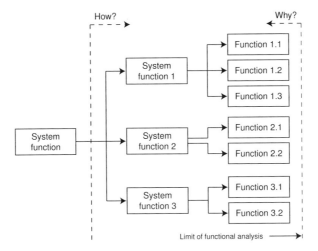

Fig. 3.5 Example of a FAST diagram.

breakdown structure is similar to the function tree in Fig. 3.3, but each box represents a physical element instead of a function. The "physical" elements may be technical items, operators, and even procedures. When each function is performed by only one physical element, the two approaches give similar results. When the system has redundancies, the trees will be different. The function "pump water" may, for example, be realized with two redundant water pumps. In the function tree, this is represented as one function, while we in the physical breakdown structure get two elements, one for each pump.

3.3.4 Functional Block Diagrams

A widely used approach to functional modeling was introduced by Douglas T. Ross of SofTech Inc. in 1973, called the structured analysis and design technique (SADT). The SADT approach is described, for example, by Lissandre (1990) and Lambert et al. (1999). In the SADT diagram each functional block is modeled according to the same structure with five main elements:

- *Function:* Definition of the function to be performed

- *Input:* The energy, materials, and information that are necessary to perform the function

- *Control:* The controls and other elements that constrain or govern how the function will be carried out.

- *Mechanism:* The people, systems, facilities or equipment necessary to carry out the function.

82 QUALITATIVE SYSTEM ANALYSIS

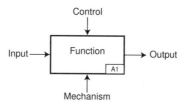

Fig. 3.6 A functional block in a SADT diagram.

- *Output:* The result of the function. The output is sometimes split in two parts: the wanted output from the function and the unwanted output.

An illustration of a functional block in a SADT diagram is shown in Fig. 3.6. The output of a functional block may be the input to another functional block, or may act as a control of another functional block. By this way we can link the functional blocks to become a functional block diagram. An illustration of an SADT diagram for subsea oil and gas stimulation is shown in Fig. 3.7. The diagram was developed as part of a student project at the Norwegian University of Science and Technology (Ødegaard 2002).

When constructing an SADT model, we use a top-down approach as illustrated in Fig. 3.8. At the top level we start with a required system function. The functions necessary to fulfill the system function are established as an SADT diagram at the next level. Each function on this level is then broken down to lower level functions, and so on, until the desired level of decomposition has been reached. The hierarchy is maintained via a numbering system that organizes parent and child diagrams.

The functional block in Fig. 3.6 is also used in the integrated definition language (IDEF), which is based on SADT and developed for the U.S. Air Force. IDEF is

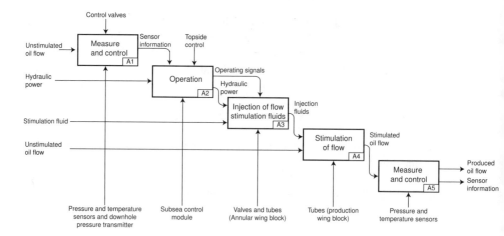

Fig. 3.7 SADT diagram.

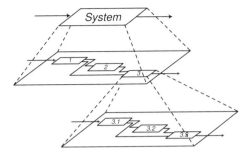

Fig. 3.8 Top-down approach to establish an SADT model.

divided into several modules. The module for modeling of system functions is called IDEF0 (U.S. Air Force 1981; Cheng Leong and Gay 1993).

For new systems, SADT and IDEF0 may be used to define the requirements and specify the functions and as a basis for suggesting a solution that meets the requirements and performs the functions. For existing systems, SADT and IDEF0 can be used to analyze the functions the system performs and to record the mechanisms (means) by which these functions are accomplished.

3.4 FAILURES AND FAILURE CLASSIFICATION

Even if we are able to identify all the required functions of a functional block, we may not be able to identify all the failure modes. This is because each function may have several failure modes. No formal procedure seems to exist that may be used to identify and classify the possible failure modes.[1]

3.4.1 Failures, Faults, and Errors

According to IEC 50(191) *failure* is the *event* when a required function is terminated (exceeding the acceptable limits), while *fault* is "the state of an item characterized by inability to perform a required function, excluding the inability during preventive maintenance or other planned actions, or due to lack of external resources." A fault is hence a state resulting from a failure.

According to IEC 50(191) an *error* is a "discrepancy between a computed, observed or measured value or condition and the true, specified or theoretically correct value or condition." An error is (yet) not a failure because it is within the acceptable limits of deviation from the desired performance (target value). An error is sometimes referred to as an *incipient failure* (e.g., see OREDA 2002; IEEE Std. 500).

---

[1]The section is based on Rausand and Øien (1996). The basic concepts of failure analysis. *Reliability Engineering and System Safety.* **53**:73–83. ©1996, with permission from Elsevier.

84 QUALITATIVE SYSTEM ANALYSIS

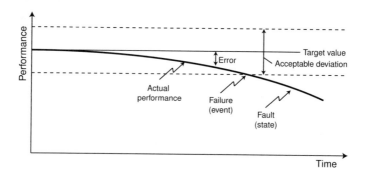

Fig. 3.9 Illustration of the difference between failure, fault, and error.

The term *failure* is sometimes confused with the terms *fault* and *error*. The relationship between these terms is illustrated in Fig. 3.9.

The distinction between failure (or fault) and error is essential in failure analysis, because this describes the borderline between what is a failure and what is not.

3.4.2 Failure Modes

A failure mode is a description of a fault, that is, how we can observe the fault. Fault mode should therefore be a more appropriate term than failure mode. IEC 50(191) recommends using the term *fault mode*, but the term *failure mode* is so widely used that a change may confuse the reader.

Identification of Failure Modes To identify the failure modes we have to study the *outputs* of the various functions. Some functions may have several outputs. Some outputs may be given a very strict definition, such that it is easy to determine whether the output requirements are fulfilled or not. In other cases the output may be specified as a target value with an acceptable deviation (see Fig. 3.9).

If we consider a process shutdown valve, it should be designed with a specified closing time, for example, 10 seconds. If the valve closes too slowly, it will not function as a safety barrier. On the other hand, if the valve closes too fast, we may get a pressure shock destroying the valve or the valve flanges. Closing times between 6 and 14 seconds may, for example, be acceptable, and we state that the valve is functioning (with respect to this particular function) as long as the closing time is within this interval. The criticality of the failure will obviously increase with the deviation from the target value.

Failure Mode Categories It is important to realize that a failure mode is a manifestation of the failure as seen from the outside, that is, the termination of one or more functions. "Internal leakage" is thus a failure mode of a shutdown valve, since the valve looses its required function to "close flow." Wear of the valve seal, however, represents a cause of failure and is hence not a failure mode of the valve.

A classification scheme for failure modes has been suggested by Blanche and Shrivastava (1994):

1. *Intermittent failures:* Failures that result in a lack of some function only for a very short period of time. The functional block will revert to its full operational standard immediately after the failure.
2. *Extended failures:* Failures that result in a lack of some function that will continue until some part of the functional block is replaced or repaired. Extended failures may be further divided into:
   (a) *Complete failures:* Failures that cause complete lack of a required function.
   (b) *Partial failures:* Failures that lead to a lack of some function but do not cause a complete lack of a required function.

Both the complete failures and the partial failures may be further classified:

   (a) *Sudden failures:* Failures that could not be forecast by prior testing or examination.
   (b) *Gradual failures:* Failures that could be forecast by testing or examination. A gradual failure will represent a gradual "drifting out" of the specified range of performance values. The recognition of gradual failures requires comparison of actual device performance with a performance specification, and may in some cases be a difficult task.

The extended failures are split into four categories; two of these are given specific names:

   (a) *Catastrophic failures:* A failure that is both sudden and complete.
   (b) *Degraded failure:* A failure that is both partial and gradual (such as the wear of the tires on a car).

The failure classification described above is illustrated in Fig. 3.10, which is adapted from Blache and Shrivastava (1994).

In some applications it may also be useful to classify failures as either primary failures, secondary failures, or command faults (e.g., see Henley and Kumamoto 1981; Villemeur 1988):

A *primary failure* is a failure caused by natural aging of the functional block. The primary failure occurs under conditions within the design envelope of the functional block. A repair action is necessary to return the functional block to a functioning state.

A *secondary failure* is a failure caused by excessive stresses outside the design envelope of the functional block. Such stresses may be shocks from thermal, mechanical, electrical, chemical, magnetic, or radioactive energy sources. The stresses may be caused by neighboring components, the environment, or by system operators/plant personnel. A repair action is necessary to return the functional block to a functioning state.

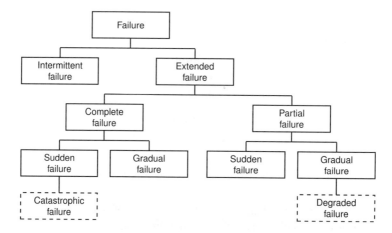

Fig. 3.10 Failure classification (adapted from Blanche and Shrivastava 1994).

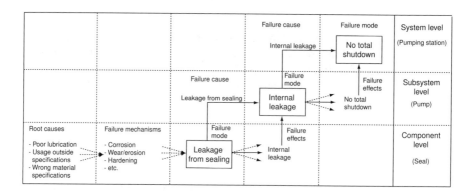

Fig. 3.11 Relationship between failure cause, failure mode, and failure effect.

A *command fault* is a failure caused by an improper control signal or noise. A repair action is usually not required to return the functional block to a functioning state. Command faults are sometimes referred to as *transient* failures.

3.4.3 Failure Causes and Failure Effects

The functions of a system may usually be split into subfunctions. Failure modes at one level in the hierarchy will often be caused by failure modes on the next lower level. It is important to link failure modes on lower levels to the main top level responses, in order to provide traceability to the essential system responses as the functional structure is refined. This is illustrated in Fig. 3.11 for a hardware structure breakdown.

Failure Causes, Mechanisms, and Root Causes According to IEC 50(191) failure cause is "the circumstances during design, manufacture or use that have led to a failure." The failure cause is a necessary information in order to avoid failures or reoccurrence of failures. Failure causes may be classified in relation to the life cycle of an functional block as illustrated in Fig. 3.12, where the various failure causes are defined as:

1. *Design failure:* A failure due to inadequate design of a functional block.

2. *Weakness failure:* A failure due to a weakness in the functional block itself when subjected to stresses within the stated capabilities of the functional block. (A weakness may be either inherent or induced.)

3. *Manufacturing failure:* A failure due to nonconformity during manufacture to the design of a functional block or to specified manufacturing processes.

4. *Ageing failure:* A failure whose probability of occurrence increases with the passage of time, as a result of processes inherent in the functional block.

5. *Misuse failure:* A failure due to the application of stresses during use that exceed the stated capabilities of the functional block.

6. *Mishandling failure:* A failure caused by incorrect handling or lack of care of the functional block.

The various failure causes in Fig. 3.12 are not necessarily disjoint. There is, for example, an obvious overlap between "weakness" failures and "design" and "manufacturing" failures.

Failure mechanisms are, in IEC 50(191), defined as the "physical, chemical or other processes that has led to a failure." A common interpretation of this term is the immediate causes to the lowest level of indenture, such as wear, corrosion, hardening, pitting, and oxidation.

This level of failure cause description is, however, not sufficient to evaluate possible remedies. Wear can, for instance, be a result of wrong material specification (design

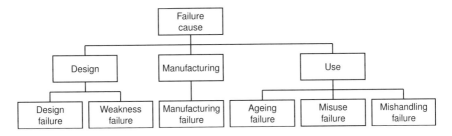

Fig. 3.12 Failure cause classification.

failure), usage outside specification limits (misuse failure), poor maintenance – inadequate lubrication (mishandling failure), and so forth. These fundamental causes are sometimes referred to as *root causes* (see Fig. 3.11), the causes upon which remedial actions can be decided.

Failure Effects A general picture of the relationship between cause and effect is that each failure mode can be caused by several different failure causes, leading to several different failure effects. To get a broader understanding of the relationship between these terms, the level of indenture being analyzed should be brought into account. This is illustrated in Fig. 3.11.

Fig. 3.11 shows that a failure mode on the lowest level of indenture is one of the failure causes on the next higher level of indenture, and the failure effect on the lowest level equals the failure mode on the next higher level. The failure mode "leakage from sealing" for the seal component is, for example, one of the possible failure causes for the failure mode "internal leakage" for the pump, and the failure effect (on the next higher level) "internal leakage" resulting from "leakage from sealing" is the same as the failure mode "internal leakage" of the pump.

For further discussions about failures and failure causes, see Rausand and Øien (1996).

3.5 FAILURE MODES, EFFECTS, AND CRITICALITY ANALYSIS

Failure mode and effects analysis (FMEA) was one of the first systematic techniques for failure analysis. It was developed by reliability engineers in the 1950s to study problems that might arise from malfunctions of military systems.

An FMEA is often the first step of a systems reliability study. It involves reviewing as many components, assemblies, and subsystems as possible to identify failure modes and causes and effects of such failures. For each component, the failure modes and their resulting effects on the rest of the system are recorded in a specific FMEA worksheet. There are numerous variations of such worksheets. An example of an FMEA worksheet is shown in Fig. 3.13.

An FMEA becomes a failure mode, effects, and criticality analysis (FMECA) if criticalities or priorities are assigned to the failure mode effects. In the following we will not distinguish between an FMEA and an FMECA and use FMECA for both.

More detailed information on how to conduct an FMECA may be found in the standards SAE-ARP 5580, IEC 60812, BS 5760-5, and MIL-STD-1629A.

3.5.1 Objectives of an FMECA

According to IEEE Std. 352, the objectives of an FMECA are to:

1. Assist in selecting design alternatives with high reliability and high safety potential during the early design phase.

System:								Performed by:			
Ref. drawing no.:								Date:		Page: of	

Description of unit			Description of failure			Effect of failure		Failure rate	Severity ranking	Risk reducing measures	Comments
Ref. no	Function	Operational mode	Failure mode	Failure cause or mechanism	Detection of failure	On the subsystem	On the system function				
(1)	(2)	(3)	(4)	(5)	(6)	(7)	(8)	(9)	(10)	(11)	(12)

Fig. 3.13 Example of an FMECA worksheet.

Fig. 3.14 FMECA activities in the various phases of product development. (Adapted from SAE-ARP 5580).

2. Ensure that all conceivable failure modes and their effects on operational success of the system have been considered.

3. List potential failures and identify the magnitude of their effects.

4. Develop early criteria for test planning and the design of the test and checkout systems.

5. Provide a basis for quantitative reliability and availability analyses.

6. Provide historical documentation for future reference to aid in analysis of field failures and consideration of design changes.

7. Provide input data for tradeoff studies.

8. Provide basis for establishing corrective action priorities.

9. Assist in the objective evaluation of design requirements related to redundancy, failure detection systems, fail-safe characteristics, and automatic and manual override.

An FMECA is mainly a qualitative analysis and should be carried out by the designers during the design stage of a system. The purpose is to identify design areas where improvements are needed to meet reliability requirements. An updated FMECA is an important basis for design reviews and inspections.

3.5.2 FMECA and the Product Development Process

If possible, the FMECA should be integrated into the product development process from the early concept phase and be updated in later development phases and in the operational phase, as illustrated in Fig. 3.14. In the conceptual design only the main functions of the new product are known. Required subfunctions are identified and may be illustrated by a function tree as shown in Fig. 3.3. The function tree is developed as a top-down approach. No, or very few, hardware solutions are known in this phase of the development process. Potential failures may be identified and evaluated for each function in the function hierarchy by a *functional* FMECA. The functional FMECA is sometimes called a *top-down* FMECA.

In the embodiment design phase, an *interface* FMECA should be carried out. The interface FMECA is performed in the same way as the functional FMECA to

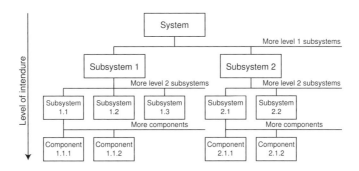

Fig. 3.15 System breakdown structure.

verify compliance to requirements. In the interface FMECA we primarily focus on interconnections between components and subsystems, and especially items designed by separate design groups.

In the detailed design and development phase hardware and software solutions are decided for the various functions, and we may establish a system breakdown structure as shown in Fig. 3.15, where a box represents a hardware or software item. To decide to which component level the analysis should be conducted is a difficult task. It is often necessary to make compromises since the workload could be overwhelming even for a system of moderate size. It is, however, a general rule to expand the analysis down to a level at which failure rate estimates are available or can be obtained. A complete list of all the components on the lowest level of indenture is usually prepared.

When the system breakdown structure is available, a *detailed* FMECA is carried out. The detailed FMECA starts by identifying all potential failure modes on the lowest level of indenture and proceeds upwards in the hierarchy. The detailed FMECA is therefore also called a *bottom-up* FMECA.

In some applications we may use the top-down approach even when the hardware and software structure has been decided. By the top-down approach, the analysis is carried out in two or more stages. The first stage is to split the system into a number of subsystems and to identify possible failure modes and failure effects of each subsystem based on knowledge of the subsystem's required functions, or from experience with similar equipment. One then proceeds to the next stage, where the components within each subsystem are analyzed. If a subsystem has no critical failure modes, no further analysis of that subsystem needs to be performed. By this screening, it is possible to save time and effort. A weakness of this top-down approach is that it does not ensure that all failure modes of a subsystem have been identified.

The FMECA worksheet in Fig. 3.13 is seen to contain a lot of information that is useful for maintenance planning and operation. The FMECA may therefore be integrated into the maintenance planning system and updated as system failures and malfunctions are detected.

3.5.3 FMECA Procedure

An FMECA is simple to conduct. It does not require any advanced analytical skills of the personnel performing the analysis. It is, however, necessary to know and understand the purpose of the system and the constraints under which it has to operate. The basic questions to be answered by an FMECA are according to IEEE Std. 352:

1. How can each part conceivably fail?
2. What mechanisms might produce these modes of failure?
3. What could the effects be if the failures did occur?
4. Is the failure in the safe or unsafe direction?
5. How is the failure detected?
6. What inherent provisions are provided in the design to compensate for the failure?

The analysis may be performed according to the following scheme:

1. Definition and delimitation of the system (which components are within the boundaries of the system and which are outside).
2. Definition of the main functions (missions) of the system.
3. Description of the operational modes of the system.
4. System breakdown into subsystems that can be handled effectively.
5. Review of system functional diagrams and drawings to determine interrelationships between the various subsystems. These interrelations may be illustrated by drawing functional block diagrams where each block corresponds to a subsystem.
6. Preparation of a complete component list for each subsystem.
7. Description of the operational and environmental stresses that may affect the system and its operation. These are reviewed to determine the adverse effects that they could generate on the system and its components.

The various entries in the FMECA worksheet are best illustrated by going through a specific worksheet column by column. We will use the FMECA worksheet in Fig. 3.13 for a detailed FMECA as an example. The changes necessary for a functional FMECA should be obvious.

Reference (column 1). The name of the item or a reference to a drawing, for example, is given in the first column.

Function (column 2). The function(s) of the item is (are) described in this column.

Operational mode (column 3). The item may have various operational modes, for example, running or standby. Operational modes for an airplane include, for example, taxi, take-off, climb, cruise, descent, approach, flare-out, and roll. In applications where it is not relevant to distinguish between operational modes, this column may be omitted.

Failure mode (column 4). For each component's function and operational mode all the failure modes are identified and recorded. Note that the failure modes should be defined as nonfulfillment of the functional requirements of the functions specified in column 2.

Failure causes and mechanisms (column 5). The possible failure mechanisms (corrosion, erosion, fatigue, etc.) that may produce the identified failure modes are recorded in this column. Other failure causes should also be recorded. To identify all potential failure causes, it may be useful to remember the interfaces illustrated in Fig. 3.1 and the inputs to the functional block in Fig. 3.6.

Detection of failure (column 6). The various possibilities for detection of the identified failure modes are then recorded. These may involve different alarms, testing, human perception, and the like. Some failure modes are called *evident failures*. Evident failures are detected instantly when they occur. The failure mode "spurious stop" of a pump with operational mode "running" is an example of an evident failure. Another type of failure is called *hidden failure*. A hidden failure is normally detected only during testing of the item. The failure mode "fail to start" of a pump with operational mode "standby" is an example of a hidden failure.

Effects on other components in the same subsystem (column 7). All the main effects of the identified failure modes on other components in the subsystem are recorded.

Effects on the function of the system (column 8). All the main effects of the identified failure mode on the function of the system are then recorded. The resulting operational status of the system after the failure may also be recorded, that is, whether the system is functioning or not, or is switched over to another operational mode.

Remark: In some applications it may be relevant to replace columns 7 and 8 by, for example, *Effect on safety* and *Effect on availability*.

Failure rate (column 9). Failure rates for each failure mode are then recorded. In many cases it is more suitable to classify the failure rate in rather broad classes. An example of such a classification is:

1	Very unlikely	Once per 1000 years or more seldom
2	Remote	Once per 100 years
3	Occasional	Once per 10 years
4	Probable	Once per year
5	Frequent	Once per month or more often

Note that the failure rate with respect to a failure mode might be different for the various operational modes. The failure mode "Leakage to the environment" for a valve may, as an example, be more likely when the valve is closed and pressurized, than when the valve is open.

Severity (column 10). By the severity of a failure mode we mean the worst potential consequence of the failure, determined by the degree of injury, property damage, or system damage that could ultimately occur. The following ranking categories (see Hammer 1972, p. 152) are often adopted:

Catastrophic	Any failure that could result in deaths or injuries or prevent performance of the intended mission.
Critical	Any failure that will degrade the system beyond acceptable limits and create a safety hazard (cause death or injury if corrective action is not immediately taken)
Major	Any failure that will degrade the system beyond acceptable limits but can be adequately counteracted or controlled by alternate means
Minor	Any failure that does not degrade the overall performance beyond acceptable limits—one of the nuisance variety

Slightly different categories are adopted in MIL-STD-882D. These categories are included as an illustration. The severity categories should be defined such that they are relevant for the practical application.

Risk reducing measures (column 11). Possible actions to correct the failure and restore the function or prevent serious consequences are then recorded. Actions that are likely to reduce the frequency of the failure modes may also be recorded.

Comments (column 12). This column may be used to record pertinent information not included in the other columns.

By combining the failure rate (column 9) and the severity (column 10) we may obtain a ranking of the criticality of the different failure modes. This ranking may be illustrated as in Fig. 3.16 by a *risk matrix*. In this example we have classified the failure rate in five classes as described under column 9. The severity is in the same way classified in four classes as described under column 10. The most critical failure modes will be represented by an ($\times$) in the upper right corner of the risk matrix, while the least critical failure modes will have ($\times$) in the lower left corner of the risk matrix.

3.5.4 Applications

Many industries require an FMECA to be integrated in the design process of technical systems and that FMECA worksheets be part of the system documentation. This is,

	Severity group			
Failure rate	Minor	Major	Critical	Catastrophic
Frequent				
Probable				
Occasional	(x)			
Remote		(x)		
Very unlikely	(x)		(x)	

Fig. 3.16 Risk matrix of the different failure modes.

for example, a common practice for suppliers to the defense, the aerospace, and the car industries. The same type of requirements are also becoming more and more usual within the offshore oil and gas industry.

Subcontractors to the car industry are usually met with requirements for both *product* and *process* FMECA. A product FMECA is a detailed FMECA of the technical items that are supplied to the car manufacturer. A process FMECA is an analysis of the producer's in-house production system, to verify that failures of the production system will not influence the quality of the products.

The FMECA is usually carried out during the design phase of a system. The main objective of the analysis is to reveal weaknesses and potential failures at an early stage, to enable the designer to incorporate corrections and barriers in the design. The results from the FMECA may also be useful during modifications of the system and for maintenance planning.

Many industries are introducing a reliability centered maintenance (RCM) program for maintenance planning. The RCM concept was introduced by the aviation industry and has formed the basis for the scheduled maintenance planning of a number of new airplane systems. The RCM concept is today applied in a wide range of industries, especially in nuclear power plants and within the offshore oil and gas industry. FMECA is one of the basic analytical tools of the RCM concept. The RCM concept is further discussed in Chapter 9.

Since all failure modes, failure mechanisms, and symptoms are documented in the FMECA, this also provides valuable information as a basis for failure diagnostic procedures and for a repairman's checklists.

An FMECA may be very effective when applied to a system where system failures most likely are the results of single component failures. During the analysis, each failure is considered individually as an independent occurrence with no relation to other failures in the system. Thus an FMECA is not suitable for analysis of systems with a fair degree of redundancy. For such systems a fault tree analysis would be a much better alternative. An introduction to fault tree analysis is given in Section 3.6. In addition, the FMECA is not well suited for analyzing systems where common cause failures are considered to be a significant problem. Common cause failures are discussed in Chapter 6.

A second limitation of FMECA is the inadequate attention generally given to human errors. This is mainly due to the concentration on hardware failures.

Perhaps the worst drawback is that all component failures are examined and documented, including those that do not have any significant consequences. For large systems, especially systems with a high degree of redundancy, the amount of unnecessary documentation work is a major disadvantage.

3.6 FAULT TREE ANALYSIS

The fault tree technique was introduced in 1962 at Bell Telephone Laboratories, in connection with a safety evaluation of the launching system for the intercontinental *Minuteman* missile. The Boeing Company improved the technique and introduced computer programs for both qualitative and quantitative fault tree analysis. Today fault tree analysis is one of the most commonly used techniques for risk and reliability studies. In particular, fault tree analysis has been used with success to analyze safety systems in nuclear power stations, such as the Reactor safety study (NUREG-0492).

A fault tree is a logic diagram that displays the interrelationships between a potential critical event (accident) in a system and the causes for this event. The causes may be environmental conditions, human errors, normal events (events that are expected to occur during the life span of the system), and specific component failures.

A fault tree analysis may be qualitative, quantitative, or both, depending on the objectives of the analysis. Possible results from the analysis may, for example, be

- A listing of the possible combinations of environmental factors, human errors, normal events, and component failures that may result in a critical event in the system.

- The probability that the critical event will occur during a specified time interval.

Only qualitative fault tree analysis is covered in this chapter. Quantitative fault tree analysis is discussed in Chapter 4. Fault tree analysis is thoroughly described in the literature; see, for example, NUREG-0492 and NASA (2002).

3.6.1 Fault Tree Construction

Fault tree analysis is a *deductive* technique where we start with a specified system failure or an accident. The system failure, or accident, is called the TOP *event* of the fault tree. The immediate causal events $A_1, A_2, \ldots$ that, either alone or in combination, may lead to the TOP event are identified and connected to the TOP event through a *logic gate*. Next, we identify all potential causal events $A_{i,1}, A_{i,2}, \ldots$ that may lead to event A_i for $i = 1, 2, \ldots$. These events are connected to event A_i through a logic gate. This procedure is continued deductively (i.e., backwards in the causal chain) until we reach a suitable level of detail. The events on the lowest level are called the *basic events* of the fault tree.

Fault tree analysis is a *binary* analysis. All events are assumed either to occur or not to occur; there are no intermediate options.

The graphical layout of the fault tree symbols are dependent on what standard we choose to follow. Table 3.1 shows the most commonly used fault tree symbols together with a brief description of their interpretation. A number of more advanced fault tree symbols are available but will not be covered in this book. A thorough description may be found in, for example, NUREG-0492 and NASA (2002). Note that the fault tree symbols used in the standard IEC 61025 are different from the symbols in Table 3.1. The meaning of the corresponding symbols are, however, the same.

A fault tree analysis is normally carried out in five steps[1]:

1. Definition of the problem and the boundary conditions.

2. Construction of the fault tree.

3. Identification of minimal cut and/or path sets.

4. Qualitative analysis of the fault tree.

5. Quantitative analysis of the fault tree.

Steps 1 to 4 are covered in this section, while step 5 is discussed in Chapter 4.

3.6.2 Definition of the Problem and the Boundary Conditions

The first activity of a fault tree analysis clearly consists of two substeps:

- Definition of the critical event (the accident) to be analyzed
- Definition of the boundary conditions for the analysis

The critical event (accident) to be analyzed is normally called the TOP event. It is very important that the TOP event is given a clear and unambiguous definition. If not, the analysis will often be of limited value. As an example, the event description "Fire in the plant" is far too general and vague. The description of the TOP event should always give answer to the questions *what*, *where*, and *when*:

What. Describes what type of critical event (accident) is occurring (e.g., fire)

Where. Describes where the critical event occurs (e.g., in the process oxidation reactor)

When. Describes when the critical event occurs (e.g., during normal operation)

---
[1] The procedure described below is strongly influenced by AIChE (1985).

Table 3.1 Fault Tree Symbols.

	Symbol	Description
Logic gates	OR-gate	
		The OR-gate indicates that the output event A occurs if any of the input events E_i occur
	AND-gate	
		The AND-gate indicates that the output event A occurs only when all the input events E_i occur at the same time
Input events	Basic event	
		The Basic event represents a basic equipment failure that requires no further development of failure causes
	Undeveloped event	
		The Undeveloped event represents an event that is not examined further because information is unavailable or because its consequence is insignificant
Description	Comment rectangle	
		The Comment rectangle is for supplementary information
Transfer symbols	Transfer-out	
		The Transfer-out symbol indicates that the fault tree is developed further at the occurrence of the corresponding Transfer-in symbol
	Transfer-in	

A more precise TOP event description is thus: "Fire in the process oxidation reactor during normal operation."

To get a consistent analysis, it is important that the boundary conditions for the analysis are carefully defined. By boundary conditions we understand the following:

- *The physical boundaries of the system.* Which parts of the system are to be included in the analysis, and which parts are not?

- *The initial conditions.* What is the operational state of the system when the TOP event is occurring? Is the system running on full/reduced capacity? Which valves are open/closed, which pumps are functioning, and so on?

- *Boundary conditions with respect to external stresses.* What type of external stresses should be included in the analysis? By external stresses we mean stresses from war, sabotage, earthquake, lightning, and so on.

- *The level of resolution.* How far down in detail should we go to identify potential reasons for a failed state? Should we, for example, be satisfied when we have identified the reason to be a "valve failure," or should we break it further down to failures in the valve housing, valve stem, actuator, and so forth. When determining the preferred level of resolution, we should remember that the detailedness in the fault tree should be comparable to the detailedness of the information available.

3.6.3 Construction of the Fault Tree

The fault tree construction always starts with the TOP event. We must thereafter carefully try to identify all fault events that are the immediate, necessary, and sufficient causes that result in the TOP event. These causes are connected to the TOP event via a logic gate. It is important that the first level of causes under the TOP event be put up in a structured way. This first level is often referred to as the TOP *structure* of the fault tree. The TOP structure causes are often taken to be failures in the prime modules of the system or in the prime functions of the system. We then proceed, level by level, until all fault events have been developed to the prescribed level of resolution. The analysis is in other words deductive and is carried out by repeatedly asking "What are the reasons for this event?"

Rules for Fault Tree Construction Let *fault event* denote any event in the fault tree, whether it is a basic event or an event higher up in the tree.

1. *Describe the fault events.* Each of the basic events should be carefully described (what, where, when) in a "comment rectangle."

2. *Evaluate the fault events.* The fault events may be different types, like technical failures, human errors, or environmental stresses. Each event should be carefully evaluated. As explained on page 85, technical failures may be divided in three groups: *primary failures*, *secondary failures* and *command faults*.

100 QUALITATIVE SYSTEM ANALYSIS

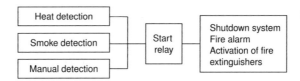

Fig. 3.17 System overview of fire detector system.

Primary failures of components are usually classified as basic events, while secondary failures and command faults are classified as intermediate events that require a further investigation to identify the prime reasons.

When evaluating a fault event, we ask the question: "Can this fault be a primary failure?" If the answer is yes, we classify the fault event as a "normal" basic event. If the answer is no, we classify the fault event as either an intermediate event that has to be further developed or as a "secondary" basic event. The secondary basic event is often called an *undeveloped* event and represents a fault event that is not examined further because information is unavailable or because its consequence is insignificant.

3. *Complete the gates.* All inputs to a specific gate should be completely defined and described before proceeding to the next gate. The fault tree should be completed in levels, and each level should be completed before beginning the next level.

Example 3.1 Fire Detector System
Let us consider a simplified version of a fire detector system located in a production room. (Observe that this system is not a fully realistic fire detector system.)

The fire detector system is divided into two parts, heat detection and smoke detection. In addition, there is an alarm button that can be operated manually. The fire detector system can be described schematically, as shown in Fig. 3.17 and 3.18.

Heat Detection
In the production room there is a closed, pneumatic pipe circuit with four identical fuse plugs, FP1, FP2, FP3, and FP4. These plugs let air out of the circuit if they are exposed to temperatures higher than 72°C. The pneumatic system has a pressure of 3 bars and is connected to a pressure switch (pressostat) PS. If one or more of the plugs are activated, the switch will be activated and give an electrical signal to the start relay for the alarm and shutdown system. In order to have an electrical signal, the direct current (DC) source must be intact.

Smoke Detection
The smoke detection system consists of three optical smoke detectors, SD1, SD2, and SD3; all are independent and have their own batteries. These detectors are very sensitive and can give warning of fire at an early stage. In order to avoid false alarms, the three smoke detectors are connected via a logical 2-out-of-3 voting unit, VU. This

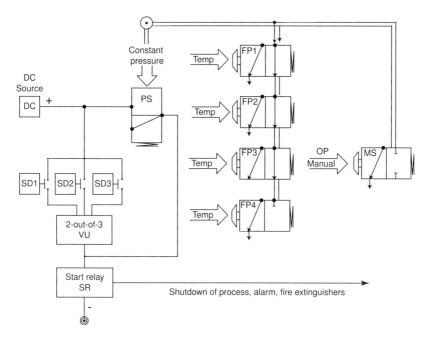

Fig. 3.18 Schematic layout of the fire detector system.

means that at least two detectors must give fire signal before the fire alarm is activated. If at least two of the three detectors are activated, the 2-out-of-3 voting unit will give an electric signal to the start relay, SR, for the alarm and shutdown system. Again the DC voltage source must be intact to obtain an electrical signal.

Manual Activation
Together with the pneumatic pipe circuit with the four fuse plugs, there is also a manual switch, MS, that can be turned to relieve the pressure in the pipe circuit. If the operator, OP, who should be continually present, notices a fire, he can activate this switch. When the switch is activated, the pressure in the pipe circuit is relieved and the pressure switch, PS, is activated and gives an electric signal to the start relay, SR. Again the *DC* source must be intact.

The Start Relay
When the start relay, SR, receives an electrical signal from the detection systems, it is activated and gives a signal to

- Shut down the process.

- Activate the alarm and the fire extinguishers.

Assume now that a fire starts. The fire detector system should detect and give warning about the fire. Let the TOP event be: *"No signals from the start relay SR*

102 QUALITATIVE SYSTEM ANALYSIS

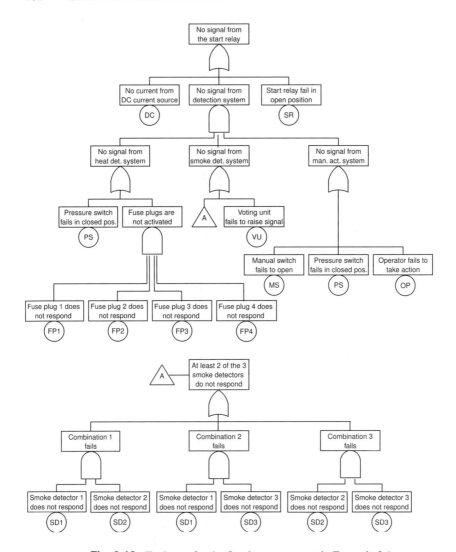

Fig. 3.19 Fault tree for the fire detector system in Example 3.1.

when a fire condition is present." (Remember *what*, *where*, and *when*.) A possible fault tree for this TOP event is presented in Fig. 3.19. □

Remark: Observe that a fault tree does not show the causes of *all* failures or accidents in a system. It only illustrates the causes of a specified failure or accident, the TOP event. The fault tree will usually also be dependent on the analyst. Two different analysts will, in most cases, construct slightly different fault trees. □

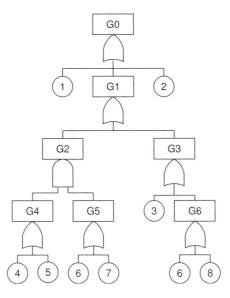

Fig. 3.20 Example of a fault tree.

3.6.4 Identification of Minimal Cut and Path Sets

A fault tree provides valuable information about possible combinations of fault events that will result in the TOP event. Such a combination of fault events is called a cut set. In the fault tree terminology, a cut set is defined as follows:

Definition 3.1 A cut set in a fault tree is a set of basic events whose occurrence (at the same time) ensures that the TOP event occurs. A cut set is said to be minimal if the set cannot be reduced without loosing its status as a cut set. □

The number of different basic events in a minimal cut set is called the *order* of the cut set. For small and simple fault trees, it is feasible to identify the minimal sets by inspection without any formal procedure/algorithm. For large or complex fault trees we need an efficient algorithm.

3.6.5 MOCUS

MOCUS (method for obtaining cut sets) is an algorithm that can be used to find the minimal cut sets in a fault tree. The algorithm is best explained by an example. Consider the fault tree in Fig. 3.20, where the gates are numbered from G0 to G6. The example fault tree is adapted from Barlow and Lambert (1975).

The algorithm starts at the G0 gate representing the TOP event. If this is an OR-gate, each input to the gate is written in separate rows. (The inputs may be new gates). Similarly, if the G0 gate is an AND-gate, the inputs to the gate are written in separate columns.

104 QUALITATIVE SYSTEM ANALYSIS

In our example, G0 is an OR-gate, hence we start with

$$\begin{matrix} 1 \\ G1 \\ 2 \end{matrix}$$

Since each of the three inputs, 1, G1 and 2 will cause the TOP event to occur, each of them will constitute a cut set.

The idea is to successively replace each gate with its inputs (basic events and new gates) until one has gone through the whole fault tree and is left with just the basic events. When this procedure is completed, the rows in the established matrix represent the cut sets in the fault tree.

Since $G1$ is an OR-gate: Since $G2$ is an AND-gate:

$$\begin{matrix} 1 \\ G2 \\ G3 \\ 2 \end{matrix} \qquad \begin{matrix} 1 \\ G4,G5 \\ G3 \\ 2 \end{matrix}$$

Since G3 is an OR-gate: Since G4 is an OR-gate:

$$\begin{matrix} 1 \\ G4,G5 \\ 3 \\ G6 \\ 2 \end{matrix} \qquad \begin{matrix} 1 \\ 4,G5 \\ 5,G5 \\ 3 \\ G6 \\ 2 \end{matrix}$$

Since G5 is an OR-gate: Since G6 is an OR-gate:

$$\begin{matrix} 1 \\ 4,6 \\ 4,7 \\ 5,6 \\ 5,7 \\ 3 \\ G6 \\ 2 \end{matrix} \qquad \begin{matrix} 1 \\ 4,6 \\ 4,7 \\ 5,6 \\ 5,7 \\ 3 \\ 6 \\ 8 \\ 2 \end{matrix}$$

We are then left with the following 9 cut sets:

{1}	{4,6}
{2}	{4,7}
{3}	{5,6}
{6}	{5,7}
{8}	

Since {6} is a cut set, {4,6} and {5,6} are not minimal. If we leave these out, we are left with the following list of minimal cut sets:

$$\{1\}, \{2\}, \{3\}, \{6\}, \{8\}, \{4,7\}, \{5,7\}$$

In other words, five minimal cut sets of order 1 and two minimal cut sets of order 2. The reason that the algorithm in this case leads to nonminimal cut sets is that basic event 6 occurs several places in the fault tree.

In some situations it may also be of interest to identify the possible combinations of components which by functioning secures that the system is functioning. Such a combination of components (basic events) is called a *path set*. In the fault tree terminology a path set is defined as follows:

Definition 3.2 A path set in a fault tree is a set of basic events whose nonoccurrence (at the same time) ensures that the TOP event does not occur. A path set is said to be minimal if the set cannot be reduced without loosing its status as a path set. □

The number of different basic events in a minimal path set is called the *order* of the path set. To find the minimal path sets in the fault tree, we may start with the so-called dual fault tree. This can be obtained by replacing all the AND-gates in the original fault tree with OR-gates, and vice versa. In addition, we let the events in the dual fault tree be complements of the corresponding events in the original fault tree. The same procedure as described above applied to the dual fault tree will now yield the minimal path sets.

For relatively "simple" fault trees one can apply the MOCUS algorithm by hand. More complicated fault trees require the use of a computer. A number of computer programs for minimal cut (path) set identification are available. Some of these are based on MOCUS, but faster algorithms have also been developed.

3.6.6 Qualitative Evaluation of the Fault Tree

A qualitative evaluation[1] of the fault tree may be carried out on the basis of the minimal cut sets. The criticality of a cut set obviously depends on the number of basic events in the cut set (i.e., the order of the cut set). A cut set of order 1 is usually more critical than a cut set of order 2, or more. When we have a cut set of order 1, the TOP event will occur as soon as the corresponding basic event occurs. When a cut set has two basic events, both of these have to occur simultaneously to cause the TOP event to occur.

Another important factor is the type of basic events of a minimal cut set. We may rank the criticality of the various cut sets according to the following ranking of basic events:

1. Human error

2. Active equipment failure

3. Passive equipment failure

---
[1] This section is strongly influenced by AIChE (1985).

Table 3.2 Criticality Ranking of Minimal Cut Sets of Order 2.

Rank	Basic event 1 (type)	Basic event 2 (type)
1	Human error	Human error
2	Human error	Active equipment failure
3	Human error	Passive equipment failure
4	Active equipment failure	Active equipment failure
5	Active equipment failure	Passive equipment failure
6	Passive equipment failure	Passive equipment failure

This ranking is based on the assumption that human errors occur more frequently than active equipment failures, and that active equipment is more prone to failure than passive equipment (e.g., an active or running pump is more exposed to failures than a passive standby pump). Based on this ranking, we get the ranking in Table 3.2 of the criticality of minimal cut sets of order 2. (Rank 1 is the most critical one.)

3.7 CAUSE AND EFFECT DIAGRAMS

The *cause and effect diagram*, also called Ishikawa diagram, was developed in 1943 by the Japanese professor Kaoru Ishikawa (1915–1989). The cause and effect diagram is used to identify and describe all the potential causes (or events) that may result in a specified event. Causes are arranged according to their level of importance or detail, resulting in a tree structure that resembles the skeleton of a fish with the main causal categories drawn as *bones* attached to the spine of the fish. The cause and effect diagram is therefore also known as a *fishbone* diagram.

A cause and effect diagram has some similarities with a fault tree but is less structured and does not have the same binary restrictions as a fault tree. To construct a cause and effect diagram, we start with a system failure or an accident that may be the same as the TOP event in a fault tree. The system failure (accident) is briefly described, enclosed in a box and placed at the right end of the diagram, as the "head of the fish." We then draw the central spine as a thick line pointing to the box (head) from the left. The major categories of potential causes are then drawn as bones to the spine. When analyzing technical systems, the following five (5M) categories are frequently used:

1. Manpower

2. Methods

3. Materials

4. Machinery

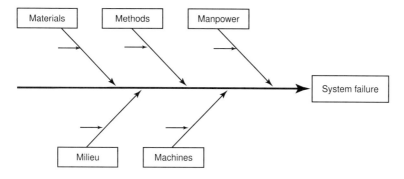

Fig. 3.21 Example of cause and effect diagram.

5. Milieu (environment)

The categories should, however, be selected to fit the actual application. It is usually recommended to use at most seven major categories. An idea-generating technique (like brainstorming) is then used by a team of experts to identify the factors within each category that may be affecting the system failure (accident) being studied. The team should ask: "What are the machine issues affecting/causing…?" This procedure is repeated for each factor under the category to produce subfactors. We continue asking why this is happening and put additional segments under each factor and subsequently under each sub-factor. We continue until we no longer get useful information when we ask: "Why may this happen?" An example of a cause and effect diagram is shown in Fig. 3.21.

When the team members agree that an adequate amount of detail has been provided under each major category, we may analyze the diagram and group the causes. One should especially look for causes that appear in more than one category. For those items identified as the "most likely causes," the team should reach consensus on listing those causes in priority order with the first cause being the "most likely cause."

The cause and effect diagram cannot be used for quantitative analyses but is generally considered to be an excellent aid for problem solving and to illustrate the potential causes of a system failure or an accident. Cause and effects diagrams are described and discussed in textbooks on quality engineering and management, for example, Ishikawa (1986) and Bergman and Klefsjö (1994).

3.8 BAYESIAN BELIEF NETWORKS

An Bayesian belief network (BBN) can be used as an alternative to fault trees and cause and effect diagrams to illustrate the relationships between a system failure or an accident and its causes and contributing factors. A BBN is more general than a fault tree since the causes do not have to be binary events. It is also more general because we do not have to connect the causes through a specified logic gate. In this way the

108 QUALITATIVE SYSTEM ANALYSIS

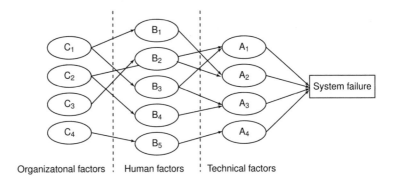

Fig. 3.22 Example of a Bayesian belief network.

BBN is rather similar to a cause and effect diagram. Contrary to the cause and effect diagram, the BBN can, however, be used as a basis for quantitative analysis.

A BBN is a directed acyclic graph. We start with the system failure or accident, the TOP event. The most immediate causes and contributing factors $A_1, A_2, \ldots$ are linked to the TOP event by arrows. Causes and contributing factors $A_{i,1}, A_{i,2}, \ldots$ that are influencing factor A_i are then linked to A_i by arrows, for $i = 1, 2, \ldots$. This procedure is continued until a desired level of resolution is reached. Dependencies between factors on different "levels" in the diagram may further be illustrated by arrows. In some applications it may be beneficial to group the causes and contributing factors in some major categories, for example, as technical factors (A factors), human factors (B factors), and organizational factors (C factors). An example of a BBN with this grouping is shown in Fig. 3.22. This application of BBNs is further discussed by Øien (2001).

Quantitative evaluation of Bayesian belief networks will not be discussed in this book. The reader is advised to consult, for example, Pearl (2000), Jensen (2001), Barlow (1998), and Bedford and Cooke (2001) for details about qualitative and quantitative assessment of BBNs.

3.9 EVENT TREE ANALYSIS

In many accident scenarios, the initiating (accidental) event, for example, a ruptured pipeline, may have a wide spectrum of possible outcomes, ranging from no consequences to a catastrophe. In most well-designed systems, a number of safety functions, or barriers, are provided to stop or mitigate the consequences of potential accidental events. The safety functions may comprise technical equipment, human interventions, emergency procedures, and combinations of these. Examples of technical safety functions are: fire and gas detection systems, emergency shutdown (ESD) systems, automatic train stop systems, fire-fighting systems, fire walls, and evacuation systems. The consequences of the accidental event are determined by how the accident progression is affected by subsequent failure or operation of these safety functions, by human errors made in responding to the accidental event, and by various factors like weather conditions and time of the day.

The accident progression is best analyzed by an inductive method. The most commonly used method is the *event tree analysis*. An event tree is a logic tree diagram that starts from a basic initiating event and provides a systematic coverage of the time sequence of event propagation to its potential outcomes or consequences. In the development of the event tree, we follow each of the possible sequences of events that result from assuming failure or success of the safety functions affected as the accident propagates. Each event in the tree will be conditional on the occurrence of the previous events in the event chain. The outcomes of each event are most often assumed to be binary (*true* or *false* – *yes* or *no*) but may also include multiple outcomes (e.g., *yes*, *partly*, and *no*).

Event tree analyses have been used in risk and reliability analyses of a wide range of technological systems. The event tree analysis is a natural part of most risk analyses but may be used as a design tool to demonstrate the effectiveness of protective systems in a plant. Event tree analyses are also used for human reliability assessment, for example, as part of the THERP technique (NUREG/CR-1278).

The event tree analysis may be qualitative, quantitative, or both, depending on the objectives of the analysis. In quantitative risk assessment (QRA) application, event trees may be developed independently or follow on from fault tree analysis.

An event tree analysis is usually carried out in six steps (AIChE 1985):

1. Identification of a relevant initiating (accidental) event that may give rise to unwanted consequences

2. Identification of the safety functions that are designed to deal with the initiating event

3. Construction of the event tree

4. Description of the resulting accident event sequences

5. Calculation of probabilities/frequencies for the identified consequences

6. Compilation and presentation of the results from the analysis

A simple event tree for a (dust) explosion is shown in Fig. 3.23. Following the initiating event explosion in Fig. 3.23, fire may or may not break out. A sprinkler system and an alarm system have been installed. These may or may not function. The quantitative analysis of the event tree is discussed on page 112.

Initiating Event Selection of a relevant initiating event is very important for the analysis. The initiating event is usually defined as the first significant deviation from the normal situation that may lead to a system failure or an accident. The initiating event may be a technical failure or some human error and may have been identified by other risk analysis techniques like FMECA, preliminary hazard analysis (PHA), or hazard and operability analysis (HAZOP). To be of interest for further analysis,

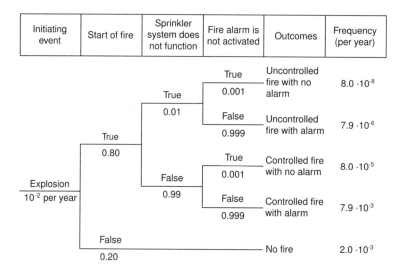

Fig. 3.23 A simple event tree for a dust explosion (adapted from IEC 60300-3-9).

the initiating event must give rise to a number of consequence sequences. If the initiating event gives rise to only one consequence sequence, fault tree analysis is a more suitable technique to analyze the problem.

The initiating event is often identified and anticipated as a possible critical event already in the design phase. In such cases, barriers and safety functions have usually been introduced to deal with the event.

Various analysts may define slightly different initiating events. For a safety analysis of, for example, an oxidation reactor, one analyst may choose "Loss of cooling water to the reactor" as a relevant initiating event. Another analyst may, for example, choose "Rupture of cooling water pipeline" as initiating event. Both of these are equally correct.

Safety Functions The safety functions (e.g., barriers, safety systems, procedures, and operator actions) that respond to the initiating event may be thought of as the system's defense against the occurrence of the initiating event. The safety functions may be classified in the following groups (AIChE 1985):

- Safety systems that automatically respond to the initiating event (e.g., automatic shutdown systems)

- Alarms that alert the operator(s) when the initiating event occurs (e.g., fire alarm systems)

- Operator procedures following an alarm

- Barriers or containment methods that are intended to limit the effects of the initiating event

The analyst must identify all barriers and safety functions that have impact on the consequences of an initiating event, in the sequence they are assumed to be activated.

The possible event chains, and sometimes also the safety functions, will be affected by various hazard contributing factors (events or states) like:

- Ignition or no ignition of a gas release
- Explosion or no explosion
- Time of the day
- Wind direction toward community or not
- Meteorological conditions
- Liquid/gas release contained or not

Event Tree Construction The event tree displays the chronological development of event chains, starting with the initiating event and proceeding through successes and/or failures of the safety functions that respond to the initiating event. The consequences are clearly defined events that result from the initiating event.

The diagram is usually drawn from left to right, starting from the initiating event. Each safety function or hazard contributing factor is called a *node* in the event tree and is formulated either as an event description or as a question, usually with two possible outcomes (*true* or *false* – *yes* or *no*). At each node the tree splits into two branches: the upper branch signifying that the event description in the box above that node is *true*, and a lower branch signifying that it is *false*. If we formulate the description of each node such that the worst outcome will always be on the upper branch, the consequences will be ranked in a descending order, with the worst consequence highest up in the list.

The outputs from one event lead to other events. The development is continued to the resulting consequences.

If the diagram is too big to be drawn on a single page, it is possible to isolate branches and draw them on different pages. The different pages may be linked together by transfer symbols.

Note that for a sequence of n events, there will be 2^n branches of the tree. The number may, however, in some cases be reduced by eliminating impossible branches.

Description of Resulting Event Sequences The last step in the qualitative part of the analysis is to describe the different event sequences arising from the initiating event. One or more of the sequences may represent a safe recovery and a return to normal operation or an orderly shutdown. The sequences of importance, from a safety point of view, are those that result in accidents.

The analyst must strive to describe the resulting consequences in a clear and unambiguous way. When the consequences are described, the analyst may rank them according to their criticality. The structure of the diagram, clearly showing the progression of the accident, helps the analyst in specifying where additional procedures or safety systems will be most effective in protecting against these accidents.

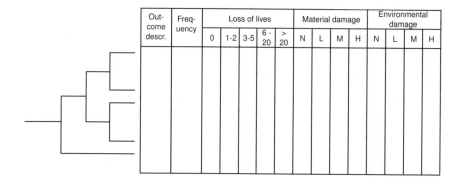

Fig. 3.24 Presentation of results from event tree analysis.

Sometimes we may find it beneficial to split the end consequences (outcomes) of the event tree analysis into various consequence categories as illustrated in Fig. 3.24. In this example the following categories are used:

- Loss of lives
- Material damage
- Environmental damage

Within each category, the consequences may be ranked. For the category "loss of lives," the subcategories 0, 1–2, 3–5, 6–20, and ≥ 21 are proposed. For the categories "material damage" and "environmental damage" the subcategories are: negligible (N), low (L), medium (M), and high (H). What is meant by these categories has to be defined in each particular case. If we are unable to put the consequences into a single group, we may give a probability distribution over the subcategories. The outcome of an event chain may, for example, be that nobody will be killed with probability 50%, 1–2 persons will be killed with probability 40%, and 3–5 persons will be killed with probability 10%. If we in addition are able to estimate the frequency of the outcome (see below) it is straightforward to estimate the fatal accident rate[2] (FAR) associated to the specified initiating event.

Quantitative Assessment If experience data are available for the initiating event and all the relevant safety functions and hazard contributing factors, a quantitative analysis of the event tree may be carried out to give frequencies or probabilities of the resulting consequences.

The occurrences of the initiating event is usually modeled by a homogeneous Poisson process with frequency λ, which is measured as the expected number of

---

[2] FAR is a commonly used measure for personnel risk and is defined as the expected number of fatalities per 10^8 hours of exposure.

occurrences per year (or some other time unit). Homogeneous Poisson processes are further discussed in Chapter 7.

For each safety function we have to estimate the conditional probability that it will function properly in the relevant context, that is, when the previous events in the event chain have occurred. Some safety functions, like ESD systems on offshore oil/gas platforms, may be very complex and will require a detailed reliability analysis.

The (conditional) reliability of a safety function will depend on a wide range of environmental and operational factors, like loads from previous events in the event chain, and the time since the last function test. In many cases it will also be difficult to distinguish between "functioning" and "nonfunctioning." A fire pump may, for example, start, but stop prematurely before the fire is extinguished.

The reliability assessment of a safety function may in most cases be performed by a fault tree analysis or an analysis based on a reliability block diagram. If the analysis is computerized, a link may be established between the reliability assessment and the appropriate node in the event tree to facilitate automatic updating of the outcome frequencies and sensitivity analyses. It may, for example, be relevant to study the effect on the outcome frequencies by changing the testing interval of a safety valve. Graphically the link may be visualized by a transfer symbol on one of the output branches from the node.

The probabilities of the various hazard contributing factors (events/states) that enter into the event tree must also be estimated for the relevant contexts. Some of these factors will be independent of the previous events in the event chain, while others are not.

It is important to note that most of the probabilities in the event tree are *conditional* probabilities. The probability that the sprinkler system in Fig. 3.23 will function is, for example, not equal to a probability that is estimated based on tests under normal conditions. We have to take into account that the sprinkler system may have been damaged during the dust explosion and the first phase of the fire (i.e., before it is activated).

Consider the event tree in Fig. 3.23. Let λ_A denote the frequency of the initiating event A, "explosion." In this example, λ_A is assumed to be equal to 10^{-2} per year, which means that an explosion on the average will occur once every 100 years. Let B denote the event "start of a fire," and let $\Pr(B) = 0.8$ be the conditional probability of this event when a dust explosion has already occurred. A more correct notation would be the conditional probability $\Pr(B \mid A)$ to make it clear that event B is considered when event A has already occurred.

In the same way, let C denote the event that the sprinkler system does not function, following the dust explosion and the outbreak of a fire. The conditional probability of C is assumed to be $\Pr(C) = 0.01$.

The fire alarm will not be activated (event D) with probability $\Pr(D) = 0.001$. In this example we have assumed that this probability is the same whether the sprinkler system is functioning or not. In most cases, however, the probability of this event would depend on the outcome of the previous event.

Let B^*, C^*, and D^* denote the negation (nonoccurrence) of the events B, C, and D, respectively. We know that $\Pr(B^*)$ is equal to $1 - \Pr(B)$, and so on.

The frequencies (per year) of the end consequences may now be calculated as follows:

1. Uncontrolled fire with no alarm:

$$\lambda_4 = \lambda_A \cdot \Pr(B) \cdot \Pr(C) \cdot \Pr(D) = 10^{-2} \cdot 0.8 \cdot 0.01 \cdot 0.001 \approx 8.0 \cdot 10^{-8}$$

2. Uncontrolled fire with alarm:

$$\lambda_3 = \lambda_A \cdot \Pr(B) \cdot \Pr(C^*) \cdot \Pr(D) = 10^{-2} \cdot 0.8 \cdot 0.01 \cdot 0.999 \approx 8.0 \cdot 10^{-5}$$

3. Controlled fire with no alarm:

$$\lambda_2 = \lambda_A \cdot \Pr(B) \cdot \Pr(C^*) \cdot \Pr(D) = 10^{-2} \cdot 0.8 \cdot 0.99 \cdot 0.001 \approx 7.9 \cdot 10^{-6}$$

4. Controlled fire with alarm:

$$\lambda_1 = \lambda_A \cdot \Pr(B) \cdot \Pr(C^*) \cdot \Pr(D^*) = 10^{-2} \cdot 0.8 \cdot 0.99 \cdot 0.999 \approx 7.9 \cdot 10^{-3}$$

5. No fire:

$$\lambda_5 = \lambda_A \cdot \Pr(B^*) = 10^{-2} \cdot 0.2 \approx 2.0 \cdot 10^{-3}$$

It is seen that the frequency of a specific outcome (consequence) simply is obtained by multiplying the frequency of the initiating event by the probabilities along the event sequence leading to the outcome in question.

If we assume that occurrences of the initiating event may be described by a homogeneous Poisson process, and that all the probabilities of the safety functions and hazard contributing factors are constant and independent of time, then the occurrences of each outcome will also follow a homogeneous Poisson process.

Example 3.2 Offshore Separator

In this example we consider a part of the processing section on an offshore oil and gas production installation. A mixture of oil, gas, and water coming from the various wells is collected in a wellhead manifold and led into two identical process trains. The gas, oil, and water are separated in several separators. The gas from the process trains is then collected in a compressor manifold and led to the gas export pipeline via compressors. The oil is loaded onto tankers and the water is cleaned and reinjected into the reservoir. Fig. 3.25 shows a simplified sketch of section of one of the process trains. The mixture of oil, gas, and water from the wellhead manifold is led into the separator, where the gas is (partly) separated from the fluids. The process is controlled by a *process control system* that is not illustrated in the figure. If the process control system fails, a separate *process safety system* should prevent a major accident. This example is limited to this process safety system. The process safety system has three protection layers:

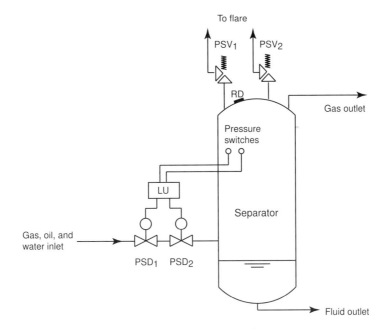

Fig. 3.25 Sketch of a first-stage gas separator.

1. On the inlet pipeline, there are installed two process shutdown (PSD) valves, PSD_1 and PSD_2 in series. The valves are fail-safe close and are held open by hydraulic (or pneumatic) pressure. When the hydraulic (pneumatic) pressure is bled off, the valves will close by the force of a precharged actuator. The system supplying hydraulic (pneumatic) pressure to the valve actuators is not illustrated in Fig. 3.25.

   Two pressure switches, PS_1, and PS_2 are installed in the separator. If the pressure in the separator increases above a set value, the pressure switches should send a signal to a logic unit (LU). If the LU receives at least one signal from the pressure switches, it will send a signal to the PSD valves to close.

2. Two pressure safety valves (PSV) are installed to relieve the pressure in the separator in case the pressure increases beyond a specified high pressure. The PCV valves, PSV_1 and PSV_2, are equipped with a spring-loaded actuator that may be adjusted to a preset pressure.

3. A rupture disc (RD) is installed on top of the separator as a last safety barrier. If the other safety systems fail, the rupture disc will open and prevent the separator from rupturing or exploding. If the rupture disc opens, the gas will blow out from the top of the separator and maybe into a blowdown system.

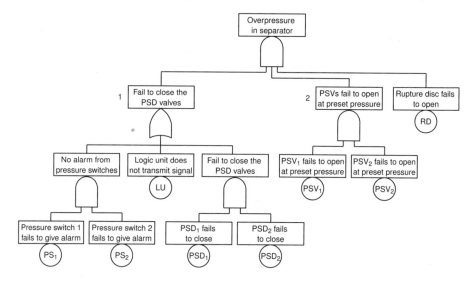

Fig. 3.26 Fault tree for the first-stage separator in Example 3.2.

The reliability of the process safety system may be analyzed by different approaches. We will here illustrate how a fault tree and an event tree analysis may be performed.

Fault Tree Analysis
The most critical situation will arise if the gas outlet line A is suddenly blocked. The pressure in the separator will then rapidly increase and will very soon reach a critical overpressure, if the process safety system does not function properly. A relevant TOP event is therefore "Critical overpressure in the first-stage separator." We assume that the critical situation occurs during normal production and that the fluid level in the separator is normal when the event occurs. We may therefore disregard the fluid outlet line from the fault tree analysis. A possible fault tree for this TOP event is presented in Fig. 3.26. In Chapter 4 we will show how to enter failure rates and other reliability parameters into the fault tree, as well as how to calculate the probability $Q_0(t)$ of the TOP event when the gas outlet is suddenly blocked.

In the construction of the fault tree in Fig. 3.26 we have made a number of assumptions. The assumptions should be recorded in a separate file and integrated in the report from the analysis. The lowest level of resolution in the fault tree in Fig. 3.26 is a failure mode of a technical item. Some of these items are rather complex, and it might be of interest to break them down into subitems and attribute failures to these. The valves may, for example, be broken down into valve body and actuator. These subitems may again be broken down to sub-subitems and so on. The failure of the pressure switches to give a signal may be split in two parts, individual failures and common cause failures that cause both pressure switches to fail at the same time. A pressure switch may fail due to an inherent component failure or due to a miscalibration by the maintenance crew. How far we should proceed depends on the objective of the analysis. Anyway, the assumptions made should be recorded.

Event Tree Analysis

The activation pressures for the three protection layers of the process safety system are illustrated in Fig. 3.27. We will get different consequences depending on whether or not the three protection systems are functioning, and the system is therefore suitable for an event tree analysis. The initiating event is "blockage of the gas outlet line." A possible event tree for this initiating event is presented in Fig. 3.28. The four outcomes are seen to give very different consequences. The most critical outcome is "rupture or explosion of separator" and may lead to total loss of the installation if the gas is ignited. The probability of this outcome is, however, very low since the rupture disc is a very simple and reliable item. The second most critical outcome is "gas flowing out of rupture disc." The criticality of this outcome depends on the design of the system, but may for some installations be very critical if the gas is ignited. The

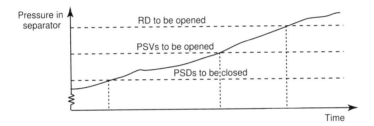

Fig. 3.27 Activation pressures for the three protection layers of the process safety system.

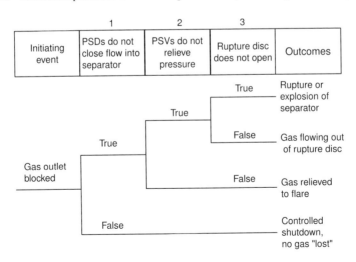

Fig. 3.28 An event tree for the initiating event "blockage of the gas outlet line."

Fig. 3.29 Component i illustrated by a block.

next outcome "gas relieved to flare" is usually a noncritical event but will lead to an economic loss (CO_2 tax) and production downtime. The last outcome is a controlled shutdown that will only lead to production downtime.

In this case the event tree analysis is seen to provide more detailed results than the fault tree analysis. The two analyses may be combined. The causes of failure of barrier 1 (PSDs do not close flow into separator) are found in branch 1 of the fault tree in Fig. 3.26. The causes of failure of barrier 2 (PSVs do not relieve pressure) are found in branch 2 of the fault tree in Fig. 3.26. If we have reliability data for all the basic events, we may use the fault tree to find the probabilities of the various branches in the event tree. □

3.10 RELIABILITY BLOCK DIAGRAMS

In this section we illustrate the structure of a system by what is known as a *reliability block diagram* (RBD). A reliability block diagram is a success-oriented network describing the *function* of the system. It shows the logical connections of (functioning) components needed to fulfill a specified system function. If the system has more than one function, each function must be considered individually, and a separate reliability block diagram has to be established for each system function.

Reliability block diagrams are suitable for systems of nonrepairable components and where the order in which failures occur does not matter. When the systems are repairable and/or the order in which failures occur is important, Markov methods will usually be more suitable. Markov methods are described in Chapter 8.

Consider a system with n different components. Each of the n components is illustrated by a block as shown is Fig. 3.29. When there is connection between the end points a and b in Fig. 3.29, we say that component i is functioning. This does not necessarily mean that component i functions in all respects. It only means that one, or a specified set of functions, is achieved [i.e., that some specified failure mode(s) do not occur]. What is meant by functioning must be specified in each case and will depend on the objectives of the study. It is also possible to put more information into the block in Fig. 3.29 and include a brief description of the required function of the component. An example is shown in Fig. 3.30, where the component is a safety shutdown valve that is installed in a pipeline. A label is used to identify the block. The label is usually a combination of three to five letters and digits.

The way the n components are interconnected to fulfill a specified system function may be illustrated by a reliability block diagram, as shown in Fig. 3.31. When we have connection between the end points a and b in Fig. 3.31, we say that the specified system function is achieved, which means that some specified system failure mode(s)

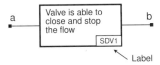

Fig. 3.30 Alternative representation of the block in Fig. 3.29.

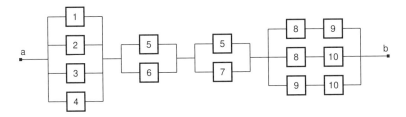

Fig. 3.31 System function illustrated by a reliability block diagram.

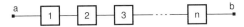

Fig. 3.32 Reliability block diagram of a series structure.

do(es) not occur. The symbols that are used in this chapter to establish a reliability block diagram are according to the standard IEC 61078.

3.10.1 Series Structure

A system that is functioning if and only if all of its n components are functioning is called a *series structure*. The corresponding reliability block diagram is shown in Fig. 3.32. We have connection between the end points a and b (the system is functioning) if and only if we have connection through all the n blocks representing the components.

3.10.2 Parallel Structure

A system that is functioning if at least one of its n components is functioning is called a *parallel structure*. The corresponding reliability block diagram is shown in Fig. 3.33. In this case we have connection between the end points a and b (i.e., the system is functioning) if we have connection through at least one of the blocks representing the components.

Example 3.3
Consider a pipeline with two independent safety valves V_1 and V_2 that are physically installed in series, as illustrated in Fig. 3.34a. The valves are supplied with a spring loaded fail-safe-close hydraulic actuator. The valves are opened and held open by hydraulic control pressure and are closed automatically by spring force whenever the

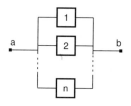

Fig. 3.33 Reliability block diagram of a parallel structure.

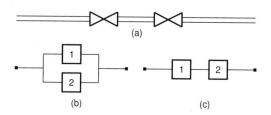

Fig. 3.34 Two safety valves in a pipeline: (a) physical layout, (b) reliability block diagram with respect to the safety barrier function, and (c) reliability block diagram with respect to spurious closure.

control pressure is removed or lost. In normal operation, both valves are held open. The main function of the valves is to act as a safety barrier, that is, to close and stop the flow in the pipeline in case of an emergency.

Since it is sufficient that one of the valves closes in order to stop the flow, the valves will form a *parallel* system with respect to the safety barrier function, as shown in Fig. 3.34b. The valves may close spuriously, that is, without a control signal, and stop the flow in the pipeline. Also in this case it is sufficient that only one of the valves fails in order to stop the flow. The valves will thus form a *series* system with respect to spurious closures, as shown in Fig. 3.34c. Notice the different meanings of the functional blocks in Fig. 3.34b and 3.34c. In Fig. 3.34b, connection through the block $\boxed{i}$ means that valve i is able to stop the flow in the pipeline, while connection through $\boxed{i}$ in Fig. 3.34c means that valve i does not close spuriously, for $i = 1, 2$.

□

3.10.3 Reliability Block Diagrams versus Fault Trees

In some practical applications, we may choose whether to model the system structure by a fault tree or by a reliability block diagram. When the fault tree is limited to only OR-gates and AND-gates, both methods may yield the same result, and we may convert the fault tree to a reliability block diagram, and vice versa.

In a reliability block diagram, connection through a block means that the component represented by the block is functioning. This again means that one or a specified

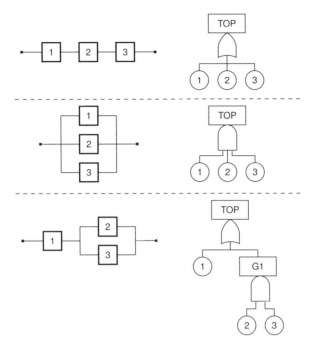

Fig. 3.35 Relationship between some simple reliability block diagrams and fault trees.

set of failure modes of the component is not occurring. In a fault tree we may let a basic event be the occurrence of the same failure mode or the same specified set of failure modes for the component. When the TOP event in the fault tree represents "system failure" and the basic events are defined as above, it is easy to see, for instance, that a series structure is equivalent to a fault tree where all the basic events are connected through an OR-gate. The TOP event occurs if either component 1 or component 2 or component 3 or ...component n fails.

In the same way a parallel structure may be represented as a fault tree where all the basic events are connected through an AND-gate. The TOP event occurs (i.e., the parallel structure fails) if component 1 and component 2 and component 3 and ...component n fail. The relationship between some simple reliability block diagrams and fault trees is illustrated in Fig. 3.35.

Example 3.2 (Cont.)
It is usually an easy task to convert a fault tree to a reliability block diagram. The reliability block diagram corresponding to the fault tree for the fire detector system in Fig. 3.19 is shown in Fig. 3.36. In this conversion we start from the TOP event and replace the gates successively. OR-gates are replaced by series structures of the "components" directly beneath the gate, and AND-gates are replaced by a parallel structure of the "components" directly beneath the gate. □

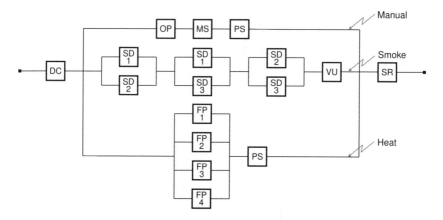

Fig. 3.36 Reliability block diagram for the fire detector system.

From Fig. 3.36 we observe that some of the components are represented in two different locations in the diagram. It should be emphasized that a reliability block diagram is not a physical layout diagram for the system. It is a logic diagram, illustrating the function of the system.

Recommendation In most practical applications it is recommended to start by constructing a fault tree instead of a reliability block diagram. In the construction of the fault tree we search for all potential causes of a specified system failure (accident). We think in terms of *failures* and will often reveal more potential failure causes than if we think in terms of *functions*, as we do by establishing a reliability block diagram. The construction of a fault tree will give the analyst a better understanding of the potential causes of failure. If the analysis is carried out in the design phase, the analyst may rethink the design and operation of the system and take actions to eliminate potential hazards.

When we establish a reliability block diagram, we think in terms of functions and will often forget auxiliary functions and equipment that is, or should be, installed to protect the equipment, people, or the environment.

For further evaluations, however, it is often more natural to base these on a reliability block diagram. A fault tree will therefore sometimes be converted to a reliability block diagram for qualitative and quantitative analyses. This is the main reason why we have chosen to focus on reliability block diagrams in the rest of this book and look upon fault trees as an alternative approach.

3.10.4 Structure Function

A system that is composed of n components will in the following be denoted a system of *order n*. The components are assumed to be numbered consecutively from 1 to n[1].

In this chapter we will confine ourselves to situations where it suffices to distinguish between only two states, a functioning state and a failed state. This applies to each component as well as to the system itself. The state of component i, $i = 1, 2, \ldots, n$ can then be described by a binary[2] variable x_i, where

$$x_i = \begin{cases} 1 & \text{if component } i \text{ is functioning} \\ 0 & \text{if component } i \text{ is in a failed state} \end{cases}$$

$\boldsymbol{x} = (x_1, x_2, \cdots, x_n)$ is called the *state vector*. Furthermore, we assume that by knowing the states of all the n components, we also know whether the system is functioning or not.

Similarly, the state of the system can then be described by a binary function

$$\phi(\boldsymbol{x}) = \phi(x_1, x_2, \ldots, x_n)$$

where

$$\phi(\boldsymbol{x}) = \begin{cases} 1 & \text{if the system is functioning} \\ 0 & \text{if the system is in a failed state} \end{cases} \quad (3.1)$$

and $\phi(\boldsymbol{x})$ is called the *structure function of the system* or just the *structure*. In the following we often talk about structures instead of systems. Examples of simple structures are given in the following sections.

3.10.5 Series Structure

A system that is functioning if and only if *all* of its n components are functioning is called a *series structure*. The structure function is

$$\phi(\boldsymbol{x}) = x_1 \cdot x_2 \cdots x_n = \prod_{i=1}^{n} x_i \quad (3.2)$$

A series structure of order n is illustrated by the reliability block diagram in Fig. 3.32. Connection between a and b is interpreted as "the structure (system) is functioning."

3.10.6 Parallel Structure

A system that is functioning if at least one of its n components is functioning is called a *parallel structure*. A parallel structure of order n is illustrated by the reliability

---

[1] The remaining sections of this chapter are influenced by Barlow and Proschan (1975).
[2] In this context a binary variable (function) is a variable (function) that can take only the two values, 0 or 1.

block diagram in Fig. 3.33. In this case the structure function can be written

$$\phi(x) = 1 - (1-x_1)(1-x_2)\cdots(1-x_n) = 1 - \prod_{i=1}^{n}(1-x_i) \qquad (3.3)$$

The expression on the right-hand side of (3.3) is often written as $\coprod_{i=1}^{n} x_i$ where $\coprod$ is read "ip."

Hence a parallel structure of order 2 has structure function

$$\phi(x_1, x_2) = 1 - (1-x_1)(1-x_2) = \coprod_{i=1}^{2} x_i$$

The right hand side may also be written: $x_1 \sqcup x_2$. Note that

$$\phi(x_1, x_2) = x_1 + x_2 - x_1 x_2 \qquad (3.4)$$

Since x_1 and x_2 are binary variables, $x_1 \sqcup x_2$ will be equal to the maximum of the x_i's. Similarly

$$\coprod_{i=1}^{n} x_i = \max_{i=1,2,\ldots,n} x_i$$

3.10.7 k-out-of-n Structure

A system that is functioning if and only if at least k of the n components are functioning is called a *k-out-of-n structure* (*k*oo*n*). A series structure is therefore an *n*-out-of-*n* (*n*oo*n*) structure, and a parallel structure is a 1-out-of-*n* (1oo*n*) structure.

The structure function of a *k*-out-of-*n* (*k*oo*n*) structure can be written

$$\phi(x) = \begin{cases} 1 & \text{if } \sum_{i=1}^{n} x_i \geq k \\ 0 & \text{if } \sum_{i=1}^{n} x_i < k \end{cases} \qquad (3.5)$$

As an example consider a 2-out-of-3 (2oo3) structure, which is illustrated in Fig. 3.37. In this case the failure of one component is tolerated, while two or more component failures lead to system failure. The reliability block diagram of the 2-out-of-3 structure may also be drawn as shown in Fig. 3.38. This representation is preferred by IEC 61078 but may be more problematic as a basis for establishing the structure function. In the rest of this book, we will therefore prefer the representation in Fig. 3.37.

A three-engined airplane which can stay in the air if and only if at least two of its three engines are functioning is an example of a 2-out-of-3 (2oo3) structure. The

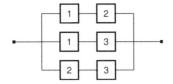

Fig. 3.37 The 2-out-of-3 structure.

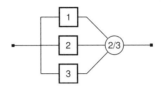

Fig. 3.38 The 2-out-of-3 structure (alternative representation).

structure function of the 2-out-of-3 structure in Fig. 3.37 may also be written

$$\begin{aligned}\phi(x) &= x_1x_2 \sqcup x_1x_3 \sqcup x_2x_3 \\ &= 1 - (1-x_1x_2)(1-x_1x_3)(1-x_2x_3) \\ &= x_1x_2 + x_1x_3 + x_2x_3 - x_1^2x_2x_3 - x_1x_2^2x_3 - x_1x_2x_3^2 + x_1^2x_2^2x_3^2 \\ &= x_1x_2 + x_1x_3 + x_2x_3 - 2x_1x_2x_3\end{aligned}$$

(Note that since x_i is a binary variable, $x_i^k = x_i$ for all i and k.)

3.11 SYSTEM STRUCTURE ANALYSIS

3.11.1 Coherent Structures

When establishing the structure of a system, it seems reasonable first to leave out all components that do not play any *direct* role for the functioning ability of the system. The components we are left with are called *relevant*. The components that are not relevant are called *irrelevant*.

If component i is irrelevant, then

$$\phi(1_i, x) = \phi(0_i, x) \quad \text{for all} \quad (\cdot_i, x) \tag{3.6}$$

where $(1_i, x)$ represents a state vector where the state of the ith component = 1, $(0_i, x)$ represents a state vector where the state of the ith component = 0, and $(\cdot_i, x)$ represents a state vector where the state of the ith component = 0 or 1. Fig. 3.39 illustrates a system of order 2, where component 2 is irrelevant.

Remark: The notation "relevant/irrelevant" is sometimes misleading, as it is easy to find examples of components of great importance for a system without being relevant

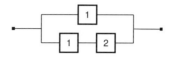

Fig. 3.39 Component 2 is irrelevant.

in the above sense. The reliability block diagram and the structure function are established for a specific system function, for example, "separate gas from oil and water" in Example 3.2. To fulfill this system function a number of components are required to function, and therefore relevant in the above sense. The shutdown function of the protection systems will be irrelevant with respect to this system function, since the production will not be influenced by the protections system's ability to shut down the process in an emergency.

When we say that a component is irrelevant, this is always with respect to a specific system function. The same component may be highly relevant with respect to another system function.

Also remember that x_i represents the state of a specific function (or a specific subset of functions) of a component. When we say that component i is irrelevant, we in fact say that the specific function i of the physical component is irrelevant. In Example 3.2, "spurious shutdowns" of the protection system will be relevant for the system function "separate gas from oil and water," while the shutdown function of the same protection system will be irrelevant. □

Now we will assume that the system will not run worse than before if we replace a component in a failed state with one that is functioning. This is obviously the same as requiring that the structure function shall be nondecreasing in each of its arguments. Let us now define what is meant by a *coherent system:*

Definition 3.3 A system of components is said to be coherent if all its components are relevant and the structure function is nondecreasing in each argument. □

All the systems that we have considered so far (except the one in Fig. 3.39) are coherent. One might get the impression that all systems of interest must be coherent, but this is not the case. It is, for example, easy to find systems where the failure of one component prevents another component from failing. This complication will be discussed later.

3.11.2 General Characteristics of Coherent Systems

Theorem 3.1 Let $\phi(x)$ be the structure function of a coherent system. Then

$$\phi(\mathbf{0}) = 0 \quad \text{and} \quad \phi(\mathbf{1}) = 1$$

In other words, Theorem 3.1 merely says that:

- If all the components in a coherent system are functioning, then the system is functioning.

- If all the components in a coherent system are in a failed state, then the system is in a failed state.

Proof: The argument uses the fact that $\phi(x)$ is binary, that is, that it can only assume the values 0 and 1.

If $\phi(\mathbf{0}) = 1$, then we must have $\phi(\mathbf{0}) = \phi(\mathbf{1}) = 1$, since $\phi(x)$ is assumed to be nondecreasing in each argument. This implies that all the components in the system are irrelevant, which contradicts the assumption that the system is coherent. Hence $\phi(\mathbf{0}) = 0$.

Similarly $\phi(\mathbf{1}) = 0$ implies that $\phi(\mathbf{0}) = 0$, that is, that all the components are irrelevant. This contradicts the assumption of coherence. Hence $\phi(\mathbf{1}) = 1$. □

Theorem 3.2 Let $\phi(x)$ be the structure function of a coherent system of order n. Then

$$\prod_{i=1}^{n} x_i \leq \phi(x) \leq \coprod_{i=1}^{n} x_i \tag{3.7}$$

Theorem 3.3 states that any coherent system is functioning at least as well as a corresponding system where all the n components are connected in series and at most as well as a system where all the n components are connected in parallel.

Proof: First note that $\prod_{i=1}^{n} x_i$ and $\coprod_{i=1}^{n} x_i$ are both binary. Assume that $\prod_{i=1}^{n} x_i = 0$. Since we already know that $\phi(x) \geq 0$, the left-hand side of (3.7) is satisfied.

Assume that $\prod_{i=1}^{n} x_i = 1$, that is, $x = \mathbf{1}$. Then according to Theorem 3.1, $\phi(x) = 1$. Hence the left hand side of (3.7) is always satisfied.

Further assume that $\coprod_{i=1}^{n} x_i = 0$, that is, $x = \mathbf{0}$. Then according to Theorem 3.1, $\phi(x) = 0$, and the right-hand side of (3.7) is satisfied. Finally assume that $\coprod_{i=1}^{n} x_i = 1$. Since we already know that $\phi(x) \leq 1$, the right-hand side of (3.7) is automatically satisfied. □

Let $x = (x_1, x_2, \ldots, x_n)$ and $y = (y_1, y_2, \ldots, y_n)$ be state vectors, and let $x \cdot y$ and $x \sqcup y$ be defined as follows:

$$x \cdot y = (x_1 y_1, x_2 y_2, \ldots, x_n y_n)$$
$$x \sqcup y = (x_1 \sqcup y_1, x_2 \sqcup y_2, \ldots, x_n \sqcup y_n)$$

We will now prove the following important result:

Theorem 3.3 Let ϕ be a coherent structure. Then

$$\phi(x \sqcup y) \geq \phi(x) \sqcup \phi(y) \tag{3.8}$$
$$\phi(x \cdot y) \leq \phi(x) \cdot \phi(y) \tag{3.9}$$

Proof: For (3.8) we know that $x_i \sqcup y_i \geq x_i$ for all i. Since ϕ is coherent, ϕ is nondecreasing in each argument and therefore

$$\phi(x \sqcup y) \geq \phi(x)$$

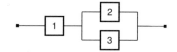

Fig. 3.40 Example system.

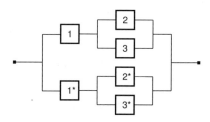

Fig. 3.41 Redundancy at system level.

For symmetrical reasons

$$\phi(x \sqcup y) \geq \phi(y)$$

Furthermore, $\phi(x)$ and $\phi(y)$ are both binary. Therefore

$$\phi(x \sqcup y) \geq \phi(x) \sqcup \phi(y)$$

For (3.9) we know that $x_i \cdot y_i \leq x_i$ for all i. Since ϕ is coherent, then

$$\phi(x \cdot y) \leq \phi(x)$$

Similarly

$$\phi(x \cdot y) \leq \phi(y)$$

Since $\phi(x)$ and $\phi(y)$ are binary, then

$$\phi(x \cdot y) \leq \phi(x) \cdot \phi(y)$$

□

Let us interpret (3.8) in common language. Consider the structure in Fig. 3.40 with structure function $\phi(x)$. Assume that we also have an identical structure $\phi(y)$ with state vector y. Fig. 3.41 illustrates a structure with "redundancy at system level". The structure function for this system is $\phi(x) \sqcup \phi(y)$.

Next consider the system we get from Fig. 3.40 when we connect each pair x_i, y_i in parallel; see Fig. 3.42. This figure illustrates a structure with "redundancy at component level."

The structure function is $\phi(x \sqcup y)$.

According to Theorem 3.3, $\phi(x \sqcup y) \geq \phi(x) \sqcup \phi(y)$. This means that:

> We obtain a "better" system by introducing redundancy at component level than by introducing redundancy at system level.

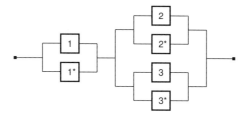

Fig. 3.42 Redundancy at component level.

This principle, which is well known to designers, is further discussed by Shooman (1968, pp. 281–289). The principle is, however, not obvious when the system has two ore more failure modes, for example, "fail to function" and "false alarm" of a fire detection system. The concept of redundancy is further discussed in Section 4.6.

3.11.3 Structures Represented by Paths and Cuts

A structure of order n consists of n components numbered from 1 to n. The set of components is denoted by

$$C = \{1, 2, \ldots, n\}$$

Definition 3.4 Path Sets, Minimal Path Sets A path set P is a set of components in C which by functioning ensures that the system is functioning. A path set is said to be minimal if it cannot be reduced without loosing its status as a path set. □

Definition 3.5 Cut Sets, Minimal Cut Sets A cut set K is a set of components in C which by failing causes the system to fail. A cut set is said to be minimal if it cannot be reduced without loosing its status as a cut set. □

Example 3.4
Consider the reliability block diagram in Fig. 3.40. The component set is $C = \{1, 2, 3\}$

Path sets:	Cut sets:
{1,2}∗	{1}∗
{1,3}∗	{2,3}∗
{1,2,3}	{1,2}
	{1,3}
	{1,2,3}

The minimal path sets and cut sets are marked with an ∗.
  In this case the minimal path sets are

$$P_1 = \{1, 2\} \quad \text{and} \quad P_2 = \{1, 3\}$$

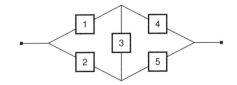

Fig. 3.43 Bridge structure.

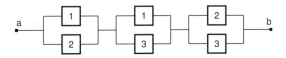

Fig. 3.44 The 2-out-of-3 structure represented as a series structure of the minimal cut parallel structures.

while the minimal cut sets are

$$K_1 = \{1\} \quad \text{and} \quad K_2 = \{2, 3\}$$

□

Example 3.5 Bridge Structure

Consider a bridge structure such as that given by the physical network in Fig. 3.43. The minimal path sets are

$$P_1 = \{1, 4\}, \quad P_2 = \{2, 5\}, \quad P_3 = \{1, 3, 5\}, \quad \text{and} \quad P_4 = \{2, 3, 4\}$$

The minimal cut sets are

$$K_1 = \{1, 2\}, \quad K_2 = \{4, 5\}, \quad K_3 = \{1, 3, 5\}, \quad \text{and} \quad K_4 = \{2, 3, 4\}$$

□

Example 3.6 2-out-of-3 Structure

Consider the 2-out-of-3 structure in Fig. 3.37. The minimal path sets are

$$P_1 = \{1, 2\}, \quad P_2 = \{1, 3\}, \quad \text{and} \quad P_3 = \{2, 3\}$$

The minimal cut sets are

$$K_1 = \{1, 2\}, \quad K_2 = \{1, 3\}, \quad \text{and} \quad K_3 = \{2, 3\}$$

The 2-out-of-3 structure may therefore be represented as a series structure of its minimal cut parallel structures as illustrated in Fig. 3.44. □

In these particular examples the number of minimal cut sets coincides with the number of minimal path sets. This will usually not be the case.

The Designer's Point of View Consider a designer who wants to ensure that a system is functioning with the least possible design effort. What the designer needs is a list of the minimal path sets from which one will be chosen for the design.

The Saboteur's Point of View Next consider a saboteur who wants to bring the system into a failed state, again with the least possible effort on his or her part. What the saboteur needs is a list of the minimal cut sets from which to choose one for the sabotage plan.

Consider an arbitrary structure with minimal path sets $P_1, P_2, \ldots, P_p$ and minimal cut sets $K_1, K_2, \ldots, K_k$. To the minimal path set P_j, we associate the binary function

$$\rho_j(x) = \prod_{i \in P_j} x_i \quad \text{for } j = 1, 2, \ldots, s \tag{3.10}$$

Note that $\rho_j(x)$ represents the structure function of a series structure composed of the components in P_j. Therefore $\rho_j(x)$ is called the jth minimal *path series structure*.

Since we know that the structure is functioning if and only if at least one of the minimal path series structures is functioning,

$$\phi(x) = \coprod_{j=1}^{p} \rho_j(x) = 1 - \prod_{j=1}^{p}(1 - \rho_j(x)) \tag{3.11}$$

Hence our structure may be interpreted as a parallel structure of the minimal path series structures.

From (3.10) and (3.11) we get

$$\phi(x) = \coprod_{j=1}^{p} \prod_{i \in P_j} x_i \tag{3.12}$$

Example 3.5 (Cont.)
In the bridge structure in Fig. 3.43, the minimal path sets were $P_1 = \{1, 4\}$, $P_2 = \{2, 5\}$, $P_3 = \{1, 3, 5\}$, and $P_4 = \{2, 3, 4\}$. The corresponding minimal path series structures are

$$\begin{aligned}
\rho_1(x) &= x_1 \cdot x_4 \\
\rho_2(x) &= x_2 \cdot x_5 \\
\rho_3(x) &= x_1 \cdot x_3 \cdot x_5 \\
\rho_4(x) &= x_2 \cdot x_3 \cdot x_4
\end{aligned}$$

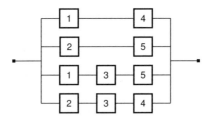

Fig. 3.45 The bridge structure represented as a parallel structure of the minimal path series structures.

Accordingly, the structure function may be written:

$$\begin{aligned}
\phi(x) &= \coprod_{j=1}^{4} \rho_j(x) = 1 - \prod_{j=1}^{4}(1 - \rho_j(x)) \\
&= 1 - (1 - \rho_1(x))(1 - \rho_2(x))(1 - \rho_3(x))(1 - \rho_4(x)) \\
&= 1 - (1 - x_1 x_4)(1 - x_2 x_5)(1 - x_1 x_3 x_5)(1 - x_2 x_3 x_4) \\
&= x_1 x_4 + x_2 x_5 + x_1 x_3 x_5 + x_2 x_3 x_4 - x_1 x_3 x_4 x_5 - x_1 x_2 x_3 x_5 \\
&\quad - x_1 x_2 x_3 x_4 - x_2 x_3 x_4 x_5 - x_1 x_2 x_4 x_5 + 2 x_1 x_2 x_3 x_4 x_5
\end{aligned}$$

(Note that since x_i is a binary variable, $x_i^k = x_i$ for all i and k.)

Hence the bridge structure can be represented by the reliability block diagram in Fig. 3.45. □

Similarly, we can associate the following binary function to the minimal cut set K_j:

$$\kappa_j(x) = \coprod_{i \in K_j} x_i = 1 - \prod_{i \in K_j}(1 - x_i) \quad \text{for } j = 1, 2, \ldots, k \quad (3.13)$$

We see that $\kappa_j(x)$ represents the structure function of a parallel structure composed of the components in K_j. Therefore $\kappa_j(x)$ is called the jth minimal *cut parallel structure*.

Since we know that the structure is failed if and only if at least one of the minimal cut parallel structures is failed, then

$$\phi(x) = \prod_{j=1}^{k} \kappa_j(x) \quad (3.14)$$

Hence we can regard this structure as a series structure of the minimal cut parallel structures. By combining (3.13) and (3.14) we get

$$\phi(x) = \prod_{j=1}^{k} \coprod_{i \in K_j} x_i \quad (3.15)$$

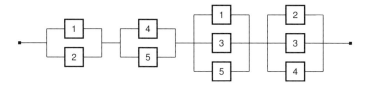

Fig. 3.46 The bridge structure represented as a series structure of the minimal cut parallel structures.

Example 3.5 (Cont.)
In the bridge structure the minimal cut sets were $K_1 = \{1, 2\}$, $K_2 = \{4, 5\}$, $K_3 = \{1, 3, 5\}$, and $K_4 = \{2, 3, 4\}$. The corresponding minimal cut parallel structures become:

$$\kappa_1(\mathbf{x}) = x_1 \sqcup x_2 = 1 - (1 - x_1)(1 - x_2)$$
$$\kappa_2(\mathbf{x}) = x_4 \sqcup x_5 = 1 - (1 - x_4)(1 - x_5)$$
$$\kappa_3(\mathbf{x}) = x_1 \sqcup x_3 \sqcup x_5 = 1 - (1 - x_1)(1 - x_3)(1 - x_5)$$
$$\kappa_4(\mathbf{x}) = x_2 \sqcup x_3 \sqcup x_4 = 1 - (1 - x_2)(1 - x_3)(1 - x_4)$$

and we may find the structure function of the bridge structure by inserting these expressions into (3.14). The bridge structure may therefore be represented by the reliability block diagram in Fig. 3.46. □

3.11.4 Structural Importance of Components

Some components in a system may obviously be more important than the others in determining whether the system is functioning or not. A component in series with the rest of the system will, for example, be at least as important as any other component in the system. It would be useful to have a quantitative measure of the importance of the individual components in the system. Before we can establish such a measure, we need to define some new concepts.

Definition 3.6 A critical path vector for component i is a state vector $(1_i, \mathbf{x})$ such that

$$\phi(1_i, \mathbf{x}) = 1 \quad \text{while} \quad \phi(0_i, \mathbf{x}) = 0$$

□

This is equivalent to requiring that

$$\phi(1_i, \mathbf{x}) - \phi(0_i, \mathbf{x}) = 1 \qquad (3.16)$$

In other words, given the states of the other components $(\cdot_i, \mathbf{x})$, the system is functioning if and only if component i is functioning. It is therefore natural to call $(1_i, \mathbf{x})$ a *critical* path vector for component i.

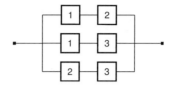

Fig. 3.47 The 2-out-of-3 structure.

Definition 3.7 A critical path set $C(1_i, x)$ corresponding to the critical path vector $(1_i, x)$ for component i is defined by:

$$C(1_i, x) = \{i\} \cup \{j; x_j = 1, \ j \neq i\} \quad (3.17)$$

□

The total number of critical path sets (path vectors) for component i is

$$\eta_\phi(i) = \sum_{(\cdot_i, x)} [\phi(1_i, x) - \phi(0_i, x)] \quad (3.18)$$

Since the x_j's are binary variables and thus can take only two possible values, 0 and 1, the total number of state vectors $(\cdot, x) = (x_1, \ldots, x_{i-1}, \cdot, x_{i+1}, \ldots, x_n)$ is 2^{n-1}.

Birnbaum's Measure of Structural Importance Birnbaum (1969) proposed the following measure for the structural importance of component i:

$$B_\phi(i) = \frac{\eta_\phi(i)}{2^{n-1}} \quad (3.19)$$

Birnbaum's measure[3] of structural importance $B_\phi(i)$ expresses the relative proportion of the 2^{n-1} possible state vectors $(\cdot_i, x)$ which are critical path vectors for component i. The components in the system can now be (partially) ranked according to the size of $B_\phi(i)$.

Example 3.7
Consider the 2-out-of-3 structure in Fig. 3.47. For component 1, we have

$(\cdot, x_2, x_3)$	$\phi(1, x_2, x_3) - \phi(0, x_2, x_3)$	$C(1_1, x_2, x_3)$
$(\cdot 00)$	0	
$(\cdot 01)$	1	$\{1, 3\}$
$(\cdot 10)$	1	$\{1, 2\}$
$(\cdot 11)$	0	

---

[3] Named after the Hungarian-American professor Zygmund William Birnbaum (1903–2000).

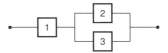

Fig. 3.48 Reliability block diagram for Example 3.8.

In this case the total number of critical path vectors for component 1 is 2:

$$\eta_\phi(1) = 2$$

while the total number of state vectors $(\cdot_1, x_2, x_3)$ is $2^{3-1} = 4$. Hence

$$B_\phi(1) = \frac{2}{4} = \frac{1}{2}$$

For symmetrical reasons

$$B_\phi(1) = B_\phi(2) = B_\phi(3) = \frac{1}{2}$$

Hence the components 1, 2 and 3 are of equal structural importance. □

Example 3.8
Consider the structure in Fig. 3.48. Here the structure function is

$$\phi(x_1, x_2, x_3) = x_1(x_2 \sqcup x_3) = x_1(1 - (1 - x_2)(1 - x_3))$$
$$= x_1 x_2 + x_1 x_3 - x_1 x_2 x_3$$

Consider component 1. Then

$(\cdot, x_2, x_3)$	$\phi(1, x_2, x_3) - \phi(0, x_2, x_3)$	$C(1_1, x_2, x_3)$
$(\cdot 00)$	0	
$(\cdot 01)$	1	$\{1,3\}$
$(\cdot 10)$	1	$\{1,2\}$
$(\cdot 11)$	1	$\{1,2,3\}$

The structural importance of component 1 is therefore

$$B_\phi(1) = \frac{\eta_\phi(1)}{2^{3-1}} = \frac{3}{4}$$

Now consider component 2.

$(x_1, \cdot, x_3)$	$\phi(x_1, 1, x_3) - \phi(x_1, 0, x_3)$	$C(x_1, 1, x_3)$
$(\cdot 00)$	0	
$(\cdot 01)$	0	
$(\cdot 10)$	1	$\{1,2\}$
$(\cdot 11)$	0	

136 QUALITATIVE SYSTEM ANALYSIS

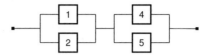

Fig. 3.49 The structure $\phi(1_3, x)$ of the bridge structure.

The structural importance of component 2 is therefore

$$B_\phi(2) = \frac{n_\phi(2)}{2^{3-1}} = \frac{1}{4}$$

For symmetrical reasons

$$B_\phi(2) = B_\phi(3) = \frac{n_\phi(2)}{2^{3-1}} = \frac{1}{4}$$

Component 1 is accordingly of greater structural importance than components 2 and 3. □

3.11.5 Pivotal Decomposition

The following identity holds for every structure function $\phi(x)$:

$$\phi(x) \equiv x_i \phi(1_i, x) + (1 - x_i)\phi(0_i, x) \quad \text{for all } x \tag{3.20}$$

We can easily see that this identity is correct from the fact that

$$x_i = 1 \Rightarrow \phi(x) = 1 \cdot \phi(1_i, x) \text{ and } x_i = 0 \Rightarrow \phi(x) = 1 \cdot \phi(0_i, x)$$

By repeated use of (3.20) we arrive at $\phi(x) = \sum_y \prod_j x_j^{y_j}(1 - x_j)^{1-y_j} \phi(y)$ where the summation is taken over all n-dimensional binary vectors.

Example 3.9 Bridge Structure
Consider the bridge structure in Fig. 3.43. The structure function $\phi(x)$ of this system can be determined by pivotal decomposition with respect to component 3.

$$\phi(x) = x_3 \phi(1_3, x) + (1 - x_3)\phi(0_3, x)$$

Here, $\phi(1_3, x)$ is the structure function of the system in Fig. 3.49:

$$\phi(1_3, x) = (x_1 \sqcup x_2)(x_4 \sqcup x_5) = (x_1 + x_2 - x_1 x_2)(x_4 + x_5 - x_4 x_5)$$

while $\phi(0_3, x)$ is the structure function of the system in Fig. 3.50:

$$\phi(0_3, x) = x_1 x_4 \sqcup x_2 x_5 = x_1 x_4 + x_3 x_5 - x_1 x_2 x_4 x_5$$

Hence the structure function of the bridge system becomes

$$\begin{aligned}\phi(x) =\ & x_3(x_1 + x_2 - x_1 x_2)(x_4 + x_5 - x_4 x_5) \\ & + (1 - x_3)(x_1 x_4 + x_2 x_5 - x_1 x_2 x_4 x_5)\end{aligned}$$

□

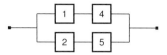

Fig. 3.50 The structure $\phi(0_3, x)$ of the bridge structure.

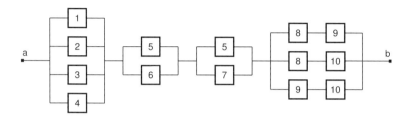

Fig. 3.51 Reliability block diagram.

Fig. 3.52 Structure of modules.

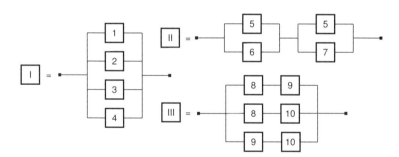

Fig. 3.53 The three substructures.

3.11.6 Modules of Coherent Structures

Consider the structure represented by the reliability block diagram in Fig. 3.51. The structure may be split in three modules as illustrated by Fig. 3.52, where the modules $\boxed{\text{I}}$, $\boxed{\text{II}}$, and $\boxed{\text{III}}$ are defined in Fig. 3.53 The modules $\boxed{\text{I}}$, $\boxed{\text{II}}$, and $\boxed{\text{III}}$ may now be analyzed individually, and the results may be put together logically. Regarding this logical connection, it is important that the partitioning into subsystems is done in such a way that *each single component never appears within more than one of the modules*.

When this partitioning is carried out in a specific way, described later, the procedure is called a *modular decomposition of the system*. In the following, we will denote a system with (C, ϕ) where C is the set of components and ϕ the structure function.

Let A represent a subset of C,

$$A \subseteq C$$

and A^c denote the complement of A with respect to C,

$$A^c = C - A$$

We denote the elements in A by $i_1, i_2, \ldots, i_\nu$, where $i_1 < i_2 < \cdots < i_\nu$. Let $\mathbf{x}^A$ be the state vector corresponding to the elements in A:

$$\mathbf{x}^A = (x_{i_1}, x_{i_2}, \ldots, x_{i_\nu})$$

and let

$$\chi(\mathbf{x}^A) = \chi(x_{i_1}, x_{i_2}, \ldots, x_{i_\nu})$$

be a binary function of $\mathbf{x}^A$. Obviously (A, χ) can be interpreted as a system (a structure).

In our example, $C = \{1, 2, \ldots, 10\}$. Let us choose $A = \{5, 6, 7\}$ and $\chi(\mathbf{x}^A) = (x_5 \sqcup x_6)(x_5 \sqcup x_7)$. The (A, χ) represents the substructure $\boxed{\text{II}}$. With this notation, a precise definition of the concept of a coherent module can be given as follows:

Definition 3.8 Coherent Modules Let the coherent structure (C, ϕ) be given and let $A \subseteq C$. Then (A, χ) is said to be a coherent module of (C, ϕ), if $\phi(\mathbf{x})$ can be written as a function of $\chi(\mathbf{x}^A)$ and $\mathbf{x}^{A^c}$, $\psi(\chi(\mathbf{x}^A), \mathbf{x}^{A^c})$ where ψ is the structure function of a coherent system. □

A is called a modular set of (C, ϕ), and if in particular $A \subset C$, (A, χ) is said to be a *proper* module of (C, ϕ).

What we actually do here is to consider all the components with index belonging to A as one "component" with state variable $\chi(\mathbf{x}^A)$. When we interpret the system in this way, the structure function will be

$$\psi(\chi(\mathbf{x}^A), \mathbf{x}^{A^c})$$

In our example we choose $A = \{5, 6, 7\}$. Then

$$\psi(\chi(\mathbf{x}^A), \mathbf{x}^{A^c}) = \chi(x_5, x_6, x_7)\left(\coprod_{i=1}^{4} x_i\right)(x_8 x_9 \sqcup x_8 x_{10} \sqcup x_9 x_{10})$$

and since $A \subset C$, (A, χ) will be a proper module of (C, ϕ). Now let us define the concept of modular decomposition.

Definition 3.9 Modular Decomposition A modular decomposition of a coherent structure (C, ϕ) is a set of disjoint modules (A_i, χ_i), $i = 1, \ldots, r$, together with an organizing structure ω, such that

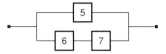

Fig. 3.54 Module II.

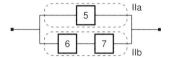

Fig. 3.55 Two prime modules.

1. $C = \cup_{i=1}^{r} A_i;$ where $A_i \cap A_j = \varnothing$ for $i \neq j$
2. $\phi(x) = \omega[\chi_1(x^{A_1}), \chi_2(x^{A_2}), \ldots, \chi_r(x^{A_r})]$ □

The "finest" partitioning into modules that we can have is obviously to let each individual component constitute one module. The "coarsest" partitioning into modules is to let the whole system constitute one module. To be of practical use, a module decomposition should, if possible, be something between these two extremes. A module that cannot be partitioned into smaller modules without letting each component represent a module is called a *prime module*.

In our example III represents a prime module. But II is not a prime module, since it may be described as in Fig. 3.54. and hence can be partitioned into two modules IIa and IIb as in Fig. 3.55. This gives no guidance on how to determine individual prime modules in a system. However, algorithms have been developed, for example, by Chatterjee (1975), that can be used to find all the prime modules in a fault tree or in a reliability block diagram.

In Chapter 4 we will justify the fact that it is natural to interpret the state vector as stochastic. In accordance with what we do in probability theory, we will from now on denote the state variables with *capital* letters from the end of the alphabet, for example, $X_1, X_2, \ldots, X_n$.

Occasionally, two or more of these can be stochastically dependent. In such situations it is advisable to try to "collect" the state variables in modules in such a way that dependency occurs only *within the modules*. If one succeeds in this, the individual modules can be considered as being independent. This will make the further analysis simpler.

PROBLEMS

3.1 Establish a function tree and a hardware breakdown structure for a dishwasher.

3.2 Consider the subsea shutdown valve in Fig. 3.56. The valve is a spring-loaded, fail-safe close gate valve which is held open by hydraulic pressure. The gate is a solid block with a cylindrical hole with the same diameter as the pipeline. To open the valve, hydraulic pressure is applied on the upper side of the piston. The pressure forces the piston, the piston rod, and the gate downwards until the hole in the gate is in line with the pipeline. When the pressure is bled off, the spring forces the piston upwards until the hole in the gate is no longer in contact with the pipeline conduct. The solid part of the gate is now pressed against the seat seal and the valve is closed.

Carry out an FMECA analysis of the shutdown valve according to the procedure described in Section 3.5.

3.3 Fig. 3.57 shows a sketch of a steam boiler system which supplies steam to a process system at a specified pressure. Water is led to the boiler through a pipeline with a regulator valve, a level indicator controller valve (LICV). Fuel (oil) is led to the burner chamber through a pipeline with a regulator valve, a pressure controller valve (PCV). The valve PCV is installed in parallel with a bypass valve V-1 together with two isolation valves to facilitate inspection and maintenance of the PCV during normal operation.

The level of the water in the boiler is surveyed by a level emitter (LE). The water level is maintained in an interval between a specified *low* level and a specified *high* level by a pneumatic control circuit connected to the water regulator valve LICV. The level indicator controller (LIC) translates the pneumatic "signal" from LE to a pneumatic "signal" controlling the valve LICV.

It is very important that the water level does not come below the specified *low* level. When the water level approaches the *low* level, a pneumatic "signal" is passed from the level indicator controller LIC to the level transmitter (LT). The LT translates the pneumatic "signal" to an electrical "signal" which is sent to the solenoid valve (SV). The solenoid valve again controls the valve PCV on the fuel inlet pipeline. This circuit is thus installed to cut off the fuel supply in case the water level comes below the specified *low* level.

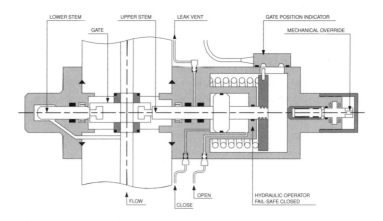

Fig. 3.56 Hydraulically operated gate valve (Problem 3.2).

PROBLEMS 141

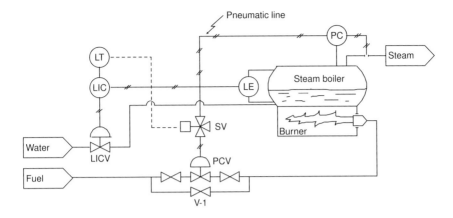

Fig. 3.57 Steam boiler system (Problem 3.3).

The pressure in the boiler and in the steam outlet pipeline is surveyed by a pressure controller PC which is connected to the solenoid valve SV, and thereby to the valve PCV on the fuel inlet pipeline. This circuit is thus installed to cut off the fuel supply in case the pressure in the boiler increases above a specified *high* pressure.

A *critical* situation occurs if the boiler is boiled dry. In this case the pressure in the vessel will increase very rapidly and the vessel may explode.

(a) Construct a fault tree where the TOP event is the *critical* situation mentioned above. Secondary failure causes shall not be included. Write down assumptions and limitations you have to make during the fault tree construction.

(b) Establish a reliability block diagram corresponding to the fault tree in (a) and determine the structure function $\phi(x)$.

(c) Determine all minimal cut sets in the fault tree (reliability block diagram).

3.4 Fig. 3.58 shows a sketch of the lubrication system on a ship engine. The separator separates water from the oil lubricant. The separator will only function satisfactory when the oil is heated to a specified temperature. When the water content in the oil is too high, the quality of the lubrication will be too low, and this may lead to damage or breakdown of the engine.

The engine will generally require

- Sufficient throughput of oil/lubricant.

- Sufficient quality of the oil/lubricant.

The oil throughput is sufficient when at least one cooler is functioning, at least one filter is open (i.e., not clogged), and the pump is functioning. In addition, all necessary pipelines must be open, no valves must be unintentionally closed, the lubrication channels in the engine must be open (not clogged), and the lubrication system must

142   QUALITATIVE SYSTEM ANALYSIS

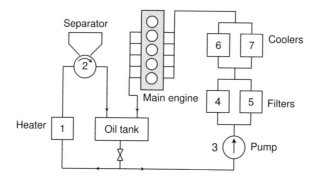

Fig. 3.58 Lubrication system on a ship engine (Problem 3.4).

not have significant leakages to the environment. We will here assume that the probabilities of these "additional" events are very low and that these events therefore may be neglected.

The quality of the oil is sufficient when

- Both coolers are functioning (with full throughput) such that the temperature of the oil to the engine is sufficiently low.
- None of the filters is clogged, and there are no holes in the filters.
- The separator system is functioning.

(a) Construct a fault tree with respect to the TOP event "Too low throughput of oil/lubricant."

(b) Construct a fault tree with respect to the TOP event "Too low quality of the oil/lubricant."

3.5 Use MOCUS to identify all the minimal cut sets of the fault tree in Fig. 3.19.

3.6 Consider the system in Fig. 3.59.

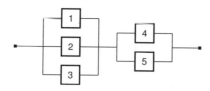

Fig. 3.59 Reliability block diagram (Problem 3.6).

(a) Derive the corresponding structure function by a direct approach.

(b) Determine the path sets and the cut sets, then the minimal path sets and the minimal cut sets.

(c) Derive the structure function of the system by using the property that the structure may be considered as a parallel structure of the minimal path series structures.

(d) Derive the structure function of the system by using the property that the structure may be considered as a series structure of the minimal cut parallel structures.

(e) Compare the results obtained in (a), (c), and (d).

3.7 Reduce the reliability block diagram in Fig. 3.60 to the simplest possible form and determine the structure function.

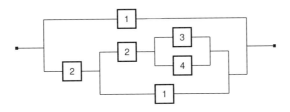

Fig. 3.60 Reliability block diagram (Problem 3.7).

3.8 Show that

(a) If ϕ represents a parallel structure, then

$$\phi(x \sqcup y) = \phi(x) \sqcup \phi(y).$$

(b) If ϕ represents a series structure, then

$$\phi(x \cdot y) = \phi(x) \cdot \phi(y)$$

3.9 The dual structure $\phi^D(x)$ to a given structure $\phi(x)$ is defined by

$$\phi^D(x) = 1 - \phi(1 - x)$$

where $(1 - x) = (1 - x_1, 1 - x_2, \ldots, 1 - x_n)$.

(a) Show that the dual structure of a k-out-of-n structure is a $(n - k + 1)$-out-of-n structure.

(b) Show that the minimal cut sets for ϕ are minimal path sets for ϕ^D, and vice versa.

3.10 Determine the structural importance of the different components of the structure studied in Problem 3.6.

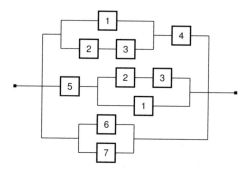

Fig. 3.61 Reliability block diagram (Problem 3.11).

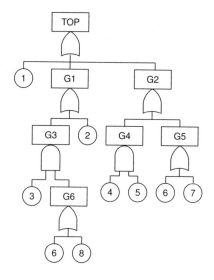

Fig. 3.62 Example fault tree (Problem 3.12).

3.11 Determine the structure function of the system in Fig. 3.61 by applying an appropriate modular decomposition.

3.12 Consider the fault tree in Fig. 3.62

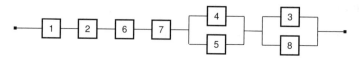

Fig. 3.63 Reliability block diagram (Problem 3.12).

(a) Use MOCUS to identify all the minimal path sets of the fault tree.

(b) Show that the system may be represented by the reliability block diagram in Fig. 3.63.

3.13 Use appropriate pivotal decompositions to determine the structure function of the system in Fig. 3.64.

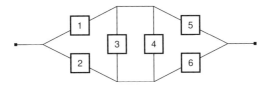

Fig. 3.64 Reliability block diagram (Problem 3.13).

3.14 Determine the structure function of the structure in Fig. 3.65 by use of appropriate pivotal decompositions.

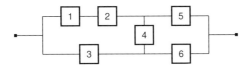

Fig. 3.65 Reliability block diagram (Problem 3.14).

3.15 Determine the structure function of the structure in Fig. 3.66.

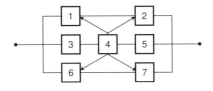

Fig. 3.66 Reliability block diagram (Problem 3.15).

3.16 Show that the structure function of the system in Fig. 3.67 may be written

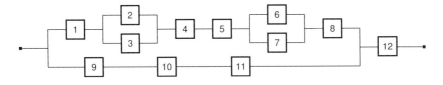

Fig. 3.67 Reliability block diagram (Problem 3.16).

$$\phi(x) = [x_1(x_2 + x_3 - x_2 x_3) x_4 x_5 (x_6 + x_7 - x_6 x_7) x_8$$
$$+ x_9 x_{10} x_{11} - x_1 (x_2 + x_3 - x_2 x_3) x_4 x_5 (x_6 + x_7 - x_6 x_7) x_8 x_9 x_{10} x_{11}] x_{12}$$

3.17 Consider the structure in Fig. 3.68.

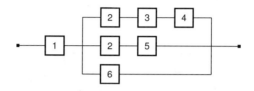

Fig. 3.68 Reliability block diagram (Problem 3.17).

(a) Find the minimal path sets $P_1, P_2, \ldots, P_p$ and the minimal cut sets $K_1, K_2, \ldots, K_k$ of the structure.

(b) Derive the structure function of this system.

4

Systems of Independent Components

4.1 INTRODUCTION

In Chapter 3 we discussed the structural relationship between a system and its components and showed how a *deterministic* model of the structure can be established, using a reliability block diagram or a fault tree. Whether or not a given component will be in a failed state after t time units can usually not be predicted with certainty. Rather, when studying the occurrence of such failures, one looks for statistical regularity. Hence it seems reasonable to interpret the state variables of the n components at time t as random variables. We denote the state variables by

$$X_1(t), X_2(t), \ldots, X_n(t)$$

Correspondingly the state *vector* and the structure function are denoted, respectively, by

$$X(t) = (X_1(t), X_2(t), \ldots, X_n(t)) \text{ and } \phi(X(t))$$

The following probabilities are of interest:

$$\Pr(X_i(t) = 1) = p_i(t) \text{ for } i = 1, 2, \ldots, n \quad (4.1)$$
$$\Pr(\phi(X(t)) = 1) = p_S(t) \quad (4.2)$$

Throughout Chapter 4 we restrict ourselves to studying systems where failures of the individual components can be interpreted as *independent* events. This implies that the state variables at time t, $X_1(t), X_2(t), \ldots, X_n(t)$ can be considered as being stochastically independent. Unfortunately, independence is often assumed just to "simplify" the analysis, even though it is unrealistic. (This is discussed in more detail in Chapter 6.)

In the main part of this chapter we consider *nonrepairable* components and systems, which are discarded the first time they fail. In that case (4.1) and (4.2) correspond to what in Chapter 2 is called the *survivor function* of component i and of the system, respectively.

Components and systems that are *replaced* or *repaired* after failure are (in this book) called *repairable*. Then (4.1) and (4.2) correspond with what in Section 1.6 we called the *availability* at time t of component i and of the system, respectively. Repairable components and systems that are considered until the first failure only are treated as nonrepairable. The reliability of repairable systems is further discussed in Chapter 9.

During most of this chapter, $p_i(t)$ will, for brevity, be called the *reliability* of component i at time t, and $q_i(t) = 1 - p_i(t)$ will be called the *unreliability* of component i at time t, for $i = 1, 2, \ldots, n$. In the same way, $p_S(t)$ will be called the *system reliability*, and $Q_0(t) = 1 - p_S(t)$ will be called the *system unreliability* at time t.

In Sections 4.1 and 4.2 we study system reliability in general. Nonrepairable systems are discussed in Section 4.3. In most of the examples we assume that the components have constant failure rates. This is done because most alternative distributions will give complicated mathematical expressions that can only be solved by using a computer. In Section 4.4 we discuss quantitative fault tree analysis and provide both exact and approximation formulas. Several types of nonrepairable standby systems are discussed in Section 4.6.

4.2 SYSTEM RELIABILITY

Since the state variables $X_i(t)$ for $i = 1, 2, \ldots, n$ are binary, then

$$\begin{aligned} E[X_i(t)] &= 0 \cdot \Pr(X_i(t) = 0) + 1 \cdot \Pr(X_i(t) = 1) \\ &= p_i(t) \quad \text{for } i = 1, 2, \ldots, n \end{aligned} \quad (4.3)$$

Similarly the system reliability (at time t) is

$$p_S(t) = E(\phi(X(t))) \quad (4.4)$$

It can be shown (see Problem 4.1) that when the components are independent, the system reliability, $p_S(t)$, will be a function of the $p_i(t)$'s only. Hence $p_S(t)$ may be written

$$p_S(t) = h(p_1(t), p_2(t), \ldots, p_n(t)) = h(\mathbf{p}(t)) \quad (4.5)$$

Unless we state otherwise, we will use the letter h to express system reliability in situations *where the components are independent*. Now let us determine the reliability of some simple structures.

4.2.1 Reliability of Series Structures

In Section 3.2 we found that a series structure of order n has the structure function

$$\phi(X(t)) = \prod_{i=1}^{n} X_i(t)$$

Since $X_1(t), X_2(t), \ldots, X_n(t)$ are assumed to be independent, the system reliability is

$$h(\boldsymbol{p}(t)) = E(\phi(X(t))) = E\left(\prod_{i=1}^{n} X_i(t)\right)$$
$$= \prod_{i=1}^{n} E(X_i(t)) = \prod_{i=1}^{n} p_i(t) \qquad (4.6)$$

Note that

$$h(\boldsymbol{p}(t)) \leq \min_{i}(p_i(t))$$

In other words, a series structure is *at most* as reliable as the *least* reliable component.

Example 4.1
Consider a series structure of three independent components. At a specified point of time t the component reliabilities are $p_1 = 0.95$, $p_2 = 0.97$, and $p_3 = 0.94$. The system reliability at time t is then, according to (4.6),

$$p_S = h(\boldsymbol{p}) = p_1 \cdot p_2 \cdot p_3 = 0.95 \cdot 0.97 \cdot 0.94 \approx 0.866$$

□

In particular, if all the components have the same reliability $p(t)$, then the system reliability of a series structure of order n is

$$p_S(t) = p(t)^n \qquad (4.7)$$

If, for example, $n = 10$ and $p(t) = 0.995$, then

$$p_S(t) = 0.995^{10} \approx 0.951$$

Hence the system reliability of a series structure is low already when $n = 10$, even if the component reliability is relatively high ($= 0.995$).

Remark: The reliability $h(\boldsymbol{p}(t))$ of a series structure may also be determined by a more direct approach, without using the structure function. Let $E_i(t)$ be the event that component i is functioning at time t. The probability of this event is $\Pr(E_i(t)) = p_i(t)$. Since a series structure is functioning if, and only if, all its components are functioning, and since the components are independent, the reliability of the series structure is

$$
\begin{aligned}
h(\boldsymbol{p}(t)) &= \Pr(E_1(t) \cap E_2(t) \cap \cdots \cap E_n(t)) \\
&= \Pr(E_1(t)) \cdot \Pr(E_2(t)) \cdots \Pr(E_n(t)) = \prod_{i=1}^{n} p_i(t)
\end{aligned}
$$

which is the same result we got in (4.6) by using the structure function. □

4.2.2 Reliability of Parallel Structures

In Section 3.3 we found that a parallel structure of order n has the structure function

$$\phi(X(t)) = \coprod_{i=1}^{n} X_i(t) = 1 - \prod_{i=1}^{n} (1 - X_i(t))$$

Hence

$$
\begin{aligned}
h(\boldsymbol{p}(t)) &= E(\phi(X(t))) = 1 - \prod_{i=1}^{n}(1 - E(X_i(t))) \\
&= 1 - \prod_{i=1}^{n}(1 - p_i(t)) = \coprod_{i=1}^{n} p_i(t) \qquad (4.8)
\end{aligned}
$$

Example 4.2
Consider a parallel structure of three independent components. At a specified point of time t the component reliabilities are $p_1 = 0.95$, $p_2 = 0.97$, and $p_3 = 0.94$. The system reliability at time t is then, according to (4.8)

$$p_S = h(\boldsymbol{p}) = 1 - (1 - p_1)(1 - p_2)(1 - p_3) = 1 - 0.05 \cdot 0.03 \cdot 0.06 \approx 0.99991$$

□

In particular, if all the components have the same reliability $p(t)$, then the system reliability of a parallel structure of order n is

$$p_S(t) = 1 - (1 - p(t))^n \qquad (4.9)$$

Remark: As for the series structure, the reliability $h(\boldsymbol{p}(t))$ of a parallel structure may be determined by a more direct approach, without using the structure function. Let $E_i^*(t)$ be the event that component i is in a failed state at time t. The probability of this event is $\Pr(E_i^*(t)) = 1 - p_i(t)$. Since a parallel structure is in a failed state if, and only if, all its components are in a failed state, and since the components are independent, we have that

$$
\begin{aligned}
1 - h(\boldsymbol{p}(t)) &= \Pr(E_1^*(t) \cap E_2^*(t) \cap \cdots \cap E_n^*(t)) \\
&= \Pr(E_1^*(t)) \cdot \Pr(E_2^*(t)) \cdots \Pr(E_n^*(t)) = \prod_{i=1}^{n}(1 - p_i(t))
\end{aligned}
$$

and, therefore in accordance with (4.8)

$$h(p(t)) = 1 - \prod_{i=1}^{n}(1 - p_i(t))$$

This direct approach is feasible for series and parallel structures, but will be complicated for more complex structures, in which case the approach by using structure functions is much more suitable. □

4.2.3 Reliability of k-out-of-n Structures

In Section 3.4 we found that a k-out-of-n structure has the structure function

$$\phi(X(t)) = \begin{cases} 1 & \text{if } \sum_{i=1}^{n} X_i(t) \geq k \\ 0 & \text{if } \sum_{i=1}^{n} X_i(t) < k \end{cases}$$

Let us for simplicity consider a k-out-of-n structure where all the n components have identical reliabilities $p_i(t) = p(t)$ for $i = 1, 2, \ldots, n$.

Since we have assumed that failures of individual components are independent events, then at a given time t, $Y(t) = \sum_{i=1}^{n} X_i(t)$ will be binomially distributed $(n, p(t))$:

$$\Pr(Y(t) = y) = \binom{n}{y} p(t)^y (1 - p(t))^{n-y} \quad \text{for } y = 0, 1, \ldots, n$$

Hence the reliability of a k-out-of-n structure of components with identical reliabilities is

$$p_S(t) = \Pr(Y(t) \geq k) = \sum_{y=k}^{n} \binom{n}{y} p(t)^y (1 - p(t))^{n-y} \tag{4.10}$$

Example 4.3
Consider a 2-out-of-4 structure with four independent components of the same type. At a specified point of time t the component reliability is $p = 0.97$. The system reliability at time t is then, according to (4.10)

$$p_S = h(p) = \binom{4}{2} 0.97^2 \, 0.03^2 + \binom{4}{3} 0.97^3 \, 0.03 + \binom{4}{4} 0.97^4 \approx 0.99989$$

□

Finally, let us see how the system reliability of a more complex structure can be determined.

Example 4.4
Fig. 4.1 shows the reliability block diagram of a simplified automatic alarm system

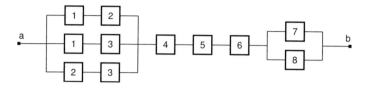

Fig. 4.1 Reliability block diagram of a simplified automatic alarm system for gas leakage.

for gas leakage. In the case of gas leakage, "connection" is established between a and b so that at least one of the alarm bells (7 and 8) will start ringing. The system has three independent gas detectors (1, 2, and 3) which are connected to a 2-out-of-3 voting unit (4); that is, at least two detectors must indicate gas leakage before an alarm is triggered off. Component 5 is a power supply unit, and component 6 is a relay.

We will consider the system at a given time t. To simplify the notation we therefore omit the explicit reference to the time t. The structure function of the system is

$$\phi(X) = (X_1 X_2 + X_1 X_3 + X_2 X_3 - 2 X_1 X_2 X_3)$$
$$\cdot (X_4 X_5 X_6)(X_7 + X_8 - X_7 X_8)$$

If the component reliability at time t_0 of component i is denoted by p_i, $i = 1, 2, \ldots, 8$, and $X_1, X_2, \ldots, X_8$ are independent, then the system reliability at time t_0 is

$$p_S = (p_1 p_2 + p_1 p_3 + p_2 p_3 - 2 p_1 p_2 p_3) \cdot p_4 p_5 p_6 \cdot (p_7 + p_8 - p_7 p_8) \quad (4.11)$$

□

4.2.4 Pivotal Decomposition

By pivotal decomposition the structure function $\phi(X(t))$ at time t can be written as

$$\begin{aligned} \phi(X(t)) &= X_i(t) \cdot \phi(1_i, X(t)) + (1 - X_i(t)) \cdot \phi(0_i, X(t)) \\ &= X_i(t) \cdot [\phi(1_i, X(t)) - \phi(0_i, X(t))] + \phi(0_i, X(t)) \quad (4.12) \end{aligned}$$

When the components are independent, the system reliability becomes

$$h(\boldsymbol{p}(t)) = p_i(t) \cdot E[\phi(1_i, X(t))] + (1 - p_i(t)) \cdot E[\phi(0_i, X(t))]$$

Let $h(1_i, \boldsymbol{p}(t)) = E[\phi(1_i, X(t))]$ and $h(0_i, \boldsymbol{p}(t)) = E[\phi(0_i, Xt))]$. Hence

$$\begin{aligned} h(\boldsymbol{p}(t)) &= p_i(t) \cdot h(1_i, \boldsymbol{p}(t)) + (1 - p_i(t)) \cdot h(0_i, \boldsymbol{p}(t)) \\ &= p_i(t) \cdot [h(1_i, \boldsymbol{p}(t)) - h(0_i, \boldsymbol{p}(t))] - h(0_i, \boldsymbol{p}(t)) \quad (4.13) \end{aligned}$$

Notice that the system reliability $h(\boldsymbol{p}(t))$ is a linear function of $p_i(t)$ when all the other component reliabilities are kept constant.

4.3 NONREPAIRABLE SYSTEMS

As explained in Section 4.1 the component reliability and the survivor function will coincide for nonrepairable components:

$$p_i(t) = R_i(t) \text{ for } i = 1, 2, \ldots, n$$

4.3.1 Nonrepairable Series Structures

According to (4.6) the survivor function of a nonrepairable series structure consisting of independent components is

$$R_S(t) = \prod_{i=1}^{n} R_i(t) \qquad (4.14)$$

Furthermore, according to (2.9)

$$R_i(t) = e^{-\int_0^t z_i(u)\,du} \qquad (4.15)$$

where $z_i(t)$ denotes the failure rate of component i at time t.

By inserting (4.15) into (4.14) we get

$$R_S(t) = \prod_{i=1}^{n} e^{-\int_0^t z_i(u)\,du} = e^{-\int_0^t \sum_{i=1}^{n} z_i(u)\,du} \qquad (4.16)$$

Hence the failure rate $z_S(t)$ of a series structure (of independent components) is equal to the sum of the failure rates of the individual components:

$$z_S(t) = \sum_{i=1}^{n} z_i(t) \qquad (4.17)$$

The mean time to failure (MTTF) of this series structure is

$$\text{MTTF} = \int_0^{\infty} R_S(t)\,dt = \int_0^{\infty} e^{-\int_0^t \sum_{i=1}^{n} z_i(u)\,du}\,dt \qquad (4.18)$$

Example 4.5
Consider a series structure on n independent components with constant failure rates λ_i, for $i = 1, 2, \ldots, n$. The survivor function of the series structure is

$$R_S(t) = e^{-(\sum_{i=1}^{n} \lambda_i)t} \qquad (4.19)$$

and the MTTF is

$$\text{MTTF} = \int_0^{\infty} e^{-(\sum_{i=1}^{n} \lambda_i)t}\,dt = \frac{1}{\sum_{i=1}^{n} \lambda_i} \qquad (4.20)$$

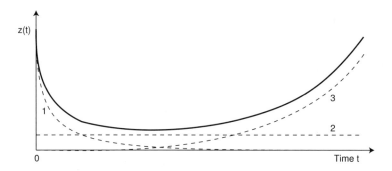

Fig. 4.2 The failure rate function of a series structure of three independent components, where component 1 has decreasing failure rate, component 2 has constant failure rate, and component 3 has increasing failure rate.

□

Example 4.6
Consider a series structure with n independent components. The time to failure of component i has a Weibull distribution with common shape parameter α and scale parameter λ_i, for $i = 1, 2, \ldots, n$. The survivor function of the series structure is from (4.15):

$$R_S(t) = \sum_{i=1}^{n} e^{-(\lambda_i t)^\alpha} = \exp\left(-\left[\left(\sum_{i=1}^{n} \lambda_i^\alpha\right)^{1/\alpha} \cdot t\right]^\alpha\right) \quad (4.21)$$

If we define $\lambda_0 = \left(\sum_{i=1}^{n} \lambda_i^\alpha\right)^{1/\alpha}$, the survivor function (4.21) can be written

$$R_S(t) = e^{-(\lambda_0 t)^\alpha} \quad (4.22)$$

The time to failure of the series structure is therefore Weibull distributed with shape parameter α and scale parameter $\lambda_0 = \left(\sum_{i=1}^{n} \lambda_i^\alpha\right)^{1/\alpha}$. □

Example 4.7
Consider a series structure of $n = 3$ independent components. Component 1 has a decreasing failure rate, for example, a Weibull distributed time to failure with shape parameter $\alpha < 1$. Component 2 has a constant failure rate, while component 3 has an increasing failure rate, for example, a Weibull distributed time to failure with shape parameter $\alpha > 2$. The failure rates of the three components are illustrated in Fig. 4.2. The failure rate function of the series structure is from (4.17) the sum of the three individual failure rate functions and is illustrated by the fully drawn line in Fig. 4.2. The failure rate function of the series structure is seen to have a bathtub shape. A bathtub-shaped failure rate of a component may therefore be obtained by replacing the component by three independent imaginary components in a series; one with decreasing failure rate function, one with constant failure rate, and one with increasing failure rate function. □

4.3.2 Nonrepairable Parallel Structures

According to (4.8) the survivor function of a nonrepairable parallel structure of independent components is

$$R_S(t) = 1 - \prod_{i=1}^{n}(1 - R_i(t)) \tag{4.23}$$

When all the components have constant failure rates $z_i(t) = \lambda_i$, for $i = 1, 2, \ldots, n$, then

$$R_S(t) = 1 - \prod_{i=1}^{n}(1 - e^{-\lambda_i t}) \tag{4.24}$$

Example 4.8
Consider a parallel structure of n independent components of the same type with constant failure rate λ. The survivor function of the parallel structure is

$$R_S(t) = 1 - \left(1 - e^{-\lambda t}\right)^n \tag{4.25}$$

By using the binomial formula (4.25) can be written

$$R_S(t) = 1 - \sum_{x=0}^{n} \binom{n}{x} \left(-e^{-\lambda t}\right)^x = \sum_{x=1}^{n} \binom{n}{x}(-1)^{x+1} e^{-\lambda x t}$$

The MTTF is

$$\text{MTTF} = \int_0^\infty R_S(t)\, dt = \frac{1}{\lambda} \sum_{x=1}^{n} \binom{n}{x} \frac{(-1)^{x+1}}{x} \tag{4.26}$$

□

Remark: In Section 4.3.5 an alternative formula for MTTF based on another argument is given. The two formulas give the same result. The MTTF of a parallel structure n independent and identical components with failure rate λ is listed in the first row of Table 4.2 for some selected values of n. □

Example 4.9
Consider a nonrepairable parallel structure of two components with lifetimes T_1 and T_2, which are assumed to be independent and exponentially distributed with failure rates λ_1 and λ_2, respectively.

The survivor function of the system is

$$\begin{aligned} R_S(t) &= 1 - (1 - e^{-\lambda_1 t})(1 - e^{-\lambda_2 t}) \\ &= e^{-\lambda_1 t} + e^{-\lambda_2 t} - e^{-(\lambda_1+\lambda_2)t} \end{aligned} \tag{4.27}$$

Note that the time to failure T of this parallel structure is *not* exponentially distributed, even if both components have exponentially distributed times to failure.

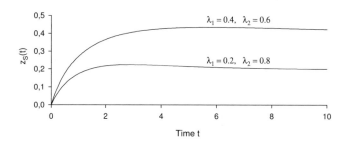

Fig. 4.3 The failure rate for a parallel structure of two independent components for selected values of λ_1 and λ_2 ($\lambda_1 + \lambda_2 = 1$).

The MTTF of the system is

$$\text{MTTF} = \int_0^\infty R_S(t)\,dt = \frac{1}{\lambda_1} + \frac{1}{\lambda_2} - \frac{1}{\lambda_1 + \lambda_2} \qquad (4.28)$$

The corresponding failure rate function is given by

$$z_S(t) = -\frac{R'_S(t)}{R_S(t)}$$

Hence

$$z_S(t) = \frac{\lambda_1 e^{-\lambda_1 t} + \lambda_2 e^{-\lambda_2 t} - (\lambda_1 + \lambda_2) e^{-(\lambda_1 + \lambda_2)t}}{e^{-\lambda_1 t} + e^{-\lambda_2 t} - e^{-(\lambda_1 + \lambda_2)t}} \qquad (4.29)$$

In Fig. 4.3, $z_S(t)$ is sketched for selected combinations of λ_1 and λ_2, such that $\lambda_1 + \lambda_2 = 1$. Notice that when $\lambda_1 \neq \lambda_2$, the failure rate function $z_S(t)$ will increase up to a maximum at a time t_0, and then decrease for $t \geq t_0$ down to $\min\{\lambda_1, \lambda_2\}$. □

This example illustrates that even if the individual components of a system have *constant* failure rates, the system itself may not have a constant failure rate.

Example 4.10
Consider a parallel structure of two independent and identical components with failure rate λ. The survivor function is

$$R_S(t) = 2e^{-\lambda t} - e^{-2\lambda t} \qquad (4.30)$$

The probability density function of the time to failure of the parallel structure is

$$f_S(t) = -R'_S(t) = 2\lambda e^{-\lambda t} - 2\lambda e^{-2\lambda t} \qquad (4.31)$$

The mode of the distribution is the value of t that maximizes $f_S(t)$

$$t_{\text{mode}} = \frac{\ln 2}{\lambda} \qquad (4.32)$$

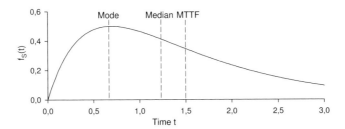

Fig. 4.4 The probability density function of a parallel structure with two independent and identical components with failure rate $\lambda = 1$, together with its mode, median, and MTTF.

The median life of the parallel structure is

$$t_m = R_S^{-1}(0.5) \approx \frac{1.228}{\lambda} \qquad (4.33)$$

The MTTF is

$$\text{MTTF} = \int_0^\infty R_S(t)\,dt = \frac{3}{2\lambda} \qquad (4.34)$$

The probability density $f_S(t)$ of the parallel structure, together with its mode, median, and MTTF, are illustrated in Fig. 4.4. The mean residual life (MRL) of the parallel structure at age t is

$$\text{MRL}(t) = \frac{1}{R_S(t)} \int_t^\infty R_S(x)\,dx = \frac{1}{2\lambda} \cdot \frac{4 - e^{-\lambda t}}{2 - e^{-\lambda t}} \qquad (4.35)$$

Notice that $\lim_{t \to \infty} \text{MRL}(t) = 1/\lambda$. Since the two components of the (nonrepairable) parallel structure are independent, one of them will fail first, and we will sooner or later be left with only one component. When one of the components has failed, the MRL of the system is equal to the MRL of the remaining component. Since the failure rate is constant, the MRL of the remaining component is equal to its MTTF $= 1/\lambda$. □

4.3.3 Nonrepairable 2-out-of-3 Structures

According to Section 3.8 the structure function of a 2-out-of-3 structure is

$$\phi(t) = X_1(t)X_2(t) + X_1(t)X_3(t) + X_2(t)X_3(t) - 2X_1(t)X_2(t)X_3(t)$$

Thus the survivor function of the 2-out-of-3 structure is

$$R_S(t) = R_1(t)R_2(t) + R_1(t)R_3(t) + R_2(t)R_3(t) - 2R_1(t)R_2(t)R_3(t)$$

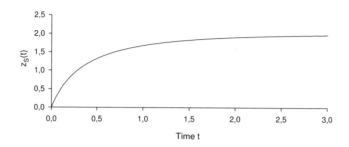

Fig. 4.5 The failure rate function $z_S(t)$ for a 2-out-of-3 structure of independent and identical components with failure rate $\lambda = 1$.

In the special case where all the three components have the common constant failure rate λ, then

$$R_S(t) = 3\,e^{-2\lambda t} - 2e^{-3\lambda t} \tag{4.36}$$

The failure rate function of this 2-out-of-3 structure is

$$z_S(t) = \frac{-R'_S(t)}{R_S(t)} = \frac{6\lambda\left(e^{-2\lambda t} - e^{-3\lambda t}\right)}{3e^{-2\lambda t} - 2e^{3\lambda t}} \tag{4.37}$$

A sketch of the failure rate function $z_S(t)$ is given in Fig. 4.5.

Note that $\lim_{t \to \infty} z_C(t) = 2\lambda$ (see Problem 4.8). The MTTF of this 2-out-of-3 structure is

$$\text{MTTF} = \int_0^\infty R_S(t)\,dt = \frac{3}{2\lambda} - \frac{2}{3\lambda} = \frac{5}{6}\frac{1}{\lambda} \tag{4.38}$$

4.3.4 A Brief Comparison

Let us now compare the three simple systems:

1. A single component
2. A parallel structure of two identical components
3. A 2-out-of-3 system with identical components

All the components are assumed to be independent with a common constant failure rate λ. A brief comparison of the three systems is presented in Table 4.1. Note that a single component has a higher MTTF than the 2-out-of-3 structure. The survivor functions of the three simple systems are also compared in Fig. 4.6. The introduction of a 2-out-of-3 structure instead of a single component hence reduces the MTTF by about 16%. The 2-out-of-3 structure has, however, a significantly higher reliability (probability of functioning) in the interval $(0, t]$ for $t < \ln 2/\lambda$.

Table 4.1 Brief Comparison of Systems (1), (2), and (3).

System	Survivor Function $R_S(t)$	Mean Time to Failure MTTF
1oo1	$e^{-\lambda t}$	$\dfrac{1}{\lambda}$
1oo2	$2e^{-\lambda t} - e^{-2\lambda t}$	$\dfrac{3}{2}\dfrac{1}{\lambda}$
2oo3	$3e^{-2\lambda t} - 2e^{-3\lambda t}$	$\dfrac{5}{6}\dfrac{1}{\lambda}$

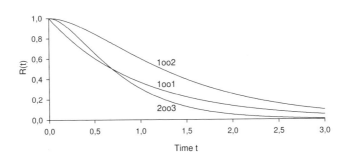

Fig. 4.6 The survivor functions of the three systems in Table 4.1 ($\lambda = 1$).

4.3.5 Nonrepairable k-out-of-n Structures

Assume that we have a k-out-of-n structure of n identical and independent components with constant failure rate λ. The survivor function of the k-out-of-n structure is from (4.10)

$$R_S(t) = \sum_{x=k}^{n} \binom{n}{x} e^{-\lambda t x} (1 - e^{-\lambda t})^{n-x} \tag{4.39}$$

According to (2.12)

$$\text{MTTF} = \sum_{x=k}^{n} \binom{n}{x} \int_0^{\infty} e^{-\lambda t x} (1 - e^{-\lambda t})^{n-x} \, dt \tag{4.40}$$

By introducing $v = e^{-\lambda t}$ we obtain (see Appendix A)

$$\begin{aligned}
\text{MTTF} &= \sum_{x=k}^{n} \binom{n}{x} \frac{1}{\lambda} \int_0^1 v^{x-1} (1-v)^{n-x} \, dv \\
&= \sum_{x=k}^{n} \binom{n}{x} \frac{1}{\lambda} \frac{\Gamma(x) \cdot \Gamma(n-x+1)}{\Gamma(n+1)} \\
&= \frac{1}{\lambda} \sum_{x=k}^{n} \binom{n}{x} \frac{(x-1)!(n-x)!}{n!} = \frac{1}{\lambda} \sum_{x=k}^{n} \frac{1}{x}
\end{aligned} \tag{4.41}$$

The MTTF of some simple k-out-of-n systems, computed by (4.41) are listed in Table 4.2. Note that a 1-out-of-n system is a parallel system, while a n-out-of-n system is a series system.

4.4 QUANTITATIVE FAULT TREE ANALYSIS

We will now see how we can analyze a structure that is modeled as a fault tree. Let n denote the number of different basic events in the fault tree. By analogy with our notation for reliability block diagrams, the fault tree is said to be of order n. The n basic events are numbered, and the following state variables are introduced:

$$Y_i(t) = \begin{cases} 1 & \text{if basic event } i \text{ occurs at time } t \\ 0 & \text{otherwise} \end{cases} \quad i = 1, 2, \ldots, n$$

Let $\mathbf{Y}(t) = (Y_1(t), Y_2(t), \ldots, Y_n(t))$ denote the state vector for the structure at time t. The purpose of a "quantitative analysis" of a fault tree usually is to determine the probability of the TOP event (system failure). The procedure is completely analogous to the one for reliability block diagrams.

The state of the TOP event at time t can be described by the binary variable $\psi(\mathbf{Y}(t))$ where

$$\psi(\mathbf{Y}(t)) = \begin{cases} 1 & \text{if the TOP event occurs at time } t \\ 0 & \text{otherwise} \end{cases}$$

Table 4.2 MTTF of some k-out-of-n Systems of Identical and Independent Components with Constant Failure Rate λ.

$k \backslash n$	1	2	3	4	5
1	$\dfrac{1}{\lambda}$	$\dfrac{3}{2\lambda}$	$\dfrac{11}{6\lambda}$	$\dfrac{25}{12\lambda}$	$\dfrac{137}{60\lambda}$
2	–	$\dfrac{1}{2\lambda}$	$\dfrac{5}{6\lambda}$	$\dfrac{13}{12\lambda}$	$\dfrac{77}{60\lambda}$
3	–	–	$\dfrac{1}{3\lambda}$	$\dfrac{7}{12\lambda}$	$\dfrac{47}{60\lambda}$
4	–	–	–	$\dfrac{1}{4\lambda}$	$\dfrac{9}{20\lambda}$
5	–	–	–	–	$\dfrac{1}{5\lambda}$

It is assumed that if we know the states of all the n basic events, we also know whether or not the TOP event occurs:

$$\psi(Y(t)) = \psi(Y_1(t), Y_2(t), \ldots, Y_n(t)) \qquad (4.42)$$

The function $\psi(Y(t))$ is called the *structure function* of the fault tree.

Let $q_i(t)$ denote the probability that basic event i occurs at time t, for $i = 1, 2, \ldots, n$. Then

$$\Pr(Y_i(t) = 1) = E(Y_i(t)) = q_i(t) \quad \text{for } i = 1, 2, \ldots, n$$

Let $Q_0(t)$ denote the probability that the TOP event (system failure) occurs at time t. Then

$$Q_0(t) = \Pr(\psi(Y(t)) = 1) = E(\psi(Y(t))) \qquad (4.43)$$

Remark: The statement "basic event i occurs at time t" may be a bit misleading. The basic events and the TOP event in a fault tree are in reality *states* and not events. When we say that a basic event (or the TOP event) occurs at time t, we mean that the corresponding *state* is present at time t. □

If the basic event i means that component i in the system is in a failed state for $i = 1, 2, \ldots, n$, then

$$\Pr(Y_i(t) = 1) = q_i(t) = 1 - p_i(t) \quad \text{for } i = 1, 2, \ldots, n$$

where $p_i(t)$ is the probability that component i is in a functioning state at time t; $q_i(t)$ is called the *unreliability* of *component* i at time t, while $Q_0(t)$ denotes the *unreliability* of the *system* at the same point of time.

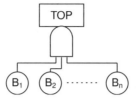

Fig. 4.7 Fault tree with a single AND-gate.

In this case

$$Q_0(t) = 1 - h(\boldsymbol{p}(t)) = 1 - h[1 - q_1(t), 1 - q_2(t), \ldots, 1 - q_n(t)]$$

where $p_i(t)$, for $i = 1, 2, \ldots, n$ and $h(\boldsymbol{p}(t))$ are defined in (4.3) and (4.5). Note that $Q_0(t)$ in this case is a function of the $q_i(t)$'s only.

In the same way as for reliability block diagrams, it can be shown that when the basic events are independent, then $Q_0(t)$ will be a function of the $q_i(t)$'s only for $i = 1, 2, \ldots, n$. Hence $Q_0(t)$ may be written

$$Q_0(t) = g(q_1(t), q_2(t), \ldots, q_n(t)) = g(\boldsymbol{q}(t)) \tag{4.44}$$

4.4.1 Fault Trees with a Single AND-Gate

Consider the fault tree in Fig. 4.7. Here the TOP event occurs if and only if all the basic events $B_1, B_2, \ldots, B_n$ occur simultaneously.

The structure function of this fault tree is

$$\psi(\boldsymbol{Y}(t)) = Y_1(t) \cdot Y_2(t) \cdots Y_n(t) = \prod_{i=1}^{n} Y_i(t)$$

Since the basic events are assumed to be independent, then

$$\begin{aligned} Q_0(t) &= E(\psi(\boldsymbol{Y}(t))) = E(Y_1(t) \cdot Y_2(t) \cdots Y_n(t)) \\ &= E(Y_1(t)) \cdot E(Y_2(t)) \cdots E(Y_n(t)) \\ &= q_1(t) \cdot q_2(t) \cdots q_n(t) = \prod_{i=1}^{n} q_i(t) \end{aligned} \tag{4.45}$$

The probability $Q_0(t)$ of the TOP event may also be determined directly by the following argument: Let $B_i(t)$ denote that basic event B_i occurs at time t; $i = 1, 2, \ldots, n$. Then

$$\begin{aligned} Q_0(t) &= \Pr(B_1(t) \cap B_2(t) \cap \cdots \cap B_n(t)) \\ &= \Pr(B_1(t)) \cdot \Pr(B_2(t)) \cdots \Pr(B_n(t)) \\ &= q_1(t) \cdot q_2(t) \cdots q_n(t) = \prod_{i=1}^{n} q_i(t) \end{aligned}$$

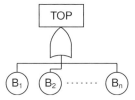

Fig. 4.8 Fault tree with a single OR-gate.

4.4.2 Fault Tree with a Single OR-Gate

Consider the fault tree in Fig. 4.8. The TOP event occurs if at least one of the basic events $B_1, B_2, \ldots, B_n$ occurs.

The structure function of this fault tree is

$$\psi(Y(t)) = Y_1(t) \sqcup Y_2(t) \sqcup \ldots \sqcup Y_n(t)$$
$$= 1 - \prod_{i=1}^{n}(1 - Y_i(t))$$

Since the basic events are assumed to be independent, then

$$Q_0(t) = E(\psi(Y(t))) = 1 - \prod_{i=1}^{n} E(1 - Y_i(t))$$
$$= 1 - \prod_{i=1}^{n}(1 - E(Y_i(t))) = 1 - \prod_{i=1}^{n}(1 - q_i(t)) \quad (4.46)$$

Then $Q_0(t)$ can also be determined directly in the following way: Let $B_i^*(t)$ denote that basic event B_i does *not* occur at time t. Then

$$\Pr(B_i^*(t)) = 1 - \Pr(B_i(t)) = 1 - q_i(t) \quad \text{for } i = 1, 2, \ldots, n$$

$$Q_0(t) = \Pr(B_1(t) \cup B_2(t) \cup \cdots \cup B_n(t))$$
$$= 1 - \Pr(B_1^*(t) \cap B_2^*(t) \cap \cdots \cap B_n^*(t))$$
$$= 1 - \Pr(B_1^*(t)) \cdot \Pr(B_2^*(t)) \cdots \Pr(B_n^*(t))$$
$$= 1 - \prod_{i=1}^{n}(1 - q_i(t))$$

4.4.3 Approximation Formula for $Q_0(t)$

Calculation of the TOP event probability by means of the structure function may in many cases be both time-consuming and cumbersome. Hence there may be a need for approximation formulas.

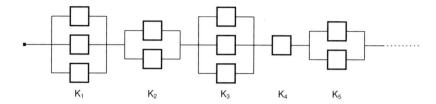

Fig. 4.9 A structure represented as a series structure of the minimal cut parallel structures.

Consider a system (fault tree) with k minimal cut sets $K_1, K_2, \ldots, K_k$. This system may be represented by a series structure of the k minimal cut parallel structures, as illustrated by the reliability block diagram in Fig. 4.9.

The TOP event occurs if at least one of the k minimal cut parallel structures fails. A minimal cut parallel structure fails if each and all the basic events in the minimal cut set occur simultaneously. Note that the same input event may enter in many different cut sets.

Let $\check{Q}_j(t)$ denote the probability that minimal cut parallel structure j fails at time t. If the basic events are assumed to be independent, then

$$\check{Q}_j(t) = \prod_{i \in K_j} q_i(t) \qquad (4.47)$$

Let $Q_0(t)$ denote the probability that the TOP event (system failure) occurs at time t. If all the k minimal cut parallel structures were independent, then

$$Q_0(t) = \coprod_{j=1}^{k} \check{Q}_j(t) = 1 - \prod_{j=1}^{k}(1 - \check{Q}_j(t)) \qquad (4.48)$$

But since the same basic event may occur in several minimal cut sets, the minimal cut parallel structures can obviously be positively dependent (or associated, see Chapter 6). In Chapter 6 it is shown that

$$Q_0(t) \leq 1 - \prod_{j=1}^{k}(1 - \check{Q}_j(t)) \qquad (4.49)$$

Hence the right-hand side of (4.49) may be used as an upper bound (conservative) for the probability of system failure.

When all the $q_i(t)$'s are very small, it can be shown that with good approximation

$$Q_0(t) \approx 1 - \prod_{j=1}^{k}(1 - \check{Q}_j(t)) \qquad (4.50)$$

This approximation is called the *upper bound approximation*, and it is used in a number of computer programs for fault tree analysis, for example, in CARA FaultTree.

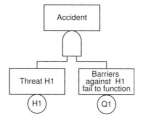

Fig. 4.10 System failure caused by fire (threat H_1)..

Equation (4.50), however, has to be used with care when at least one of the $q_i(t)$'s is of order 10^{-2} or larger.

Assume now that all the $\check{Q}_j(t)$'s are so small that we can disregard their products. In this case (4.50) may be approximated by

$$Q_0(t) \approx 1 - \prod_{j=1}^{k}(1 - \check{Q}_j(t)) \approx \sum_{j=1}^{k} \check{Q}_j(t) \qquad (4.51)$$

It is straightforward to verify that the last approximation is more conservative than the first one:

$$Q_0(t) \leq 1 - \prod_{j=1}^{k}(1 - \check{Q}_j(t)) \leq \sum_{j=1}^{k} \check{Q}_j(t) \qquad (4.52)$$

Example 4.11 TOP Event Frequency

In risk analyzes we are sometimes interested in the *frequency* of the TOP event in a specific situation. Consider a system that is exposed to a set of threats or hazards $H_1, H_2, \ldots, H_m$. The threats may be extreme loads (caused by fire, explosion, earthquake, lightning, etc.), component failures or operational or maintenance errors. Some of these threats (hazards) have been identified during system design, and barriers, and/or protective systems may have been established to withstand the threats.

To have a system failure (accident), one of the threats must manifest itself *and* the protective system must fail. As an example, assume that the threat H_1 denotes a fire in a specified system module. The expected frequency of such fires has been estimated to be λ_{H_1}.

The fire protection and extinguisher systems that have been installed to withstand the fire, may be studied by a fault tree analysis with TOP event "Failure of the fire protection or extinguisher system." Assume that we find the TOP event probability $Q_{H_1}(t)$. The expected frequency of system failures (accidents) caused by this type of fires is thus (see Section 7.2) equal to $\lambda_{H_1} Q_{H_1}(t)$. This situation is illustrated by the fault tree in Fig. 4.10.

The total expected frequency of system failures may now be determined by combining all the identified threats $H_1, H_2, \ldots, H_m$. Protection systems may be available for only a limited number of the threats. Some of the threats may, on the other hand,

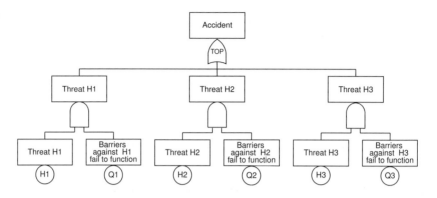

Fig. 4.11 System failure caused by the threats H_1, H_2, and H_3.

have a number of redundant protection systems. For some of the threats, it may be impossible to establish protection systems or barriers.

A fault tree showing the total expected frequency of system failures (accidents) is illustrated in Fig. 4.11.

Let $Q_{H_j}(t)$ denote the probability that all the protection systems against the threat H_j fails, for $j = 1, 2, \ldots, m$. If the threat H_j has n_j redundant and independent protection systems, then $Q_{H_j}(t) = \prod_{i=1}^{n_j} Q_{H_{j,i}}$, where $Q_{H_{j,i}}(t)$ denotes the probability that protection system i against threat H_j fails at time t. When no protection system is available for threat H_j, then $Q_{H_j}(t) = 1$.

The total expected frequency of system failures (accidents) λ_S may now be approximated by

$$\lambda_S \approx \sum_{j=1}^{m} \lambda_{H_j} Q_{H_j}(t)$$

Note that for this approach to be valid, the fault tree in Fig. 4.11 must have one and only one threat in each minimal cut set. This application of fault tree analysis is further discussed in AIChE (1989). □

4.5 EXACT SYSTEM RELIABILITY

In this section we describe some different methods for calculation of exact system reliability at a given time t_0 when the n components are *independent*. To simplify the notation we replace the component reliabilities $p_j(t_0)$ by p_j for $j = 1, 2, \ldots, n$.

4.5.1 Computation Based on the Structure Function

The most straightforward method for computation of the exact system reliability is illustrated in Example 4.4. The structure function of the system is established, and powers of the X_i's are deleted ($i = 1, 2, \ldots, n$). The exact system reliability is then obtained by replacing the X_i's by the corresponding p_i's.

4.5.2 Computation Based on Pivotal Decomposition

By repeated pivotal decomposition (see Section 3.11) a structure function can always be written as

$$\phi(X) = \sum_{y} \prod_{j=1}^{n} X_j^{y_j} (1 - X_j)^{1-y_j} \phi(y) \tag{4.53}$$

where the summation over y is a summation over all n-dimensional binary vectors y, and $0^0 \equiv 1$. If $X_1, X_2, \ldots, X_n$ are assumed to be independent, $X_j^{y_j}(1 - X_j)^{1-y_j}$ are also independent for all j. Since y_j can only take the values 0 and 1, then

$$E[X_j^{y_j}(1 - X_j)^{1-y_j}] = p_j^{y_j}(1 - p_j)^{1-y_j} \quad \text{for } j = 1, 2, \ldots, n$$

Therefore the system reliability may be written

$$p_S = E[\phi(X)] = \sum_{y} \prod_{j=1}^{n} p_j^{y_j}(1 - p_j)^{1-y_j} \phi(y) \tag{4.54}$$

The approach is the following:

1. First, determine by experiment the value of $\phi(y)$ for all 2^n possible y vectors.

2. Next put the obtained values of $\phi(y)$ for the different vectors y into (4.54). [$\phi(y)$ is set equal to 1 if the system is in a functioning state, and equal to 0 otherwise.]

Thereby an expression is obtained for the desired system reliability.

4.5.3 Computation Based on Minimal Cut (Path) Sets

When all the minimal cut sets $K_1, K_2, \ldots, K_k$ and/or all the minimal path sets $P_1, P_2, \ldots, P_P$ are determined, the structure function may be written

$$\phi(X) = \prod_{j=1}^{k} \coprod_{i \in K_j} X_i \tag{4.55}$$

alternatively

$$\phi(X) = \coprod_{j=1}^{p} \prod_{i \in P_j} X_i \tag{4.56}$$

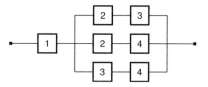

Fig. 4.12 Reliability block diagram for Example 4.12.

The structure function is thus written on a multilinear form. (Remember that since $X_i, i = 1, 2, \ldots, n$ are binary, all exponents can be omitted.) Since $X_1, X_2, \ldots, X_n$ are assumed to be independent, the system reliability is obtained by replacing all the X_i's in the structure function by the corresponding p_i's.

Example 4.12
The structure in Fig. 4.12 has the following three minimal path sets:

$$P_1 = \{1, 2, 3\}, \quad P_2 = \{1, 2, 4\}, \quad P_1 = \{1, 3, 4\}$$

Hence

$$\phi(X) = \coprod_{j=1}^{3} \prod_{i \in P_j} X_i = X_1 X_2 X_3 \sqcup X_1 X_2 X_4 \sqcup X_1 X_3 X_4$$

$$= 1 - (1 - X_1 X_2 X_3)(1 - X_1 X_2 X_4)(1 - X_1 X_3 X_4)$$

$$= X_1 X_2 X_3 + X_1 X_2 X_4 + X_1 X_3 X_4 - X_1^2 X_2^2 X_3 X_4$$
$$- X_1^2 X_2 X_3^2 X_4 - X_1^2 X_2 X_3 X_4^2 + X_1^3 X_2^2 X_3^2 X_4^2$$

$$= X_1 X_2 X_3 + X_1 X_2 X_4 + X_1 X_3 X_4 - 2 X_1 X_2 X_3 X_4$$

Since the components are independent, the system reliability is

$$h(p) = p_1 p_2 p_3 + p_1 p_2 p_4 + p_1 p_3 p_4 - 2 p_1 p_2 p_3 p_4$$

Time is reintroduced by replacing p_j by $p_j(t_0)$ for $j = 1, 2, \ldots, n$ in the above expression. □

4.5.4 The Inclusion-Exclusion Principle

In this section we study how the inclusion-exclusion principle can be applied to determine the *unreliability* of a system. The same approach can also be used to determine the system *reliability*. This is shown at the end of this section.

A system of n independent components has the minimal cut sets $K_1, K_2, \ldots, K_k$. Let E_j denote the event that the components of the minimal cut set K_j are all in a failed state, that is, that the jth minimal cut parallel structure has failed at time t. According to (4.47),

$$\Pr(E_j) = \check{Q}_j = \prod_{i \in K_j} q_i$$

where q_i denotes the unreliability of component i at time t_0 for $i = 1, 2, \ldots, n$.

Since the system fails as soon as one of its minimal cut parallel structures fails, the system unreliability may be expressed by

$$Q_0 = \Pr\left(\bigcup_{j=1}^{k} E_j\right) \tag{4.57}$$

In general, the individual events E_j, $j = 1, 2, \ldots, k$ are not disjoint. Hence, the probability $\Pr(\bigcup_{j=1}^{k} E_j)$ is determined by using the general addition theorem in probability theory (e.g., see Dudewicz and Mishra, 1988, p. 45).

$$\begin{aligned} Q_0 &= \sum_{j=1}^{k} \Pr(E_j) - \sum_{i<j} \Pr(E_i \cap E_j) + \cdots \\ &\quad + (-1)^{j+1} \Pr(E_1 \cap E_2 \cap \cdots \cap E_k) \end{aligned} \tag{4.58}$$

By introducing

$$W_1 = \sum_{j=1}^{k} \Pr(E_j)$$

$$W_2 = \sum_{i<j} \Pr(E_i \cap E_j)$$

$$\vdots$$

$$W_k = \Pr(E_1 \cap E_2 \cap \cdots \cap E_k)$$

equation (4.58) may be written

$$\begin{aligned} Q_0 &= W_1 - W_2 + W_3 - \cdots + (-1)^{k+1} W_k \\ &= \sum_{j=1}^{k} (-1)^{j+1} W_j \end{aligned} \tag{4.59}$$

Example 4.13

The minimal cut sets of the bridge structure in Fig. 4.13 are

$$K_1 = \{1, 2\}, \quad K_2 = \{4, 5\}, \quad K_3 = \{1, 3, 5\}, \quad K_4 = \{2, 3, 4\}$$

Let B_i denote that component i is failed, $i = 1, 2, 3, 4, 5$.

According to (4.59) the unreliability Q_0 of the bridge structure is

$$Q_0 = W_1 - W_2 + W_3 - W_4$$

where

$$\begin{aligned} W_1 &= \sum_{j=1}^{4} \Pr(E_j) \\ &= \Pr(B_1 \cap B_2) + \Pr(B_4 \cap B_5) + \Pr(B_1 \cap B_3 \cap B_5) + \Pr(B_2 \cap B_3 \cap B_4) \\ &= q_1 q_2 + q_4 q_5 + q_1 q_3 q_5 + q_2 q_3 q_4 \end{aligned}$$

170 SYSTEMS OF INDEPENDENT COMPONENTS

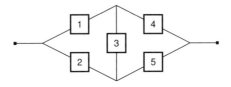

Fig. 4.13 The bridge structure.

$$\begin{aligned}
W_2 = \sum_{i<j} \Pr(E_i \cap E_j) &= \Pr(E_1 \cap E_2) + \Pr(E_1 \cap E_3) + \Pr(E_1 \cap E_4) \\
&\quad + \Pr(E_2 \cap E_3) + \Pr(E_2 \cap E_4) + \Pr(E_3 \cap E_4) \\
&= \Pr(B_1 \cap B_2 \cap B_4 \cap B_5) \\
&\quad + \Pr(B_1 \cap B_2 \cap B_1 \cap B_3 \cap B_5) \\
&\quad + \Pr(B_1 \cap B_2 \cap B_2 \cap B_3 \cap B_4) \\
&\quad + \Pr(B_4 \cap B_5 \cap B_1 \cap B_3 \cap B_5) \\
&\quad + \Pr(B_4 \cap B_5 \cap B_2 \cap B_3 \cap B_4) \\
&\quad + \Pr(B_1 \cap B_3 \cap B_5 \cap B_2 \cap B_3 \cap B_4) \\
&= q_1 q_2 q_4 q_5 + q_1 q_2 q_3 q_5 + q_1 q_2 q_3 q_4 + q_1 q_3 q_4 q_5 \\
&\quad + q_2 q_3 q_4 q_5 + q_1 q_2 q_3 q_4 q_5
\end{aligned}$$

Similarly

$$W_3 = 4 q_1 q_2 q_3 q_4 q_5$$

and

$$W_4 = q_1 q_2 q_3 q_4 q_5$$

Hence the system unreliability is

$$\begin{aligned}
Q_0 &= W_1 - W_2 + W_3 - W_4 \\
&= q_1 q_2 + q_4 q_5 + q_1 q_3 q_5 + q_2 q_3 q_4 - q_1 q_2 q_4 q_5 - q_1 q_2 q_3 q_5 \\
&\quad - q_1 q_2 q_3 q_4 - q_1 q_3 q_4 q_5 - q_2 q_3 q_4 q_5 + 2 q_1 q_2 q_3 q_4 q_5
\end{aligned}$$

□

Example 4.13 shows that, when using the general addition theorem (4.58) we have to calculate the probability of a large number of terms that later cancel each other. An alternative approach has been proposed by Satyanarayana and Prabhakar (1978). The idea behind their method is—with the help of graph-theoretic arguments—to leave out the canceling terms at an early stage without having to calculate them. Their method has been computerized in a program called TRAP (topological reliability analysis program).

A number of alternatives to the inclusion-exclusion principle have been proposed. One of these alternatives is the ERAC algorithm (exact reliability/availability calculation) which was developed by Aven (1986). The ERAC algorithm has been implemented in CARA FaultTree.

Calculating the exact value of a system's unreliability Q_0 by means of (4.59) may be cumbersome and time-consuming, even when the system is relatively simple. In such cases one may sometimes be content with an approximative value of the system's unreliability.

One way of determining approximate values of the system unreliability Q_0 utilizes the following result based on inclusion-exclusion:

$$\begin{aligned} Q_0 &\leq W_1 \\ W_1 - W_2 \leq Q_0 & \\ Q_0 &\leq W_1 - W_2 + W_3 \\ &\vdots \end{aligned} \qquad (4.60)$$

It can be shown (see Feller, 1968, pp. 98–102) that

$$(-1)^{j-1} Q_0 \leq (-1)^{j-1} \sum_{v=1}^{j} (-1)^{v-1} W_v \ , \quad j = 1, 2, \ldots, k \qquad (4.61)$$

Equation (4.61) may give the impression that the differences between the consecutive upper and lower bounds are monotonically decreasing. This is, however, not true in general.

In practice (4.61) is used in the following way: Successively one determines upper and lower bounds for Q_0, proceeding downwards in (4.61) until one obtains bounds that are sufficiently close. The first upper bound for the system unreliability at time t, $Q_0(t)$, is according to (4.61)

$$Q_0(t) \leq W_1 = \sum_{j=1}^{k} \check{Q}_j(t) \qquad (4.62)$$

According to (4.49)

$$Q_0(t) \leq 1 - \prod_{j=1}^{k} (1 - \check{Q}_j(t)) \qquad (4.63)$$

and from (4.52) we have that

$$1 - \prod_{j=1}^{k} (1 - \check{Q}_j(t)) \leq \sum_{j=1}^{k} \check{Q}_j(t) \qquad (4.64)$$

Hence the right-hand side of (4.63) is a more accurate approximation to the true value of $Q_0(t)$ than (4.62).

Example 4.14
Reconsider the bridge structure in Example 4.13 and assume that all the component unreliabilities q_i are equal to 0.05. By introducing these q_i's in the expression for the W_i's in Example 4.13, we obtain

$$\begin{aligned} W_1 &= 5250 \cdot 10^{-6} \\ W_2 &= 3156 \cdot 10^{-6} \\ W_3 &= 1.25 \cdot 10^{-6} \\ W_4 &= 0.31 \cdot 10^{-6} \end{aligned}$$

From (4.61) we get:

$$\begin{aligned} Q_0 &\leq W_1 \approx 5250 \cdot 10^{-6} = 0.5250\% \\ Q_0 &\geq W_1 - W_2 \approx 5218.4 \cdot 10^{-6} = 0.5218\% \end{aligned}$$

Hence from the first two inequalities of (4.61) we know that

$$0.5218\% \leq Q_0 \leq 0.5250\%$$

For many applications this precision may be sufficient. If not, we proceed and calculate the next inequality:

$$Q_0 \leq W_1 - W_2 + W_3 \approx 5219.69 \cdot 10^{-6} = 0.5220\%$$

Now we know that Q_0 is bounded by

$$0.5218\% \leq Q_0 \leq 0.5220\%$$

The exact value is

$$Q_0 = W_1 - W_2 + W_3 - W_4 = 5219.38 \cdot 10^{-6} \approx 0.5219\%$$

By comparison, the upper bound obtained by (4.63) is equal to

$$1 - \prod_{j=1}^{k}(1 - \check{Q}_j) = 0.00524249 \approx 0.5242\%$$

□

A number of computer programs for reliability and fault tree analysis is based on the inclusion-exclusion principle. Among these is the fundamental KITT code (Kinetic Tree Theory) (see Vesely and Narum, 1970).

The inclusion-exclusion principle may also be applied to the minimal path sets $P_1, P_2, \ldots, P_p$. Let F_j denote the event that the components in the minimal path set P_j are all functioning; $j = 1, 2, \ldots, p$.

In this case the system reliability $p_S = 1 - Q_0$ is

$$p_S = \Pr\left(\bigcup_{j=1}^{p} F_j\right)$$

and

$$\Pr(F_j) = \prod_{i \in P_j} p_i$$

where p_i is the reliability of component i for $i = 1, 2, \ldots, n$. Successive upper and lower bounds of the system reliability p_S may be derived in the same way as we dealt with (4.57).

4.6 REDUNDANCY

In some structures, single items (components, subsystems) may be of much greater importance for the system's ability to function than others. If, for example, a single item is operating in series with the rest of the system, failure of this single item implies that the system fails. Two ways of ensuring higher system reliability in such situations are (1) use items with very high reliability in these critical places in the system, or (2) introduce *redundancy* in these places (i.e., introduce one or more reserve items). The type of redundancy obtained by replacing the important item with two or more items operating in parallel, is called *active redundancy*. These items then share the load right from the start until one of them fails.

The reserve items can also be kept in standby in such a way that the first of them is activated when the ordinary item fails, the second is activated when the first reserve item fails, and so on. If the reserve items carry no load in the waiting period before activation (and therefore cannot fail in this period), the redundancy is called *passive*. In the waiting period such an item is said to be in *cold* standby. If the standby items carry a weak load in the waiting period (and therefore might fail in this period), the redundancy is called *partly loaded*. In the following sections we will illustrate these types of redundancy by considering some simple examples.

4.6.1 Passive Redundancy, Perfect Switching, No Repairs

Consider the standby system in Fig. 4.14. The system functions in the following way: Item 1 is put into operation at time $t = 0$. When it fails, item 2 is activated. When it fails, item 3 is activated, and so forth. The item that is in operation is called the *active* item, while the items that are standing by ready to take over are called *standby* or *passive* items. When item n fails, the system fails.

Here we assume that the switch S functions perfectly and that items cannot fail while they are passive. Let T_i denote the time to failure of item i, for $i = 1, 2, \ldots, n$.

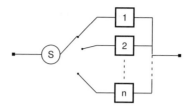

Fig. 4.14 Standby system with n items.

The lifetime, T, of the whole standby system is then

$$T = \sum_{i=1}^{n} T_i$$

The mean time to system failure, MTTF_S, is obviously

$$\mathrm{MTTF}_S = \sum_{i=1}^{n} \mathrm{MTTF}_i$$

where MTTF_i denotes the mean time to failure of item i, $i = 1, 2, \ldots, n$.

The exact distribution of the lifetime T can only be determined in some very special cases. Such a special case occurs when $T_1, T_2, \ldots, T_n$ are independent and exponentially distributed with failure rate λ. According to (2.40) T is gamma (Erlangian) distributed with parameters n and λ. The survivor function of the system is then

$$R_S(t) = \sum_{k=0}^{n-1} \frac{(\lambda t)^k}{k!} e^{-\lambda t} \qquad (4.65)$$

If we have only one standby item, such that $n = 2$, the survivor function is

$$R_S(t) = e^{-\lambda t} + \frac{\lambda t}{1!} e^{-\lambda t} = (1 + \lambda t) e^{-\lambda t} \qquad (4.66)$$

If we have two standby items (i.e., $n = 3$), the survivor function is

$$R_S(t) = e^{-\lambda t} + \frac{\lambda t}{1!} e^{-\lambda t} + \frac{(\lambda t)^2}{2!} e^{-\lambda t} = \left(1 + \lambda t + \frac{(\lambda t)^2}{2}\right) e^{-\lambda t} \qquad (4.67)$$

If we are unable to determine the exact distribution of T, we have to be content with an approximate expression for the distribution. Assume, for example, that the lifetimes $T_1, T_2, \ldots, T_n$ are independent and identically distributed with mean time to failure μ and variance σ^2. According to Lindeberg-Levy's central limit theorem (Dudewicz and Mishra, 1988, p. 316), when $n \to \infty$, T will be asymptotically normally distributed with mean $n\mu$ and variance $n\sigma^2$.

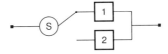

Fig. 4.15 Standby system with 2 items.

In this case the survivor function of the system may be approximated by

$$R_S(t) = \Pr\left(\sum_{i=1}^{n} T_i > t\right) = 1 - \Pr\left(\sum_{i=1}^{n} T_i \leq t\right)$$

$$= 1 - \Pr\left(\frac{\sum_{i=1}^{n} T_i - n\mu}{\sigma\sqrt{n}} \leq \frac{t - n\mu}{\sigma\sqrt{n}}\right) \approx \Phi\left(\frac{n\mu - t}{\sigma\sqrt{n}}\right)$$

where $\Phi(\cdot)$ denotes the distribution function of the standard normal distribution $\mathcal{N}(0, 1)$.

4.6.2 Cold Standby, Imperfect Switch, No Repairs

Here we will restrict ourselves to considering the simplest case with $n = 2$ items. Fig. 4.15 shows a standby system with an active item (item 1) and an item in *cold* standby (item 2). The active item is under surveillance by a switch, which activates the standby item when the active item fails.

Let us furthermore assume that the active item has constant failure rate λ_1. When the active item fails, the switch will activate the standby item. The probability that this switching is successful is denoted by $1 - p$. The failure rate of item 2 in standby position is assumed to be negligible. When the standby item is activated, its failure rate is λ_2. The three items operate independently. No repairs are carried out. In addition we assume that the only way in which the switch S can fail is by not activating the standby item when the active item fails. In many practical applications the switching will be performed by a human operator. The probability p of unsuccessful activation of the standby item will often include the probability of not being able to start the standby item.

The system is able to survive the interval $(0, t]$ in two *disjoint* ways.

1. Item 1 does *not* fail in $(0, t]$ (i.e., $T_1 > t$).

2. Item 1 fails in a time interval $(\tau, \tau + d\tau]$ where $0 < \tau < t$. The switch S is able to activate item 2. Item 2 is activated at time τ and does not fail in the time interval $(\tau, t]$.

Let T denote the time to system failure. Events 1 and 2 are clearly disjoint. Hence the survivor function of the system $R_S(t) = \Pr(T > t)$ will be the sum of the probability of the two events.

The probability of event 1 is

$$\Pr(T_1 > t) = e^{-\lambda_1 t}$$

Next consider event 2: Item 1 fails in $(\tau, \tau + d\tau]$ with probability $f_1(\tau)\, d\tau = \lambda_1 e^{-\lambda_1 \tau}\, d\tau$. The switch S is able to activate item 2 with probability $(1 - p)$.

Item 2 does *not* fail in $(\tau, t]$ with probability $e^{-\lambda_2(t-\tau)}$. Since item 1 may fail at any point of time τ in $(0, t]$, the survivor function of the system is when $\lambda_1 \neq \lambda_2$.

$$\begin{aligned}
R_S(t) &= e^{-\lambda_1 t} + \int_0^t (1-p)\, e^{-\lambda_2(t-\tau)} \lambda_1 e^{-\lambda_1 \tau}\, d\tau \\
&= e^{-\lambda_1 t} + (1-p)\lambda_1 e^{-\lambda_2 t} \int_0^t e^{-(\lambda_1-\lambda_2)\tau}\, d\tau \\
&= e^{-\lambda_1 t} + \frac{(1-p)\lambda_1}{\lambda_1 - \lambda_2} e^{-\lambda_2 t} - \frac{(1-p)\lambda_1}{\lambda_1 - \lambda_2} e^{-\lambda_1 t}
\end{aligned} \quad (4.68)$$

When $\lambda_1 = \lambda_2 = \lambda$, we get

$$\begin{aligned}
R_S(t) &= e^{-\lambda t} + \int_0^t (1-p) e^{-\lambda(t-\tau)} \lambda e^{-\lambda \tau}\, d\tau \\
&= e^{-\lambda t} + (1-p)\lambda e^{-\lambda t} \int_0^t d\tau \\
&= e^{-\lambda t} + (1-p)\lambda t e^{-\lambda t}
\end{aligned} \quad (4.69)$$

The MTTF$_S$ for the system is

$$\begin{aligned}
\text{MTTF}_S &= \int_0^\infty R_S(t)\, dt = \frac{1}{\lambda_1} + \frac{(1-p)\lambda_1}{\lambda_1 - \lambda_2}\left(\frac{1}{\lambda_2} - \frac{1}{\lambda_1}\right) \\
&= \frac{1}{\lambda_1} + (1-p)\frac{1}{\lambda_2}
\end{aligned} \quad (4.70)$$

This result applies for all values of λ_1 and λ_2.

Example 4.15
Consider the standby system in Fig. 4.15, composed of two identical pumps each with constant failure rate $\lambda = 10^{-3}$ failures/hour. The probability p that the switch S will fail to activate (switch over and start) the standby pump has been estimated to 1.5%.

The survivor function of the pump system at time t is, according to (4.69)

$$R_S(t) = (1 + (1-p)\lambda t) e^{-\lambda t} \quad (4.71)$$

Hence, the probability that this system survives 1.000 hours is

$$R_S(1.000) = 0.7302$$

The mean time to system failure is

$$\text{MTTF}_S = \frac{1}{\lambda}(1 + (1-p)) = 1985 \text{ hours}$$

□

4.6.3 Partly Loaded Redundancy, Imperfect Switch, No Repairs

Consider the same standby system as the one in Fig. 4.15, but change the assumptions so that item 2 carries a certain load before it is activated. Let λ_0 denote the failure rate of item 2 while in partly loaded standby. The system is able to survive the interval $(0, t]$ in two disjoint ways.

1. Item 1 does *not* fail in $(0, t]$ (i.e., $T_1 > t$).

2. Item 1 fails in a time interval $(\tau, \tau + d\tau)$, where $0 < \tau < t$. The switch S is able to activate item 2. Item 2 does not fail in $(0, \tau]$, is activated at time τ, and does not fail in $(\tau, t]$.

Let T denote the time to system failure. The survivor function of the system, $R_S(t) = \Pr(T > t)$, will be the sum of the probabilities for the two events, since they are disjoint.

Consider event 2: Item 1 fails in $(\tau, \tau+d\tau]$ with probability $f_1(\tau) d\tau = \lambda_1 e^{-\lambda_1 \tau} d\tau$. The switch S is able to activate item 2 with probability $1 - p$. Item 2 does not fail in $(0, \tau]$ in partly loaded standby with probability $e^{-\lambda_0 \tau}$, and item 2 does not fail in $(\tau, t]$ in active state with probability $e^{-\lambda_2(t-\tau)}$.

Since item 1 may fail at any point of time t in $(0, \tau]$, the survivor function of the system becomes

$$R_S(t) = e^{-\lambda_1 t} + \int_0^t (1-p) e^{-\lambda_0 \tau} e^{-\lambda_2(t-\tau)} \lambda_1 e^{-\lambda_1 \tau} d\tau$$

$$= e^{-\lambda_1 t} + \frac{(1-p)\lambda_1}{\lambda_0 + \lambda_1 - \lambda_2} \left(e^{-\lambda_2 t} - e^{-(\lambda_0+\lambda_1)t} \right) \quad (4.72)$$

where we have assumed that $(\lambda_1 + \lambda_0 - \lambda_2) \neq 0$.

When $(\lambda_1 + \lambda_0 - \lambda_2) = 0$, the survivor function becomes

$$R_S(t) = e^{-\lambda_1 t} + (1-p)\lambda_1 t e^{-\lambda_2 t} \quad (4.73)$$

The mean time to system failure is

$$\text{MTTF}_S = \frac{1}{\lambda_1} + \frac{(1-p)\lambda_1}{\lambda_1 + \lambda_0 - \lambda_2} \left(\frac{1}{\lambda_2} - \frac{1}{\lambda_1 + \lambda_0} \right)$$

$$= \frac{1}{\lambda_1} + (1-p)\frac{\lambda_1}{\lambda_2(\lambda_1 + \lambda_0)} \quad (4.74)$$

This result applies for all values of λ_0, λ_1, and λ_2. In this section we have tacitly made certain assumptions about independence. These assumptions will not be discussed thoroughly here. An introduction to standby redundancy is given by Trivedi (1982), Billinton and Allen (1983), and Endrenyi (1978). A more detailed discussion is presented by Ravichandran (1990). The concept of redundancy is also discussed in Chapter 8, where we use Markov models to study repairable as well as nonrepairable standby systems.

PROBLEMS

4.1 Show that when the components are independent, the system reliability $p_S(t)$ may be written as (4.5), that is, a function of the component reliabilities, $p_i(t)$ ($i = 1, 2, \ldots, n$), only.

4.2 An old-fashioned string of Christmas tree lights has 10 bulbs connected in series. The 10 identical bulbs are assumed to have independent life times with constant failure rate λ. Determine λ such that the probability that the string survives 3 weeks is at least 99%.

4.3 Consider three identical items in parallel. What is the system reliability if each item has a reliability of 98%?

4.4 A system consists of five identical components connected in parallel. Determine the reliability of the components such that the system reliability is 97%.

4.5 A system must have a reliability of 99%. How many components are required in parallel when each component has a reliability of 65%?

4.6 A plant has two identical and parallel process streams A and B. Each process stream has a transfer pump and a rotary filter as shown in Fig. 4.16. Both process streams have to be functioning to secure full production. [This problem is adapted from an example in Henley and Kumamoto (1981, p. 305)] To improve the sys-

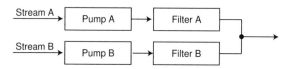

Fig. 4.16 Two parallel process streams (system 1).

tem availability it has been proposed to install an extra process stream, as shown in Fig. 4.17.

System 2 has full production when at least two of the three process streams are functioning (i.e., a 2-out-of-3 system). It is assumed that the pumps and the filters are functioning and repaired independent of each other. The average availability of a pump has been estimated to be 99.2% while the average availability of a filter is 96.8%.

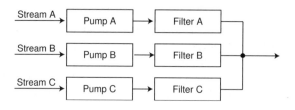

Fig. 4.17 Three parallel process streams (system 2).

(a) Determine the average availability with respect to full production for the two systems. (Note that full production is achieved when at least two process streams are functioning).

(b) Assume that the total cost of a pump is $15 per day (including installation, operation, and maintenance). The total cost of a filter is estimated to $60 per day. The company gets a penalty of $10000 per day when the system is not able to give full production. Which of the two systems would you choose to minimize the cost?

4.7 Consider a 2-out-of-3 system of independent and identical components with constant failure rate λ.

(a) Find the probability density function of the time to failure T of the 2-out-of-3 system.

(b) Find the mode, the median, and the MTTF for the time to failure T.

(c) Make a sketch of the probability density function, and note where the mode, the median, and the MTTF are located.

(d) Find the MRL of the 2-out-of-3 system at age t. Make a sketch of the function $g(t) = \text{MRL}(t)/\text{MTTF}$. Find $\lim_{t \to \infty} g(t)$ and give a physical interpretation of this limit.

4.8 Consider the failure rate function $z_S(t)$ for a 2-out-of-3 structure of independent and identical components with constant failure rate λ that is illustrated in Fig. 4.5. Show that $\lim_{t \to \infty} z_S(t) = 2\lambda$, and give a physical explanation of why this is a realistic limit.

4.9 Consider a coherent structure of n independent components with system survivor function $R_S(t) = h(R_1(t), R_2(t), \ldots, R_n(t))$. Assume that all the n components have life distributions with increasing failure rate (IFR), and that the mean time to failure of component i is $\text{MTTF}_i = \mu_i$, for $i = 1, 2, \ldots, n$. Show that

$$R_S(t) \geq h\left(e^{-t/\mu_1}, e^{-t/\mu_2}, \ldots, e^{-t/\mu_n}\right) \quad \text{for } 0 < t < \min\{\mu_1, \mu_2, \ldots, \mu_n\}$$

Hint: Use the result in Problem 2.32.

4.10 Consider a series structure with two "independent" components B and C with constant failure rates $\lambda_B = 5.0 \cdot 10^{-4}$ failures per hour and $\lambda_C = 3.0 \cdot 10^{-3}$ failures per hour, respectively. The system is put into operation and is functioning at time $t = 0$. Each time the system fails, the component which is responsible for the failure is repaired to an "as good as new" condition. When one of the components fails, the load on the other component will disappear, and this component will consequently not fail when the failed component is "down." The mean repair time of component B is $\tau_B = 5$ hours, while the mean repair time of component C is $\tau_C = 10$ hours.

(a) Show that the mean repair time of the *system* is

$$\text{MTTR} = \frac{\lambda_B \tau_B + \lambda_C \tau_C}{\lambda_B + \lambda_C}$$

(b) Determine the average availability A of the system.

(c) Since the failure of one component prevents the other component from failing, the states of the components are not fully independent of each other. Determine the average availability A_I of the system when you assume (wrongly) that the components are fully independent, and compare with the average availability A which was found in (a). Discuss the difference between A and A_I.

4.11 Consider a series structure of 10 independent and identical components. Each component has an MTTF = 5000 hours.

(a) Determine MTTF$_{\text{system}}$ when the components have constant failure rates.

(b) Assume next that the life lengths of the components are Weibull distributed with shape parameter $\alpha = 2.0$. Determine MTTF$_{\text{system}}$ and compare with (a).

(c) Assume now that you have a parallel structure of two independent and identical components with MTTF = 5000 hours. Repeat problems (a) and (b) and discuss the results.

4.12 A process plant needs a regular supply of high-pressure steam. If the steam supply is shut down, the process must also be shut down immediately. The start-up procedures are rather time-consuming. A shutdown of the steam supply may therefore imply significant consequences. The steam producing system comprises three identical steam vessels. The three vessels are physically separated and have separate control and supply systems. The three vessels may thus be considered as independent items.

Consider first only one of the vessels. Assume that the vessel has a constant failure rate $2.02 \cdot 10^{-4}$ failures per hour when it is operated with normal capacity. Failure is in this context defined to be a spurious shutdown of the steam supply from the vessel.

(a) Find the MTTF of the vessel.

(b) Find the probability that the vessel will survive a period of 4 months without failure.

The failure rate of the vessel depends on the capacity at which the vessel is operated. It has been estimated that

$$\lambda_p = \lambda_0 + 2p(p - 1.6)\lambda_1 \quad \text{for } 0 < p \leq 1$$

where λ_p denotes the failure rate when the vessel is operated at $100p\%$ capacity, and

$$\lambda_0 = 6.5 \cdot 10^{-4} \text{ failures per hour}$$
$$\lambda_1 = 3.5 \cdot 10^{-4} \text{ failures per hour}$$

(c) Make a sketch of the failure rate λ_p as a function of p. In particular, write down the failure rate when the capacity is 40, 80, and 100%.

Each vessel is equipped with four independent burner elements. When a burner has been started (ignited), it has a constant failure rate $\lambda_b = 5.0 \cdot 10^{-5}$ failure per hour. When a vessel is operated at 80% capacity, all the four burner elements are normally active. It is, however, sufficient that three of the four burner elements are functioning.

(d) Consider a vessel which is operated at 80% capacity where all the four burner elements are active at time $t = 0$. Determine the probability that the burner system survives a period of 2 months without any burner element failures, when no repair is carried out.

The probability that a passive burner cannot be started has been estimated to be 3%.

(e) Determine the probability that a standby (passive) vessel may be started and survives a period of 2 months, when the vessel is operated at 80% capacity, and no repair is carried out.

The process needs a steam supply corresponding to 80% capacity of *one* vessel. Based on regularity arguments, two vessels are, however, normally active at 40% capacity each. The third vessel remains in *cold* standby. The changeover from passive to active operation is normally carried out without other problems than those connected to the start-up of the burner elements. When a vessel is operated at 40% capacity, only two of the four burner elements are used. Two burners are also necessary for the operation. If one of the two active vessels fails, the following procedure is used:

- The capacity of the functioning vessel is increased to 80% if possible.

- The standby vessel is started, if possible, by starting two burner elements (40% capacity).

(f) Assume that one of the active vessels fails. Determine the probability that the capacity of the other active vessel can be increased from 40 to 80%, when you know that the two other burner elements in the same vessel have been passive since the last major overhaul of the vessel.

5
Component Importance

5.1 INTRODUCTION

From the preceding chapters it should be obvious that some components in a system are more important for the system reliability than other components. A component in series with the rest of the system is a cut set of order 1, and is generally more important than a component that is a member of a cut set of higher order. In this chapter a number of component importance measures are defined and discussed. The importance measures may be used to *rank* the components, that is, to arrange the components in order of increasing or decreasing importance. The measures may also be used for *classification* of importance, that is, to allocate the components into two or more groups, according to some preset criteria.

Throughout this chapter we will consider a system of n independent components with component reliabilities $p_i(t)$, for $i = 1, 2, \ldots, n$. The system reliability with respect to a specified system function is denoted $h(\boldsymbol{p}(t))$. When we discuss *component importance*, the importance is always seen in relation to the specified system function. As an example, consider a process shutdown system. The essential function of the system is to shut down the process on demand in case of a significant process disturbance or an emergency. Another function for the shutdown system is to distinguish real demand signals from false signals and prevent spurious process shutdowns. When we talk about the importance of a component in the process safety system, we have to specify which of the two functions we are considering. A component may be very important for the essential shutdown function but may have little, or no, importance for the other system functions.

The following component importance measures are defined and discussed in this chapter:

1. Birnbaum's measure (and some variants)
2. The improvement potential measure (and some variants)
3. Risk achievement worth
4. Risk reduction worth
5. The criticality importance measure
6. Fussell-Vesely's measure

Several other measures are defined and described by Lambert (1975) and Henley and Kumamoto (1981).

The various measures are based on slightly different interpretations of the concept *component importance*. Intuitively, the importance of a component should depend on two factors:

- The location of the component in the system
- The reliability of the component in question

and, perhaps, also the uncertainty in our estimate of the component reliability.

How a component importance measure is applied depends on the phase in the system's life cycle. In the system design phase, the importance measure may be used to identify weak points (bottlenecks) and components that should be improved to improve the system reliability. The reliability of a component may be improved by using a higher quality component, by introducing redundant components, by reducing the operational and environmental loads on the component, or by improving the maintainability of the component. To chose the optimal improvement is a complex task and will not be discussed any further in this chapter. The objective of the component importance measure is to help the designer to identify the components that should be improved and rank these components in order of importance. In the operational phase, the component importance measure may be used to allocate inspection and maintenance resources to the most important components. The measure may also be used to identify components that should be modified or replaced with higher quality components.

Component importance measures are commonly used in risk assessments and especially within probabilistic risk assessments of nuclear power plants (e.g., see EPRI 1995). In these applications the component importance measures are often called *risk importance measures* and are mainly used to identify components and subsystems that should be improved to reduce the risk and to identify components and subsystems for risk-based in service inspection and testing (e.g., see Cheok et al. 1998; Blakey et al. 1998). The risk importance measures are similar to the component importance measures but are defined within a risk analysis terminology that is slightly different from the terminology used in this book.

The various component importance measures are defined and discussed in the following sections. Some numerical examples are presented in Section 5.8 together with a brief comparison of the measures.

5.2 BIRNBAUM'S MEASURE

Birnbaum (1969) proposed the following measure of the reliability importance of a component:

Definition 5.1 Birnbaum's measure of importance[1] of component i at time t is

$$I^B(i \mid t) = \frac{\partial h(\boldsymbol{p}(t))}{\partial p_i(t)} \quad \text{for } i = 1, 2, \ldots, n. \tag{5.1}$$

□

Birnbaum's measure is thus obtained by partial differentiation of the system reliability with respect to $p_i(t)$. This approach is well known from classical sensitivity analysis. If $I^B(i \mid t)$ is large, a small change in the reliability of component i will result in a comparatively large change in the system reliability at time t.

By using the fault tree notation introduced in Section 4.4,

$$\begin{aligned} q_i(t) &= 1 - p_i(t) \quad \text{for } i = 1, 2, \ldots, n \\ Q_0(t) &= 1 - h(\boldsymbol{p}(t)) \end{aligned} \tag{5.2}$$

we observe that (5.1) may be written

$$I^B(i \mid t) = \frac{\partial Q_0(t)}{\partial q_i(t)} \quad \text{for } i = 1, 2, \ldots, n \tag{5.3}$$

In Section 4.2 we used *pivotal decomposition* to show that the system reliability $h(\boldsymbol{p}(t))$ may be written as a linear function of $p_i(t)$ for $i = 1, 2, \ldots, n$ when the n components are independent.

$$\begin{aligned} h(\boldsymbol{p}(t)) &= p_i(t) \cdot h(1_i, \boldsymbol{p}(t)) + (1 - p_i(t)) \cdot h(0_i, \boldsymbol{p}(t)) \\ &= p_i(t) \cdot [h(1_i, \boldsymbol{p}(t)) - h(0_i, \boldsymbol{p}(t))] - h(0_i, \boldsymbol{p}(t)) \end{aligned} \tag{5.4}$$

where $h(1_i, \boldsymbol{p}(t))$ denotes the (conditional) probability that the system is functioning when it is known that component i is functioning at time t, and $h(0_i, \boldsymbol{p}(t))$ denotes the (conditional) probability that the system is functioning when component i is in a failed state at time t.

Birnbaum's measure of the reliability importance of component i at time t can thus be written as

$$I^B(i \mid t) = \frac{\partial h(\boldsymbol{p}(t))}{\partial p_i(t)} = h(1_i, \boldsymbol{p}(t)) - h(0_i, \boldsymbol{p}(t)) \tag{5.5}$$

---

[1] Named after the Hungarian-American professor Zygmund William Birnbaum (1903–2000).

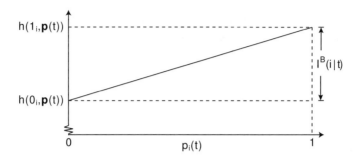

Fig. 5.1 Illustration of Birnbaum's measure of reliability importance.

This alternative expression for Birnbaum's measure is illustrated in Fig. 5.1.

Remark: Note that Birnbaum's measure $I^B(i \mid t)$ of component i only depends on the structure of the system and the reliabilities of the other components. $I^B(i \mid t)$ is independent of the actual reliability $p_i(t)$ of component i. This may be regarded as a weakness of Birnbaum's measure. □

Example 5.1
Consider a series structure of two independent components 1 and 2. Let p_i denote the reliability of component i, and assume that p_i is not a function of the time t, for $i = 1, 2$. When component 1 is functioning ($X_1 = 1$), the system is functioning if and only if component 2 is functioning, in which case the reliability is $h(1_1, \boldsymbol{p}) = p_2$. When component 1 is in a failed state, the system will be in a failed state irrespective of the state of component 2, in which case $h(0_1, \boldsymbol{p}) = 0$. Birnbaum's measure of component 1 is therefore from (5.5)

$$I^B(1) = h(1_1, \boldsymbol{p}) - h(0_1, \boldsymbol{p}) = p_2$$

□

This procedure of determining Birnbaum's measure is in many cases more simple to calculate than (5.1) and is therefore used in some computer programs.

We will now show a third way of expressing Birnbaum's measure. In Section 4.2 we saw that $h(\cdot_i, \boldsymbol{p}(t)) = E[\phi(\cdot_i, \boldsymbol{X}(t))]$, such that (5.5) can be written

$$\begin{aligned} I^B(i \mid t) &= E[\phi(1_i, \boldsymbol{X}(t))] - E[\phi(0_i, \boldsymbol{X}(t))] \\ &= E[\phi(1_i, \boldsymbol{X}(t)) - \phi(0_i, \boldsymbol{X}(t))] \end{aligned}$$

When $\phi(\boldsymbol{X}(t))$ is a coherent structure, $[\phi(1_i, \boldsymbol{X}(t)) - \phi(0_i, \boldsymbol{X}(t))]$ can only take on the values 0 and 1. Birnbaum's measure (5.5) can therefore be written as

$$I^B(i \mid t) = \Pr(\phi(1_i, \boldsymbol{X}(t)) - \phi(0_i, \boldsymbol{X}(t)) = 1) \tag{5.6}$$

This is to say that $I^B(i \mid t)$ is equal to the probability that $(1_i, \boldsymbol{X}(t))$ is a *critical path vector* for component i at time t (see Definition 3.6).

When $(1_i, \boldsymbol{X}(t))$ is a critical path vector for component i, we often, for the sake of brevity, say that component i is *critical* for the system. The state of component i will, in other words, be decisive for whether or not the system functions at time t.

Note that the fact that component i is critical for the system, tells nothing about the state of component i. The statement concerns the other components of the system only. If component i is critical for the system, then component i must either be a cut set of order 1, or be a member of a cut set where all the other components in the same cut set have failed.

An alternative definition of Birnbaum's measure is thus:

Definition 5.2 Birnbaum's measure of reliability importance of component i at time t is equal to the probability that the system is in such a state at time t that component i is critical for the system. □

Note that this definition of reliability importance also can be used when the components are *dependent*.

Example 5.2
Reconsider the series structure in Example 5.4 where component 1 is *critical* for the system if, and only if, component 2 is functioning. Birnbaum's measure of component 1 is therefore from Definition 5.2:

$$\begin{aligned} I^B(1) &= \Pr(\text{Component 1 is critical for the system}) \\ &= \Pr(\text{Component 2 is functioning}) = p_2 \end{aligned}$$

□

Birnbaum's Measure of Structural Importance Birnbaum's measure of the structural importance $B_\phi(i)$ of component i was defined in Section 3.5. We will now show how $B_\phi(i)$ can be found from Birnbaum's measure of reliability importance.

Let the reliabilities $p_j(t) = \frac{1}{2}$ for all $j \neq i$. The different realizations of the stochastic vector

$$(\cdot_i, \boldsymbol{X}(t)) = (X_1(t), \ldots, X_{i-1}(t), \cdot, X_{i+1}(t), \ldots, X_n(t))$$

will then all have the probability

$$\frac{1}{2^{n-1}}$$

COMPONENT IMPORTANCE

since the $X_i(t)$'s are assumed to be independent. Then

$$\begin{aligned}
I^B(t) &= E(\phi(1_i, \boldsymbol{X}(t)) - \phi(0_i, \boldsymbol{X}(t))) \\
&= \sum_{(\cdot_i, \boldsymbol{x})} (\phi(1_i, \boldsymbol{x}) - (\phi(0_i, \boldsymbol{x})) \cdot \Pr(\boldsymbol{X}(t) = \boldsymbol{x}) \\
&= \frac{1}{2^{n-1}} \sum_{(\cdot_i, \boldsymbol{x})} (\phi(1_i, \boldsymbol{x}) - \phi(0_i, \boldsymbol{x})) \\
&= \frac{\eta_\phi}{2^{n-1}} = B_\phi(i)
\end{aligned} \qquad (5.7)$$

where $\eta_\phi(i)$ is defined as in Section 3.11.

Thus we have shown that when all the component reliabilities $p_j(t) = \frac{1}{2}$ for $j \neq i$, then Birnbaum's measure of reliability importance of component i and his measure of structural importance for component i coincide.

$$B_\phi(i) = I^B(t)\big|_{p_j(t)=1/2,\ j\neq i} = \frac{\partial h(\boldsymbol{p}(t))}{\partial p_i(t)}\bigg|_{p_j(t)=1/2, j\neq i} \qquad (5.8)$$

Equation (5.8) is often used to calculate structural importance.

Remarks

1. Assume that component i has failure rate λ_i. In some situations we may be interested in measuring how much the system reliability will change by making a small change to the failure rate λ_i. The sensitivity of the system reliability with respect to changes in λ_i can obviously be measured by

$$\frac{\partial h(\boldsymbol{p}(t))}{\partial \lambda_i} = \frac{\partial h(\boldsymbol{p}(t))}{\partial p_i(t)} \cdot \frac{\partial p_i(t)}{\partial \lambda_i} = I^B(i \mid t) \cdot \frac{\partial p_i(t)}{\partial \lambda_i} \qquad (5.9)$$

   A similar measure can be used for all parameters related to the component reliability $p_i(t)$, for $i = 1, 2, \ldots, n$. In some cases, several components in a system will have the same failure rate λ. To find the sensitivity of the system reliability with respect to changes in λ, we can still use $\partial h(\boldsymbol{p}(t))/\partial \lambda$.

2. Consider a system where component i has reliability $p_i(t)$ that is a function of a parameter θ_i. The parameter θ_i may be the failure rate, the repair rate, or the test frequency, of component i. To improve the system reliability, we may want to change the parameter θ_i (by buying a higher quality component or changing the maintenance strategy). Assume that we are able to determine the cost of the improvement as a function of θ_i, that is, $c_i = c(\theta_i)$, and that this function is strictly increasing or decreasing such that we can find its inverse function. The effect of an extra investment related to component i may now be measured by

$$\frac{\partial h(\boldsymbol{p}(t))}{\partial c_i} = \frac{\partial h(\boldsymbol{p}(t))}{\partial \theta_i} \cdot \frac{\partial \theta_i}{\partial c_i} = I^B(i \mid t) \cdot \frac{\partial p_i(t)}{\partial \theta_i} \cdot \frac{\partial \theta_i}{\partial c_i}$$

3. In a practical reliability study of a complex system, one of the most time-consuming tasks is to find adequate estimates for the input parameters (failure rates, repair rates, etc.). In some cases, we may start with rather rough estimates, calculate Birnbaum's measure of importance for the various components, or the parameter sensitivities (5.9), and then spend the most time finding high-quality data for the most important components. Components with a very low value of Birnbaum's measure will have a negligible effect on the system reliability, and extra efforts finding high-quality data for such components may be considered a waste of time.

5.3 IMPROVEMENT POTENTIAL

Consider a system with reliability $h(\boldsymbol{p}(t))$ at time t. In some cases it may be of interest to know how much the system reliability increases if component i ($i = 1, 2, \ldots, n$) is replaced by a perfect component, that is, a component such that $p_i(t) = 1$. The difference between $h(1_i, \boldsymbol{p}(t))$ and $h(\boldsymbol{p}(t))$ is called the *improvement potential* (IP) with respect to component i and denoted by $I^{\text{IP}}(i \mid t)$.

Definition 5.3 The improvement potential with respect to component i at time t is

$$I^{\text{IP}}(i \mid t) = h(1_i, \boldsymbol{p}(t)) - h(\boldsymbol{p}(t)) \quad \text{for } i = 1, 2, \ldots, n \tag{5.10}$$

□

Birnbaum's measure of importance, $I^{\text{B}}(i \mid t)$ is from (5.5) seen to be the slope of the line in Fig. 5.1 and can alternatively be expressed as

$$I^{\text{B}}(i \mid t) = \frac{h(1_i, \boldsymbol{p}(t)) - h(\boldsymbol{p}(t))}{1 - p_i(t)} \quad \text{for } i = 1, 2, \ldots, n \tag{5.11}$$

The improvement potential with respect to component i is then

$$I^{\text{IP}}(i \mid t) = I^{\text{B}}(i \mid t) \cdot (1 - p_i(t)) \tag{5.12}$$

or, by using the fault tree notation (5.2)

$$I^{\text{IP}}(i \mid t) = I^{\text{B}}(i \mid t) \cdot q_i(t) \tag{5.13}$$

Remark: The improvement potential of component i is the difference between the system reliability with a *perfect* component i, and the system reliability with the actual component i. In practice, it is not possible to improve the reliability $p_i(t)$ of component i to 100% reliability. Let us assume that it is possible to improve $p_i(t)$ to new value $p_i^{(n)}(t)$ representing, for example, the state of the art for this type of components. We may then calculate the realistic or *credible improvement potential* (CIP) of component i at time t, defined by

$$I^{\text{CIP}}(i \mid t) = h(p_i^{(n)}(t), \boldsymbol{p}(t)) - h(\boldsymbol{p}(t)) \tag{5.14}$$

where $h(p_i^{(n)}(t), \boldsymbol{p}(t))$ denotes the system reliability when component i is replaced by a new component with reliability $p_i^{(n)}(t)$. Because the system reliability $h(\boldsymbol{p}(t))$ is a linear function of $p_i(t)$, and because Birnbaum's measure is the slope of the line in Fig. 5.1, we can write (5.14) as

$$I^{\mathrm{CIP}}(i \mid t) = I^{\mathrm{B}}(i \mid t) \cdot \left(p_i^{(n)}(t) - p_i(t)\right) \qquad (5.15)$$

Schmidt et al. (1985) introduced the concept of a *generalized importance measure* (GIM) that may be written as

$$I^{\mathrm{GIM}}(i \mid t) = \frac{I^{\mathrm{CIP}}(i \mid t)}{1 - h(\boldsymbol{p}(t))}$$

Cheok et al. (1998) rearranged this expression and introduced a new *general importance measure* given by (with our notation):

$$I^{\mathrm{GIM-2}}(i \mid t) = \frac{1 - h(p_i^{(n)}(t), \boldsymbol{p}(t))}{1 - h(\boldsymbol{p}(t))}$$

which may be written as

$$I^{\mathrm{GIM-2}}(i \mid t) = 1 + I^{\mathrm{GIM}}(i \mid t)$$

The three importance measures $I^{\mathrm{CIP}}(i \mid t)$, $I^{\mathrm{GIM}}(i \mid t)$, and $I^{\mathrm{GIM-2}}(i \mid t)$ will give approximately the same information but different numerical values. All the three measures can be considered to be *general* since the new component reliability $p_i^{(n)}(t)$ is not restricted to a value of 0 or 1 as was the case for the improvement potential (and also for the measures risk reducing worth and the risk achievement worth that are introduced in the following sections). It is therefore possible to plot the measures as a function of $p_i^{(n)}(t)$ to illustrate the effect of changes in $p_i^{(n)}(t)$. Such a curve is called a *risk importance curve* by Cheok et al. (1998). □

5.4 RISK ACHIEVEMENT WORTH

Risk achievement worth (RAW) has been introduced as a risk importance measure in probabilistic safety assessments of nuclear power stations (EPRI 1995). The RAW is defined by using risk analysis terminology (e.g., see Cheok et al. 1998). Here we present the measure using reliability terminology.

Definition 5.4 The importance measure *risk achievement worth* (RAW) of component i at time t is

$$I^{\mathrm{RAW}}(i \mid t) = \frac{1 - h(0_i, \boldsymbol{p}(t))}{1 - h(\boldsymbol{p}(t))} \quad \text{for } i = 1, 2, \ldots, n \qquad (5.16)$$

□

The RAW, $I^{\text{RAW}}(i \mid t)$, is the ratio of the (conditional) system unreliability if component i is not present (or if component i is always failed) with the actual system unreliability. The RAW presents a measure of the *worth* of component i in achieving the present level of system reliability and indicates the importance of maintaining the current level of reliability for the component.

Example 5.3
Consider a process safety system with reliability $h(\mathbf{p})$ at time t. In many applications we study the long-term, or steady-state, situation such that $h(\mathbf{p})$ is independent of time. Let us assume that the demands for the safety system occurs at random with frequency ν_0. An accident occurs if the safety system does not function when a demand occurs. The accident frequency is therefore $\nu_{\text{acc}} = \nu_0 \cdot (1 - h(\mathbf{p}))$ (see Section 8.2 for a detailed explanation). Let us now assume that component i in the safety system is known not to be functioning. This may, for example, be because component i is disconnected for maintenance or some other reason. The (conditional) system unreliability is now $1 - h(0_i, \mathbf{p})$ and the accident frequency is $\nu_{\text{acc}}^i = \nu_0 \cdot (1 - h(0_i, \mathbf{p}))$. We may then write

$$\nu_{\text{acc}}^i = \frac{1 - h(0_i, \mathbf{p})}{1 - h(\mathbf{p})} \cdot \nu_{\text{acc}} = I^{\text{RAW}}(i) \cdot \nu_{\text{acc}}$$

If we disconnect component (or subsystem) i, the accident frequency will increase with a factor of $I^{\text{RAW}}(i)$ [and the unreliability of the safety system will increase with a factor of $I^{\text{RAW}}(i)$].

In nuclear applications, we are primarily interested in the core damage frequency (CDF). If we disconnect item i from the main safety system, the core damage frequency becomes

$$\text{CDF}^i = I^{\text{RAW}}(i) \cdot \text{CDF}_0$$

where CDF_0 denotes the base core damage frequency. $\square$

5.5 RISK REDUCTION WORTH

Risk reduction worth (RRW) is, like RAW introduced in Section 5.4, mainly used as a risk importance measure in probabilistic safety assessments of nuclear power stations (EPRI 1995).

Definition 5.5 The importance measure *risk reduction worth* (RRW) of component i at time t is

$$I^{\text{RRW}}(i \mid t) = \frac{1 - h(\mathbf{p}(t))}{1 - h(1_i, \mathbf{p}(t))} \quad \text{for } i = 1, 2, \ldots, n \tag{5.17}$$

$\square$

The RRW, $I^{\text{RRW}}(i \mid t)$, is the ratio of the actual system unreliability with the (conditional) system unreliability if component i is replaced by a perfect component with

$p_i(t) \equiv 1$. In some applications, failure of a "component" may be an operator error or some external event. If such "components" can be removed from the system, for example, by canceling an operator intervention, this may be regarded as replacement with a perfect component.

We notice that $I^{\text{RRW}}(i \mid t) \geq 1$, and that

$$I^{\text{RRW}}(i \mid t) = \left(1 - \frac{I^{\text{IP}}(i \mid t)}{1 - h(p(t))}\right)^{-1}$$

By fault tree notation and (5.13) we get

$$I^{\text{RRW}}(i \mid t) = \left(1 - \frac{I^{\text{IP}}(i \mid t)}{Q_0(t)}\right)^{-1} = \left(1 - \frac{I^{\text{B}}(i \mid t) \cdot q_i(t)}{Q_0(t)}\right)^{-1} \quad (5.18)$$

The RRW may therefore be expressed by the improvement potential or by Birnbaum's measure.

Example 5.3 (Cont.)
Reconsider the process safety system in Example 5.3 with reliability $h(p)$. Let us assume that we contemplate improving component i and would like to know the maximum potential improvement by replacing component i with a perfect component with reliability $p_i \equiv 1$. The (conditional) system reliability will then be $1 - h(1_i, p)$, which we can use (5.17) to express as

$$1 - h(1_i, p) = \frac{1 - h(p)}{I^{\text{RRW}}(i)}$$

If we, as an example, find that $I^{\text{RRW}}(i) = 2$, than the system unreliability we would obtain by replacing component i with a perfect component would be 50% of the initial unreliability $1 - h(p)$. □

5.6 CRITICALITY IMPORTANCE

Criticality importance (CI) is a component importance measure that is particularly suitable for prioritizing maintenance actions. Criticality importance is related to Birnbaum's measure. As a motivation for the definition of criticality importance, we recall from page 187 that component i is *critical* for the system if the other components of the system are in such states that the system is functioning if and only if component i is functioning. To say that component i is critical is thus a statement about the other components in the system, and not a statement about component i.

Let $C(1_i, X(t))$ denote the event that the system at time t is in a state where component i is critical. According to (5.6), the probability of this event is equal to Birnbaum's measure of component i at time t.

$$\Pr(C(1_i, X(t))) = I^{\text{B}}(i \mid t) \quad (5.19)$$

Since the components of the system are independent, the event $C(1_i, X(t))$ will be independent of the state of component i at time t.

The probability that component i is critical for the system and at the same time is failed at time t is then

$$\Pr(C(1_i, X(t)) \cap (X_i(t) = 0)) = I^{\mathrm{B}}(i \mid t) \cdot (1 - p_i(t)) \tag{5.20}$$

Let us now assume that we know that the system is in a failed state at time t, that is, $\phi(X(t)) = 0$. The conditional probability (5.19) when we know that the system is in a failed state at time t is then

$$\Pr(C(1_i, X(t)) \cap (X_i(t) = 0) \mid \phi(X(t)) = 0) \tag{5.21}$$

Since the event $C(1_i, X(t)) \cap (X(t) = 0)$ implies that $\phi(X(t)) = 0$, we can use (5.20) to get

$$\frac{\Pr(C(1_i, X(t)) \cap (X_i(t) = 0))}{\Pr(\phi(X(t)) = 0)} = \frac{I^{\mathrm{B}}(i \mid t) \cdot (1 - p_i(t))}{1 - h(\boldsymbol{p}(t))} \tag{5.22}$$

This result is called the criticality importance, and we give the formal definition as:

Definition 5.6 The component importance measure *criticality importance* $I^{\mathrm{CR}}(i \mid t)$ of component i at time t is the probability that component i is critical for the system and is failed at time t, when we know that the system is failed at time t.

$$I^{\mathrm{CR}}(i \mid t) = \frac{I^{\mathrm{B}}(i \mid t) \cdot (1 - p_i(t))}{1 - h(\boldsymbol{p}(t))} \tag{5.23}$$

□

By using the fault tree notation (5.2), $I^{\mathrm{CR}}(i \mid t)$ may be written

$$I^{\mathrm{CR}}(i \mid t) = \frac{I^{\mathrm{B}}(i \mid t) \cdot q_i(t)}{Q_0(t)} \tag{5.24}$$

The criticality importance $I^{\mathrm{CR}}(i \mid t)$ is in other words the probability that component i has *caused* system failure, when we know that the system is failed at time t. For component i to cause system failure, component i must be critical, and then fail. Component i will then, by failing, cause the system to fail. When component i is repaired, the system will start functioning again. This is why the criticality importance measure may be used to prioritize maintenance actions in complex systems.

Remark: We notice from (5.24) that the criticality importance $I^{\mathrm{CR}}(i \mid t)$ is close to a linear function of $q_i(t)$, at least for systems with a high level of redundancy. This is because Birnbaum's measure is not a function of $q_i(t)$, and because $q_i(t)$ will have a rather low influence on $Q_0(t)$ in highly redundant systems. As seen from Examples 5.4 and 5.5, the linearity is not adequate for very simple systems with two components.

From (5.18) we observe that the RRW can be expressed as a function of criticality importance:

$$I^{\text{RRW}}(i \mid t) = \left(1 - I^{\text{CR}}(i \mid t)\right)^{-1} = \frac{1}{1 - I^{\text{CR}}(i \mid t)} \qquad (5.25)$$

□

5.7 FUSSELL-VESELY'S MEASURE

J. B. Fussell and W. Vesely suggested the following measure of the importance of component i (see Fussell 1975):

Definition 5.7 Fussell-Vesely's measure of importance, $I^{\text{FV}}(i \mid t)$ is the probability that at least one minimal cut set that contains component i is failed at time t, given that the system is failed at time t. □

We say that a minimal cut set is failed when all the components in the minimal cut set are failed.

Fussell-Vesely's measure takes into account the fact that a component may contribute to system failure without being critical. The component contributes to system failure when a minimal cut set, containing the component, is failed.

Consider a system with k minimal cut sets $K_1, K_2, \ldots, K_k$. According to (3.12), at time t, the system can then be represented logically by a series structure of k minimal cut parallel structures $\kappa_1(X(t)), \kappa_2(X(t)), \ldots, \kappa_k(X(t))$. The system is failed if and only if at least one of the k minimal cuts is failed. Note that the same component may be a member of *several* different minimal cut sets.

We introduce the following notation:

$D_i(t)$ At least one minimal cut set which contains component i is failed at time t
$C(t)$ The system is failed at time t
m_i The number of minimal cut sets which contain component i
$E^i_j(t)$ Minimal cut set j among those containing component i is failed at time t for $i = 1, 2, \ldots, n$ and $j = 1, 2, \ldots, m_i$

We may then write Fussell-Vesely's measure as

$$I^{\text{FV}}(i \mid t) = \Pr(D_i(t) \mid C(t)) = \frac{\Pr(D_i(t) \cap C(t))}{\Pr(C(t))}$$

but since $D_i(t)$ implies $C(t)$

$$I^{\text{FV}}(i \mid t) = \frac{\Pr(D_i(t))}{\Pr(C(t))} \qquad (5.26)$$

Since the event $D_i(t)$ occurs if at least one of the events $E^i_j(t)$ occurs, $j = 1, 2, \ldots, m_i$, $i = 1, 2, \ldots, n$:

$$D_i(t) = E^i_1(t) \cup E^i_2(t) \cup \cdots \cup E^i_{m_i}(t)$$

Since the components are assumed to be independent, then

$$\Pr(C(t)) = \Pr(\phi(X(t)) = 0) = 1 - h(p(t))$$

and

$$\Pr(E^i_j(t)) = \Pr(\kappa^i_j(X(t)) = 0) = \prod_{\ell \in K^i_j} (1 - p_\ell) \tag{5.27}$$

where K^i_j denotes the jth minimal cut set that contains component i, and $\kappa^i_j(X(t))$ is the corresponding cut parallel structure.

Since the same component may be a member of several cut sets, the events $E^i_j(t)$, $j = 1, 2, \ldots, m_i$ are usually not disjoint. For the same reason, the events $E^i_j(t)$, $j = 1, 2, \ldots, m_i$, will not in general be independent, even if all the components are independent.

If the components are independent or associated (see Chapter 6), the cut parallel structures $\kappa^i_j(X(t))$ will also be associated. It can then be shown that the following inequality is valid [compare with (4.49))]:

$$\Pr(D_i(t)) \leq 1 - \prod_{j=1}^{m_i} (1 - \Pr(E^i_j(t))) \tag{5.28}$$

When the component reliabilities are high, (5.28)—with equality sign—will be approximately correct. Then

$$I^{\text{FV}}(i \mid t) \approx \frac{1 - \prod_{j=1}^{m_i} (1 - \Pr(E^i_j(t)))}{1 - h(p(t))} \tag{5.29}$$

A somewhat cruder approximation is

$$I^{\text{FV}}(i \mid t) \approx \frac{\sum_{j=1}^{m_i} \Pr(E^i_j(t))}{1 - h(p(t))} \tag{5.30}$$

By using the fault tree notation (5.2) and

$$\check{Q}^i_j(t) = \Pr(E^i_j(t)) = \prod_{\ell \in K^i_j} q_\ell(t) \tag{5.31}$$

then formulas (5.29) and (5.30) become

$$I^{\text{FV}}(i \mid t) \approx \frac{1 - \prod_{j=1}^{m_i}(1 - (\check{Q}^i_j(t))}{Q_0(t)} \tag{5.32}$$

and

$$I^{FV}(i \mid t) \approx \frac{\sum_{j=1}^{m_i} \check{Q}^i_j(t)}{Q_0(t)} \tag{5.33}$$

For complex systems, Fussell-Vesely's measure is considerably easier to calculate by hand than Birnbaum's measure and the criticality importance measure. When Fussell-Vesely's measure is to be calculated by hand, the formula (5.33) is normally used. The formula is simple to use and at the same time gives a good approximation when the component reliabilities are high.

In equation 5.33, $\check{Q}^i_j(t)$ denotes the probability that minimal cut set j which contains component i is failed at time t. From (5.31) we have that $\check{Q}^i_j(t) = \prod_{\ell \in K^i_j} q_\ell(t)$. We can put $q_i(t)$ outside the product and get

$$\check{Q}^i_j(t) = q_i(t) \cdot \prod_{\ell \in K^i_j, \ell \neq i} q_\ell(t) = q_i(t) \cdot \check{Q}^{i-}_j(t) \tag{5.34}$$

where $\check{Q}^{i-}_j(t)$ denotes the probability that minimal cut set j which contains component i, but where component i is removed, is failed at time t. We may now rewrite (5.33) and get

$$I^{FV}(i \mid t) \approx \frac{q_i(t)}{Q_0(t)} \cdot \sum_{j=1}^{m_i} \check{Q}^{i-}_j(t) \tag{5.35}$$

The system unreliability $Q_0(t)$ may, according to (4.64), be approximated by

$$Q_0(t) \approx \sum_{j=1}^{k} \check{Q}_j(t) \tag{5.36}$$

where $\check{Q}_j(t)$ denotes the probability that minimal cut set j is failed at time t for $j = 1, 2, \ldots, k$. We will use (5.36) to find an approximation to Birnbaum's measure (5.3) of component i. We therefore have to take the partial derivative of $Q_0(t)$ with respect to $q_i(t)$. The partial derivative of $\check{Q}_j(t)$ will be zero for all minimal cut sets where i is not a member. The partial derivative of a $\check{Q}^i_j(t)$ where i is a member is easily found from (5.34) and we get

$$I^B(i \mid t) = \frac{\partial Q_0(t)}{\partial q_i(t)} \approx \sum_{j=1}^{m_i} \check{Q}^{i-}_j(t)$$

The criticality importance measure is, from (5.24)

$$I^{CR}(i \mid t) = \frac{q_i(t)}{Q_0(t)} \cdot I^B(i \mid t) \approx \frac{q_i(t)}{Q_0(t)} \cdot \sum_{j=1}^{m_i} \check{Q}^{i-}_j(t) \tag{5.37}$$

Fig. 5.2 Series structure.

By comparing with (5.35) we see that

$$I^{FV}(i \mid t) \approx I^{CR}(i \mid t) \tag{5.38}$$

for systems where the approximation (5.36) is adequate.

Remark: Consider a system with minimal cut sets $C_1, C_2, \ldots, C_k$. A necessary criterion for component i to be critical for the system is that all the components, except for component i, in at least one minimal cut set containing component i are in a failed state. This is, however, not a sufficient criterion for component i to be critical, since we have to require the remaining cut sets to be functioning. This fact highlights the similarity and the difference between the definitions of criticality importance $I^{CR}(i \mid t)$, and Fussell-Vesely's measure $I^{FV}(i \mid t)$. We realize that we always have that

$$I^{CR}(i \mid t) \lesssim I^{FV}(i \mid t) \tag{5.39}$$

□

5.8 EXAMPLES

In this section we illustrate the use of the six component importance measures through three simple examples, a series structure with two components, a parallel structure with two components, and a 2-out-of-3 structure. The components are assumed to be independent, with reliabilities p_i, for $i = 1, 2, 3$, that are independent of time.

Example 5.4 Series Structure
Consider the series structure in Fig. 5.2 with two independent components with reliabilities:

$$p_1 = 0.98$$
$$p_2 = 0.96$$

The system reliability at time t is

$$h(p_1, p_2) = p_1 \cdot p_2 = 0.9408$$

Birnbaum's measure Birnbaum's measure for components 1 and 2, respectively, is

$$I^B(1) = \frac{\partial h(p_1, p_2)}{\partial p_1} = p_2 = 0.96$$

$$I^B(2) = \frac{\partial h(p_1, p_2)}{\partial p_2} = p_1 = 0.98$$

198 COMPONENT IMPORTANCE

According to Birnbaum's measure, the component with the lowest reliability in a series structure is the most important. This corresponds well with our intuition. A series structure can be compared with a chain. We know that a chain is never stronger than the weakest link in the chain. The weakest link is therefore the most important.

Improvement potential The improvement potential (IP) for components 1 and 2, respectively, is

$$I^{IP}(1) = I^{B}(1) \cdot (1 - p_1) = 0.0192$$
$$I^{IP}(2) = I^{B}(2) \cdot (1 - p_2) = 0.0392$$

The improvement potential thus gives the same ranking as Birnbaum's measure for a series structure. The weakest component is the most important.

Risk achievement worth The RAW for components 1 and 2, respectively, is

$$I^{RAW}(1) = \frac{1 - h(0_1, \boldsymbol{p})}{1 - h(\boldsymbol{p})} = \frac{1}{1 - p_1 p_2} = 16.9$$
$$I^{RAW}(2) = \frac{1 - h(0_2, \boldsymbol{p})}{1 - h(\boldsymbol{p})} = \frac{1}{1 - p_1 p_2} = 16.9$$

In a series structure, all components will have the same importance according to the RAW measure.

Risk reduction worth The RRW for components 1 and 2, respectively, is

$$I^{RRW}(1) = \frac{1 - h(\boldsymbol{p})}{1 - h(1_1, \boldsymbol{p})} = \frac{1 - p_1 p_2}{1 - p_2} = 1.480$$
$$I^{RRW}(2) = \frac{1 - h(\boldsymbol{p})}{1 - h(1_2, \boldsymbol{p})} = \frac{1 - p_1 p_2}{1 - p_1} = 2.960$$

In a series structure, the least reliable component is therefore the most important according to the RRW measure.

Criticality importance The criticality importance for components 1 and 2, respectively, is

$$I^{CR}(1) = \frac{I^{B}(1) \cdot (1 - p_1)}{1 - p_1 p_2} = 0.3243$$
$$I^{CR}(2) = \frac{I^{B}(2) \cdot (1 - p_2)}{1 - p_1 p_2} = 0.6622$$

This agrees with the ranking we got by using Birnbaum's measure. The weakest component in a series structure is the most important.

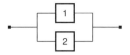

Fig. 5.3 Parallel structure.

Fussell-Vesely's measure Since there is only one minimal cut set containing components 1 and 2, we get

$$\Pr(D_1) = 1 - p_1 = 0.02$$
$$\Pr(D_2) = 1 - p_2 = 0.04$$

while

$$\Pr(C) = 1 - h(p_1, p_2) = 1 - p_1 p_2 = 0.0592$$

Fussell-Vesely's measure is then for component 1 and 2, respectively:

$$I^{\mathrm{FV}}(1) = \frac{\Pr(D_1)}{\Pr(C)} = \frac{1 - p_1}{1 - p_1 p_2} = 0.3378$$
$$I^{\mathrm{FV}}(2) = \frac{\Pr(D_2)}{\Pr(C)} = \frac{1 - p_2}{1 - p_1 p_2} = 0.6757$$

This agrees with the ranking we got by using Birnbaum's measure. The weakest component in a series structure is the most important.

Conclusion All the measures, except the RAW, gave the same ranking of the components of a series structure; the component with the lowest reliability is the most important. According to the risk achievement worth, all components in a series structure are equally important. □

Example 5.5 Parallel Structure
Consider the parallel structure with two independent components illustrated in Fig. 5.3. Let the component reliabilities at time t be as in Example 5.4. Then the system reliability at time t is

$$h(p_1, p_2) = p_1 + p_2 - p_1 \cdot p_2 = 0.9992$$

Birnbaum's measure Birnbaum's measure for components 1 and 2, respectively, is

$$I^{\mathrm{B}}(1) = \frac{\partial h(p_1, p_2)}{\partial p_1} = 1 - p_2 = 0.04$$
$$I^{\mathrm{B}}(2) = \frac{\partial h(p_1, p_2)}{\partial p_2} = 1 - p_1 = 0.02$$

According to Birnbaum's measure, the component with the highest reliability is the most important in a parallel structure. A parallel structure will function as long as at least one of its components is functioning. It makes therefore sense to say that the most reliable component is most important.

200 COMPONENT IMPORTANCE

Improvement potential The improvement potential for components 1 and 2, respectively, is

$$I^{IP}(1) = I^B(1) \cdot (1-p_1) = 0.0008$$
$$I^{IP}(2) = I^B(2) \cdot (1-p_2) = 0.0008$$

All the components in a parallel structure are equally important and have the same improvement potential.

Risk achievement worth The RAW for components 1 and 2, respectively, is

$$I^{RAW}(1) = \frac{1-h(0_1,\boldsymbol{p})}{1-h(\boldsymbol{p})} = \frac{1-p_2}{1-(p_1+p_2-p_1p_2)} = 50.00$$

$$I^{RAW}(2) = \frac{1-h(0_2,\boldsymbol{p})}{1-h(\boldsymbol{p})} = \frac{1-p_1}{1-(p_1+p_2-p_1p_2)} = 25.00$$

In a parallel structure, the most reliable component is the most important component, according to the RAW measure.

Risk reduction worth The RRW for components 1 and 2, respectively, is

$$I^{RRW}(1) = \frac{1-h(\boldsymbol{p})}{1-h(1_1,\boldsymbol{p})} = \infty$$

$$I^{RRW}(2) = \frac{1-h(\boldsymbol{p})}{1-h(1_2,\boldsymbol{p})} = \infty$$

since the denominator is zero. In a parallel structure, all components will have the same, high importance according to the RRW measure.

Criticality importance The criticality importance of components 1 and 2, respectively, is

$$I^{CR}(1) = \frac{I^B(1) \cdot (1-p_1)}{1-p_1-p_2+p_1p_2} = \frac{(1-p_1)(1-p_2)}{(1-p_1)(1-p_2)} = 1.000$$

$$I^{CR}(2) = \frac{I^B(2) \cdot (1-p_2)}{1-p_1-p_2+p_1p_2} = \frac{(1-p_1)(1-p_2)}{(1-p_1)(1-p_2)} = 1.000$$

All components in a parallel structure have the same criticality importance. This result seems reasonable. If a parallel structure is failed, it will start functioning again irrespective of which of the components we repair.

Fussell-Vesely's measure In the parallel structure the system itself constitutes a minimal cut set, that is, $D_1(t) = D_2(t) = C(t)$. Hence

$$I^{FV}(1 \mid t) = I^{FV}(2 \mid t) = 1$$

All components in a parallel structure are equally important according to Fussell-Vesely's measure.

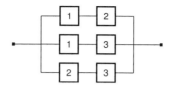

Fig. 5.4 The 2-out-of-3 structure.

Conclusion According to Birnbaum's measure and the RAW, the most reliable component in a parallel structure is the most important. The other four measures say that all components in a parallel structure are equally important. □

Example 5.6 The 2-out-of-3 Structure

Consider the 2-out-of-3 structure of three independent components in Fig. 5.4 with component reliabilities:

$$p_1 = 0.98$$
$$p_2 = 0.96$$
$$p_3 = 0.94$$

The system reliability is

$$h(\boldsymbol{p}) = p_1 p_2 + p_1 p_3 + p_2 p_3 - 2 p_1 p_2 p_3 = 0.9957$$

Birnbaum's measure Birnbaum's measure for components 1, 2, and 3, respectively, is

$$I^B(1) = \frac{\partial h(\boldsymbol{p})}{\partial p_1} = p_2 + p_3 - 2 p_2 p_3 = 0.0952$$

$$I^B(2) = \frac{\partial h(\boldsymbol{p})}{\partial p_2} = p_1 + p_3 - 2 p_1 p_3 = 0.0776$$

$$I^B(3) = \frac{\partial h(\boldsymbol{p})}{\partial p_3} = p_1 + p_2 - 2 p_1 p_2 = 0.0584$$

Hence in this particular case:

$$I^B(1) > I^B(2) > I^B(3)$$

According to Birnbaum's measure the component importance decreases with decreasing reliability in a 2-out-of-3 structure. The most important is the component with the highest reliability.

Improvement potential For the 2-out-of-3 system we have

$$I^{IP}(1) = I^B(1) \cdot (1 - p_1(t)) = 0.0019$$
$$I^{IP}(2) = I^B(2) \cdot (1 - p_2(t)) = 0.0031$$
$$I^{IP}(3) = I^B(3) \cdot (1 - p_3(t)) = 0.0035$$

such that

$$I^{\text{IP}}(1) < I^{\text{IP}}(2) < I^{\text{IP}}(3)$$

According to the improvement potential the component importance increases with decreasing reliability in a 2-out-of-3 structure. The most important is the component with lowest reliability.

Risk achievement worth The RAW of components 1, 2 and 3, respectively, is

$$I^{\text{RAW}}(1) = \frac{1 - h(0_1, \boldsymbol{p})}{1 - h(\boldsymbol{p})} = \frac{1 - p_2 p_3}{1 - (p_1 p_2 + p_1 p_3 + p_2 p_3 - 2 p_1 p_2 p_3)} = 22.7$$

$$I^{\text{RAW}}(2) = \frac{1 - h(0_2, \boldsymbol{p})}{1 - h(\boldsymbol{p})} = \frac{1 - p_1 p_3}{1 - (p_1 p_2 + p_1 p_3 + p_2 p_3 - 2 p_1 p_2 p_3)} = 18.3$$

$$I^{\text{RAW}}(3) = \frac{1 - h(0_3, \boldsymbol{p})}{1 - h(\boldsymbol{p})} = \frac{1 - p_1 p_2}{1 - (p_1 p_2 + p_1 p_3 + p_2 p_3 - 2 p_1 p_2 p_3)} = 13.8$$

and hence

$$I^{\text{RAW}}(1) > I^{\text{RAW}}(2) > I^{\text{RAW}}(3)$$

We thus get the same result as for Birnbaum's measure. The RAW of the components in a 2-out-of-3 structure decreases with decreasing component reliability.

Risk reduction worth The RRW of components 1, 2, and 3, respectively, is

$$I^{\text{RRW}}(1) = \frac{1 - h(\boldsymbol{p})}{1 - h(1_1, \boldsymbol{p})} = \frac{1 - (p_1 p_2 + p_1 p_3 + p_2 p_3 - 2 p_1 p_2 p_3)}{1 - (p_2 + p_3 - p_2 p_3 - 2 p_2 p_3)} = 1.793$$

$$I^{\text{RRW}}(2) = \frac{1 - h(\boldsymbol{p})}{1 - h(1_2, \boldsymbol{p})} = \frac{1 - (p_1 p_2 + p_1 p_3 + p_2 p_3 - 2 p_1 p_2 p_3)}{1 - (p_1 + p_3 - p_1 p_3 - 2 p_1 p_3)} = 3.587$$

$$I^{\text{RRW}}(3) = \frac{1 - h(\boldsymbol{p})}{1 - h(1_3, \boldsymbol{p})} = \frac{1 - (p_1 p_2 + p_1 p_3 + p_2 p_3 - 2 p_1 p_2 p_3)}{1 - (p_1 + p_2 - p_1 p_2 - 2 p_1 p_2)} = 5.380$$

and hence

$$I^{\text{RRW}}(1) < I^{\text{RRW}}(2) < I^{\text{RRW}}(3)$$

We thus get the same result as for the improvement potential measure. The most important component of a 2-out-of-3 structure is, according to RRW, the component with the lowest reliability.

Criticality importance The criticality importance of components 1, 2, and 3, respectively, is

$$I^{\text{CR}}(1) = \frac{I^{\text{B}}(1) \cdot (1 - p_1)}{1 - p_1 p_2 - p_1 p_3 - p_2 p_3 + 2 p_1 p_2 p_3} = 0.4428$$

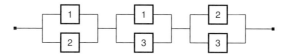

Fig. 5.5 The 2-out-of-3 structure.

Correspondingly

$$I^{CR}(2) = 0.7219$$
$$I^{CR}(3) = 0.8149$$

and hence

$$I^{CR}(1) < I^{CR}(2) < I^{CR}(3)$$

We thus get the same result as for the improvement potential measure. The most important component of a 2-out-of-3 structure is, according to the criticality importance measure, the component with the lowest reliability.

Fussell-Vesely's measure The 2-out-of-3 structure represented as a series structure of the minimal cut parallel structures can be illustrated by the reliability block diagram in Fig. 5.5. This structure has three minimal cut sets where each component enters in two cut sets.

Thus we have

$$\begin{aligned}
\Pr(D_1) &= \Pr(E_1^1 \cup E_2^1) \\
&= \Pr(E_1^1) + \Pr(E_2^1) - \Pr(E_1^1 \cap E_2^1) \\
&= q_1 q_2 + q_1 q_3 - q_1 q_2 q_3 \approx 0.0020 \\
\Pr(D_2) &= \Pr(E_1^2 \cup E_2^2) \\
&= \Pr(E_1^2) + \Pr(E_2^2) - \Pr(E_1^2 \cap E_2^2) \\
&= q_1 q_2 + q_2 q_3 - q_1 q_2 q_3 \approx 0.0032 \\
\Pr(D_3) &= \Pr(E_1^3 \cup E_2^3) \\
&= \Pr(E_1^3) + \Pr(E_2^3) - \Pr(E_1^3 \cap E_2^3) \\
&= q_1 q_3 + q_2 q_3 - q_1 q_2 q_3 \approx 0.0036
\end{aligned}$$

In Example 5.6 we found that $\Pr(C) = 1 - h(\boldsymbol{p}) = 1 - 0.9957 = 0.0043$. Fussell-Vesely's measure of reliability importance of components 1, 2, and 3, respectively then become

$$\begin{aligned}
I^{FV}(1) &= \frac{\Pr(D_1)}{\Pr(C)} = \frac{0.0020}{0.0043} \approx 0.4651 \\
I^{FV}(2) &= \frac{\Pr(D_2)}{\Pr(C)} = \frac{0.0032}{0.0043} \approx 0.7442 \\
I^{FV}(3) &= \frac{\Pr(D_3)}{\Pr(C)} = \frac{0.0036}{0.0043} \approx 0.8372
\end{aligned}$$

and hence

$$I^{\text{FV}}(1) < I^{\text{FV}}(2) < I^{\text{FV}}(3)$$

We thus get the same result as for the criticality importance measure. The most important component of a 2-out-of-3 structure is, according to Fussell-Vesely's measure, the component with the lowest reliability.

We have here determined Fussell-Vesely's measure exactly. One of the approximation formulas could also be used.

As an illustration we will calculate Fussell-Vesely's measure for components 1, 2, and 3, respectively, according to the approximation formula (5.33):

$$I^{\text{FV}}(1) \approx \frac{\check{Q}_1^1 + \check{Q}_2^1}{Q_0} = \frac{q_1 q_2 + q_1 q_3}{Q_0} \approx 0.4651$$

$$I^{\text{FV}}(2) \approx \frac{\check{Q}_1^2 + \check{Q}_2^2}{Q_0} = \frac{q_1 q_2 + q_2 q_3}{Q_0} \approx 0.7442$$

$$I^{\text{FV}}(3) \approx \frac{\check{Q}_1^3 + \check{Q}_2^3}{Q_0} = \frac{q_1 q_3 + q_2 q_3}{Q_0} \approx 0.8372$$

As we see the four first decimals are correct. The approximation seems very good, at least for this simple example.

Conclusion According to Birnbaum's measure and risk achievement worth, the component importance of a 2-out-of-3 structure decreases with decreasing component reliability. The most important component is the component with the highest reliability. The four other measures lead to the opposite conclusion: The component importance increases with decreasing component reliability. The most important component is the component with the lowest reliability. □

PROBLEMS

5.1 Let $p_S(t) = 1 - Q_0(t)$ denote the system reliability, and let $p_i(t) = 1 - q_i(t)$ denote the reliability of component i, for $i = 1, 2, \ldots, n$. Verify that

$$\frac{dp_S(t)}{dp_i(t)} = \frac{dQ_0(t)}{dq_i(t)}$$

5.2 Show that if a 2-out-of-3 structure of independent components has component reliabilities $p_1 \geq p_2 \geq p_3$, then

(a) if $p_0 \geq 0.5$ $I^{\text{B}}(1 \mid t) \geq I^{\text{B}}(2 \mid t) \geq I^{\text{B}}(3 \mid t)$

(b) if $p_1 \leq 0.5$ $I^{\text{B}}(1 \mid t) \leq I^{\text{B}}(2 \mid t) \leq I^{\text{B}}(3 \mid t)$

5.3 Find the structural importance of component 1 in the 2-out-of-3 structure in Example 5.6 by using (5.5).

5.4 Find the reliability importance and structural importance of component 7 in Example 4.3.

5.5 Find the criticality importance for component 7 in Example 4.3.

5.6 Find the reliability importance for component 7 in Example 4.3 by using Fussell-Vesely's measure.

5.7 Consider the nonrepairable structure in Fig. 5.6

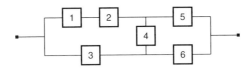

Fig. 5.6 Reliability block diagram (Problem 5.8).

(a) Determine the structure function.

(b) Assume the components to be independent. Determine the reliability importance according to Birnbaum's measure of components 2 and 4 when $p_i = 0.99$ for $i = 1, 2, \ldots, 6$.

5.8 Consider the nonrepairable structure in Fig. 5.6 . Assume that the six components are independent, and let the reliability at time t of component i be denoted by $p_i(t)$, for $i = 1, 2, \ldots, 6$.

(a) Determine Birnbaum's measure of importance of component 3.

(b) Determine the criticality importance of component 3.

(c) Determine Fussell-Vesely's measure of component 3.

(d) Select realistic values for the component reliabilities and discuss the difference between criticality importance and Fussell-Vesely's measure for this particular system. Show that the relation (5.39) is fulfilled.

5.9 Consider the nonrepairable structure in Fig. 5.7.

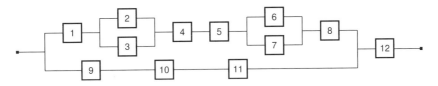

Fig. 5.7 Reliability block diagram (Problem 5.9).

(a) Show that the corresponding structure function may be written:

$$\phi(X) = [X_1 \cdot (X_2 + X_3 - X_2 X_3) X_4 X_5 (X_6 + X_7 - X_6 X_7) X_8$$
$$+ X_9 X_{10} X_{11} - X_1 (X_2 + X_3 - X_2 X_3) X_4 X_5 (X_6 + X_7 - X_6 X_7)$$
$$X_8 X_9 X_{10} X_{11}] X_{12}$$

(b) Determine the system reliability when the different component reliabilities are given as:

$$p_1 = 0.970 \quad p_5 = 0.920 \quad p_9 = 0.910$$
$$p_2 = 0.960 \quad p_6 = 0.950 \quad p_{10} = 0.930$$
$$p_3 = 0.960 \quad p_7 = 0.959 \quad p_{11} = 0.940$$
$$p_4 = 0.940 \quad p_8 = 0.900 \quad p_{12} = 0.990$$

(c) Determine the reliability importance of component 8 by using Birnbaum's measure and the criticality importance measure.

(d) Similarly determine the reliability importance of component 11, using the same measures as in (c). Compare and comment on the results obtained.

5.10 Let (C, ϕ) be a coherent structure of n independent components with state variables $X_1, X_2, \ldots, X_n$. Consider the following modular decomposition of (C, ϕ):

(i) $C = \bigcup_{j=1}^{r} A_j$ where $A_i \cap A_j = \emptyset$ for $i \neq j$

(ii) $\phi(x) = \omega\left(\chi_1(x^{A_1}), \chi_2(x^{A_2}), \ldots, \chi_r(x^{A_r})\right)$

Assume that $k \in A_j$ and show that

- the Birnbaum measure of importance of component k is equal to the product of;
- the Birnbaum measure of importance of module j relative to the system, and
- the Birnbaum measure of importance of component k relative to module j.

Is the same relation valid for the other measures?

6

Dependent Failures

6.1 INTRODUCTION

In Chapter 4, we studied situations where the n components of a system fail independently of each other. This was modeled by assuming that the state variables of the n components $X_1(t), X_2(t), \ldots, X_n(t)$ are independent random variables. This assumption considerably simplifies the modeling as well as the statistical analysis.

However, when the components of a system fail, they do not necessarily have to fail independently of each other. We may distinguish between two main types of dependence: *positive* and *negative*. If a failure of one component leads to an increased tendency for another component to fail, the dependence is said to be *positive*. If, on the other hand, the failure of one component leads to a reduced tendency for another component to fail, the dependence is called *negative*.

Example 6.1
Consider a system of two components, 1 and 2. Let A_i denote the event that component i is in a failed state ($i = 1, 2$). The probability that both components are failed is

$$\Pr(A_1 \cap A_2) = \Pr(A_1 \mid A_2) \cdot \Pr(A_2) = \Pr(A_2 \mid A_1) \cdot \Pr(A_1)$$

The components are *independent* when $\Pr(A_1 \mid A_2) = \Pr(A_1)$, and $\Pr(A_2 \mid A_1) = \Pr(A_2)$, such that

$$\Pr(A_1 \cap A_2) = \Pr(A_1) \cdot \Pr(A_2)$$

The components have a *positive dependency* when $\Pr(A_1 \mid A_2) > \Pr(A_1)$, and $\Pr(A_2 \mid A_1) > \Pr(A_2)$, such that

$$\Pr(A_1 \cap A_2) > \Pr(A_1) \cdot \Pr(A_2)$$

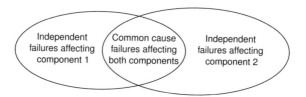

Fig. 6.1 Relationship between independent and common cause failures of a system with two components.

and the components have a *negative dependency* when $\Pr(A_1 \mid A_2) < \Pr(A_1)$, and $\Pr(A_2 \mid A_1) < \Pr(A_2)$, such that

$$\Pr(A_1 \cap A_2) < \Pr(A_1) \cdot \Pr(A_2)$$

□

In reliability applications, positive dependence is usually the most relevant type of dependence. Negative dependence may, however, also occur in practice.

Dependent failures may be classified in three main groups:

1. *Common cause failures.* A common cause event is, according to NUREG/CR-6268, a "dependent failure in which two or more component fault states exist simultaneously, or within a short time interval, and are a direct result of a shared cause". Common cause failures may be caused by (see NASA 2002):

    - A common *design or material deficiency* that results in multiple components failing to perform a function or to withstand a design environment
    - A common *installation error* that results in multiple components being misaligned or being functionally inoperable
    - A common *maintenance error* that results in multiple components being misaligned or being functionally inoperable
    - A common *harsh environment* such as vibration, radiation, moisture, or contamination that causes multiple components to fail

    The relationship between independent and common cause failures of a system of two components is illustrated in Fig. 6.1. The number of components that fail due to the common, or root, cause is called the *multiplicity* of the failures.

    When we establish models for common cause failures in systems consisting of several redundant components, we should carefully distinguish between:

    (a) Simultaneous failures of a set of components of the system, due to a common dependency, for example, on a support function. If such dependencies are well understood, they should be explicitly identified in the reliability model and handled as components. Such an explicit model is illustrated by the fault tree of a parallel system in Fig. 6.2.

INTRODUCTION

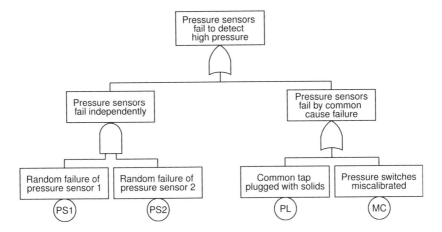

Fig. 6.2 Fault tree of parallel system with an explicitly modeled common cause failures. [Adapted from Summers and Raney (1999)].

(b) Simultaneous failures of a set of components of the system due to shared causes that are not explicitly represented in the system logic model. Such events are sometimes called "residual" common cause failures. These events should be incorporated into the reliability analysis through specific models that are described in this chapter.

2. *Cascading failures.* Cascading failures are multiple failures initiated by the failure of one component in the system that results in a chain reaction or "domino effect." When several components share a common load, failure of one component may lead to increased load on the remaining components and consequently to an increased likelihood of failure. Cascading failures are especially important in electro-power distribution systems. A well-known example is the power-line failure in Oregon that caused a massive cascading failure throughout the western United States and Canada on 10 August 1996.

Components may also influence each other through the internal environment. Malfunction of a component may, for example, lead to a more hostile working environment for the other components through increased pressure, temperature, humidity, and so on. Cascading failures are sometimes called *propagating failures*.

Some types of cascading failures may be adequately modeled and analyzed by event trees (see Section 3.9). Cascading failures are not discussed any further in this book.

3. *Negative dependencies.* Negative dependency failures are single failures that reduce the likelihood of failures of other components. If, for example, an electrical fuse fails open such that the "downstream" circuit is disconnected,

the load on the electrical devices in this circuit is removed and their likelihood of failure is reduced.

When a system is "down" for repair of a specific component, the load on other components is often removed and their likelihood of failure is consequently reduced during the system downtime.

Remark: Components that fail due to a shared cause will normally fail in the same functional mode. The term *common mode failures* was therefore used in the early literature. It is, however, not considered to be a precise term for communicating the characteristics that describe a common cause failure, and the term should therefore not be used. □

Example 6.2
One of the best known accidents resulting from a common cause failure is the fire at the Browns Ferry nuclear power plant near Decatur, Alabama, on March 22, 1975 (Rippon 1975). The fire started when two of the operators used a candle to check for air leaks between the cable room and one of the reactor buildings, which was kept at a negative air pressure. The candle's flame was drawn out along the conduit, and the urethane seal used where the cables penetrate the wall caught fire. The fire continued until the insulation of about 2000 cables was damaged. Among these were all the cables to the automatic emergency shutdown (ESD) systems and also the cables to all the "manually" operated valves, apart from four relief valves. With these four valves it was possible to close down the reactor so that a nuclear meltdown was avoided. This accident resulted in new instructions requiring that the cables to the different emergency shutdown systems be put in separate conduits and prohibit the use of combustible filling (e.g., urethane foam). □

A taxonomy and classification of the root causes for common cause failures is presented in Table 6.1 which is adapted from Edwards and Watson (1979). This taxonomy may help the analyst systematically to concentrate his attention on the possible root causes and to judge them one at a time. Common cause failures may greatly reduce the reliability of a system, especially of systems with a high degree of redundancy. A significant research activity has therefore been devoted to this problem. A wide range of technical reports and papers describing various aspects of dependent failures have been written during the last 10 to 20 years. Valuable references include Edwards and Watson (1979), Fleming et al. (1986), NUREG/CR-4780, Bodsberg and Hokstad (1988), Mosleh (1991), and Hokstad (1993).

6.2 HOW TO OBTAIN RELIABLE SYSTEMS

As for any reliability analysis, failure mode, effects, and criticality analysis (FMECA) also provides the basic framework for the identification and investigation of dependent failures. In IEEE Std. 352 two extensions of the conventional FMECA are described in order to encompass potential interdependencies between the system components:

Table 6.1 Classification of Reasons for Common Cause Failures.

Engineering				Operations			
Design		Manufacture	Construction	Procedural		Environmental	
Functional Deficiencies	Realization Faults		Installation & Commissioning	Maintenance & Test	Operation	Normal Extremes	Energetic Events
Hazard undetectable	Channel dependency	Inadequate quality control	Inadequate quality control	Imperfect repair	Operator errors	Temperature	Fire
Inadequate instrumentation	Common operation & protection components	Inadequate standards	Inadequate standards	Imperfect testing	Inadequate procedures	Pressure	Flood
Inadequate control	Operational deficiencies	Inadequate inspection	Inadequate inspection	Imperfect calibration	Inadequate supervision	Humidity	Weather
	Inadequate components	Inadequate testing	Inadequate testing & commissioning	Imperfect procedures	Communication error	Vibration	Earthquake
	Design errors			Inadequate supervision		Acceleration	Explosion
	Design limitations					Stress	Missiles
						Corrosion	Electrical power
						Contamination	Radiation
						Interference	Chemical sources
						Radiation	
						Static charge	

Source: Adapted from Edwards and Watson (1979). Reproduced with the permission of AEA Technology.

1. A common cause failure analysis to identify root causes and mechanisms of failures of components normally considered to be redundant

2. A cascading failure analysis to identify failures that can lead to "chain-type" events affecting several different areas or systems in a plant

Detailed procedures for the analyses are given in IEEE Std. 352.

The most important defense against accidental component failures is the use of redundancy (see Section 4.5). The fire at Browns Ferry shows, however, that redundancy itself is not enough. Even if there had been a higher number of redundant shutdown systems, all these would have been made inactive by a single error—the flame from a candle. Common cause failures can be prevented by a variety of defenses. A number of procedures and checklists have been developed to assist the engineer in creating a design that is robust against common cause failures; see, for example, EPRI NP-5777, Edwards and Watson (1979), and Bourne et al. (1981).

Some general defensive tactics to avoid common cause failures are presented by Parry (1991)[1]:

1. *Barriers.* Any physical impediment that tends to confine and/or restrict a potentially damaging condition.

2. *Personnel training.* A program to assure that the operators and maintainers are familiar with procedures and are able to follow them correctly during all conditions of operation.

3. *Quality control.* A program to assure that the product is in conformance with the documented design and that its operation and maintenance are according to approved procedures, standards, and regulatory requirements.

4. *Redundancy.* Additional, identical, redundant components added to a system solely for the purpose of increasing the likelihood that a sufficient number of components will survive exposure to a given cause of failure and are available to perform a given function. This is a tactic to improve system availability, but, by definition, common cause failures decrease the positive impact of this particular tactic. However, increased redundancy will generally still have value.

5. *Preventive maintenance.* A program of applicable and effective preventive maintenance tasks designed to prevent premature failure or degradation of components.

6. *Monitoring, surveillance testing, and inspection.* Monitoring via alarms, frequent tests, and/or inspections so that unannounced failures from any detectable causes are not allowed to accumulate. This includes special tests performed on redundant components in response to observed failures.

---

[1] Reprinted from *Reliability Engineering & System Safety*, Volume 34, G. W. Parry, "Common Cause Failure Analysis: A Critique and Some Suggestions". pp. 320. Copyright 1991, with kind permission from Elsevier Science Ltd, The Boulevard, Langford Lane, Kidlington OX5 1GB, UK.

7. *Procedures review.* A review of operational, maintenance, and calibration/test procedures to eliminate incorrect or inappropriate actions resulting in component or system unavailability.

8. *Diversity.* The mixture of interchangeable components made by different manufacturers (equipment diversity) or the introduction of a totally redundant system with an entirely different principle of operation (functional diversity) for the express purpose of reducing the likelihood of a total loss of function that might occur because all like components are vulnerable to the same cause(s) of failure. Diversity in staff is another form of applying this concept, whereby different teams are used to maintain and test redundant trains. This is a tactic that specifically addresses common cause failures.

An analysis of common cause failures constitutes an essential part of reliability studies of complex high reliability systems. A systematic approach is needed. A general approach for treating common cause failures in safety and reliability studies is presented in NUREG/CR-4780. A more brief presentation is given by Mosleh (1991).

The first step in an analysis of common cause failures is to identify *common cause candidates*, that is, components for which the independence assumption is suspected to be incorrect. Several checklists have been developed to aid the analyst in this task; see, for example, Summers and Raney (1999). Common cause candidates will usually have one or more *coupling factors*. A coupling factor is a property of a set of components that identifies them as being susceptible to the same mechanisms of failure. Examples of coupling factors include the same, or similar:

- Design
- Hardware
- Software
- Installation staff
- Maintenance or operation crew
- Procedures
- Environment
- Location

Even if common cause failures are caused by a common cause, they do not need to occur at the same time. A rather long time between failures does not necessarily mean that there is no dependency between the failure events. Whenever the common cause is not a "fatal" shock, common cause failures will usually occur in a sequence. An efficient failure detection system will therefore be an important defense against common cause failures in a redundant system. This defense is listed as tactic 6 on page 212.

A systematic engineering study has to be performed on how to mitigate against each dependency. Several techniques have been proposed to check whether system defenses have been considered for each potential common cause. NUREG/CR-5460 suggests an approach based on a cause-defense matrix, where all potential causes are listed in the first column, and the measures that are taken in the plant against each failure cause are listed in the other columns.

Data on the occurrence of common cause failures must be collected, analyzed statistically, and saved for later engineering use, in particular, for the following situations:

1. Future designs of the system

2. Assessment of system reliability

3. Working out rules for system operators

An international common cause database has been established for the nuclear sector by the Nuclear Energy Agency (NEA), through the International Common Cause Failure Data Exchange (ICDE) project that was launched in 1994 (ICDE 1994).

6.3 MODELING OF DEPENDENT FAILURES

In situations where we have good reasons to believe that calculations based on the assumption of independent failures are adequate, the models and methods described in Chapter 4, can be used. Cascading failures and negative dependencies may, however, often be taken into account in these analyses, because they arise from functional or physical interrelationships between components and subsystems that may explicitly be modeled in a fault tree/reliability block diagram or a Markov model. Nevertheless, in a number of situations an analysis based on the assumption of independence will lead to unrealistic results and may be of limited value for practical purposes.

At the moment large efforts are made to develop suitable models that take into account different types of dependence. In the following sections, we will give examples of some such models. In these models, there are sometimes built-in approximations, but the results obtained in this way are often far more realistic than the ones obtained by assuming independence.

When analyzing reliability, one usually deals with technical systems. In some cases it may be possible to express typical features of the system mathematically and derive a model from this (mechanistic model). In other situations when sufficient data (information) is at hand, an *empirical* model based on a sample of the data, may be adequate for the analysis. In both cases the analyst must confront the possible models with data to see if they reflect the specific design features of the system. He also has to be aware that the validity of a model may be limited.

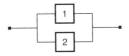

Fig. 6.3 Parallel system.

6.4 SPECIAL MODELS

In this section we will, as an illustration, discuss three models for dependent failures:

1. The square-root method
2. The β-factor model
3. The binomial failure rate (BFR) model.

Several other, and more detailed, models have been developed; for example, see Apostolakis and Moieni (1987), Hokstad (1988, 1993), Mosleh (1991), NUREG/CR-4780 (1989).

6.4.1 The Square-Root Method

In the Reactor safety study (NUREG-75/014), a simple bounding technique was used to estimate the effect of common cause failures on a system. We will illustrate this technique by a simple example.

Consider the parallel structure in Fig. 6.3 with two components 1 and 2, both of which may fail as a result of a common cause.

Let A_i represent the situation where component i is in a failed state at time t, and let $q_i = \Pr(A_i)$ denote the unavailability of component i, for $i = 1, 2$. The probability that the parallel structure is not functioning at time t, that is, the unavailability of the system, is then $Q_0 = \Pr(A_1 \cap A_2)$.

Since $(A_1 \cap A_2) \subseteq A_i$,

$$\Pr(A_1 \cap A_2) \leq \Pr(A_i) \quad \text{for } i = 1, 2$$

Therefore

$$\Pr(A_1 \cap A_2) \leq \min\{\Pr(A_1), \Pr(A_2)\} \tag{6.1}$$

If the events A_1 and A_2 are independent, by definition

$$\Pr(A_1 \cap A_2) = \Pr(A_1) \cdot \Pr(A_2)$$

If the events are positively dependent,

$$\Pr(A_1 \mid A_2) \geq \Pr(A_1)$$

and hence

$$\Pr(A_1 \cap A_2) = \Pr(A_1 \mid A_2) \cdot \Pr(A_2) \geq \Pr(A_1) \cdot \Pr(A_2) \tag{6.2}$$

Combining (6.1) and (6.2) we get the following result when A_1 and A_2 are positively dependent:

$$\Pr(A_1) \cdot \Pr(A_2) \leq \Pr(A_1 \cap A_2) \leq \min\{\Pr(A_1), \Pr(A_2)\} \tag{6.3}$$

Equation (6.3) can be written

$$q_L \leq \Pr(A_1 \cap A_2) \leq q_U$$

where

$$\begin{aligned} q_L &= \Pr(A_1) \cdot \Pr(A_2) = q_1 \cdot q_2 \\ q_U &= \min\{\Pr(A_1), \Pr(A_2)\} = \min\{q_1, q_2\} \end{aligned} \tag{6.4}$$

In the Reactor safety study (NUREG-75/014), the unavailability of the parallel structure, $Q_0 = \Pr(A_1 \cap A_2)$, is approximated by the geometric mean Q_0^* of the lower bound q_L and the upper bound q_U:

$$Q_0^* = \sqrt{q_L \cdot q_U} \tag{6.5}$$

and is called the *square-root method*.

There is, however, no proper theoretical foundation for the choice of geometric averaging of the two limits. Further, the result depends heavily on this somewhat arbitrary averaging.

In this section the square-root method has been given a very simple formulation to clarify the main principle of the method. A more general formulation is presented by Edwards and Watson (1979) and Harris (1986).

A weakness of the square-root method is that it does not take into account the various degrees of coupling between the components. Attempts have therefore been made to develop generalizations of this method. See Harris (1986) for a thorough discussion. The square-root method is now seldom used.

Example 6.3
Consider a parallel structure of n components with a common unavailability q at a specified time t. Let A_i denote the situation that component i is in a failed state at time t. Thus we have $\Pr(A_i) = q$ for $i = 1, \ldots, n$.

If the n components are all independent, the unavailability Q_0 of the parallel system becomes $Q_0 = q^n$. If the components are positively dependent, we can apply the square-root method with lower bound

$$q_L = \prod_{i=1}^{n} \Pr(A_i) = q^n$$

Table 6.2 Unavailability of Parallel Structure of n Identical Components with Unavailability $q = 0.01$ Modeled by the Square-Root Method.

n	Independent Components $Q_0 = q^n$	Square-Root Method $Q_0^* = q^{(n+1)/2}$
1	10^{-2}	10^{-2}
2	10^{-4}	10^{-3}
3	10^{-6}	10^{-4}
4	10^{-8}	10^{-5}
5	10^{-10}	10^{-6}

and upper bound

$$q_U = \min\{\Pr(A_1), \ldots, \Pr(A_n)\} = q$$

The geometric mean is thus

$$Q_0^* = \sqrt{q_L \cdot q_U} = q^{(n+1)/2}$$

The effect of the dependency modeled by the square-root method is illustrated in Table 8.2 for some values of n and $q = 0.01$. □

6.4.2 The β-Factor Model

The β-factor model was introduced by Fleming (1974) and is today the most commonly used model for common cause failures. Let a system be composed of n identical components, each with constant failure rate λ. The situation is furthermore assumed to be such that the failure of a component may be due to one of two possible causes:

1. Circumstances that concern only the component (independent of the condition of the remaining components)

2. Occurrence of an external event (independent of the system) whereby *all* the components fail at the same time

Let $\lambda^{(i)}$ denote the failure rate due to failure cause of type 1, and let $\lambda^{(c)}$ denote the failure rate due to failure cause of type 2. Assuming independence of the two failure causes, the total failure rate λ of the component, can be written as the sum of the two failure rates

$$\lambda = \lambda^{(i)} + \lambda^{(c)} \tag{6.6}$$

The failure rate $\lambda^{(i)}$ is called the failure rate due to "independent" failures and also sometimes the failure rate due to "individual" failures.

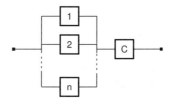

Fig. 6.4 Parallel structure with common cause "component" C.

Now introduce β as the "common cause factor":

$$\beta = \frac{\lambda^{(c)}}{\lambda^{(i)} + \lambda^{(c)}} = \frac{\lambda^{(c)}}{\lambda} \tag{6.7}$$

Then

$$\lambda^{(c)} = \beta\lambda \tag{6.8}$$

and

$$\lambda^{(i)} = (1 - \beta)\lambda \tag{6.9}$$

The β-factor thus denotes the relative proportion of common cause failures among all the failures of a component. The β-factor may also be considered as the conditional probability that a failure is a common cause failure, that is,

$$\text{Pr(Common cause failure} \mid \text{Failure)} = \beta$$

Remark: For a fixed β, the rate of common cause failures $\lambda^{(c)} = \beta\lambda$ in the β-factor model is seen to increase with the total failure rate λ. Items with many failures will hence also have more common cause failures. Since repair and maintenance is often claimed to be a prime cause of common cause failures, it is relevant to assume that items requiring a lot of repair and maintenance will also have many common cause failures. □

Example 6.4 Parallel System of Identical Components
Consider a parallel structure of n identical components with failure rate λ. An external event may occur that causes failure in each and every component of the system. This external event can be represented by a "hypothetical" component (C) that is in series with the rest of the system. This is illustrated in Fig. 6.4. The failure rate of component C is, according to (6.8), $\lambda^{(c)} = \beta\lambda$, while the n components in the parallel structure in Fig. 6.4 may be considered as independent with failure rate $\lambda^{(i)} = (1-\beta)\lambda$.

We assume the system to be nonrepairable, and let $R_I(t)$ denote the survivor function of the identical components while $R_C(t)$ denotes the survivor function of the "hypothetical" component C. The survivor function of the system can now be written

$$\begin{aligned} R(t) &= \left(1 - (1 - R_I(t))^n\right) \cdot R_C(t) \\ &= \left(1 - (1 - e^{-(1-\beta)\lambda t})^n\right) \cdot e^{-\beta\lambda t} \end{aligned} \tag{6.10}$$

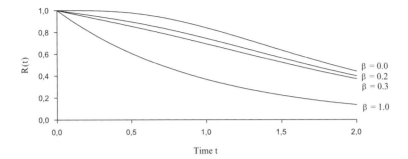

Fig. 6.5 The survivor function of a parallel system of four components for some selected values of the common cause factor β. All components have constant failure rate $\lambda = 1$.

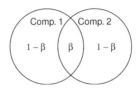

Fig. 6.6 Fraction of different types of failures for a system with two components when using a β-factor model.

Fig. 8.4 shows the survivor function $R(t)$ of a parallel system of $n = 4$ components for selected values of β. Note that when the common cause factor β is increased, the reliability of the system declines. □

Example 6.5 System with Two Components
Consider a system with $n = 2$ components with the same constant failure rate λ. By using the β-factor model, the rate of single failures becomes $\lambda_{2:1} = 2 \cdot (1 - \beta) \cdot \lambda$. The rate of double (common cause) failures is $\lambda_{2:2} = \beta \lambda$. The fractions of single and double failures are illustrated in Fig. 6.6. □

Example 6.6 System with Three Components
Consider a system with $n = 3$ components with the same constant failure rate λ. By using the β-factor model, the rate of single failures becomes $\lambda_{3:1} = 3 \cdot (1 - \beta) \cdot \lambda$. The rate of triple (common cause) failures is $\lambda_{3:3} = \beta \lambda$. Double failures are not possible when using the β-factor model. The fractions of single, double, and triple failures are illustrated in Fig. 6.7. □

Example 6.7 Comparison of Simple Systems
Consider the following three simple systems

1. A single component

2. A parallel structure of two identical components

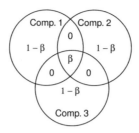

Fig. 6.7 Fraction of different types of failures for a system with three components when using a β-factor model.

3. A 2-out-of-3 system with identical components

All the components are assumed to have the same constant failure rate λ. The systems are exposed to common cause failures that may be modeled by a β-factor model. Since common cause failures are not relevant for a single component, the survivor function and the mean time to failure (MTTF) are

$$R_1(t) = e^{-\lambda t} \quad \text{and} \quad \text{MTTF}_1 = \frac{1}{\lambda}$$

The survivor function of the parallel system is, according to (6.10),

$$\begin{aligned} R_2(t) &= \left(2e^{-(1-\beta)\lambda t} - e^{-2(1-\beta)\lambda t}\right) \cdot e^{-\beta\lambda t} \\ &= 2e^{-\lambda t} - e^{-(2-\beta)\lambda t} \end{aligned}$$

Its mean time to failure is hence

$$\text{MTTF}_2 = \frac{2}{\lambda} - \frac{1}{(2-\beta)\lambda}$$

The survivor function of the 2-out-of-3 system is

$$\begin{aligned} R_3(t) &= \left(3e^{-2(1-\beta)\lambda t} - 2e^{-3(1-\beta)\lambda t}\right) \cdot e^{-\beta\lambda t} \\ &= 3e^{-(2-\beta)\lambda t} - 2e^{-(3-2\beta)\lambda t} \end{aligned}$$

Its mean time to failure is hence

$$\text{MTTF}_3 = \frac{3}{(2-\beta)\lambda} - \frac{2}{(3-2\beta)\lambda}$$

The MTTFs of the three simple systems are illustrated in Fig. 6.8 for $\lambda = 1$. It is obvious that all three systems have the same MTTF when $\beta = 1$, that is, total dependence. □

The β-factor model is very simple, and it is easy to understand the practical interpretation of the factor β. β is related to the degree of protection against common

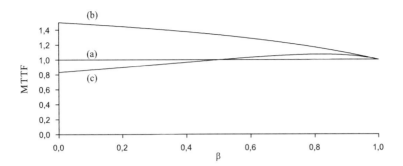

Fig. 6.8 The MTTF as a function of β for (1) a single component, (2) a parallel system of two identical components, and (3) a 2-out-of-3 system of identical components. All components have constant failure rate $\lambda = 1$.

cause failures. A number of checklist methods for the assessment of β have been proposed. One of these methods is developed by Humphreys (1987). He is basing the assessment of β on an evaluation of eight criteria: separation, similarity, complexity, analysis, procedures, training, control, and tests. Each criterion is evaluated based on a checklist and ranked according to a predefined system. The ranks allocated to the eight criteria are then added, and a simple formula is used to estimate β.

When using the β-factor model, we assume that all system failures are either single, independent failures, or common cause failures where *all* the components of the system fail simultaneously. A serious limitation of the β-factor model is that it does not allow that only a certain fraction of the components fails. The model seems quite adequate for parallel systems with two components but may not be adequate for more complex systems. NUREG/CR-4780 states that: "Although historical data collected from the operation of nuclear power plants indicate that common cause events do not always fail all redundant components, experience from using this simple model shows that, in many cases, it gives reasonably accurate (or slightly conservative) results for redundancy levels up to about three or four items. However, beyond such redundancy levels, this model generally yields results that are conservative."

To use the β-factor model we need an estimate of the failure rate λ, or $\lambda^{(i)}$, and an estimate of β. As mentioned above, β may be estimated based on a procedure similar to the one suggested by Humphreys (1987) or based on sound engineering judgement. Failure rates may be found in a variety of data sources (see Chapter 12). Some of the data sources present the total failure rate λ, while other sources present the independent failure rate $\lambda^{(i)}$. It is sometimes difficult to decide whether a specific data source presents λ or $\lambda^{(i)}$. To decide, we have to study carefully how the data have been collected, from which sources, and so on. The data in MIL-HDBK 217F are, for example, mainly compiled from laboratory testing of single components. Hence, only independent failure rates, $\lambda^{(i)}$, are presented. The data in OREDA (2002) are, on the other hand, field data collected from maintenance files. The maintenance records normally do not distinguish between independent failures and common cause failures. Hence, OREDA presents the total failure rates. Reliability data sources are discussed in Chapter 14.

The β-factor model is further discussed in Chapter 10 where the model is applied to safety instrumented systems.

6.4.3 Binomial Failure Rate Model and Its Extensions

The binomial failure rate (BFR) model was introduced by Vesely (1977) and is a simple special case of the Marshall-Olkin's multivariate exponential model (see Marshall and Olkin 1967). The situation under study is the following: A system is composed of n identical components. Each component can fail at random times, independently of each other, and they are all supposed to have the same "individual" failure rate $\lambda^{(i)}$.

Furthermore a common cause shock can hit the system with occurrence rate ν. Whenever a shock occurs, each of the n individual components is assumed to fail with probability p, independent of the states of the other components. The number Z of individual components failing as a consequence of the shock, is thus binomially distributed (n, p). The probability that the multiplicity Z of failures is equal to z is thus

$$\Pr(Z = z) = \binom{n}{z} p^z (1-p)^{n-z} \quad \text{for } z = 0, 1, 2, \ldots.$$

The mean number of components that fail in one shock is $E(Z) = np$. Two conditions are furthermore assumed:

1. Shocks and individual failures occur independently of each other.

2. All failures are immediately discovered and repaired, with the repair time being negligible.

As a consequence the time between individual failures, in the absence of shocks, will be exponentially distributed with failure rate $\lambda^{(i)}$, and the time between shocks will be exponentially distributed with failure rate ν. The number of individual failures in any time period of length t_0 is Poisson distributed ($\lambda^{(i)} t_0$). Similarly the number of shocks in any time period of length t_0 is Poisson distributed (νt_0).

The component failure rate caused by shocks thus equals $p \cdot \nu$, and the total failure rate of one component equals

$$\lambda = \lambda^{(i)} + p \cdot \nu$$

By using this model, we have to estimate the independent failure rate $\lambda^{(i)}$ and the two parameters ν and p. The parameter ν relates to the degree of "stress" on the system, while p is a function of the built-in component protection against external shocks. Note that the BFR model is identical to the β-factor model when the system has only two components.

This is Vesely's original BFR model. The statistical analysis of such models is discussed by Atwood (1986). Several aspects of the situation must be clarified to make the analysis possible. It may, for example, happen that ν cannot be estimated in a direct way from failure data, because shocks may occur unnoticed when no component fails.

Several extensions of Vesely's BFR model have been studied. It may, for example, happen that p varies from shock to shock. One way of modeling this is to assume p to be beta distributed, that is, it has the probability density function

$$f(p) = \frac{\Gamma(r+s)}{\Gamma(r)\Gamma(s)} p^{r-1}(1-p)^{s-1} \quad \text{for } r > 0, \ s > 0.$$

Such a Bayesian approach is discussed, for example, by Hokstad (1988). He introduces a re-parameterization of the beta distribution and is able to interpret the new parameters in the context of how well the components and the system are protected against the shocks.

The assumption that the components will fail independently of each other, given that a shock has occurred, represents a rather serious limitation, and this assumption is often not satisfied in practice. The problem can to some extent be remedied by defining one fraction of the shocks as being "lethal" shocks, that is, shocks that automatically cause all the components to fail ($p = 1$). If all the shocks are "lethal," one is back to the β-factor model. Observe that this case ($p = 1$) corresponds to the situation that there is no built-in protection against these shocks.

Situations where individual failures may occur together with nonlethal as well as lethal shocks are often realistic. Such models are, however, rather complicated, even if the nonlethal and the lethal shocks occur independently of each other.

Further extensions are presented, for example, by Apostolakis and Moieni (1987).

6.5 ASSOCIATED VARIABLES

Esary et al. (1967) introduced a broad class of state variables, which they call *associated*. This class includes independent variables as well as variables with a certain type of *positive* dependence that may be appropriate for reliability analysis. In this section we present some main results for systems where the state variables are assumed to be associated. The results are presented without any proofs. For proofs and more detailed results, the reader should consult Barlow and Proschan (1975).

Throughout this section we will consider a coherent structure $\phi(X)$. The reliability of component i is denoted $p_i = \Pr(X_i = 1)$ for $i = 1, 2, \ldots, n$, and the system reliability is denoted $p_S = \Pr(\phi(X) = 1)$. The structure ϕ is assumed to have k minimal cut sets $K_i, K_2, \ldots, K_k$ with associated minimal cut parallel structures $\kappa_1(X), \kappa_2(X), \ldots, \kappa_k(X)$, and p minimal path sets $P_1, P_2, \ldots, P_p$ with associated minimal path series structures $\rho_1(X), \rho_2(X), \ldots, \rho_p(X)$.

Associated variables are defined by:

Definition 6.1 The random variables $Y = [Y_1, Y_2, \ldots, Y_m]$ (not necessarily binary) are said to be *associated* if, for all increasing functions f and g, we have that

$$E(f(X) \cdot g(X)) \geq E(f(X)) \cdot E(g(X)) \tag{6.11}$$

When the state variables $X_1, X_2, \ldots, X_n$ are associated, we say that the components of the system are associated.

The following results apply:

1. If $Y_1, Y_2, \ldots, Y_m$ are independent variables, then they are also associated.

2. Any subset of an associated set is itself associated.

3. Increasing functions of associated variables are associated.

4. If $Y_1, Y_2, \ldots, Y_m$ are associated and binary variables, then $(1 - Y_1), (1 - Y_2), \ldots, (1 - Y_m)$ are also associated and binary.

5. If the binary state variables $X_1, X_2, \ldots, X_n$ of a coherent structure ϕ are associated, then the minimal path structures $\rho_j(X)$ for $j = 1, 2, \ldots, p$, are associated, and the minimal cut parallel structures $\kappa_j(X)$ for $j = 1, 2, \ldots, k$, are associated.

For a *series* structure of associated components we have that

$$p_S = \Pr\left(\prod_{i=1}^{n} X_i = 1\right) \geq \prod_{i=1}^{n} \Pr(X_i = 1) = \prod_{i=1}^{n} p_i \qquad (6.12)$$

Note that the right-hand side of (6.12) is the reliability of a series structure of n *independent* components. If the components of a series structure are associated, we underestimate the reliability of the structure when we use the formula for independent components. We say that we obtain a *conservative* bound for the reliability p_S of the series structure.

For a *parallel* structure of associated components we have that

$$p_S = \Pr\left(\coprod_{i=1}^{n} X_i = 1\right) \leq \coprod_{i=1}^{n} \Pr(X_i = 1) = 1 - \prod_{i=1}^{n}(1 - p_i) \qquad (6.13)$$

Note that the right-hand side of (6.13) is the reliability of a parallel structure of n *independent* components. If the components in a parallel structure are associated, we overestimate the reliability of the structure when we use the formula for independent components, and we get a nonconservative bound for the reliability p_S of the parallel structure.

For a system of associated components that is neither a series nor a parallel system, it is not possible to predict whether or not we over- or underestimate the reliability by proceeding as if the state variables are independent. In such cases, however, rough upper and lower bounds for the system reliability can be determined.

Any structure ϕ of order n is at least as "strong" as a series structure of its n components, and at most as "strong" as a parallel structure of its n components.

$$\prod_{i=1}^{n} X_i \leq \phi(X) \leq \coprod_{i=1}^{n} X_i$$

We can therefore use (6.12) and (6.13) to conclude that for any system of n associated components, the system reliability p_S is bounded by

$$\prod_{i=1}^{n} p_i \leq p_S \leq 1 - \prod_{i=1}^{n}(1 - p_i) \tag{6.14}$$

The bounds obtained in (6.14) are, however, normally rather wide and therefore not very useful for practical purposes.

In order to find more narrow bounds for the system reliability we will look for bounds based on the minimal path series structures and the minimal cut parallel structures. Since the structure ϕ can be represented as a series structure of the k minimal cut parallel structures, and since the minimal cut parallel structures are associated, we can use (6.12) to conclude that

$$p_S = \Pr(\phi(X) = 1) \geq \prod_{j=1}^{k} \Pr(\kappa_j(X) = 1)$$

In the same way, since the structure ϕ can be represented as a parallel structure of the p minimal path series structures, and since the minimal path series structures are associated, we can use (6.13) to conclude that

$$p_S = \Pr(\phi(X) = 1) \leq \coprod_{j=1}^{p} \Pr(\rho_j(X) = 1)$$

The reliability p_S of a system of n associated components is therefore bounded by

$$\prod_{j=1}^{k} \Pr(\kappa_j(X) = 1) \leq p_S \leq \coprod_{j=1}^{p} \Pr(\rho_i(X) = 1) \tag{6.15}$$

Let us now consider a system of n *independent* components. In this case we have

$$\Pr(\rho_j(X) = 1) = \prod_{i \in P_j} p_i$$

$$\Pr(\kappa_j(X) = 1) = \coprod_{i \in C_j} p_i = 1 - \prod_{i \in C_j}(1 - p_i)$$

Since independent components are also associated, and we can use (6.15) to find the following bounds for the reliability p_S of the system:

$$\prod_{j=1}^{k} \coprod_{i \in K_j} p_i \leq p_S \leq \coprod_{j=1}^{p} \prod_{i \in P_j} p_i \tag{6.16}$$

$$\prod_{j=1}^{k}\left(1 - \prod_{i \in K_j}(1 - p_i)\right) \leq p_S \leq 1 - \prod_{j=1}^{p}\left(1 - \prod_{i \in P_j} p_i\right) \tag{6.17}$$

Like the bounds obtained in (6.14), the bounds obtained in (6.15) are sometimes too wide to be of practical use.

By using the fault tree terminology that we introduced in Chapter 4, we have that $Q_0 = 1 - p_S$, $q_i = 1 - p_i$ for $i = 1, 2, \ldots, n$, and

$$\check{Q}_j = 1 - \Pr(\kappa_j(X) = 1)$$

We can now use (6.15) to show that

$$Q_0 \leq 1 - \prod_{j=1}^{k} \left(1 - \check{Q}_j\right) \tag{6.18}$$

We have hence shown that the *upper bound approximation* (4.50) that is used to calculate the TOP event probability Q_0 in a fault tree is valid when the basic events (components) are associated.

The reliability p_S of a system of n associated components is also bounded by

$$\max_{1 \leq j \leq p} \prod_{i \in P_j} p_i \leq p_S \leq \min_{1 \leq j \leq k} \coprod_{i \in K_j} p_i \tag{6.19}$$

and these bounds are at least as narrow as the bounds obtained in (6.14).

Example 6.8
Reconsider the gas detector system in Example 4.1. Let the component reliabilities at time t_0 be

$$\begin{aligned} p_1 = p_2 = p_3 &= 0.997 \\ p_4 &= 0.999 \\ p_5 = p_6 &= 0.998 \\ p_7 = p_8 &= 0.995 \end{aligned} \tag{6.20}$$

If the components are *independent*, we find the system reliability at time t_0 by inserting (6.21) in (4.50) and thus get

$$p_S \approx 0.9950$$

Suppose that the state variables of the system are associated but not necessarily independent.

Let us first determine the bounds for p_S by using (6.18).
The lower bound becomes

$$\prod_{i=1}^{8} p_i = (0.997)^3 \cdot 0.999 \cdot (0.998)^2 \cdot (0.995)^2 \approx 0.9762$$

The upper bound becomes

$$1 - \prod_{i=1}^{8} (1 - p_i) = 1 - (0.003)^3 \cdot (0.001) \cdot (0.002)^2 \cdot (0.005)^2 \approx 1.00$$

Hence

$$0.9762 \leq p_S \leq 1.00$$

Let us next calculate the bounds for p_S by using (6.19). The minimal path and cut sets were found in Example 4.1.

Minimal path sets: Minimal cut sets:

$S_1 = \{1,2,4,5,6,7\}$ $K_1 = \{1,2\}$
$S_2 = \{1,2,4,5,6,8\}$ $K_2 = \{1,3\}$
$S_3 = \{1,3,4,5,6,7\}$ $K_3 = \{2,3\}$
$S_4 = \{1,3,4,5,6,8\}$ $K_4 = \{4\}$
$S_5 = \{2,3,4,5,6,7\}$ $K_5 = \{5\}$
$S_6 = \{2,3,4,5,6,8\}$ $K_6 = \{6\}$
$K_7 = \{7,8\}$

Using the data in (6.21) we observe that all paths S_1 to S_6 have the same reliability

$$\prod_{i \in P_j} p_i = (0.997)^2 \cdot 0.999 \cdot (0.998)^2 \cdot 0.995 \approx 0.9841 \text{ for all } j$$

Hence the lower bound is

$$\max_{1 \leq j \leq 6} \prod_{i \in P_j} p_i \approx 0.9841$$

In order to find the upper bound $\coprod_{i \in K_j} p_i$ must be calculated for K_1 to K_7:

Cut 1	$1 - (0.003)^2$	≈ 0.999991
Cut 2	$1 - (0.003)^2$	≈ 0.999991
Cut 3	$1 - (0.003)^2$	≈ 0.999991
Cut 4	p_4	≈ 0.999
Cut 5	p_5	≈ 0.998
Cut 6	p_6	≈ 0.998
Cut 7	$1 - (0.005)^2$	≈ 0.999975

Hence the upper bound becomes

$$\min_{1 \leq j \leq 7} \coprod_{i \in K_j} p_i \approx 0.998$$

Hence application of (6.19) leads to the following bounds for p_S:

$$0.9841 \leq p_S \leq 0.998$$

PROBLEMS

6.1 Consider a parallel structure of two identical components. Let q denote the unavailability of each component.

(a) Make a sketch of the system unavailability as a function of q when the two components are independent.

(b) Determine the system unavailability by the square-root method and make a sketch of this unavailability as a function of q in the same coordinate system as in (a).

(c) Determine the difference between the system unavailability when the components are independent and the unavailability determined by the square-root method when $q = 0.15$.

6.2 Consider a 2-out-of-3 system of identical components. Let q denote the unavailability of each component. Edwards and Watson (1979) states that the square-root method gives the system unavailability $\sqrt{3 \cdot q^3}$. Discuss this statement. Is it correct? See, for example, Harris (1986) for a thorough discussion.

6.3 Consider a 2-out-of-3 system of identical components with constant failure rate λ. The system is exposed to common cause failures that may be modeled by a β-factor model. In Fig. 6.8 it is shown that the MTTF of the system has a minimum for $\beta = 0$. Determine the value of β for which MTTF attains its maximum. Explain why MTTF as a function of β has this particular shape.

6.4 Reconsider the bridge structure in Example 4.8. Assume that all the five components are identical and have constant failure rate λ. The system is exposed to common cause failures that may be modeled by a β-factor model. Determine the MTTF of the bridge structure as a function of β, and make a sketch of MTTF as a function of β when $\lambda = 5 \cdot 10^{-4}$ failures per hour, and no repair is carried out.

6.5 Consider a 2-out-of-3 structure of identical components. The system is exposed to common cause failures that may be modeled by a binomial failure rate (BFR) model. The "individual" failure rate of the components is $\lambda^{(i)} = 5 \cdot 10^{-5}$ failures per hour. Nonlethal shocks occur with frequency $\nu = 10^{-5}$ nonlethal shocks per hour. When a non-lethal shock occurs, the components may fail independently with probability $p = 0.20$. Lethal shocks occur with frequency $\omega = 10^{-7}$ lethal shocks per hour. When a lethal shock occurs, all the three components will fail simultaneously. The lethal and the nonlethal shocks are assumed to be independent.

(a) Determine the mean time between system failures, $\text{MTBF}^{(i)}$, caused by individual component failures, when you assume that the system is only repaired when a system failure occurs. In such a case the system is repaired to an "as good as new" condition.

(b) Determine the mean time between system failures, MTBF_{NL} when you assume that the only cause of system failures is the nonlethal shocks.

(c) Determine the mean time between system failures, $MTBF_L$, when you assume that the only cause of system failures is the lethal shocks.

(d) Try to find the total mean time between system failures. Discuss the problems you meet during this assessment.

7
Counting Processes

7.1 INTRODUCTION

In this chapter and in Chapter 8 we will study the reliability of a repairable system as a function of time. We are interested in finding system reliability measures like the availability of the system, the mean number of failures during a specified time interval, the mean time to the first system failure, and the mean time between system failures. For this purpose, we study the system by using stochastic processes. A *stochastic process* $\{X(t), t \in \Theta\}$ is a collection of random variables. The set Θ is called the *index set* of the process. For each *index t* in Θ, $X(t)$ is a random variable. The index t is often interpreted as time, and $X(t)$ is called the *state* of the process at time t. When the index set Θ is countable, we say that the process is a discrete-time stochastic process. When Θ is a continuum, we say that it is a continuous-time stochastic process. In this chapter and in Chapter 8 we only look at continuous-time stochastic processes. The presentation of the various processes in this book is very brief and limited, as we have focused on results that can be applied in practice instead of mathematical rigor. The reader should therefore consult a textbook on stochastic processes for more details. An excellent introduction to stochastic processes may be found in, for example, Ross (1996) and Cocozza-Thivent (1997).

In this chapter we consider a repairable system that is put into operation at time $t = 0$. When the system fails, it will be repaired to a functioning state. The repair time is assumed to be negligible. When the second failure occurs, the system will again be repaired, and so on. We thus get a sequence of failure times. We will primarily be interested in the random variable $N(t)$, the number of failures in the time interval $(0, t]$. This particular stochastic process $\{N(t), t \geq 0\}$ is called a *counting process*.

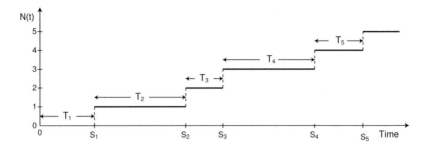

Fig. 7.1 Relation between the number of events $N(t)$, the interoccurrence times (T_i), and the calendar times (S_i).

In Chapter 8 we study the various *states* of a repairable system. A multicomponent, repairable system will have a number of possible states, depending on how many of its components are in operation. The state of the system at time t is denoted $X(t)$, and we are interested in finding the probability that the system is in a specific state at time t. We also find the steady-state probabilities, or the average proportion of time the system is in the various states. The presentation will be limited to a special class of stochastic processes $\{X(t), t \geq 0\}$ having the *Markov property*. Such a stochastic process is called a *Markov process* and is characterized by its lack of memory. If a Markov process is in state j at time t, we will get no more knowledge about its future states by knowing the history of the process up to time t.

7.1.1 Counting Processes

Consider a repairable system that is put into operation at time $t = 0$. The first failure (event) of the system will occur at time S_1. When the system has failed, it will be replaced or restored to a functioning state. The repair time is assumed to be so short that it may be neglected. The second failure will occur at time S_2 and so on. We thus get a sequence of failure times $S_1, S_2, \ldots$. Let T_i be the time between failure $i - 1$ and failure i for $i = 1, 2, \ldots$, where S_0 is taken to be 0. T_i will be called the *interoccurrence time i* for $i = 1, 2, \ldots$. T_i may also be called the *time between failures*, and the *interarrival time*. In general, counting processes are used to model sequences of *events*. In this book, most of the events considered are failures, but the results presented will apply for more general events.

Throughout this chapter t denotes a specified point of time, irrespective whether t is *calendar time* (a realization of S_i) or *local time* (a realization of an interoccurrence time T_i. We hope that this convention will not confuse the reader. The time concepts are illustrated in Fig. 7.1.

The sequence of interoccurrence times, $T_1, T_2, \ldots$ will generally not be independent and identically distributed—unless the system is replaced upon failure or restored to an "as good as new" condition, and the environmental and operational conditions remain constant throughout the whole period.

INTRODUCTION

A precise definition of a counting process is given below (from Ross 1996, p. 59).

Definition 7.1 A stochastic process $\{N(t), t \geq 0\}$ is said to be a *counting process* if $N(t)$ satisfies:

1. $N(t) \geq 0$.

2. $N(t)$ is integer valued.

3. If $s < t$, then $N(s) \leq N(t)$.

4. For $s < t$, $[N(t) - N(s)]$ represents the number of failures that have occurred in the interval $(s, t]$. □

A counting process $\{N(t), t \geq 0\}$ may alternatively be represented by the sequence of failure (calendar) times $S_1, S_2, \ldots$, or by the sequence of interoccurrence times $T_1, T_2, \ldots$. The three representations contain the same information about the counting process.

Example 7.1
The following failure times (calendar time in days) are presented by Ascher and Feingold (1984, p. 79). The data set is recorded from time $t = 0$ until 7 failures have been recorded during a total time of 410 (days). The data come from a single system, and the repair times are assumed to be negligible. This means that the system is assumed to be functioning again almost immediately after a failure is encountered.

Number of failures $N(t)$	Calendar time S_j	Interoccurrence time T_j
0	0	0
1	177	177
2	242	65
3	293	51
4	336	43
5	368	32
6	395	27
7	410	15

The data are illustrated in Fig. 7.2. The interoccurrence times are seen to become shorter with time. The system seems to be deteriorating, and failures tend to become more frequent. A system with this property is called a *sad* system by Ascher and Feingold (1984), for obvious reasons. A system with the opposite property, where failures become less frequent with operating time, is called a *happy* system.

The number of failures $N(t)$ may also be illustrated as a function of (calendar) time t as illustrated in Fig. 7.3. Note that $N(t)$ by definition is constant between failures and jumps (a height of 1 unit) at the failure times S_i for $i = 1, 2, \ldots$. It is thus sufficient to plot the jumping points $(S_i, N(S_i))$ for $i = 1, 2, \ldots$. The plot is called an $N(t)$ plot, or a Nelson-Aalen plot (see Section 7.4.3).

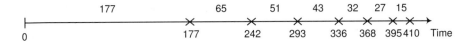

Fig. 7.2 The data set in Example 7.1.

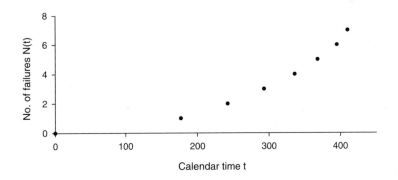

Fig. 7.3 Number of failures $N(t)$ as a function of time for the data in Example 7.1.

Note that $N(t)$ as a function of t will tend to be convex when the system is *sad*. In the same way, $N(t)$ will tend to be concave when the system is *happy*.[1] If $N(t)$ is (approximately) linear, the system is steady, that is, the interoccurrence times will have the same expected length. In Fig. 7.3 $N(t)$ is clearly seen to be convex. Thus the system is *sad*. □

Example 7.2 Compressor Failure Data

Failure time data for a specific compressor at a Norwegian process plant have been collected as part of a student thesis at the Norwegian University of Science and Technology. All compressor failures in the time period from 1968 until 1989 have been recorded. In this period a total of 321 failures occurred, 90 of which were critical failures and 231 were noncritical. In this context, a critical failure is defined to be a failure causing compressor downtime. Noncritical failures may be corrected without having to close down the compressor. The majority of the noncritical failures were instrument failures and failures of the seal oil system and the lubrication oil system.

As above, let $N(t)$ denote the number of compressor failures in the time interval $(0, t]$. From a production regularity point of view, the critical failures are the most important, since these failures are causing process shutdown. The operating times (in days) at which the 90 critical failures occurred are listed in Table 7.1. Here the time t denotes the *operating* time, which means that the downtimes caused by compressor

---

[1] Notice that we are using the terms *convex* and *concave* in a rather inaccurate way here. What we mean is that the observed points $\{t_i, N(t_i)\}$ for $i = 1, 2, \ldots$ approximately follow a convex/concave curve.

Table 7.1 Failure Times (Operating Days) in Chronological Order.

1.0	4.0	4.5	92.0	252.0	277.0
277.5	284.5	374.0	440.0	444.0	475.0
536.0	568.0	744.0	884.0	904.0	1017.5
1288.0	1337.0	1338.0	1351.0	1393.0	1412.0
1413.0	1414.0	1546.0	1546.5	1575.0	1576.0
1666.0	1752.0	1884.0	1884.2	1884.4	1884.6
1884.8	1887.0	1894.0	1907.0	1939.0	1998.0
2178.0	2179.0	2188.5	2195.5	2826.0	2847.0
2914.0	3156.0	3156.5	3159.0	3211.0	3268.0
3276.0	3277.0	3321.0	3566.5	3573.0	3594.0
3640.0	3663.0	3740.0	3806.0	3806.5	3809.0
3886.0	3886.5	3892.0	3962.0	4004.0	4187.0
4191.0	4719.0	4843.0	4942.0	4946.0	5084.0
5084.5	5355.0	5503.0	5545.0	5545.2	5545.5
5671.0	5939.0	6077.0	6206.0	6206.5	6305.0

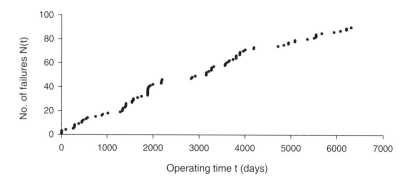

Fig. 7.4 Number of critical compressor failures $N(t)$ as a function of time (days), (totaling 90 failures).

failures and process shutdowns are not included. An $N(t)$ plot with respect to the 90 critical failures is presented in Fig. 7.4. In this case the $N(t)$ plot is slightly concave, which indicates a *happy* system. The time between critical failures hence seems to increase with the time in operation. Also note that several failures have occurred within short intervals. This indicates that the failures may be dependent, or that the maintenance crew has not been able to correct the failures properly at the first attempt.
□

An analysis of life data from a repairable system should always be started by establishing an $N(t)$ plot. If $N(t)$ as a function of the time t is nonlinear, methods based on the assumption of independent and identically distributed times between failures are obviously not appropriate. It is, however, not certain that such methods are appropriate even if the $N(t)$ plot is very close to a straight line. The interoccurrence times may be strongly correlated. Methods to check whether the interoccurrence times are correlated or not are discussed, for example, by Ascher and Feingold (1984) and Bendell and Walls (1985). The $N(t)$ plot is further discussed in Section 7.4.

7.1.2 Some Basic Concepts

A number of concepts associated with counting processes are defined in the following. Throughout this section we assume that the events that are counted are *failures*. In some of the applications later in this chapter we also study other types of events, like repairs. Some of the concepts must be reformulated to be meaningful in these applications. We hope that this will not confuse the reader.

- *Independent increments.* A counting process $\{N(t), t \geq 0\}$ is said to have independent increments if for $0 < t_1 < \cdots < t_k$, $k = 2, 3, \ldots$ $[N(t_1) - N(0)], [N(t_2) - N(t_1)], \ldots, [N(t_k) - N(t_{k-1})]$ are all independent random variables. In that case the number of failures in an interval is not influenced by the number of failures in any strictly earlier interval (i.e., with no overlap). This means that even if the system has experienced an unusual high number of failures in a certain time interval, this will not influence the distribution of future failures.

- *Stationary increments.* A counting process is said to have stationary increments if for any two disjoint time points $t > s \geq 0$ and any constant $c > 0$, the random variables $[N(t) - N(s)]$ and $[N(t+c) - N(s+c)]$ are identically distributed. This means that the distribution of the number of failures in a time interval depends only on the length of the interval and not on the interval's distance from the origin.

- *Stationary process.* A counting process is said to be stationary (or homogeneous) if it has stationary increments.

- *Nonstationary process.* A counting process is said to be nonstationary (or nonhomogeneous) if it is neither stationary nor eventually becomes stationary.

- *Regular process.* A counting process is said to be regular (or orderly) if

$$\Pr(N(t + \Delta t) - N(t) \geq 2) = o(\Delta t) \tag{7.1}$$

when Δt is small, and $o(\Delta t)$ denotes a function of Δt with the property that $\lim_{\Delta t \to 0} o(\Delta t)/\Delta t = 0$. In practice this means that the system will not experience two or more failures simultaneously.

- *Rate of the process.* The rate of the counting process at time t is defined as:

$$w(t) = W'(t) = \frac{d}{dt} E(N(t)) \tag{7.2}$$

INTRODUCTION

where $W(t) = E(N(t))$ denotes the mean number of failures (events) in the interval $(0, t]$. Thus

$$w(t) = W'(t) = \lim_{\Delta t \to 0} \frac{E(N(t + \Delta t) - N(t))}{\Delta t} \quad (7.3)$$

and when Δt is small,

$$\begin{aligned} w(t) &\approx \frac{E(N(t + \Delta t) - N(t))}{\Delta t} \\ &= \frac{\text{Mean no. of failures in } (t, t + \Delta t]}{\Delta t} \end{aligned}$$

Thus a natural estimator of $w(t)$ is

$$\hat{w}(t) = \frac{\text{Number of failures in } (t, t + \Delta t]}{\Delta t} \quad (7.4)$$

for some suitable Δt. It follows that the rate $w(t)$ of the counting process may be regarded as the mean number of failures (events) per time unit at time t.

When we are dealing with a *regular* process, the probability of two or more failures in $(t, t + \Delta t]$ is negligible when Δt is small. Thus for small Δt we may assume that

$$N(t + \Delta t) - N(t) = 0 \text{ or } 1$$

Thus the mean number of failures in $(t, t + \Delta t]$ is approximately equal to the probability of failure in $(t, t + \Delta t]$, and

$$w(t) \approx \frac{\text{Probability of failure in } (t, t + \Delta t]}{\Delta t} \quad (7.5)$$

Hence $w(t) \, \Delta t$ can be interpreted as the probability of failure in the time interval $(t, t + \Delta t]$.

Some authors use (7.5) written as

$$w(t) = \lim_{\Delta t \to 0} \frac{\Pr(N(t + \Delta t) - N(t) = 1)}{\Delta t}$$

as definition of the rate of the process. Observe also that

$$E(N(t_0)) = W(t_0) = \int_0^{t_0} w(t) \, dt \quad (7.6)$$

- ROCOF. When the events of a counting process are failures, the rate $w(t)$ of the process is often called the *rate of occurrence of failures* (ROCOF).

- *Time between failures.* We have denoted the time T_i between failure $i - 1$ and failure i, for $i = 1, 2, \ldots$, the interoccurrence times. For a general counting

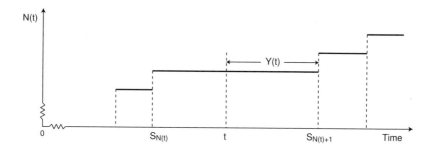

Fig. 7.5 The forward recurrence time $Y(t)$.

process the interoccurrence times will neither be identically distributed nor independent. Hence the mean time between failures, $\text{MTBF}_i = E(T_i)$, will in general be a function of i and $T_1, T_2, \ldots, T_{i-1}$.

- *Forward recurrence time.* The forward recurrence time $Y(t)$ is the time to the next failure measured from an arbitrary point of time t. Thus $Y(t) = S_{N(t)+1} - t$. The forward recurrence time is also called the *residual lifetime*, the *remaining lifetime* or the *excess life*. The forward recurrence time is illustrated in Fig. 7.5.

Many of the results in this chapter are only valid for *nonlattice* distributions. The definition of a lattice distribution follows.

Definition 7.2 A nonnegative random variable is said to have a *lattice* (or periodic) distribution if there exists a number $d \geq 0$ such that

$$\sum_{n=0}^{\infty} \Pr(X = nd) = 1$$

In words, X has a lattice distribution if X can only take on values that are integral multiples of some nonnegative number d. □

7.1.3 Martingale Theory

Martingale theory can be applied to counting processes to make a record of the *history* of the process. Let $\mathcal{H}_t$ denote the history of the process up to, but not including, time t. Usually we think of $\mathcal{H}_t$ as $\{N(s), 0 \leq s < t\}$ which keeps records of all failures before time t. It could, however, contain more specific information about each failure.

We may define a conditional rate of failures as

$$w_C(t \mid \mathcal{H}_t) = \lim_{\Delta t \to \infty} \frac{\Pr(N(t + \Delta t) - N(t) = 1 \mid \mathcal{H}_t)}{\Delta t} \tag{7.7}$$

Thus, $w_C(t \mid \mathcal{H}_t) \cdot \Delta t$ is approximately the probability of failure in the interval $[t, t + \Delta t)$ conditional on the failure history up to, but not including time t. Note that the rate of the process (ROCOF) defined in (7.2) is the corresponding unconditional rate of failures.

Usually the process depends on the history through random variables and $w_C(t \mid \mathcal{H}_t)$ will consequently be *stochastic*. It should, however, be noted that $w_C(t \mid \mathcal{H}_t)$ is stochastic only through the history: for a fixed history (i.e., for a given state just before time t), $w_C(t \mid \mathcal{H}_t)$ is not stochastic. To simplify the notation, we will in the following omit the explicit reference to the history $\mathcal{H}_t$ and let $w_C(t)$ denote the conditional ROCOF.

The martingale approach for modeling counting processes requires rather sophisticated mathematics. We will therefore avoid using this approach during the main part of the chapter but will touch upon martingales on page 254 and in Section 7.5 where we discuss imperfect repair models.

A brief, but clear, introduction to martingales used in counting processes is given by Hokstad (1997). A more thorough description is given by Andersen et al. (1993).

7.1.4 Four Types of Counting Processes

In this chapter four types of counting processes are discussed:

1. Homogeneous Poisson processes (HPP)

2. Renewal processes

3. Nonhomogeneous Poisson processes (NHPP)

4. Imperfect repair processes

The Poisson process got its name after the French mathematician Siméon Denis Poisson (1781–1840).

The HPP was introduced in Section 2.10. In the HPP model all the interoccurrence times are independent and exponentially distributed with the same parameter (failure rate) λ.

The renewal process as well as the NHPP are generalizations of the HPP, both having the HPP as a special case. A renewal process is a counting process where the interoccurrence times are independent and identically distributed with an arbitrary life distribution. Upon failure the component is thus replaced or restored to an "as good as new" condition. This is often called a *perfect repair*. Statistical analysis of observed interoccurrence times from a renewal process is discussed in detail in Chapter 11.

The NHPP differs from the HPP in that the rate of occurrences of failures varies with time rather than being a constant. This implies that for an NHPP model the interoccurrence times are neither independent nor identically distributed. The NHPP is often used to model repairable systems that are subject to a *minimal repair* strategy, with negligible repair times. Minimal repair means that a failed system is restored just back to functioning state. After a minimal repair the system continues as if nothing had happened. The likelihood of system failure is the same immediately before and after a failure. A minimal repair thus restores the system to an "as bad as

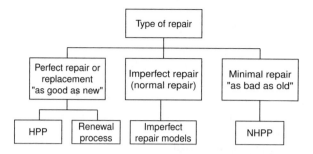

Fig. 7.6 Types of repair and stochastic point processes covered in this book.

old" condition. The minimal repair strategy is discussed, for example, by Ascher and Feingold (1984) and Akersten (1991) who gives a detailed list of relevant references on this subject.

The renewal process and the NHPP represent two extreme types of repair: replacement to an "as good as new" condition and replacement to "as bad as old" (minimal repair), respectively. Most repair actions are, however, somewhere between these extremes and are often called *imperfect repair* or *normal repair*. A number of different models have been proposed for imperfect repair. A survey of some of these models is given in Section 7.5.

The various types of repair and the models covered in this book are illustrated in Fig. 7.6.

7.2 HOMOGENEOUS POISSON PROCESSES

The homogeneous Poisson process was introduced in Section 2.10. The HPP may be defined in a number of different ways. Three alternative definitions of the HPP are presented in the following to illustrate different features of the HPP. The first two definitions are from Ross (1996, pp. 59–60).

Definition 7.3 The counting process $\{N(t), t \geq 0\}$ is said to be an HPP having rate λ, for $\lambda > 0$, if

1. $N(0) = 0$.

2. The process has independent increments.

3. The number of events in any interval of length t is Poisson distributed with mean λt. That is, for all $s, t > 0$,

$$\Pr(N(t+s) - N(s) = n) = \frac{(\lambda t)^n}{n!} e^{-\lambda t} \quad \text{for } n = 0, 1, 2, \ldots \qquad (7.8)$$

□

Note that it follows from property 3 that an HPP has stationary increments and also that $E(N(t)) = \lambda t$, which explains why λ is called the rate of the process.

Definition 7.4 The counting process $\{N(t), t \geq 0\}$ is said to be an HPP having rate λ, for $\lambda > 0$, if

1. $N(0) = 0$.

2. The process has stationary and independent increments.

3. $\Pr(N(\Delta t) = 1) = \lambda \Delta t + o(\Delta t)$.

4. $\Pr(N(\Delta t) \geq 2) = o(\Delta t)$.

□

These two alternative definitions of the HPP are presented to clarify the analogy to the definition of the NHPP which is presented in Section 7.4.

A third definition of the HPP is given, for example, by Cocozza-Thivent (1997, p. 24):

Definition 7.5 The counting process $\{N(t), t \geq 0\}$ is said to be an HPP having rate λ, for $\lambda > 0$, if $N(0) = 0$, and the interoccurrence times $T_1, T_2, \ldots$ are independent and exponentially distributed with parameter λ. □

7.2.1 Main Features of the HPP

The main features of the HPP can be easily deduced from the three alternative definitions:

1. The HPP is a regular (orderly) counting process with independent and stationary increments.

2. The ROCOF of the HPP is constant and independent of time,
$$w(t) = \lambda \quad \text{for all } t \geq 0 \tag{7.9}$$

3. The number of failures in the interval $(t, t+v]$ is Poisson distributed with mean λv,
$$\Pr(N(t+v) - N(t) = n) = \frac{(\lambda v)^n}{n!} e^{-\lambda v}$$
$$\text{for all } t \geq 0, \ v > 0 \tag{7.10}$$

4. The mean number of failures in the time interval $(t, t+v]$ is
$$W(t+v) - W(t) = E(N(t+v) - N(t)) = \lambda v \tag{7.11}$$

Especially note that $E(N(t)) = \lambda t$, and $\text{var}(N(t)) = \lambda t$.

5. The interoccurrence times $T_1, T_2, \ldots$ are independent and identically distributed exponential random variables having mean $1/\lambda$.

6. The time of the nth failure $S_n = \sum_{i=1}^{n} T_i$ has a gamma distribution with parameters (n, λ). Its probability density function is

$$f_{S_n}(t) = \frac{\lambda}{(n-1)!} (\lambda t)^{n-1} e^{-\lambda t} \quad \text{for } t \geq 0 \tag{7.12}$$

Further features of the HPP are presented and discussed, for example, by Ross (1996), Thompson (1988), and Ascher and Feingold (1984).

Remark: Consider a counting process $\{N(t), t \geq 0\}$ where the interoccurrence times $T_1, T_2, \ldots$ are independent and exponentially distributed with parameter λ (i.e., Definition 7.5). The arrival time S_n has, according to (7.12), a gamma distribution with parameters (n, λ).

Since $N(t) = n$ if and only if $S_n \leq t < S_{n+1}$, and the interoccurrence time $T_{n+1} = S_{n+1} - S_n$, we can use the law of total probability to write

$$\begin{aligned}
\Pr(N(t) = n) &= \Pr(S_n \leq t < S_{n+1}) \\
&= \int_0^t \Pr(T_{n+1} > t - s \mid S_n = s) f_{S_n}(s) \, ds \\
&= \int_0^t e^{-\lambda(t-s)} \frac{\lambda}{(n-1)!} (\lambda s)^{n-1} e^{-\lambda s} \, ds \\
&= \frac{(\lambda t)^n}{n!} e^{-\lambda t}
\end{aligned} \tag{7.13}$$

We have thus shown that $N(t)$ has a Poisson distribution with mean λt, in accordance with Definition 7.3. □

7.2.2 Asymptotic Properties

The following asymptotic results apply:

$$\frac{N(t)}{t} \to \lambda \quad \text{with probability 1, when } t \to \infty$$

and

$$\frac{N(t) - \lambda t}{\sqrt{\lambda t}} \xrightarrow{\mathcal{L}} \mathcal{N}(0, 1)$$

such that

$$\Pr\left(\frac{N(t) - \lambda t}{\sqrt{\lambda t}} \leq t\right) \approx \Phi(t) \quad \text{when } t \to \infty \tag{7.14}$$

where $\Phi(t)$ denotes the distribution function of the standard normal (Gaussian) distribution $\mathcal{N}(0, 1)$.

7.2.3 Estimate and Confidence Interval

An obvious estimator for λ is

$$\hat{\lambda} = \frac{N(t)}{t} \qquad (7.15)$$

The estimator is unbiased, $E(\hat{\lambda}) = \lambda$, with variance, $\text{var}(\hat{\lambda}) = \lambda/t$.

A $1 - \varepsilon$ confidence interval for λ, when $N(t) = n$ events (failures) are observed during a time interval of length t, is given by (e.g., see Cocozza-Thivent 1997, p. 63)

$$\left(\frac{1}{2t} z_{1-\varepsilon/2,\, 2n},\ \frac{1}{2t} z_{\varepsilon/2,\, 2(n+1)} \right) \qquad (7.16)$$

where $z_{\varepsilon,\nu}$ denotes the upper $100\varepsilon\%$ percentile of the chi-square (χ^2) distribution with ν degrees of freedom. A table of $z_{\varepsilon,\nu}$ for some values of ε and ν is given in Appendix F.

In some situations it is of interest to give an upper $(1 - \varepsilon)$ confidence limit for λ. Such a limit is obtained through the one-sided confidence interval given by

$$\left(0,\ \frac{1}{2t} z_{\varepsilon,\, 2(n+1)} \right) \qquad (7.17)$$

Note that this interval is applicable even if no failures ($N(t) = 0$) are observed during the interval $(0, t)$.

7.2.4 Sum and Decomposition of HPPs

Let $\{N_1(t), t \geq 0\}$ and $\{N_2(t), t \geq 0\}$ be two independent HPPs with rates λ_1 and λ_2, respectively. Further, let $N(t) = N_1(t) + N_2(t)$. It is then easy to verify that $\{N(t), t \geq 0\}$ is an HPP with rate $\lambda = \lambda_1 + \lambda_2$.

Suppose that in an HPP $\{N(t), t \geq 0\}$ we can classify each event as type 1 and type 2 that are occurring with probability p and $(1 - p)$, respectively. This is, for example, the case when we have a sequence of failures with two different failure modes (1 and 2), and p equals the relative number of failure mode 1. Then the number of events, $N_1(t)$ of type 1, and $N_2(t)$ of type 2, in the interval $(0, t]$ also give rise to HPPs, $\{N_1(t), t \geq 0\}$ and $\{N_2(t), t \geq 0\}$ with rates $p\lambda$ and $(1 - p)\lambda$, respectively. Furthermore, the two processes are independent. For a formal proof, see, for example, Ross (1996, p. 69). These results can be easily generalized to more than two cases.

Example 7.3

Consider an HPP $\{N(t), t \geq 0\}$ with rate λ. Some failures develop into a consequence C, others do not. The failures developing into a consequence C are denoted a C-failure. The consequence C may, for example, be a specific failure mode. The probability that a failure develops into consequence C is denoted p and is constant for each failure. The failure consequences are further assumed to be independent of

244 COUNTING PROCESSES

each other. Let $N_C(t)$ denote the number of C-failures in the time interval $(0, t]$. When $N(t)$ is equal to n, $N_C(t)$ will have a binomial distribution:

$$\Pr(N_C(t) = y \mid N(t) = n) = \binom{n}{y} p^y (1-p)^{n-y} \quad \text{for } y = 0, 1, 2, \ldots, n$$

The marginal distribution of $N_C(t)$ is

$$\begin{aligned}
\Pr(N_C(t) = y) &= \sum_{n=y}^{\infty} \binom{n}{y} p^y (1-p)^{n-y} \frac{(\lambda t)^n}{n!} e^{-\lambda t} \\
&= \frac{p^y e^{-\lambda t}}{y!} (\lambda t)^y \sum_{n=y}^{\infty} \frac{[\lambda t (1-p)]^{n-y}}{(n-y)!} \\
&= \frac{(p\lambda t)^y e^{-\lambda t}}{y!} \sum_{x=0}^{\infty} \frac{[\lambda t (1-p)]^x}{x!} \\
&= \frac{(p\lambda t)^y e^{-\lambda t}}{y!} e^{\lambda t (1-p)} \\
&= \frac{(p\lambda t)^y}{y!} e^{-p\lambda t} \qquad (7.18)
\end{aligned}$$

Thus $\{N_C(t), t \geq 0\}$ is an HPP with rate $p\lambda$, and the mean number of C-failures in the time interval $(0, t]$ is

$$E(N_C(t)) = p\lambda t$$

□

7.2.5 Conditional Distribution of Failure Time

Suppose that exactly one failure of an HPP with rate λ is known to have occurred some time in the interval $(0, t_0]$. We want to determine the distribution of the time T_1 at which the failure occurred:

$$\begin{aligned}
\Pr(T_1 \leq t \mid N(t_0) = 1) &= \frac{\Pr(T_1 \leq t \cap N(t_0) = 1)}{\Pr(N(t_0) = 1)} \\
&= \frac{\Pr(1 \text{ failure in } (0, t] \cap 0 \text{ failures in } (t, t_0])}{\Pr(N(t_0) = 1)} \\
&= \frac{\Pr(N(t) = 1) \cdot \Pr(N(t_0) - N(t) = 0)}{\Pr(N(t_0) = 1)} \\
&= \frac{\lambda t e^{-\lambda t} \, e^{-\lambda (t_0 - t)}}{\lambda t_0 e^{-\lambda t_0}} \\
&= \frac{t}{t_0} \quad \text{for } 0 < t \leq t_0 \qquad (7.19)
\end{aligned}$$

When we know that exactly one failure (event) takes place in the time interval $(0, t_0]$, the time at which the failure occurs is *uniformly* distributed over $(0, t_0]$. Thus each interval of equal length in $(0, t_0]$ has the same probability of containing the failure. The expected time at which the failure occurs is

$$E(T_1 \mid N(t_0) = 1) = \frac{t_0}{2} \qquad (7.20)$$

7.2.6 Compound HPPs

Consider an HPP $\{N(t), t \geq 0\}$ with rate λ. A random variable V_i is associated to failure event i, for $i = 1, 2, \ldots$. The variable V_i may, for example, be the consequence (economic loss) associated to failure i. The variables $V_1, V_2, \ldots$ are assumed to be independent with common distribution function

$$F_V(v) = \Pr(V \leq v)$$

The variables $V_1, V_2, \ldots$ are further assumed to be independent of $N(t)$. The cumulative consequence at time t is

$$Z(t) = \sum_{i=1}^{N(t)} V_i \quad \text{for } t \geq 0 \qquad (7.21)$$

The process $\{Z(t), t \geq 0\}$ is called a *compound Poisson process*. Compound Poisson processes are discussed, for example, by Ross (1996, p. 82) and Taylor and Karlin (1984, p. 200). The same model is called a *cumulative damage model* by Barlow and Proschan (1975, p. 91). To determine the mean value of $Z(t)$, we need the following important theorem:

Theorem 7.1 *(Wald's Equation)* Let $X_1, X_2, X_3, \ldots$ be independent and identically distributed random variables with finite mean μ. Further let N be a stochastic integer variable so that the event $(N = n)$ is independent of $X_{n+1}, X_{n+2}, \ldots$ for all $n = 1, 2, \ldots$. Then

$$E\left(\sum_{i=1}^{N} X_i\right) = E(N) \cdot \mu \qquad (7.22)$$

□

A proof of Wald's equation may be found, for example, in Ross (1996, p. 105). The variance of $\sum_{i=1}^{N} X_i$ is (see Ross 1996, pp. 22–23):

$$\text{var}\left(\sum_{i=1}^{N} X_i\right) = E(N) \cdot \text{var}(X_i) + [E(X_i)]^2 \cdot \text{var}(N) \qquad (7.23)$$

Let $E(V_i) = \nu$ and $\text{var}(V_i) = \tau^2$. From (7.22) and (7.23) we get

$$E(Z(t)) = \nu \lambda t \quad \text{and} \quad \text{var}(Z(t)) = \lambda(\nu^2 + \tau^2)t$$

Assume now that the consequences V_i are all positive, that is, $\Pr(V_i > 0) = 1$ for all i, and that a total system failure occurs as soon as $Z(t) > c$ for some specified critical value c. Let T_c denote the time to system failure. Note that $T_c > t$ if and only if $Z(t) \leq c$.

Let $V_0 = 0$, then

$$\begin{aligned}
\Pr(T_c > t) &= \Pr(Z(t) \leq c) = \Pr\left(\sum_{i=0}^{N(t)} V_i \leq c\right) \\
&= \sum_{n=0}^{\infty} \Pr\left(\sum_{i=0}^{n} V_i \leq c \mid N(t) = n\right) \frac{(\lambda t)^n}{n!} e^{-\lambda t} \\
&= \sum_{n=0}^{\infty} \frac{(\lambda t)^n}{n!} e^{-\lambda t} F_V^{(n)}(c) \qquad (7.24)
\end{aligned}$$

where $F_V^{(n)}(v)$ denotes the distribution function of $\sum_{i=0}^{n} V_i$, and the last equality is due to the fact that $N(t)$ is independent of $V_1, V_2, \ldots$.

The mean time to total system failure is thus

$$\begin{aligned}
E(T_c) &= \int_0^{\infty} \Pr(T_c > t)\, dt \\
&= \sum_{n=0}^{\infty} \left(\int_0^{\infty} \frac{(\lambda t)^n}{n!} e^{-\lambda t}\, dt\right) F_V^{(n)}(c) \\
&= \frac{1}{\lambda} \sum_{n=0}^{\infty} F_V^{(n)}(c) \qquad (7.25)
\end{aligned}$$

Example 7.4
Consider a sequence of failure events that can be described as an HPP $\{N(t), t \geq t\}$ with rate λ. Failure i has consequence V_i, where $V_1, V_2, \ldots$ are independent and exponentially distributed with parameter ρ. The sum $\sum_{i=1}^{n} V_i$ therefore has a gamma distribution with parameters (n, ρ) [see Section 2.11 (2.45)]:

$$F_V^{(n)}(v) = 1 - \sum_{k=0}^{n-1} \frac{(\rho v)^k}{k!} e^{-\rho v} = \sum_{k=n}^{\infty} \frac{(\rho v)^k}{k!} e^{-\rho v}$$

Total system failure occurs as soon as $Z(t) = \sum_{i=1}^{N(t)} V_i > c$. The mean time to total system failure is given by (7.25) where

$$\begin{aligned}
\sum_{n=0}^{\infty} F_V^{(n)}(c) &= \sum_{n=0}^{\infty} \sum_{k=n}^{\infty} \frac{(\rho c)^k}{k!} e^{-\rho c} = \sum_{k=0}^{\infty} \sum_{n=0}^{k} \frac{(\rho c)^k}{k!} e^{-\rho c} \\
&= \sum_{k=0}^{\infty} (1+k) \frac{(\rho c)^k}{k!} e^{-\rho c} = 1 + \rho c
\end{aligned}$$

Hence when the consequences $V_1, V_2, \ldots$ are exponentially distributed with parameter ρ, the mean time to total system failure is

$$E(T_c) = \frac{1 + \rho c}{\lambda} \qquad (7.26)$$

□

The distribution of the time T_c to total system failure is by Barlow and Proschan (1975, p. 94) shown to be an increasing failure rate average (IFRA) distribution for *any* distribution $F_V(v)$. (IFRA distributions are discussed in Section 2.19).

7.3 RENEWAL PROCESSES

Renewal theory had its origin in the study of strategies for replacement of technical components, but later it was developed as a general theory within stochastic processes. As the name of the process indicates, it is used to model *renewals*, or replacement of equipment. This section gives a summary of some main aspects of renewal theory which are of particular interest in reliability analysis. This includes formulas for calculation of exact availability and mean number of failures within a given time interval. The latter can, for example, be used to determine optimal allocation of spare parts.

Example 7.5
A component is put into operation and is functioning at time $t = 0$. When the component fails at time T_1, it is replaced by a new component of the same type, or restored to an "as good as new" condition. When this component fails at time $T_1 + T_2$, it is again replaced, and so on. The replacement time is assumed to be negligible. The life lengths $T_1, T_2, \ldots$ are assumed to be independent and identically distributed. The number of failures, and *renewals*, in a time interval $(0, t]$ is denoted $N(t)$. □

7.3.1 Basic Concepts

A *renewal process* is a counting process $\{N(t), t \geq 0\}$ with interoccurrence times $T_1, T_2, \ldots$ that are independent and identically distributed with distribution function

$$F_T(t) = \Pr(T_i \leq t) \quad \text{for } t \geq 0, \ i = 1, 2, \ldots$$

The events that are observed (mainly failures) are called *renewals*, and $F_T(t)$ is called the underlying distribution of the renewal process. We will assume that $E(T_i) = \mu$ and $\text{var}(T_i) = \sigma^2 < \infty$ for $i = 1, 2, 3, \ldots$. Note that the HPP discussed in Section 7.2 is a renewal process where the underlying distribution is exponential with parameter λ. A renewal process may thus be considered as a generalization of the HPP.

The concepts that were introduced for a general counting process in Section 7.1.2 are also relevant for a renewal process. The theory of renewal processes has, however, been developed as a specific theory, and many of the concepts have therefore been given specific names. We will therefore list the main concepts of renewal processes and introduce the necessary terminology.

1. The time until the nth renewal (the nth arrival time), S_n:

$$S_n = T_1 + T_2 + \cdots + T_n = \sum_{i=1}^{n} T_i \qquad (7.27)$$

2. The number of renewals in the time interval $(0, t]$:

$$N(t) = \max\{n : S_n \leq t\} \qquad (7.28)$$

3. The renewal function:

$$W(t) = E(N(t)) \qquad (7.29)$$

Thus $W(t)$ is the mean number of renewals in the time interval $(0, t]$.

4. The renewal density:

$$w(t) = \frac{d}{dt} W(t) \qquad (7.30)$$

Note that the renewal density coincides with the rate of the process defined in (7.2), which is called the rate of occurrence of failures when the renewals are failures. The mean number of renewals in the time interval $(t_1, t_2]$ is

$$W(t_2) - W(t_1) = \int_{t_1}^{t_2} w(t)\, dt \qquad (7.31)$$

The relation between the renewal periods T_i and the number of renewals $N(t)$, the renewal process is illustrated in Fig. 7.1. The properties of renewal processes are discussed in detail by Cox (1962), Ross (1996), and Cocozza-Thivent (1997).

7.3.2 The Distribution of S_n

To find the exact distribution of the time to the nth renewal S_n is often very complicated. We will outline an approach that may be used, at least in some cases. Let $F^{(n)}(t)$ denote the distribution function of $S_n = \sum_{i=1}^{n} T_i$.

Since S_n may be written as $S_n = S_{n-1} + T_n$, and S_{n-1} and T_n are independent, the distribution function of S_n is the *convolution* of the distribution functions of S_{n-1} and T_n, respectively,

$$F^{(n)}(t) = \int_0^t F^{(n-1)}(t - x)\, dF_T(x) \qquad (7.32)$$

The convolution of two (life) distributions F and G is often denoted $F * G$, meaning that $F * G(t) = \int_0^t G(t-x) \, dF(x)$. Equation (7.32) can therefore be written $F^{(n)} = F_T * F^{(n-1)}$.

When $F_T(t)$ is absolutely continuous with probability density function $f_T(t)$, the probability density function $f^{(n)}(t)$ of S_n may be found from

$$f^{(n)}(t) = \int_0^t f^{(n-1)}(t-x) f_T(x) \, dx \tag{7.33}$$

By successive integration of (7.32) for $n = 2, 3, 4, \ldots$, the probability distribution of S_n for a specified value of n can, in principle, be found.

It may also sometimes be relevant to use Laplace transforms to find the distribution of S_n. The Laplace transform of Equation (7.33) is (see Appendix B)

$$f^{*(n)}(s) = (f_T^*(s))^n \tag{7.34}$$

The probability density function of S_n can now, at least in principle, be determined from the inverse Laplace transform of (7.34).

In practice it is often very time-consuming and complicated to find the exact distribution of S_n from formulas (7.32) and (7.34). Often an approximation to the distribution of S_n is sufficient.

From the strong law of large numbers, that is, with probability 1,

$$\frac{S_n}{n} \to \mu \quad \text{as} \quad n \to \infty \tag{7.35}$$

According to the central limit theorem, $S_n = \sum_{i=1}^n T_i$ is asymptotically normally distributed:

$$\frac{S_n - n\mu}{\sigma\sqrt{n}} \xrightarrow{\mathcal{L}} \mathcal{N}(0, 1)$$

and

$$F^{(n)}(t) = \Pr(S_n \leq t) \approx \Phi\left(\frac{t - n\mu}{\sigma\sqrt{n}}\right) \tag{7.36}$$

where $\Phi(\cdot)$ denotes the distribution function of the standard normal distribution $\mathcal{N}(0, 1)$.

Example 7.6 IFR Interoccurrence Times

Consider a renewal process where the interoccurrence times have an increasing failure rate (IFR) distribution $F_T(t)$ (see Section 2.19) with mean time to failure μ. In this case, Barlow and Proschan (1965, p. 27) have shown that the survivor function, $R_T(t) = 1 - F_T(t)$, satisfies

$$R_T(t) \geq e^{-t/\mu} \quad \text{when} \quad t < \mu \tag{7.37}$$

The right-hand side of (7.37) is the survivor function of a random variable U_j with exponential distribution with failure rate $1/\mu$. Let us assume that we have n independent random variable $U_1, U_2, \ldots, U_n$ with the same distribution. The distribution of $\sum_{j=1}^{n} U_j$ has then a gamma distribution with parameters $(n, 1/\mu)$ (see Section 2.11), and we therefore get

$$\begin{aligned} 1 - F^{(n)}(t) &= \Pr(S_n > t) = \Pr(T_1 + T_2 + \cdots + T_n > t) \\ &\geq \Pr(U_1 + U_2 + \cdots + U_n > t) = \sum_{j=0}^{n-1} \frac{(t/\mu)^j}{j!} e^{-t/\mu} \end{aligned}$$

Hence

$$F^n(t) \leq 1 - \sum_{j=0}^{n-1} \frac{(t/\mu)^j}{j!} e^{-t/\mu} \quad \text{for } t < \mu \tag{7.38}$$

For a renewal (failure) process where the interoccurrence times have an IFR distribution with mean μ, equation (7.38) provides a conservative bound for the probability that the nth failure will occur before time t, when $t < \mu$. □

7.3.3 The Distribution of $N(t)$

From the strong law of large numbers, that is, with probability 1,

$$\frac{N(t)}{t} \to \frac{1}{\mu} \quad \text{as } t \to \infty \tag{7.39}$$

When t is large, $N(t) \approx t/\mu$. This means that $N(t)$ is approximately a linear function of t when t is large. In Fig. 7.7 the number of renewals $N(t)$ is plotted as a function of t for a simulated renewal process where the underlying distribution is Weibull with parameters $\lambda = 1$ and $\alpha = 3$.

From the definition of $N(t)$ and S_n, it follows that

$$\Pr(N(t) \geq n) = \Pr(S_n \leq t) = F^{(n)}(t) \tag{7.40}$$

and

$$\begin{aligned} \Pr(N(t) = n) &= \Pr(N(t) \geq n) - \Pr(N(t) \geq n+1) \\ &= F^{(n)}(t) - F^{(n+1)}(t) \end{aligned} \tag{7.41}$$

For large values of n we can apply (7.36) and obtain

$$\Pr(N(t) \leq n) \approx \Phi\left(\frac{(n+1)\mu - t}{\sigma}\right) \tag{7.42}$$

and

$$\Pr(N(t) = n) \approx \Phi\left(\frac{t - n\mu}{\sigma\sqrt{n}}\right) - \Phi\left(\frac{t - (n+1)\mu}{\sigma\sqrt{n+1}}\right) \tag{7.43}$$

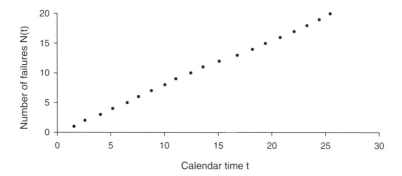

Fig. 7.7 Number of renewals $N(t)$ as a function of t for a simulated renewal process where the underlying distribution is Weibull with parameters $\lambda = 1$ and $\alpha = 3$.

Takács (1956) derived the following alternative approximation formula which is valid when t is large:

$$\Pr(N(t) \leq n) \approx \Phi\left(\frac{n - (t/\mu)}{\sigma\sqrt{t/\mu^3}}\right) \qquad (7.44)$$

A proof of (7.44) is provided in Ross (1996, p. 109).

7.3.4 The Renewal Function

Since $N(t) \geq n$ if and only if $S_n \leq t$, we get that (see Problem 7.4)

$$W(t) = E(N(t)) = \sum_{n=1}^{\infty} \Pr(N(t) \geq n) = \sum_{n=1}^{\infty} \Pr(S_n \leq t) = \sum_{n=1}^{\infty} F^{(n)}(t) \qquad (7.45)$$

An integral equation for $W(t)$ may be obtained by combining (7.45) and (7.32):

$$\begin{aligned}
W(t) &= F_T(t) + \sum_{r=2}^{\infty} F^{(r)}(t) = F_T(t) + \sum_{r=1}^{\infty} F^{(r+1)}(t) \\
&= F_T(t) + \sum_{r=1}^{\infty} \int_0^t F^{(r)}(t-x)\, dF_T(x) \\
&= F_T(t) + \int_0^t \sum_{r=1}^{\infty} F^{(r)}(t-x)\, dF_T(x) \\
&= F_T(t) + \int_0^t W(t-x)\, dF_T(x) \qquad (7.46)
\end{aligned}$$

This equation is known as the *fundamental renewal equation* and can sometimes be solved for $W(t)$.

Equation (7.46) can also be derived by a more direct argument. By conditioning on the time T_1 of the first renewal, we obtain

$$W(t) = E(N(t)) = E(E(N(t) \mid T_1))$$
$$= \int_0^\infty E(N(t) \mid T_1 = x) \, dF_{T_1}(x) \quad (7.47)$$

where

$$E(N(t) \mid T_1 = x) = \begin{cases} 0 & \text{when } t < x \\ 1 + W(t - x) & \text{when } t \geq x \end{cases} \quad (7.48)$$

If the first renewal occurs at time x for $x \leq t$, the process starts over again from this point in time. The mean number of renewals in $(0, t]$ is thus 1 plus the mean number of renewals in $(x, t]$, which is $W(t - x)$.

Combining equations (7.47) and (7.48) yields

$$W(t) = \int_0^t (1 + W(t - x)) \, dF_T(x)$$
$$= F_T(t) + \int_0^t W(t - x) \, dF_T(x)$$

and thus an alternative derivation of (7.46) is provided.

The exact expression for the renewal function $W(t)$ is often difficult to determine from (7.46). Approximation formulas and bounds may therefore be useful.

Since $W(t)$ is the expected number of renewals in the interval $(0, t]$, the average length μ of each renewal is approximately $t/W(t)$. We should therefore expect that when $t \to \infty$, we get

$$\lim_{t \to \infty} \frac{W(t)}{t} = \frac{1}{\mu} \quad (7.49)$$

This result is known as the *elementary renewal theorem* and is valid for a general renewal process. A proof may, for example, be found in Ross (1996, p. 107).

When the renewals are component failures, the mean number of failures in $(0, t]$ is thus approximately

$$E(N(t)) = W(t) \approx \frac{t}{\mu} = \frac{t}{\text{MTBF}} \quad \text{when } t \text{ is large}$$

where $\mu = \text{MTBF}$ denotes the mean time between failures.

From the elementary renewal theorem (7.49), the mean number of renewals in the interval $(0, t]$ is

$$W(t) \approx \frac{t}{\mu} \quad \text{when } t \text{ is large}$$

The mean number of renewals in the interval $(t, t+u]$ is

$$W(t+u) - W(t) \approx \frac{u}{\mu} \quad \text{when } t \text{ is large, and } u > 0 \tag{7.50}$$

and the underlying distribution $F_T(t)$ is nonlattice. This result is known as *Blackwell's theorem*, and a proof may be found in Feller (1968, Chapter XI).

Blackwell's theorem (7.50) has been generalized by Smith (1958), who showed that when the underlying distribution $F_T(t)$ is nonlattice, then

$$\lim_{t \to \infty} \int_0^t Q(t-x) \, dW(x) = \frac{1}{\mu} \int_0^\infty Q(u) \, du \tag{7.51}$$

where $Q(t)$ is a nonnegative, nonincreasing function which is Riemann integrable over $(0, \infty)$. This result is known as the *key renewal theorem*.

By introducing $Q(t) = \alpha^{-1}$ for $0 < t \le \alpha$ and $Q(t) = 0$ otherwise, in (7.51), we get Blackwell's theorem (7.50).

Let

$$F_e(t) = \frac{1}{\mu} \int_0^t (1 - F_T(u)) \, du \tag{7.52}$$

where $F_e(t)$ is a distribution function with a special interpretation that is further discussed on page 267. By using $Q(t) = 1 - F_e(t)$ in (7.51) we get

$$\lim_{t \to \infty} \left(W(t) - \frac{t}{\mu} \right) = \frac{E(T_i^2)}{2\mu^2} - 1 = \frac{\sigma^2 + \mu^2}{2\mu^2} - 1 = \frac{1}{2}\left(\frac{\sigma^2}{\mu^2} - 1 \right)$$

if $E(T_i^2) = \sigma^2 + \mu^2 < \infty$. We may thus use the following approximation when t is large

$$W(t) \approx \frac{t}{\mu} + \frac{1}{2}\left(\frac{\sigma^2}{\mu^2} - 1 \right) \tag{7.53}$$

Upper and lower bounds for the renewal function are supplied on page 262.

7.3.5 The Renewal Density

When $F_T(t)$ has density $f_T(t)$, we may differentiate (7.45) and get

$$w(t) = \frac{d}{dt} W(t) = \frac{d}{dt} \sum_{n=1}^\infty F_T^{(n)}(t) = \sum_{n=1}^\infty f_T^{(n)}(t) \tag{7.54}$$

This formula can sometimes be used to find the renewal density $w(t)$. Another approach is to differentiate (7.46) with respect to t

$$w(t) = f_T(t) + \int_0^t w(t-x) f_T(x) \, dx \tag{7.55}$$

Yet another approach is to use Laplace transforms. From Appendix B the Laplace transform of (7.55) is

$$w^*(s) = f_T^*(s) + w^*(s) \cdot f_T^*(s)$$

Hence

$$w^*(s) = \frac{f_T^*(s)}{1 - f_T^*(s)} \qquad (7.56)$$

Remark: According to (7.5) the probability of a failure (renewal) in a short interval $(t, t+\Delta t]$ is approximately $w(t) \cdot \Delta t$. Since the probability that the *first* failure occurs in $(t, t + \Delta t]$ is approximately $f_T(t) \cdot \Delta t$, we can use (7.55) to conclude that a "later" failure (i.e., not the first) will occur in $(t, t + \Delta t]$ with probability approximately equal to $\left(\int_0^t w(t - x) f_T(x) \, dx \right) \cdot \Delta t$. □

The exact expression for the renewal density $w(t)$ is often difficult to determine from (7.54), (7.55), and (7.56). In the same way as for the renewal function, we therefore have to suffice with approximation formulas and bounds.

From (7.49) we should expect that

$$\lim_{t \to \infty} w(t) = \frac{1}{\mu} \qquad (7.57)$$

Smith (1958) has shown that (7.57) is valid for a renewal process with underlying probability density function $f_T(t)$ when there exists a $p > 1$ such that $|f_T(t)|^p$ is Riemann integrable. The renewal density $w(t)$ will therefore approach a constant $1/\mu$ when t is large.

Consider a renewal process where the renewals are component failures. The interoccurrence times $T_1, T_2, \ldots$ then denote the times to failure, and $S_1, S_2, \ldots$ are the times when the failures occur. Let $z(t)$ denote the failure rate [force of mortality (FOM)] function of the time to the first failure T_1. The conditional renewal density (ROCOF) $w_C(t)$ in the interval $(0, T_1)$ must equal $z(t)$ (see page 238). When the first failure has occurred, the component will be renewed or replaced and started up again with the same failure rate (FOM) as for the initial component. The conditional renewal rate (ROCOF) may then be expressed as

$$w_C(t) = z\left(t - S_{N(t-)}\right)$$

where $t - S_{N(t-)}$ is the time since the last failure strictly before time t. The conditional ROCOF is illustrated in Fig. 7.8 when the interoccurrence times are Weibull distributed with scale parameter $\lambda = 1$ and shape parameter $\alpha = 3$. The plot is based on simulated interoccurrence times from this distribution.

Example 7.7
Consider a renewal process where the renewal periods $T_1, T_2, \ldots$ are independent and gamma distributed with parameters $(2, \lambda)$, with probability density function

$$f_T(t) = \lambda^2 t \, e^{-\lambda t} \quad \text{for } t > 0, \lambda > 0$$

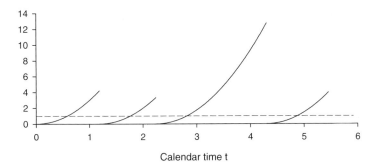

Fig. 7.8 Illustration of the conditional ROCOF (fully drawn line) for simulated data from a Weibull distribution with parameters $\alpha = 3$ and $\lambda = 1$. The corresponding asymptotic renewal density is drawn by a dotted line.

The mean renewal period is $E(T_i) = \mu = 2/\lambda$, and the variance is $\text{var}(T_i) = \sigma^2 = 2/\lambda^2$. The time until the nth renewal, S_n, is gamma distributed (see Section 2.11) with probability density function

$$f^{(n)}(t) = \frac{\lambda}{(2n-1)!} (\lambda t)^{2n-1} e^{-\lambda t} \quad \text{for } t > 0$$

The renewal density is according to (7.54)

$$w(t) = \sum_{n=1}^{\infty} f^{(n)}(t) = \lambda e^{-\lambda t} \sum_{n=1}^{\infty} \frac{(\lambda t)^{2n-1}}{(2n-1)!}$$

$$= \lambda e^{-\lambda t} \cdot \frac{e^{\lambda t} - e^{-\lambda t}}{2} = \frac{\lambda}{2} (1 - e^{-2\lambda t})$$

The renewal function is

$$W(t) = \int_0^t w(x)\, dx = \frac{\lambda}{2} \int_0^t (1 - e^{-2\lambda x})\, dx = \frac{\lambda t}{2} - \frac{1}{4}(1 - e^{-2\lambda t}) \quad (7.58)$$

The renewal density $w(t)$ and the renewal function $W(t)$ are illustrated in Fig. 7.9 for $\lambda = 1$. Note that when $t \to \infty$, then

$$W(t) \to \frac{\lambda t}{2} = \frac{t}{\mu}$$

$$w(t) \to \frac{\lambda}{2} = \frac{1}{\mu}$$

in accordance with (7.49) and (7.57), respectively. We may further use (7.53) to find a better approximation for the renewal function $W(t)$. From (7.58) we get the left-hand

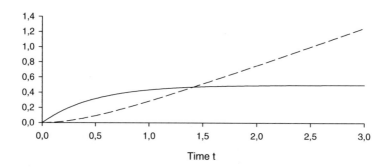

Fig. 7.9 Renewal density $w(t)$ (fully drawn line) and renewal function $W(t)$ (dotted line) for Example 7.7, with ($\lambda = 1$).

side of (7.53):

$$W(t) - \frac{t}{\mu} = W(t) - \frac{\lambda t}{2} \to -\frac{1}{4} \quad \text{when } t \to \infty$$

The right hand side of (7.53) is (with $\mu = 2/\mu$ and $\sigma^2 = 2/\lambda^2$)

$$\frac{t}{\mu} + \frac{1}{2}\left(\frac{\sigma^2}{2\mu^2} - 1\right) = \frac{t}{\mu} - \frac{1}{4}$$

We can therefore use the approximation

$$W(t) \approx \frac{\lambda t}{2} - \frac{1}{4} \quad \text{when } t \text{ is large}$$

$\square$

Example 7.8

Consider a renewal process where the renewal periods $T_1, T_2, \ldots$ are independent and Weibull distributed with shape parameter α and scale parameter λ. In this case the renewal function $W(t)$ cannot be deduced directly from (7.45). Smith and Leadbetter (1963) have, however, shown that $W(t)$ can be expressed as an infinite, absolutely convergent series where the terms can be found by a simple recursive procedure. They show that $W(t)$ can be written

$$W(t) = \sum_{k=1}^{\infty} \frac{(-1)^{k-1} \cdot A_k \cdot (\lambda t)^{k\alpha}}{\Gamma(k\alpha + 1)} \tag{7.59}$$

By introducing this expression for $W(t)$ in the fundamental renewal equation, the constants A_k; $k = 1, 2, \ldots$ can be determined. The calculation, which is quite

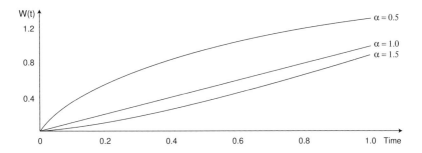

Fig. 7.10 The renewal function for Weibull distributed renewal periods with $\lambda = 1$ and $\alpha = 0.5$, $\alpha = 1$ and $\alpha = 1.5$. (The figure is adapted from Smith and Leadbetter, 1963).

comprehensive, leads to the following *recursion* formula:

$$
\begin{aligned}
A_1 &= \gamma_1 \\
A_2 &= \gamma_2 - \gamma_1 A_1 \\
A_3 &= \gamma_3 - \gamma_1 A_2 - \gamma_2 A_1 \\
&\vdots \\
A_n &= \gamma_n - \sum_{j=1}^{n-1} \gamma_j A_{n-j} \\
&\vdots
\end{aligned}
\qquad (7.60)
$$

where

$$\gamma_n = \frac{\Gamma(n\alpha + 1)}{n!} \quad \text{for } n = 1, 2, \ldots$$

For $\alpha = 1$, the Weibull distribution is an exponential distribution with parameter λ. In this case

$$\gamma_n = \frac{\Gamma(n+1)}{n!} = 1 \quad \text{for } n = 1, 2, \ldots$$

This leads to

$$
\begin{aligned}
A_1 &= 1 \\
A_n &= 0 \quad \text{for } n \geq 2
\end{aligned}
$$

The renewal function is thus according to (7.59)

$$W(t) = \frac{(-1)^0 A_1 \lambda t}{\Gamma(2)} = \lambda t$$

The renewal function $W(t)$ is illustrated in Fig. 7.10 for $\lambda = 1$ and three values of α. □

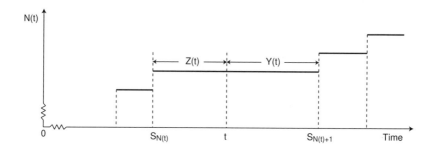

Fig. 7.11 The age $Z(t)$ and the remaining lifetime $Y(t)$.

7.3.6 Age and Remaining Lifetime

The *age* $Z(t)$ of an item operating at time t is defined as

$$Z(t) = \begin{cases} t & \text{for } N(t) = 0 \\ t - S_{N(t)} & \text{for } N(t) > 0 \end{cases} \qquad (7.61)$$

The *remaining lifetime* $Y(t)$ of an item that is in operation at time t is given as

$$Y(t) = S_{N(t)+1} - t \qquad (7.62)$$

The age $Z(t)$ and the remaining lifetime $Y(t)$ are illustrated in Fig. 7.11. The remaining lifetime is also called the residual life, the excess life, or the forward recurrence time (e.g., see Ross 1996, and Ascher and Feingold 1984). Note that $Y(t) > y$ is equivalent to no renewal in the time interval $(t, t + y]$.

Consider a renewal process where the renewals are component failures, and let T denote the time from start-up to the first failure. The distribution of the remaining life $Y(t)$ of the component at time t is given by

$$\Pr(Y(t) > y) = \Pr(T > y + t \mid T > t) = \frac{\Pr(T > y + t)}{\Pr(T > t)}$$

and the mean remaining lifetime at time t is

$$E(Y(t)) = \frac{1}{\Pr(T > t)} \int_t^\infty \Pr(T > u)\, du$$

See also Section 2.7, where $E(Y(t))$ was called the mean residual life (MRL) at time t. When T has an exponential distribution with failure rate λ, the mean remaining lifetime at time t is $1/\lambda$ which is an obvious result because of the memoryless property of the exponential distribution.

Limiting distribution Consider a renewal process with a nonlattice underlying distribution $F_T(t)$. We observe the process at time t. The time till the next failure is the remaining lifetime $Y(t)$. The limiting distribution of $Y(t)$ when $t \to \infty$ is (see Ross 1996, p. 116)

$$\lim_{t \to \infty} \Pr(Y(t) \le t) = F_e(t) = \frac{1}{\mu} \int_0^t (1 - F_T(u))\,du \qquad (7.63)$$

which is the same distribution as we used in (7.52). The mean of the limiting distribution $F_e(t)$ of the remaining lifetime is

$$\begin{aligned}
E(Y) &= \int_0^\infty \Pr(Y > y)\,dy = \int_0^\infty (1 - F_e(y))\,dy \\
&= \frac{1}{\mu} \int_0^\infty \int_y^\infty \Pr(T > t)\,dt\,dy = \frac{1}{\mu} \int_0^\infty \int_0^t \Pr(T > t)\,dy\,dt \\
&= \frac{1}{\mu} \int_0^\infty t\,\Pr(T > t)\,dt = \frac{1}{2\mu} \int_0^\infty \Pr(T > \sqrt{x})\,dx \\
&= \frac{1}{2\mu} \int_0^\infty \Pr(T^2 > x)\,dx = \frac{E(T^2)}{2\mu} = \frac{\sigma^2 + \mu^2}{2\mu}
\end{aligned}$$

where $E(T) = \mu$ and $\text{var}(T) = \sigma^2$, and we assume that $E(T^2) = \sigma^2 + \mu^2 < \infty$. We have thus shown that the limiting mean remaining life is

$$\lim_{t \to \infty} E(Y(t)) = \frac{\sigma^2 + \mu^2}{2\mu} \qquad (7.64)$$

Example 7.7 (Cont.)
Again, consider he renewal process in Example 7.7 where the underlying distribution was a gamma distribution with parameters $(2, \lambda)$, with mean time between renewals $E(T_i) = \mu = 2/\lambda$ and variance $\text{var}(T_i) = 2/\lambda^2$. The mean remaining life of an item that is in operation at time t far from now is from (7.64)

$$E(Y(t)) \approx \frac{\sigma^2 + \mu^2}{2\mu} = \frac{3}{2\lambda} \quad \text{when } t \text{ is large}$$

□

The distribution of the age $Z(t)$ of an item that is in operation at time t can be derived by starting with

$$\begin{aligned}
Z(t) > z &\iff \text{no renewals in } (t - z, t) \\
&\iff Y(t - z) > z
\end{aligned}$$

Therefore

$$\Pr(Z(t) > z) = \Pr(Y(t - z) > z)$$

When the underlying distribution $F_T(t)$ is nonlattice, we can show that the limiting distribution of the age $Z(t)$ when $t \to \infty$ is

$$\lim_{t \to \infty} \Pr(Z(t) \leq t) = F_e(t) = \frac{1}{\mu} \int_0^t (1 - F_T(u)) \, du \qquad (7.65)$$

that is, the same distribution as (7.63). When $t \to \infty$, both the remaining lifetime $Y(t)$ and the age $Z(t)$ at time t will have the same distribution. When t is large, then

$$E(Y(t)) \approx E(Z(t)) \approx \frac{\sigma^2 + \mu^2}{2\mu} \qquad (7.66)$$

Let us now assume that a renewal process with a nonlattice underlying distribution has been "running" for a long time, and that we observe the process at a random time, which we denote $t = 0$. The time T_1 to the first renewal after time $t = 0$ is equal to the remaining lifetime of the item that is in operation at time $t = 0$. The distribution of T_1 is equal to (7.63) and the mean time to the first renewal is given by (7.64). Similarly, the age of the item that is in operation at time $t = 0$ has the same distribution and the same mean as the time to the first renewal. For a formal proof, see Ross (1996, p. 131) or Bon (1995, p. 136).

Remark: This result may seem a bit strange. When we observe a renewal process that has been "running" for a long time at a random time t, the length of the corresponding interoccurrence time is $S_{N(t)+1} - S_{N(t)}$, as illustrated in Fig. 7.15, and the mean length of the interoccurrence time is μ. We obviously have that $S_{N(t)+1} - S_{N(t)} = Z(t) + Y(t)$, but $E(Z(t) + E(Y(t)) = (\sigma^2 + \mu^2)/\mu$ is greater than μ. This rather surprising result is known as the *inspection paradox*, and is further discussed by Ross (1996, p. 117), and Bon (1995, p 141). □.

If the underlying distribution function $F_T(t)$ is new better than used (NBU) or new worse than used (NWU) (see Section 2.19), bounds may be derived for the distribution of the remaining lifetime $Y(t)$ of the item that is in operation at time t. Barlow and Proschan (1975, p. 169) have shown that the following apply:

$$\text{If } F_T(t) \text{ is NBU, then } \Pr(Y(t) > y) \leq \Pr(T > y) \qquad (7.67)$$

$$\text{If } F_T(t) \text{ is NWU, then } \Pr(Y(t) > y) \geq \Pr(T > y) \qquad (7.68)$$

Intuitively, these results are obvious. If an item has an NBU life distribution, then a new item should have a higher probability of surviving the interval $(0, y]$ than a used item. The opposite should apply for an item with an NWU life distribution.

When the distributions of $Z(t)$ and $Y(t)$ are to be determined, the following lemma is useful:

Lemma 7.1 If

$$g(t) = h(t) + \int_0^t g(t - x) \, dF(x) \qquad (7.69)$$

where the functions h and F are known, while g is unknown, then

$$g(t) = h(t) + \int_0^t h(t-x)\,dW_F(x) \tag{7.70}$$

where

$$W_F(x) = \sum_{r=1}^{\infty} F^{(r)}(x)$$

□

Note that equation (7.70) is a generalization of the fundamental renewal equation (7.46).

Example 7.9
Consider a renewal process with underlying distribution $F_T(t)$. The distribution of the remaining lifetime $Y(t)$ of an item that is in operation at time t can be given by (e.g., see Bon 1995, p. 129)

$$\Pr(Y(t) > y) = \Pr(T > y+t) + \int_0^t \Pr(T > y+t-u)\,dW_F(u) \tag{7.71}$$

By introducing the survivor function $R(t) = 1 - F_T(t)$, and assuming that the renewal density $w_F(t) = dW_F(t)/dt$ exists, (7.71) may be written

$$\Pr(Y(t) > y) = R(y+t) + \int_0^t R(y+t-u)w_F(u)\,du \tag{7.72}$$

If the probability density function $f(t) = dF_T(t)/dt = -dR(t)/dt$ exists, we have from the definition of $f(t)$ that

$$R(t) - R(t+y) \approx f(t) \cdot y \quad \text{when } y \text{ is small}$$

Equation (7.72) may in this case be written

$$\begin{aligned}
\Pr(Y(t) > y) &\approx R(t) - f(t) \cdot y + \int_0^t (R(t-u) - f(t-u) \cdot y)w_F(u)\,du \\
&= R(t) + \int_0^t R(t-u)w_F(u)\,du \\
&\quad -y\left(f(t) + \int_0^t f(t-u)w_F(u)\,du\right) \\
&= \Pr(Y(t) > 0) - w_F(t) \cdot y
\end{aligned} \tag{7.73}$$

The last line in (7.73) follows from Lemma 7.1. Since $\Pr(Y(t) > 0) = 1$, we have the following approximation:

$$\Pr(Y(t) > y) \approx 1 - w_F(t) \cdot y \quad \text{when } y \text{ is small} \tag{7.74}$$

If we observe a renewal process at a random time t, the probability of having a failure (renewal) in a short interval of length y after time t is, from (7.74), approximately $w_F(t) \cdot y$, and it is hence relevant to call $w_F(t)$ the ROCOF.

□

7.3.7 Bounds for the Renewal Function

We will now establish some bounds for the renewal function $W(t)$. For this purpose consider a renewal process with interarrival times $T_1, T_2, \ldots$. We stop observing the process at the first renewal after time t, that is, at renewal $N(t) + 1$. Since the event $N(t) + 1 = n$ only depends on $T_1, T_2, \ldots, T_n$, we can use Wald's equation to get

$$E\left(S_{N(t)+1}\right) = E\left(\sum_{i=1}^{N(t)+1}\right) = E(T) \cdot E(N(t) + 1) = \mu\left[W(t) + 1\right] \qquad (7.75)$$

Since $S_{N(t)+1}$ is the first renewal after t, it can be expressed as

$$S_{N(t)+1} = t + Y(t)$$

The mean value is from (7.75)

$$\mu[W(t) + 1] = t + E(Y(t))$$

such that

$$W(t) = \frac{t}{\mu} + \frac{E(Y(t))}{\mu} - 1 \qquad (7.76)$$

When t is large and the underlying distribution is nonlattice, we can use (7.64) to get

$$W(t) - \frac{t}{\mu} \to \frac{1}{2}\left(\frac{\sigma^2}{\mu^2} - 1\right) \quad \text{when } t \to \infty \qquad (7.77)$$

which is the same result as we found in (7.53).

Lorden (1970) has shown that the renewal function $W(t)$ of a *general* renewal process is bounded by

$$\frac{t}{\mu} - 1 \leq W(t) \leq \frac{t}{\mu} + \frac{\sigma^2}{\mu^2} \qquad (7.78)$$

For a proof, see Cocozza-Thivent (1997, p. 170).

In section 2.19 we introduced several families of life distribution. A distribution was said to be "new better than used in expectation" (NBUE) when the mean remaining lifetime of a used item was less, or equal to, the mean life of a new item. In the same way, a distribution was said to be "new worse than used in expectation" (NWUE) when the mean remaining life of a used item was greater, or equal to, the mean life of a new item.

If we have an NBUE distribution, then $E(Y(t)) \leq \mu$, and

$$W(t) = \frac{t + E(Y(t))}{\mu} - 1 \leq \frac{t}{\mu} \quad \text{for } t \geq 0$$

and

$$\frac{t}{\mu} - 1 \leq W(t) \leq \frac{t}{\mu} \qquad (7.79)$$

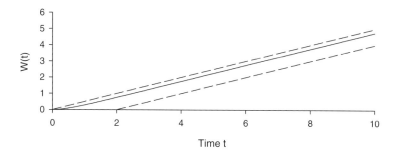

Fig. 7.12 The renewal function $W(t)$ of a renewal process with underlying distribution that is gamma $(2, \lambda)$, together with the bounds for $W(t)$, for $\lambda = 1$.

If we have an NWUE distribution, then $E(Y(t)) \geq \mu$, and

$$W(t) = \frac{t + E(Y(t))}{\mu} - 1 \geq \frac{t}{\mu} \quad \text{for } t \geq 0 \tag{7.80}$$

Further bounds for the renewal function are given by Dohi et al. (2002).

Example 7.7 (Cont.)
Reconsider the renewal process where the underlying distribution has a gamma distribution with parameters $(2, \lambda)$. This distribution has an increasing failure rate and is therefore also NBUE. We can therefore apply the bounds in (7.79). In Fig. 7.12 the renewal function (7.58)

$$W(t) = \frac{\lambda t}{2} - \frac{1}{4}(1 - e^{-2\lambda t})$$

is plotted together with the bounds in (7.79)

$$\frac{\lambda t}{2} - 1 \leq W(t) \leq \frac{\lambda t}{2}$$

□

7.3.8 Superimposed Renewal Processes

Consider a series structure of n independent components that are put into operation at time $t = 0$. All the n components are assumed to be new at time $t = 0$. When a component fails, it is replaced with a new component of the same type or restored to an "as good as new" condition. Each component will thus produce a renewal process. The n components will generally be different, and the renewal processes will therefore have different underlying distributions.

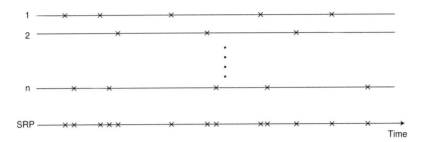

Fig. 7.13 Superimposed renewal process.

The process formed by the union of all the failures is called a *superimposed renewal process* (SRP). The n individual renewal processes and the SRP are illustrated in Fig. 7.13.

In general, the SRP will *not* be a renewal process. However, it has been shown, for example, by Drenick (1960), that superposition of an infinite number of independent *stationary* renewal processes is an HPP. Many systems are composed of a large number of components in series. Drenick's result is often used as a justification for assuming the time between system failures to be exponentially distributed.

Example 7.10
Consider a series structure of two components. When a component fails, it is replaced or repaired to an "as good as new" condition. Each component will therefore produce an ordinary renewal process. The time required to replace or repair a component is considered to be negligible. The components are assumed to fail and be repaired independent of each other. Both components are put into operation and are functioning at time $t = 0$. The series system fails as soon as one of its components fails, and the system failures will produce a superimposed renewal process. Times to failure for selected life distributions with increasing failure rates for the two components and the series system have been simulated on a computer and are illustrated in Fig. 7.14. The conditional ROCOF (when the failure times are given) is also shown in the figure. As illustrated in Fig. 7.14 the system is not restored to an "as good as new" state after each system failure. The system is subject to *imperfect repairs* (see Section 7.5) and the process of system failures is not a renewal process since the times between system failures do not have a common distribution. □

The superimposed renewal process is further discussed, for example, by Cox and Isham (1980), Ascher and Feingold (1984), and Thompson (1988).

7.3.9 Renewal Reward Processes

Consider a renewal process $\{N(t), t \geq 0\}$, and let $(S_{i-1}, S_i]$ be the duration of the ith renewal cycle, with interoccurrence time $T_i = S_i - S_{i-1}$. Let V_i be a reward associated to renewal T_i, for $i = 1, 2, \ldots$. The rewards $V_1, V_2, \ldots$ are assumed to be independent random variables with the common distribution function $F_V(v)$, and

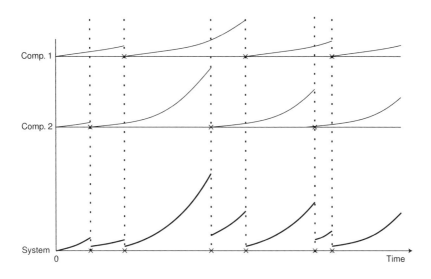

Fig. 7.14 Superimposed renewal process. Conditional ROCOF $w_C(t)$ of a series system with two components that are renewed upon failure.

with $E(T_i) < \infty$. This model is comparable with the compound Poisson process that was described on page 245. The accumulated reward in the time interval $(0, t]$ is

$$V(t) = \sum_{i=1}^{N(t)} V_i \qquad (7.81)$$

Let $E(T_i) = \mu_T$ and $E(V_i) = \mu_V$. According to Wald's equation (7.22) the mean accumulated reward is

$$E(V(t)) = \mu_V \cdot E(N(t)) \qquad (7.82)$$

According to the elementary renewal theorem (7.49), when $t \to \infty$,

$$\frac{W(t)}{t} = \frac{E(N(t))}{t} \to \frac{1}{\mu_T}$$

Hence

$$\frac{E(V(t))}{t} = \frac{\mu_V \cdot E(N(t))}{t} \to \frac{\mu_V}{\mu_T} \qquad (7.83)$$

The same result is true even if the reward V_i is allowed to depend on the associated interoccurrence time T_i for $i = 1, 2, \ldots$. The pairs (T_i, V_i) for $i = 1, 2, \ldots$ are, however, assumed to be independent and identically distributed (for proof, see Ross

1996, p. 133). The reward V_i in renewal cycle i may, for example, be a function of the interoccurrence time T_i, for $i = 1, 2, \ldots$. When t is very large, then

$$V(t) \approx \mu_V \cdot \frac{t}{\mu_T}$$

which is an obvious result.

7.3.10 Delayed Renewal Processes

Sometimes we consider counting processes where the first interoccurrence time T_1 has a distribution function $F_{T_1}(t)$ that is different from the distribution function $F_T(t)$ of the subsequent interoccurrence times. This may, for example, be the case for a failure process where the component at time $t = 0$ is not new. Such a renewal process is called a *delayed* renewal process, or a *modified* renewal process. To specify that the process is not delayed, we sometimes say that we have an *ordinary* renewal process.

Several of the results presented earlier in this section can be easily extended to delayed renewal processes:

The Distribution of $N(t)$ Analogous with (7.41) we get

$$\Pr(N(t) = n) = F_{T_1}^* * F_T^{*(n-1)} - F_{T_1}^* * F_T^{*(n)} \tag{7.84}$$

The Distribution of S_n The Laplace transform of the density of S_n is from (7.34):

$$f^{*(n)}(s) = f_{T_1}^*(s) \left(f_T^*(s) \right)^{n-1} \tag{7.85}$$

The Renewal Function The integral equation (7.46) for the renewal function $W(t)$ becomes

$$W(t) = F_{T_1}(t) + \int_0^t W(t-x) \, dF_T(x) \tag{7.86}$$

and the Laplace transform is

$$W^*(s) = \frac{f_{T_1}^*(s)}{s(1 - f_T^*(s))} \tag{7.87}$$

The Renewal Density Analogous with (7.55) we get

$$w(t) = f_{T_1}(t) + \int_0^t w(t-x) f_T(x) \, dx \tag{7.88}$$

and the Laplace transform is

$$w^*(s) = \frac{f_{T_1}^*(s)}{1 - f_T^*(s)} \tag{7.89}$$

All the limiting properties for ordinary renewal processes, when $t \to \infty$, will obviously also apply for delayed renewal processes.

For more detailed results see, for example, Cocozza-Thivent (1997, section 6.2). We will here briefly discuss a special type of a delayed renewal process, the *stationary renewal process*.

Definition 7.6 A stationary renewal process is a delayed renewal process where the first renewal period has distribution function

$$F_{T_1}(t) = F_e(t) = \frac{1}{\mu} \int_0^t (1 - F_T(x)) \, dx \qquad (7.90)$$

while the underlying distribution of the other renewal periods is $F_T(t)$. □

Remarks

1. Note that $F_e(t)$ is the same distribution as we found in (7.63).

2. When the probability density function $f_T(t)$ of $F_T(t)$ exists, the density of $F_e(t)$ is

$$f_e(t) = \frac{dF_e(t)}{dt} = \frac{1 - F_T(t)}{\mu} = \frac{R_T(t)}{\mu}$$

3. As pointed out by Cox (1962, p. 28) the stationary renewal process has a simple physical interpretation. Suppose a renewal process is started at time $t = -\infty$, but that the process is not observed before time $t = 0$. Then the first renewal period observed, T_1, is the remaining lifetime of the component in operation at time $t = 0$. According to (7.63) the distribution function of T_1 is $F_e(t)$. A stationary renewal process is called an *equilibrium renewal process* by Cox (1962). This is the reason why we use the subscript e in $F_e(t)$. In Ascher and Feingold (1984) the stationary renewal process is called a renewal process with *asynchronous sampling*, while an ordinary renewal process is called a renewal process with *synchronous sampling*. □

Let $\{N_S(t), t \geq 0\}$ be a stationary renewal process, and let $Y_S(t)$ denote the remaining life of an item at time t. The stationary renewal process has the following properties (e.g., see Ross 1996, p. 131):

$$W_S(t) = t/\mu \qquad (7.91)$$
$$\Pr(Y_S(t) \leq y) = F_e(y) \quad \text{for all } t \geq 0 \qquad (7.92)$$
$$\{N_S(t), t \geq 0\} \quad \text{has stationary increments} \qquad (7.93)$$

where $F_e(y)$ is defined by equation (7.90).

268   COUNTING PROCESSES

Remark: A homogeneous Poisson process is a stationary renewal process, because of the memoryless property of exponential distribution. The HPP is seen to fulfill all the three properties (7.91), (7.92), and (7.93). □

Example 7.11
Reconsider the renewal process in Example 7.7 where the interoccurrence times had a gamma distribution with parameters $(2, \lambda)$. The underlying distribution function is then

$$F_T(t) = 1 - e^{-\lambda t} - \lambda t\, e^{-\lambda t}$$

and the mean interoccurrence time is $E(T_i) = 2/\lambda$. Let us now assume that the process has been running for a long time and that when we start observing the process at time $t = 0$, it may be considered as a stationary renewal process.

According to (7.91), the renewal function for this stationary renewal process is $W_S(t) = \lambda t/2$, and the distribution of the remaining life, $Y_S(t)$ is (see 7.92)

$$\Pr(Y_S(t) \leq y) = \frac{\lambda}{2} \int_0^y (e^{-\lambda u} + \lambda u\, e^{-\lambda u})\, du$$

$$= 1 - \left(1 + \frac{\lambda y}{2}\right) e^{-\lambda y}$$

The mean remaining lifetime of an item at time t is

$$E(Y_S(t)) = \int_0^\infty \Pr(Y_S(t) > y)\, dy = \int_0^\infty \left(1 + \frac{\lambda y}{2}\right) e^{-\lambda y}\, dy = \frac{3}{2\lambda}$$

□

Delayed renewal processes are used in the next section to analyze alternating renewal processes.

7.3.11 Alternating Renewal Processes

Consider a system that is activated and functioning at time $t = 0$. Whenever the system fails, it is repaired. Let $U_1, U_2, \ldots$ denote the successive times to failure (up-times) of the system. Let us assume that the times to failure are independent and identically distributed with distribution function $F_U(t) = \Pr(U_i \leq t)$ and mean $E(U) = $ MTTF (mean time to failure). Likewise we assume the corresponding repair times $D_1, D_2, \ldots$ to be independent and identically distributed with distribution function $F_D(d) = \Pr(D_i \leq d)$ and mean $E(D) = $ MTTR (mean time to repair). MTTR denotes the total mean downtime following a failure and will usually involve much more that the active repair time. We therefore prefer to use the term MDT (mean downtime) instead of MTTR[2].

---

[2] In the rest of this book we are using T to denote time to failure. In this chapter we have already used T to denote interoccurrence time (renewal period), and we will therefore use U to denote the time to failure (up-time). We hope that this will not confuse the reader.

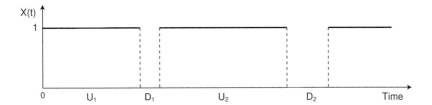

Fig. 7.15 Alternating renewal process.

If we define the completed repairs to be the renewals, we obtain an ordinary renewal process with renewal periods (interoccurrence times) $T_i = U_i + D_i$ for $i = 1, 2, \ldots$. The mean time between renewals is $\mu_T = \text{MTTF} + \text{MDT}$. This resulting process is called an *alternating renewal process* and is illustrated in Fig. 7.15. The underlying distribution function, $F_T(t)$, is the convolution of the distribution functions $F_U(t)$ and $F_D(t)$,

$$F_T(t) = \Pr(T_i \leq t) = \Pr(U_i + D_i \leq t) = \int_0^t F_U(t-x)\,dF_D(x) \qquad (7.94)$$

If instead we let the renewals be the events when a failure occurs, we get a *delayed renewal process* where the first renewal period T_1 is equal to U_1 while $T_i = D_{i-1} + U_i$ for $i = 2, 3, \ldots$.

In this case the distribution function $F_{T_1}(t)$ of the first renewal period is given by

$$F_{T_1}(t) = \Pr(T_1 \leq t) = \Pr(U_1 \leq t) = F_U(t) \qquad (7.95)$$

while the distribution function $F_T(t)$ of the other renewal periods is given by (7.94).

Example 7.12
Consider the alternating renewal process described above, and let the renewals be the completed repairs such that we have an ordinary renewal process. Let a reward V_i be associated to the ith interoccurrence time, and assume that this reward is defined such that we earn one unit per unit of time the system is functioning in the time period since the last failure. When the reward is measured in time units, then $E(V_i) = \mu_V = \text{MTTF}$. The average availability $A_{\text{av}}(0, t)$ of the component in the time interval $(0, t)$ has been defined as the mean fraction of time in the interval $(0, t)$ where the system is functioning. From (7.83) we therefore get

$$A_{\text{av}}(0, t) \to \frac{\mu_V}{\mu_T} = \frac{\text{MTTF}}{\text{MTTF} + \text{MDT}} \quad \text{when } t \to \infty \qquad (7.96)$$

which is the same result we obtain in Section 9.4 based on heuristic arguments. □

Availability The availability $A(t)$ of an item (component or system) was defined as the probability that the item is functioning at time t, that is, $A(t) = \Pr(X(t) = 1)$, where $X(t)$ denotes the state variable of the item.

As above, consider an alternating renewal process where the renewals are completed repairs, and let $T = U_1 + D_1$. The availability of the item is then

$$A(t) = \Pr(X(t) = 1) = \int_0^\infty \Pr(X(t) = 1 \mid T = x) \, dF_T(x)$$

Since the component is assumed to be "as good as new" at time $T = U_1 + D_1$, the process repeats itself from this point in time and

$$\Pr(X(t) = 1 \mid T = x) = \begin{cases} A(t - x) & \text{for } t > x \\ \Pr(U_1 > t \mid T = x) & \text{for } t \leq x \end{cases}$$

Therefore

$$A(t) = \int_0^t A(t - x) \, dF_T(x) + \int_t^\infty \Pr(U_1 > t \mid T = x) \, dF_T(x)$$

But since $D_1 > 0$, then

$$\int_t^\infty \Pr(U_1 > t \mid U_1 + D_1 = x) \, dF_T(x) = \int_0^\infty \Pr(U_1 > t \mid T = x) \, dF_T(x)$$
$$= \Pr(U_1 > t) = 1 - F_U(t)$$

Hence

$$A(t) = 1 - F_U(t) + \int_0^t A(t - x) \, dF_T(x) \tag{7.97}$$

We may now apply Lemma 7.1 and get

$$A(t) = 1 - F_T(t) + \int_0^t (1 - F_T(t - x)) \, dW_{F_T}(x) \tag{7.98}$$

where

$$W_{F_T}(t) = \sum_{n=1}^\infty F_T^{(n)}(t)$$

is the renewal function for a renewal process with underlying distribution $F_T(t)$.

When $F_U(t)$ is a nonlattice distribution, the key renewal theorem (7.51) can be used with $Q(t) = 1 - F_U(t)$ and we get

$$\int_0^t (1 - F_U(t - x)) \, dW_{F_T}(x) \xrightarrow[t \to \infty]{} \frac{1}{E(T)} \int_0^\infty (1 - F_U(t)) \, dt = \frac{E(U)}{E(U) + E(D)}$$

Since $F_T(t) \to 1$ when $t \to \infty$, we have thus shown that

$$A = \lim_{t \to \infty} A(t) = \frac{E(U)}{E(U) + E(D)} = \frac{\text{MTTF}}{\text{MTTF+MDT}} \tag{7.99}$$

Notice that this is the same result as we got in (7.96) by using results from renewal reward processes.

Example 7.13
Consider a parallel structure of n components that fail and are repaired independent of each other. Component i has a time to failure (up-time) U_i which is exponentially distributed with failure rate λ_i and a time to repair (downtime) D_i which is also exponentially distributed with repair rate μ_i, for $i = 1, 2, \ldots$. The parallel structure will fail when all the n components are in a failed state at the same time. Since the components are assumed to be independent, a system failure must occur in the following way: Just prior to the system failure, $(n - 1)$ components must be in a failed state, and then the functioning component must fail.

Let us now assume that the system has been in operation for a long time, such that we can use limiting (average) availabilities. The probability that component i is in a failed state is then approximately:

$$\bar{A}_i \approx \frac{\text{MDT}}{\text{MTTF} + \text{MDT}} = \frac{1/\mu_i}{1/\lambda_i + 1/\mu_i} = \frac{\lambda_i}{\lambda_i + \mu_i}$$

Similarly, the probability that component i is functioning is approximately

$$A_i \approx \frac{\mu_i}{\lambda_i + \mu_i}$$

The probability that a functioning component i will fail within a very short time interval of length Δt is approximately

$$\Pr(\Delta t) \approx \lambda_i \, \Delta t$$

The probability of system failure in the interval $(t, t + \Delta t)$, when t is large, is

$$\Pr(\text{System failure in}(t, t + \Delta t)) = \sum_{i=1}^{n} \left[\frac{\mu_i}{\lambda_i + \mu_i} \prod_{j \neq i} \frac{\lambda_j}{\lambda_j + \mu_j} \right] \cdot \lambda_i \, \Delta t + o(\Delta t)$$

$$= \sum_{i=1}^{n} \left[\frac{\lambda_i}{\lambda_i + \mu_i} \prod_{j \neq i} \frac{\lambda_j}{\lambda_j + \mu_j} \right] \cdot \mu_i \, \Delta t + o(\Delta t)$$

$$= \prod_{j=1}^{n} \frac{\lambda_j}{\lambda_j + \mu_j} \sum_{i=1}^{n} \mu_i \Delta t + o(\Delta t)$$

Since Δt is assumed to be very small, no more than one system failure will occur in the interval. We can therefore use Blackwell's theorem (7.50) to conclude that the above expression is just Δt times the reciprocal of the mean time between system failures, MTBF_S, that is,

$$\text{MTBF}_S = \left[\prod_{j=1}^{n} \frac{\lambda_j}{\lambda_j + \mu_j} \sum_{i=1}^{n} \mu_i \right]^{-1} \qquad (7.100)$$

When the system is in a failed state, all the n components are in a failed state. Since the repair times (downtimes) are assumed to be independent with repair rates μ_i for $i = 1, 2, \ldots, n$, the system downtime will be exponential with repair rate $\sum_{i=1}^{n} \mu_i$, and the mean downtime to repair the system is

$$\text{MDT}_S = \frac{1}{\sum_{i=1}^{n} \mu_i}$$

The mean up-time, or the mean time to failure, MTTF_S, of the system is equal to $\text{MTBF}_S - \text{MDT}_S$:

$$\text{MTTF}_S = \left[\prod_{j=1}^{n} \frac{\lambda_j}{\lambda_j + \mu_j} \sum_{i=1}^{n} \mu_i \right]^{-1} - \frac{1}{\sum_{i=1}^{n} \mu_i} \qquad (7.101)$$

$$= \frac{1 - \prod_{j=1}^{n} \lambda_j/(\lambda_j + \mu_j)}{\prod_{j=1}^{n} \lambda_j/(\lambda_j + \mu_j) \sum_{i=1}^{n} \mu_i} \qquad (7.102)$$

To check that the above calculations are correct, we may calculate the average unavailability:

$$\bar{A}_S = \frac{\text{MDT}_S}{\text{MTTF}_S + \text{MDT}_S} = \prod_{j=1}^{n} \frac{\lambda_j}{\lambda_j + \mu_j}$$

which is in accordance with the results obtained in Chapter 9. [Example 7.13 is adapted from Example 3.5(B) in Ross (1996)]. □

Mean Number of Failures/Repairs First, let the renewals be the events where a repair is completed. Then we have an ordinary renewal process with renewal periods $T_1, T_2, \ldots$ which are independent and identically distributed with distribution function (7.94).

Assume that U_i and D_i both are continuously distributed with densities $f_U(t)$ and $f_D(t)$, respectively. The probability density function of the T_i's is then

$$f_T(t) = \int_0^t f_U(t-x) f_D(x) \, dx \qquad (7.103)$$

According to Appendix B the Laplace transform of (7.103) is

$$f_T^*(s) = f_U^*(s) \cdot f_D^*(s)$$

Let $W_1(t)$ denote the renewal function, that is, the mean number of completed repairs in the time interval $(0, t]$. According to (7.84)

$$W_1^*(s) = \frac{f_U^*(s) \cdot f_D^*(s)}{s(1 - f_U^*(s) \cdot f_D^*(s))} \qquad (7.104)$$

In this case both the U_i's and the D_i's were assumed to be continuously distributed. This, however, turns out *not* to be essential. Equation (7.104) is also valid for discrete distributions, or for a mixture of discrete and continuous distributions. In this case we may use that

$$f_U^*(s) = E(e^{-sU_i})$$
$$f_D^*(s) = E(e^{-sD_i})$$

The mean number of completed repairs in $(0, t]$ can now, at least in principle, be determined for any choice of life- and repair time distributions.

Next, let the renewals be the events where a failure occurs. In this case we get a delayed renewal process. The renewal periods $T_1, T_2, \ldots$ are independent and $F_{T_1}(t)$ is given by (7.95) while the distribution of $T_2, T_3, \ldots$ is given by (7.94).

Let $W_2(t)$ denote the renewal function, that is, the mean number of failures in $(0, t]$ under these conditions. According to (7.87) the Laplace transform is

$$W_2^*(s) = \frac{f_U^*(s)}{s(1 - f_U^*(s) \cdot f_D^*(s))} \qquad (7.105)$$

which, at least in principle, can be inverted to obtain $W_2(t)$.

Availability at a Given Point of Time By taking Laplace transforms of (7.98) we get

$$A^*(s) = \frac{1}{s} - F_U^*(s) + \left(\frac{1}{s} - F_U^*(s)\right) \cdot w_{F_T}^*(s)$$

Since

$$F^*(s) = \frac{1}{s} f^*(s)$$

then

$$A^*(s) = \frac{1}{s}(1 - f_U^*(s)) \cdot (1 + w_{F_T}^*(s))$$

If we have an ordinary renewal process (i.e., the renewals are the events where a repair is completed), then

$$w_{F_T}^*(s) = s W_1^*(s)$$

Hence

$$A^*(s) = \frac{1}{s}(1 - f_U^*(s)) \cdot \left(1 + \frac{f_U^*(s) \cdot f_D^*(s)}{1 - f_U^*(s) \cdot f_D^*(s)}\right)$$

that is,

$$A^*(s) = \frac{1 - f_U^*(s)}{s\left(1 - f_U^*(s) \cdot f_D^*(s)\right)} \qquad (7.106)$$

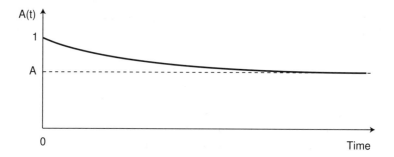

Fig. 7.16 Availability of a component with exponential life and repair times.

The availability $A(t)$ can in principle be determined from (7.106) for any choice of life and repair time distributions.

Example 7.14 Exponential Lifetime–Exponential Repair Time
Consider an alternating renewal process where the component up-times $U_1, U_2, \ldots$ are independent and exponentially distributed with failure rate λ. The corresponding downtimes are also assumed to be independent and exponentially distributed with the repair rate $\mu = 1/\text{MDT}$.

Then

$$f_U(t) = \lambda e^{-\lambda t} \text{ for } t > 0$$
$$f_U^*(s) = \frac{\lambda}{\lambda + s}$$

and

$$f_D(t) = \mu e^{-\mu t} \text{ for } t > 0$$
$$f_D^*(s) = \frac{\mu}{\mu + s}$$

The availability $A(t)$ is then obtained from (7.106):

$$A^*(s) = \frac{1 - \lambda/(\lambda + s)}{s[1 - (\lambda/(\lambda + s)) \cdot (\mu/(\mu + s))]}$$
$$= \frac{\mu}{\lambda + \mu} \cdot \frac{1}{s} + \frac{\lambda}{\lambda + \mu} \cdot \frac{1}{s + (\lambda + \mu)} \quad (7.107)$$

Equation (7.107) can be inverted (see Appendix B) and we get

$$A(t) = \frac{\mu}{\lambda + \mu} + \frac{\lambda}{\lambda + \mu} e^{-(\lambda + \mu)t} \quad (7.108)$$

which is the same result as we get in Chapter 8. The availability $A(t)$ is illustrated in Fig. 7.16.

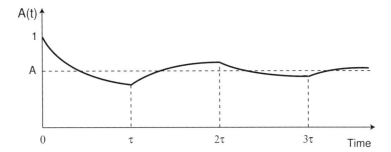

Fig. 7.17 The availability of a component with exponential lifetimes and constant repair time (τ).

The limiting availability is

$$A = \lim_{t \to \infty} A(t) = \frac{\mu}{\lambda + \mu} = \frac{1/\lambda}{1/\lambda + 1/\mu} = \frac{\text{MTTF}}{\text{MTTF+MDT}}$$

By inserting $f_U^*(s)$ and $f_D^*(s)$ into (7.104) we get the Laplace transform of the mean number of renewals $W(t)$,

$$\begin{aligned} W^*(s) &= \frac{(\lambda/(\lambda+s)) \cdot (\mu/(\mu+s))}{s[1 - (\lambda/(\lambda+s)) \cdot (\mu/(\mu+s))]} \\ &= \frac{\lambda\mu}{\lambda+\mu} \cdot \frac{1}{s^2} - \frac{\lambda\mu}{(\lambda+\mu)^2} \cdot \frac{1}{s} + \frac{\lambda\mu}{(\lambda+\mu)^2} \cdot \frac{1}{s+(\lambda+\mu)} \end{aligned}$$

By inverting this expression we get the mean number of completed repairs in the time interval $(0, t]$

$$W(t) = \frac{\lambda\mu}{\lambda+\mu} t - \frac{\lambda\mu}{(\lambda+\mu)^2} + \frac{\lambda\mu}{(\lambda+\mu)^2} e^{-(\lambda+\mu)t} \qquad (7.109)$$

□

Example 7.15 Exponential Lifetime–Constant Repair Time

Consider an alternating renewal process where the system up-times $U_1, U_2, \ldots$ are independent and exponentially distributed with failure rate λ. The downtimes are assumed to be constant and equal to τ with probability 1: $\Pr(D_i = \tau) = 1$ for $i = 1, 2, \ldots$. The corresponding Laplace transforms are

$$f_U^*(s) = \frac{\lambda}{\lambda+s}$$

$$f_D^*(s) = E(e^{-sD}) = e^{-s\tau} \cdot \Pr(D = \tau) = e^{-s\tau}$$

Hence the Laplace transform of the availability (7.106) becomes

$$\begin{aligned}
A^*(s) &= \frac{1 - \lambda/(\lambda + s)}{s[1 - (\lambda/(\lambda + s))\, e^{-s\tau}]} = \frac{1}{s + \lambda - \lambda e^{-s\tau}} \\
&= \frac{1}{\lambda + s} \cdot \left[\frac{1}{1 - (\lambda/(\lambda + s))\, e^{-s\tau}} \right] = \frac{1}{\lambda + s} \sum_{\nu=0}^{\infty} \left(\frac{\lambda}{\lambda + s} \right)^{\nu} e^{-s\nu\tau} \\
&= \frac{1}{\lambda} \sum_{\nu=0}^{\infty} \left(\frac{\lambda}{\lambda + s} \right)^{\nu+1} e^{-s\nu\tau}
\end{aligned} \qquad (7.110)$$

The availability then becomes

$$A(t) = \mathcal{L}^{-1}(A^*(s)) = \sum_{\nu=0}^{\infty} \frac{1}{\lambda} \mathcal{L}^{-1}\left[\left(\frac{\lambda}{\lambda + s} \right)^{\nu+1} e^{-s\nu\tau} \right]$$

According to Appendix B

$$\mathcal{L}^{-1}\left[\left(\frac{\lambda}{\lambda + s} \right)^{\nu+1} \right] = \frac{\lambda^{\nu+1}}{\nu!} t^{\nu} e^{-\lambda t} = f(t)$$

$$\mathcal{L}^{-1}(e^{-s\nu\tau}) = \delta(t - \nu\tau)$$

where $\delta(t)$ denotes the Dirac delta-function. Thus

$$\begin{aligned}
\mathcal{L}^{-1}\left[\left(\frac{\lambda}{\lambda + s} \right)^{\nu+1} \cdot e^{-s\nu\tau} \right] &= \mathcal{L}^{-1}\left[\left(\frac{\lambda}{\lambda + s} \right)^{\nu+1} \right] * \mathcal{L}^{-1}(e^{-s\nu\tau}) \\
&= \int_0^{\infty} \delta(t - \nu\tau - x) f(x)\, dx \\
&= f(t - \nu\tau) \cdot u(t - \nu\tau)
\end{aligned}$$

where

$$u(t - \nu\tau) = \begin{cases} 1 & \text{if } t \geq \nu\tau \\ 0 & \text{if } t < \nu\tau \end{cases}$$

Hence the availability is

$$A(t) = \sum_{\nu=0}^{\infty} \frac{\lambda^{\nu}}{\nu!} (t - \nu\tau)^{\nu} e^{-\lambda(t - \nu\tau)} u(t - \nu\tau) \qquad (7.111)$$

The availability $A(t)$ is illustrated in Fig. 7.17.
The limiting availability is then according to (7.99)

$$A = \lim_{t \to \infty} A(t) = \frac{\text{MTTF}}{\text{MTTF+MDT}} = \frac{1/\lambda}{(1/\lambda) + \tau} = \frac{1}{1 + \lambda\tau} \qquad (7.112)$$

The Laplace transform for the renewal density is

$$w^*(s) = \frac{f_T^*(s) \cdot f_D^*(s)}{1 - f_T^*(s) \cdot f_D^*(s)} = \frac{\lambda e^{-s\tau}/(\lambda + s)}{1 - \lambda e^{-s\tau}/(\lambda + s)}$$

$$= \frac{1}{\lambda + s - \lambda e^{-s\tau}} \lambda e^{-s\tau} = \lambda \cdot A^*(s) \, e^{-s\tau}$$

where $A^*(s)$ is given by (7.110).

Then the renewal density becomes

$$w(t) = \lambda \mathcal{L}^{-1}(A^*(s) \cdot e^{-s\tau}) = \lambda \int_0^\infty \delta(t - \tau - x) A(x) \, dx$$

that is,

$$w(t) = \begin{cases} \lambda \cdot A(t - \tau) & \text{if } t \geq \tau \\ 0 & \text{if } t < \tau \end{cases} \qquad (7.113)$$

Hence the mean number of completed repairs in the time interval $(0, t]$ for $t > \tau$ is

$$W(t) = \int_0^t w(u) \, du = \lambda \int_\tau^t A(u - \tau) \, du = \lambda \int_0^{t-\tau} A(u) \, du \qquad (7.114)$$

□

7.4 NONHOMOGENEOUS POISSON PROCESSES

In this section the homogeneous Poisson process is generalized by allowing the rate of the process to be a function of time.

7.4.1 Introduction and Definitions

Definition 7.7 A counting process $\{N(t), t \geq 0\}$ is a nonhomogeneous (or nonstationary) Poisson process with rate function $w(t)$ for $t \geq 0$, if

1. $N(0) = 0$.
2. $\{N(t), t \geq 0\}$ has independent increments.
3. $\Pr(N(t + \Delta t) - N(t) \geq 2) = o(\Delta t)$, which means that the system will not experience more than one failure at the same time.
4. $\Pr(N(t + \Delta t) - N(t) = 1) = w(t)\Delta t + o(\Delta t)$. □

The basic "parameter" of the NHPP is the ROCOF function $w(t)$. This function is also called the *peril rate* of the NHPP. The *cumulative rate* of the process is

$$W(t) = \int_0^t w(u) \, du \qquad (7.115)$$

This definition of course covers the situation in which the rate is a function of some observed explanatory variable that is a function of the time t.

It is important to note that the NHPP model does not require stationary increments. This means that failures may be more likely to occur at certain times than others, and hence the interoccurrence times are generally neither independent nor identically distributed. Consequently, statistical techniques based on the assumption that the data are independent and identically distributed cannot be applied to an NHPP.

The NHPP is often used to model trends in the interoccurrence times, that is, improving (*happy*) or deteriorating (*sad*) systems. It seems intuitive that a happy system will have a decreasing ROCOF function, while a sad system will have an increasing ROCOF function. Several studies of failure data from practical systems have concluded that the NHPP was an adequate model, and that the systems that were studied approximately satisfied the properties of the NHPP listed in Definition 7.7.

Due to the assumption of independent increments, the number of failures in a specified interval $(t_1, t_2]$ will be independent of the failures and interoccurrence times prior to time t_1. When a failure has occurred at time t_1, the conditional ROCOF $w_C(t \mid \mathcal{H}_t)$ (see page 254) in the next interval will be $w(t)$ and independent of the history $\mathcal{H}_{t_1}$ up to time t_1, and especially the case when no failure has occurred before t_1, in which case $w(t) = z(t)$, that is, the failure rate function (FOM) for $t < t_1$. A practical implication of this assumption is that the conditional (ROCOF), $w_C(t)$, is the same just before a failure and immediately after the corresponding repair. This assumption has been termed *minimal repair* (see Ascher and Feingold, 1984, p. 51). When replacing failed parts that may have been in operation for a long time, by new ones, an NHPP clearly is not a realistic model. For the NHPP to be realistic, the parts put into service should be identical to the old ones, and hence should be aged outside the system under identical conditions for the same period of time.

Now consider a system consisting of a large number of components. Suppose that a critical component fails and causes a system failure and that this component is immediately replaced by a component of the same type, thus causing a negligible system downtime. Since only a small fraction of the system is replaced, it seems natural to assume that the systems's reliability after the repair essentially is the same as immediately before the failure. In other words, the assumption of *minimal repair* is a realistic approximation. When an NHPP is used to model a repairable system, the system is treated as a *black box* in that there is no concern about how the system "looks inside."

A car is a typical example of a repairable system. Usually the operating time of a car is expressed in terms of the mileage indicated on the speedometer. Repair actions will usually not imply any extra mileage. The repair "time" is thus negligible. Many repairs are accomplished by adjustments or replacement of single components. The minimal repair assumption is therefore often applicable and the NHPP may be accepted as a realistic model, at least as a first order approximation.

Consider an NHPP with ROCOF $w(t)$, and suppose that failures occur at times $S_1, S_2, \ldots$. An illustration of $w(t)$ is shown in Fig. 7.18.

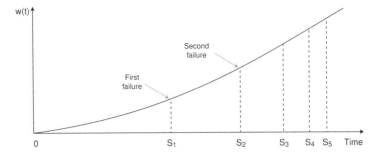

Fig. 7.18 The ROCOF $w(t)$ of an NHPP and random failure times.

7.4.2 Some Results

From the definition of the NHPP it is straightforward to verify (e.g., see Ross 1996, p. 79) that the number of failures in the interval $(0, t]$ is Poisson distributed:

$$\Pr(N(t) = n) = \frac{[W(t)]^n}{n!} e^{-W(t)} \quad \text{for } n = 0, 1, 2, \ldots \tag{7.116}$$

The mean number of failures in $(0, t]$ is therefore

$$E(N(t)) = W(t)$$

and the variance is $\mathrm{var}(N(t)) = W(t)$. The cumulative rate $W(t)$ of the process (7.115) is therefore the mean number of failures in the interval $(0, t]$ and is sometimes called the *mean value function* of the process. When n is large, $\Pr(N(t) \leq n)$ may be determined by normal approximation:

$$\begin{aligned}
\Pr(N(t) \leq n) &= \Pr\left(\frac{N(t) - W(t)}{\sqrt{W(t)}} \leq \frac{n - W(t)}{\sqrt{W(t)}}\right) \\
&= \Phi\left(\frac{n - W(t)}{\sqrt{W(t)}}\right)
\end{aligned} \tag{7.117}$$

From (7.116) it follows that the number of failures in the interval $(v, t + v]$ is Poisson distributed:

$$\Pr(N(t+v) - N(v) = n) = \frac{[W(t+v) - W(v)]^n}{n!} e^{-[W(t+v) - W(v)]}$$
$$\text{for } n = 0, 1, 2, \ldots$$

and that the mean number of failures in the interval $(v, t + v]$ is

$$E(N(t+v) - N(v)) = W(t+v) - W(v) = \int_v^{t+v} w(u)\, du \tag{7.118}$$

The probability of no failure in the interval (t_1, t_2) is

$$\Pr(N(t_2) - N(t_1) = 0) = e^{-\int_{t_1}^{t_2} w(t)\, dt}$$

Let S_n denote the time until failure n for $n = 0, 1, 2, \ldots$, where $S_0 = 0$. The distribution of S_n is given by:

$$\Pr(S_n > t) = \Pr(N(t) \leq n-1) = \sum_{k=0}^{n-1} \frac{W(t)^k}{k!} e^{-W(t)} \qquad (7.119)$$

When $W(t)$ is small, this probability may be determined from standard tables of the Poisson distribution. When $W(t)$ is large, the probability may be determined by normal approximation; see (7.117):

$$\begin{aligned}\Pr(S_n > t) &= \Pr(N(t) \leq n-1) \\ &\approx \Phi\left(\frac{n-1-W(t)}{\sqrt{W(t)}}\right)\end{aligned} \qquad (7.120)$$

Time to First Failure Let T_1 denote the time from $t = 0$ until the first failure. The survivor function of T_1 is

$$R_1(t) = \Pr(T_1 > t) = \Pr(N(t) = 0) = e^{-W(t)} = e^{-\int_0^t w(t)\,dt} \qquad (7.121)$$

Hence the failure rate (FOM) function $z_{T_1}(t)$ of the first interoccurrence time T_1 is equal to the ROCOF $w(t)$ of the process. Note, however, the different meaning of the two expressions. $z_{T_1}(t)\Delta t$ approximates the (conditional) probability that the *first* failure occurs in $(t, t + \Delta t]$, while $w(t)\Delta t$ approximates the (unconditional) probability that a failure, not necessarily the first, occurs in $(t, t + \Delta t]$.

A consequence of (7.121) is that the distribution of the first interoccurrence time, that is, the time from $t = 0$ until the system's first failure, will determine the ROCOF of the entire process. Thompson (1981) claims that this is a nonintuitive fact which is casting doubt on the NHPP as a realistic model for repairable systems. Use of an NHPP model implies that if we are able to estimate the failure rate (FOM) function of the time to the *first* failure, such as for a specific type of automobiles, we at the same time have an estimate of the ROCOF of the entire life of the automobile.

Time Between Failures Assume that the process is observed at time t_0, and let $Y(t_0)$ denote the time until the next failure. In the previous sections $Y(t_0)$ was called the remaining lifetime, or the forward recurrence time. By using (7.116), we can express the distribution of $Y(t_0)$ as

$$\begin{aligned}\Pr(Y(t_0) > t) &= \Pr(N(t+t_0) - N(t_0) = 0) = e^{-[W(t+t_0)-W(t_0)]} \\ &= e^{-\int_{t_0}^{t+t_0} w(u)\,du} = e^{-\int_0^t w(u+t_0)\,du}\end{aligned} \qquad (7.122)$$

Note that this result is independent of whether t_0 denotes a failure time or an arbitrary point in time.

Assume that t_0 is the time, S_{n-1}, at failure $n-1$. In this case $Y(t_0)$ denotes the time between failure $n-1$ and failure n (i.e., the nth interoccurrence time $T_n = S_n - S_{n-1}$). The failure rate (FOM) function of the nth interoccurrence time T_n is from (7.122):

$$z_{t_0}(t) = w(t + t_0) \quad \text{for } t \geq 0 \qquad (7.123)$$

Notice that this is a conditional failure rate, given that $S_{n-1} = t_0$. The mean time between failure $n - 1$ (at time t_0) and failure n, MTBF$_n$, is

$$\text{MTBF}_n = E(T_n) = \int_0^\infty \Pr(Y_{t_0} > t)\,dt$$

$$= \int_0^\infty e^{-\int_0^t w(u+t_0)\,du}\,dt \qquad (7.124)$$

Example 7.16
Consider an NHPP with ROCOF $w(t) = 2\lambda^2 t$, for $\lambda > 0$ and $t \geq 0$. The mean number of failures in the interval $(0, t)$ is $W(t) = E(N(t)) = \int_0^t w(u)\,du = (\lambda t)^2$. The distribution of the time to the first failure, T_1, is given by the survivor function

$$R_1(t) = e^{-W(t)} = e^{-(\lambda t)^2} \quad \text{for } t \geq 0$$

that is, a Weibull distribution with scale parameter λ and shape parameter $\alpha = 2$. If we observe the process at time t_0, the distribution of the time $Y(t_0)$ till the next failure is from (7.122):

$$\Pr(Y(t_0) > t) = e^{-\int_0^t w(u+t_0)\,du} = e^{-\lambda^2(t^2 + 2t_0 t)}$$

If t_0 is the time of failure $n - 1$, the time to the next failure, $Y(t_0)$, is the nth interoccurrence time T_n and the failure rate (FOM) function of T_n is

$$z_{t_0}(t) = 2\lambda^2(t + t_0)$$

which is linearly increasing with the time t_0 of failure $n - 1$. Notice again that this is a conditional rate, given that failure $n - 1$ occurred at time $S_{n-1} = t_0$. The mean time between failure $n - 1$ and failure n is from (7.124):

$$\text{MTBF}_n = \int_0^\infty e^{-\lambda^2(t^2 + 2t_0 t)}\,dt$$

□

Relation to the Homogeneous Poisson Process Let $\{N(t), t \geq 0\}$ be an NHPP with ROCOF $w(t) > 0$ such that the inverse $W^{-1}(t)$ of the cumulative rate $W(t)$ exists, and let $S_1, S_2, \ldots$ be the times when the failures occur.

Consider the time-transformed occurrence times $W(S_1), W(S_2), \ldots$, and let $\{N^*(t), t \geq 0\}$ denote the associated counting process. The distribution of the (transformed) time $W(S_1)$ till the first failure is from (7.121)

$$\Pr(W(S_1) > t) = \Pr(S_1 > W^{-1}(t)) = e^{-W(W^{-1}(t))} = e^{-t}$$

that is, an exponential distribution with parameter 1.
The new counting process is defined by

$$N(t) = N^*(W(t)) \quad \text{for } t \geq t$$

hence

$$N^*(t) = N(W^{-1}(t)) \quad \text{for } t \geq 0$$

and we get from (7.116)

$$\Pr(N^*(t) = n) = \Pr(N(W^{-1}(t)) = n)$$
$$= \frac{[W(W^{-1}(t))]^n}{n!} e^{-W(W^{-1}(t))} = \frac{1^n}{n!} e^{-t}$$

that is, the Poisson distribution with rate 1. We have thereby shown that an NHPP with cumulative rate $W(t)$ (where the inverse of $W(t)$ exists) can be transformed into an HPP with rate 1, by time-transforming the failure occurrence times $S_1, S_2, \ldots$ to $W(S_1), W(S_2), \ldots$.

7.4.3 The Nelson-Aalen Estimator

Let $\{N(t), t \geq 0\}$ be an NHPP with ROCOF $w(t)$. We want to find an estimate of the mean number of failures $W(t) = E(N(t)) = \int_0^t w(u)\,du$ in the interval $(0, t)$. An obvious estimate is $\widehat{W}(t) = N(t)$.

Assume that we have n different and independent NHPPs $\{N_i(t), t \geq 0\}$ for $i = 1, 2, \ldots, n$ with a common ROCOF $w(t)$. This is, for example, the situation when we observe failures of the same type of repairable equipment installed in different places. Each installation will then produce a separate NHPP with rate $w(t)$. An obvious estimator for $W(t)$ is now

$$\widehat{W}(t) = \frac{1}{n} \sum_{i=1}^n N_i(t) = \frac{\text{Total number of failures in}(0, t]}{\text{Total number of processes in}(0, t]}$$

This estimator may be written in an alternative way:

$$\widehat{W}(t) = \frac{1}{n} \sum_{i=1}^n N_i(t) = \sum_{\{i;\, T_i \leq t\}} \frac{1}{n} = \sum_{i \geq 1} \frac{1}{n} 1_{\{T_i \leq t\}}$$

where $T_1, T_2, \ldots$ denotes the failure times (for all the processes), and $1_{\{T_i \leq t\}}$ is an indicator that is equal to 1 when $T_i \leq t$, and 0 otherwise.

In practice, the various processes will not be observed in the same interval. As an illustration, let us assume that n_1 processes are observed in the interval $(0, t_1]$, and that n_2 processes are observed for $t \geq t_1$. Let $N(t)$ denote the total number of failures in $(0, t]$, irrespective of how many processes are active. It seems now natural to estimate $W(t)$ by

$$\widehat{W}(t) = \frac{N(t)}{n_1} = \sum_{\{i;\, T_i \leq t_1\}} \frac{1}{n_1} \quad \text{when } 0 \leq t \leq t_1$$

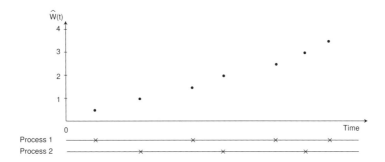

Fig. 7.19 The Nelson-Aalen estimator for two simultaneous processes.

When $t > t_1$, an estimator for $\int_{t_1}^{t} w(u)\, du = W(t) - W(t_1)$ is

$$\frac{N(t) - N(t_1)}{n_2} = \sum_{\{i;\ t_1 < T_i \leq t\}} \frac{1}{n_2} \quad \text{when } t > t_1$$

The total estimator will then be

$$\widehat{W}(t) = \sum_{\{i;\ T_i \leq t_1\}} \frac{1}{n_1} + \sum_{\{i;\ t_1 < T_i \leq t\}} \frac{1}{n_2}$$

Assume now that we have a set of independent NHPPs with a common ROCOF $w(t)$. Let $Y(s)$ denote the number of active processes immediately before time s. From the arguments above, it seems natural to use the following estimator for $W(t)$,

$$\widehat{W}(t) = \sum_{\{i;\ T_i \leq t\}} \frac{1}{Y(T_i)} = \sum_{i \geq 1} \frac{1}{Y(T_i)} 1_{\{T_i \leq t\}} \tag{7.125}$$

This nonparametric estimator is called the Nelson-Aalen estimator for $W(t)$. Note that when there is only one sample, then the Nelson-Aalen estimator coincides with $N(t)$, which is plotted in Fig. 7.3 A simple example of the Nelson-Aalen estimator for two simultaneous processes is illustrated in Fig. 7.19.

The estimator (7.125) was introduced by Aalen (1975, 1978) for counting processes in general and generalizes the Nelson (1969) estimator which is further discussed in Chapter 8. It may be shown (see the discussion in Andersen and Borgan, 1985) that $\widehat{W}(t)$ is an approximately unbiased estimator of $W(t)$ and that the variance can be estimated (almost unbiasedly) by

$$\text{var}(\widehat{W}(t)) \approx \hat{\sigma}^2(t) = \sum_{\{i;\ T_i \leq t\}} \frac{1}{Y(T_i)^2} \tag{7.126}$$

$\widehat{W}(t)$ may further be shown (see Andersen and Borgan, 1985) to be asymptotically normally distributed with mean $W(t)$ and a variance estimated by $\hat{\sigma}^2(t)$. Hence an approximate $100(1 - \alpha)\%$ pointwise confidence interval for $W(t)$, is given by

$$\widehat{W}(t) - u_{\alpha/2}\hat{\sigma}(t) \leq W(t) \leq \widehat{W}(t) + u_{\alpha/2}\hat{\sigma}(t)$$

where u_α denotes the upper $100\alpha\%$ percentile of the standard normal distribution $\mathcal{N}(0, 1)$.

7.4.4 Parametric NHPP Models

Several parametric models have been established to describe the ROCOF of an NHPP. We will discuss some of these models:

1. The power law model
2. The linear model
3. The log-linear model

All three models may be written in the common form (see Atwood, 1992)

$$w(t) = \lambda_0 \, g(t; \vartheta) \tag{7.127}$$

where λ_0 is a common multiplier, and $g(t; \vartheta)$ determines the shape of the ROCOF $w(t)$. The three models may be parameterized in various ways. In this section we shall use the parameterization of Crowder et al. (1991), although the parametrization of Atwood (1992) may be more logical.

The Power Law Model In the power law model the ROCOF of the NHPP is defined as

$$w(t) = \lambda \beta t^{\beta-1} \quad \text{for } \lambda > 0, \ \beta > 0 \text{ and } t \geq 0 \tag{7.128}$$

This NHPP is sometimes referred to as a *Weibull process*, since the ROCOF has the same functional form as the failure rate (FOM) function of the Weibull distribution. Also note that the first arrival time T_1 of this process is Weibull distributed with shape parameter β and scale parameter λ. However, according to Ascher and Feingold (1984), one should avoid the name Weibull process in this situation, since it gives the wrong impression that the Weibull distribution can be used to model trend in interoccurrence times of a repairable system. Hence such notation may lead to confusion.

A repairable system modeled by the power law model is seen to be improving (happy) if $0 < \beta < 1$, and deteriorating (sad) if $\beta > 1$. If $\beta = 1$ the model reduces to an HPP. The case $\beta = 2$ is seen to give a linearly increasing ROCOF. This model was studied in Example 7.16.

The power law model was first proposed by Crow (1974) based on ideas of Duane (1964). A goodness-of-fit test for the power law model based on total time on test (TTT) plots (see Chapter 11) is proposed and discussed by Klefsjö and Kumar (1992).

Assume that we have observed an NHPP in a time interval $(0, t_0]$ and that failures have occurred at times $s_1, s_2, \ldots, s_n$. Maximum likelihood estimates $\hat{\beta}$ and $\hat{\lambda}$ of β and λ, respectively, are given by

$$\hat{\beta} = \frac{n}{n \ln t_0 - \sum_{i=1}^{n} \ln s_i} \tag{7.129}$$

and
$$\hat{\lambda} = \frac{n}{t_0^{\hat{\beta}}} \quad (7.130)$$

The estimates are further discussed by Crowder et al. (1991, p. 171) and Cocozza-Thivent (1997, p. 64).

A $(1 - \varepsilon)$ confidence interval for β is given by (see Cocozza-Thivent 1997, p. 65)

$$\left(\frac{\hat{\beta}}{2n} z_{(1-\varepsilon/2),2n}, \frac{\hat{\beta}}{2n} z_{(1+\varepsilon/2),2n} \right) \quad (7.131)$$

where $z_{\varepsilon,\nu}$ denotes the upper $100\varepsilon\%$ percentile of the chi-square (χ^2) distribution with ν degrees of freedom (tables are given in Appendix F).

The Linear Model In the linear model the ROCOF of the NHPP is defined by

$$w(t) = \lambda(1 + \alpha t) \quad \text{for } \lambda > 0 \text{ and } t \geq 0 \quad (7.132)$$

The linear model has been discussed by Vesely (1991) and Atwood (1992). A repairable system modeled by the linear model is deteriorating if $\alpha > 0$, and improving when $\alpha < 0$. When $\alpha < 0$, then $w(t)$ will sooner or later become less than zero. The model should only be used in time intervals where $w(t) > 0$.

The Log-Linear Model In the log-linear model or *Cox-Lewis* model, the ROCOF of the NHPP is defined by

$$w(t) = e^{\alpha + \beta t} \quad \text{for } -\infty < \alpha, \beta < \infty \text{ and } t \geq 0 \quad (7.133)$$

A repairable system modeled by the log-linear model is improving (happy) if $\beta < 0$, and deteriorating (sad) if $\beta > 0$. When $\beta = 0$ the log-linear model reduces to an HPP.

The log-linear model was proposed by Cox and Lewis (1966) who used the model to investigate trends in the interoccurrence times between failures in air-conditioning equipment in aircrafts. The first arrival time T_1 has failure rate (FOM) function $z(t) = e^{\alpha + \beta t}$ and hence has a truncated Gumbel distribution of the smallest extreme (i.e., a Gompertz distribution; see Section 2.17).

Assume that we have observed an NHPP in a time interval $(0, t_0]$ and that failures have occurred at times $s_1, s_2, \ldots, s_n$. Maximum likelihood estimates $\hat{\alpha}$ and $\hat{\beta}$ of α and β, respectively, are found by solving

$$\sum_{i=1}^{n} s_i + \frac{n}{\beta} - \frac{nt_0}{1 - e^{-\beta t_0}} = 0 \quad (7.134)$$

to give $\hat{\beta}$, and then taking

$$\hat{\alpha} = \ln \left(\frac{n\hat{\beta}}{e^{\hat{\beta} t_0} - 1} \right) \quad (7.135)$$

The estimates are further discussed by Crowder et al. (1991, p. 167).

7.4.5 Statistical Tests of Trend

The simple graph in Fig. 7.3 clearly indicates an increasing rate of failures, that is, a deteriorating or *sad* system. The next step in an analysis of the data may be to perform a *statistical test* to find out whether the observed trend is *statistically significant* or just accidental. A number of tests have been developed for this purpose, that is, for testing the null hypothesis

H_0: "No trend" (or more precisely that the interoccurrence times are independent and identically exponentially distributed, that is, an HPP)

against the alternative hypothesis

H_1: "Monotonic trend" (i.e., the process is an NHPP that is either *sad* or *happy*)

Among these are two nonparametric tests that we will discuss:

1. The Laplace test
2. The military handbook test

These two tests are discussed in detail by Ascher and Feingold (1984) and Crowder et al. (1991). It can be shown that the Laplace test is optimal when the true failure mechanism is that of a log-linear NHPP model (see Cox and Lewis, 1966), while the military handbook test is optimal when the true failure mechanism is that of a power law NHPP model (see Bain et al. 1985).

The Laplace Test The test statistic for the case where the system is observed until n failures have occurred is

$$U = \frac{\frac{1}{n-1}\sum_{j=1}^{n-1} S_j - (S_n/2)}{S_n/\sqrt{12(n-1)}} \qquad (7.136)$$

where $S_1, S_2, \ldots$ denote the failure times. For the case where the system is observed until time t_0, the test statistic is

$$U = \frac{\frac{1}{n}\sum_{j=1}^{n} S_j - (t_0/2)}{t_0/\sqrt{12n}} \qquad (7.137)$$

In both cases, the test statistic U is approximately standard normally $\mathcal{N}(0, 1)$ distributed when the null hypothesis H_0 is true. The value of U is seen to indicate the direction of the trend, with $U < 0$ for a *happy* system and $U > 0$ for a *sad* system. Optimal properties of the Laplace test have, for example, been investigated by Gaudoin (1992).

Military Handbook Test The test statistic of the so-called military handbook test (see MIL-HDBK-189) for the case where the system is observed until n failures have occurred is

$$Z = 2\sum_{i=1}^{n-1} \ln \frac{S_n}{S_i} \qquad (7.138)$$

For the case where the system is observed until time t_0, the test statistic is

$$Z = 2 \sum_{i=1}^{n} \ln \frac{t_0}{S_i} \tag{7.139}$$

The asymptotic distribution of Z is in the two cases a χ^2 distribution with $2(n-1)$ and $2n$ degrees of freedom, respectively.

The hypothesis of no trend (H_0) is rejected for *small* or *large* values of Z. Low values of Z correspond to deteriorating systems, while large values of Z correspond to improving systems.

7.5 IMPERFECT REPAIR PROCESSES

In the previous sections we studied two main categories of models that can be used to describe the occurrence of failures of repairable systems: renewal processes and nonhomogeneous Poisson processes. The homogeneous Poisson process may be considered a special case of both models. When using a renewal process, the repair action is considered to be *perfect*, meaning the the system is "as good as new" after the repair action is completed. When we use the NHPP, we assume the the repair action is *minimal*, meaning that the reliability of the system is the same immediately after the repair action as it was immediately before the failure occurred. In this case we say the the system is "as bad as old" after the repair action. The renewal process and the NHPP may thus be considered as two extreme cases. Systems subject to normal repair will be somewhere between these two extremes. Several models have been suggested for the *normal*, or *imperfect*, repair situation, a repair that is somewhere between a minimal repair and a renewal.

In this section we will consider a system that is put into operation at time t. The initial failure rate (FOM) function of the system is denoted $z(t)$, and the conditional ROCOF of the system is denoted $w_C(t)$. The conditional ROCOF was defined by (7.7).

When the system fails, a repair action is initiated. The repair action will bring the system back to a functioning state and may involve a repair or a replacement of the system component that produced the system failure. The repair action may also involve maintenance and upgrading of the rest of the system and even replacement of the whole system. The time required to perform the repair action is considered to be negligible. Preventive maintenance, except for preventive maintenance carried out during a repair action, is disregarded.

A high number of models have been suggested for modeling imperfect repair processes. Most of the models may be classified in two main groups: (i) models where the repair actions reduce the rate of failures (ROCOF), and (ii) models where the repair actions reduce the (virtual) age of the system. A survey of available models are provided, for example, by Pham and Wang (1996), Hokstad (1997), and Akersten (1998).

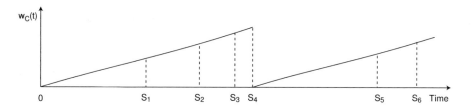

Fig. 7.20 An illustration of a possible shape of the conditional ROCOF of Brown and Proschan's imperfect repair model.

7.5.1 Brown and Proschan's Model

One of the best known imperfect repair models is described by Brown and Proschan (1983). Brown and Proschan's model is based on the following repair policy: A system is put into operation at time $t = 0$. Each time the system fails, a repair action is initiated, that with probability p is a *perfect* repair that will bring the system back to an "as good as new" condition. With probability $1 - p$ the repair action will be a *minimal* repair, leaving the system in an "as bad as old" condition. The renewal process and the NHPP are seen to be special cases of Brown and Proschan's model, when $p = 1$ and $p = 0$, respectively. Brown and Proschan's model may therefore be regarded as a mixture of the renewal process and the NHPP. Note that the probability p of a perfect repair is independent of the time elapsed since the previous failure and also of the age of the system. Let us, as an example, assume that $p = 0.02$. This means that we for most failures will make do with a minimal repair, and on the average renew (or replace) the system once for every 50 failures. This may be a realistic model, but the problem is that the renewals will come at random, meaning that we have the same probability of renewing a rather new system as an old system. Fig. 7.20 illustrates a possible shape of the conditional ROCOF.

The data obtained from a repairable system is usually limited to the times between failures, $T_1, T_2, \ldots$. The detailed repair modes associated to each failure are in general not recorded. Bases on this "masked" data set, Lim (1998) has developed a procedure for estimating p and the other parameters of Brown and Proschan's model.

Brown and Proschan's model was extended by Block et al. (1985) to age-dependent repair, that is, when the item fails at time t, a perfect repair is performed with probability $p(t)$ and a minimal repair is performed with probability $1 - p(t)$. Let Y_1 denote the time from $t = 0$ until the first perfect repair. When a perfect repair is carried out, the process will start over again, and we get a sequence of times between perfect repairs $Y_1, Y_2, \ldots$ that will form a renewal process. Assume that $F(t)$ is the distribution of the time to the first failure T_1, and let $f(t)$ and $R(t) = 1 - F(t)$ be the corresponding probability density function and the survivor function, respectively. The failure rate (FOM) function of T_1 is then $z(t) = f(t)/R(t)$, and we know from Chapter 2 that the distribution function may be written as

$$F(t) = 1 - e^{-\int_0^t z(x)\,dx} = 1 - e^{-\int_0^t [f(x)/R(x)]\,dx}$$

The distribution of Y_i is given by (see Block et al. 1985)

$$F_p(t) = \Pr(Y_i \leq t) = 1 - e^{-\int_0^t [p(x)f(x)/R(x)]dx} = 1 - e^{-\int_0^t z_p(x)dx} \quad (7.140)$$

Hence, the time between renewals, Y has failure rate (FOM) function

$$z_p(t) = \frac{p(t)f(t)}{R(t)} = p(t)z(t) \quad (7.141)$$

Block et al. (1985) also supply an explicit formula for the renewal function and discuss the properties of of $F_p(t)$.

Failure Rate Reduction Models Several models have been suggested where each repair action results in a reduction of the conditional ROCOF. The reduction may be a fixed reduction, a certain percentage of the actual value of the rate of failures, or a function of the history of the process. Models representing the first two types were proposed by Chan and Shaw (1993). Let $z(t)$ denote the failure rate (FOM) function of the time to the first failure. If all repairs were minimal repairs, the ROCOF of the process would be $w_1(t) = z(t)$. Consider the failure at time S_i, and let S_{i-} denote the time immediately before time S_i. In the same way, let S_{i+} denote the time immediately after time S_i. The models suggested by Chan and Shaw (1993) may then be expressed by the conditional ROCOF as

$$w_C(S_{i+}) = \begin{cases} w_C(S_{i-}) - \Delta & \text{for a fixed reduction } \Delta \\ w_C(S_{i-})(1-\rho) & \text{for a proportional reduction } 0 \leq \rho \leq 1 \end{cases} \quad (7.142)$$

Between two failures, the conditional ROCOF is assumed to be vertically parallel to the initial ROCOF $w_1(t)$. The parameter ρ in (7.142) is an index representing the efficiency of the repair action. When $\rho = 0$, we have minimal repair, and the NHPP is therefore a special case of Chan and Shaw's proportional reduction model. When $\rho = 1$, the repair action will bring the conditional ROCOF down to zero, but this will not represent a renewal process since the interoccurrence times will not be identically distributed, except for the special case when $w_1(t)$ is a linear function. The conditional ROCOF of Chan and Shaw's proportional reduction model is illustrated in Fig. 7.21 for some possible failure times and with $\rho = 0.30$.

Chan and Shaw's model (7.142) has been generalized by Doyen and Gaudoin (2002a,b). They propose a set of models where the proportionality factor ρ depends on the history of the process. In their models the conditional ROCOF is expressed as

$$w_C(S_{i+}) = w_C(S_{i-}) - \varphi(i, S_1, S_2, \ldots, S_i) \quad (7.143)$$

where $\varphi(i, S_1, S_2, \ldots, S_i)$ is the reduction of the conditional ROCOF resulting from the repair action. Between two failures they assume that the conditional ROCOF is vertically parallel to the initial ROCOF $w_1(t)$. These assumptions lead to the conditional ROCOF

$$w_C(t) = w_1(t) - \sum_{i=1}^{N(t)} \varphi(i, S_1, S_2, \ldots, S_i) \quad (7.144)$$

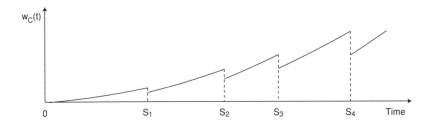

Fig. 7.21 The conditional ROCOF of Chan and Shaw's proportional reduction model for some possible failure times ($\rho = 0.30$).

When we, as in Chan and Shaw's model (7.142), assume a proportional reduction after each repair action, the conditional ROCOF in the interval $(0, S_1)$ becomes $w_C(t) = w_1(t)$. In the interval $[S_1, S_2)$ the conditional ROCOF is $w_C(t) = w_1(t) - \rho w_1(S_1)$. In the third interval $[S_2, S_3)$ the conditional ROCOF is

$$\begin{aligned} w_C(t) &= w_1(t) - \rho w_1(S_1) - \rho \left[w_1(S_2) - \rho w_1(S_1)\right] \\ &= w_1(t) - \rho \left[(1-\rho)^0 w_1(S_2) + (1-\rho)^1 w_1(S_1)\right] \end{aligned}$$

It is now straightforward to continue this derivation and show that the conditional ROCOF of Chan and Shaw's proportional reduction model (7.142) may be written as

$$w_C(t) = w_1(t) - \rho \sum_{i=0}^{N(t)} (1-\rho)^i \, w_1(S_{N(t)-i}) \tag{7.145}$$

This model is called arithmetic reduction of intensity with infinite memory (ARI_∞) by Doyen and Gaudoin (2002a).

In the model (7.142) the reduction is proportional to the conditional ROCOF just before time t. Another approach is to assume that a repair action can only reduce a proportion of the wear that has accumulated since the previous repair action. This can be formulated as:

$$w_C(S_{i+}) = w_C(S_{i-}) - \rho \left[w_C(S_{i-}) - w_C(S_{i-1+})\right] \tag{7.146}$$

The conditional ROCOF of this model is

$$w_C(t) = w_1(t) - \rho w_1(S_{N(t)}) \tag{7.147}$$

This model is called arithmetic reduction of intensity with memory one (ARI_1) by Doyen and Gaudoin (2002a). If $\rho = 0$, the system is "as bad as old" after the repair action and the NHPP is thus a special case of the ARI_1 model. If $\rho = 1$, the conditional ROCOF is brought down to zero by the repair action, but the process is not a renewal process, since the interoccurrence times are not identically distributed. For the ARI_1 model, there exists a deterministic function $w_{\min}(t)$ that is always smaller than the

IMPERFECT REPAIR PROCESSES   291

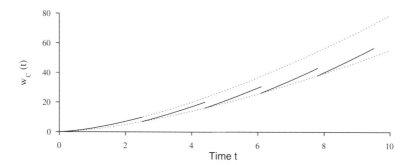

Fig. 7.22 The ARI$_1$ model for some possible failure times. The "underlying" ROCOF $w_1(t)$ is a power law model with shape parameter $\beta = 2.5$, and the parameter $\rho = 0.30$. The upper dotted curve is $w_1(t)$, and the lower dotted curve is the minimal wear intensity $(1-\rho)w_1(t)$.

conditional ROCOF such that there is a nonzero probability that the ROCOF will be excessively close to $w_{\min}(t)$.

$$w_{\min}(t) = (1-\rho)\, w_1(t)$$

This intensity is a minimal wear intensity, that is to say, a maximal lower boundary for the conditional ROCOF. The ARI$_1$ model is illustrated in Fig. 7.22 for some possible failure times.

The two models ARI$_\infty$ and ARI$_1$ may be considered as two extreme cases. To illustrate the difference, we may consider the conditional ROCOF as an index representing the wear of the system. By the ARI$_\infty$ model, every repair action will reduce, by a specified percentage ρ, the total accumulated wear of the system since the system was installed. By the ARI$_1$ model the repair action will only reduce, by a percentage ρ, the wear that has been accumulated since the previous repair action. This is why Doyen and Gaudoin (2002a) say that the ARI$_\infty$ has infinite memory, while the ARI$_1$ has memory one (one period).

Doyen and Gaudoin (2002a) have also introduced a larger class of models in which only the first m terms of the sum in (7.145) are considered. They call this model the arithmetic reduction of intensity model of memory m (ARI$_m$), and the corresponding conditional ROCOF is

$$w_C(t) = w_1(t) - \rho \sum_{i=0}^{\min\{m-1, N(t)\}} (1-\rho)^i\, w_C(S_{N(t)-i}) \qquad (7.148)$$

The ARI$_m$ model has a minimal wear intensity:

$$w_{\min}(t) = (1-\beta)^m w_1(t)$$

In all these models we note that the parameter ρ may be regarded as an index of the efficiency of the repair action.

- $0 < \rho < 1$: The repair action is efficient.

- $\rho = 1$: Optimal repair. The conditional ROCOF is put back to zero (but the repair effect is different from the "as good as new" situation.

- $\rho = 0$: The repair action has no effect on the wear of the system. The system state after the repair action is "as bad as old."

- $\rho < 0$: The repair action is harmful to the system, and will introduce extra problems.

Age Reduction Models Malik (1979) proposed a model where each repair action reduces the *age* of the system by a time that is proportional to the operating time elapsed from the previous repair action. The age of the system is hence considered as a virtual concept.

To establish a model, we assume that a system is put into operation at time $t = 0$. The initial ROCOF $w_1(t)$ is equal to the failure rate (FOM) function $z(t)$ of the interval until the first system failure. $w_1(t)$ is then the ROCOF of a system where all repairs are minimal repairs. The first failure occurs at time S_1, and the conditional ROCOF just after the repair action is completed is

$$w_C(S_{1+}) = w_1(S_1 - \vartheta)$$

where $S_1 - \vartheta$ is the new virtual age of the system. After the next failure, the conditional ROCOF will be $w_C(S_{2+}) = w_1(S_2 - 2\vartheta)$ and so on. The conditional ROCOF at time t is

$$w_C(t) = w_1(t - N(t)\vartheta)$$

We may now let ϑ be a function of the history and get

$$w_C(t) = w_1\left(t - \sum_{i=1}^{N(t)} \vartheta(i, S_1, S_2, \ldots S_i)\right) \qquad (7.149)$$

Between two consecutive failures we assume that the conditional ROCOF is horizontally parallel with the initial ROCOF $w_1(t)$.

Doyen and Gaudoin (2002a) propose an age reduction model where the repair action reduces the virtual age of the system with an amount proportional to its age just before the repair action. Let ρ denote the percentage of reduction of the virtual age. In the interval $(0, S_1)$ the conditional ROCOF is $w_C(t) = w_1(t)$. Just after the first failure (when the repair is completed) the virtual age is $S_1 - \rho S_1$, and in the interval (S_1, S_2) the conditional ROCOF is $w_C(t) = w_1(t - \rho S_1)$. Just before the second failure at time S_2, the virtual age is $S_2 - \rho S_1$, and just after the second failure the virtual age is $S_2 - \rho S_1 - \rho(S_2 - \rho S_1)$. In the interval (S_2, S_3) the conditional ROCOF is $w_C(t) = w_1(t - \rho S_1 - \rho(S_2 - \rho S_1))$ which may be written as $w_C(t) =$

$w_1(t - \rho(1-\rho)^0 S_2 - \rho(1-\rho)^1 S_1)$. By continuing this argument, it is easy to realize that the conditional ROCOF of this age reduction model is

$$w_C(t) = w_1\left(t - \rho \sum_{i=0}^{N(t)} (1-\rho)^i S_{N(t)-i}\right) \quad (7.150)$$

This model is by Doyen and Gaudoin (2002a) called arithmetic reduction of age with infinite memory (ARA$_\infty$). The same model has also been introduced by Yun and Choung (1999). We note that when $\rho = 0$, we get $w_C(t) = w_1(t)$ and we have an NHPP. When $\rho = 1$, we get $w_C(t) = w_1(t - S_{N(t)})$ which represents that the repair action leaves the system in an "as good as new" condition. The NHPP and the renewal process are therefore special cases of the ARA$_{-\infty}$ model.

Malik (1979) introduced a model in which the repair action at time S_i reduces the last operating time from $S_i - S_{i-1}$ to $\rho(S_i - S_{i-1})$ where as before, $0 \leq \rho \leq 1$. Using this model, Shin et al. (1996) developed an optimal maintenance policy and derived estimates for the various parameters. The corresponding conditional ROCOF is

$$w_C(t) = w_1(t - \rho S_{N(t)})$$

The minimal wear intensity is equal to $w_1((1-\rho)t)$. This model is by Doyen and Gaudoin (2002a) called arithmetic reduction of age with memory one (ARA$_1$).

In analogy with the failure rate reduction models, we may define a model called arithmetic reduction of age with memory m by

$$w(t) = w_1\left(t - \rho \sum_{i=0}^{\min\{m-1, N(t)\}} (1-\rho)^i S_{N(t)-i}\right)$$

The minimal wear intensity is

$$w_{\min}(t) = w_1((1-\beta)^m t)$$

Kijima and Sumita (1986) introduced a model which they called the generalized renewal process for modeling the imperfect repair process. This model has later been extended by Kaminskiy and Krivstov (1998). The model is an age reduction model that is similar to the models described by Doyen and Gaudoin (2002a). Estimation of the parameters of the generalized renewal model is discussed by Yañes et al. (2002).

Trend Renewal Process Let $S_1, S_2, \ldots$ denote the times when failure occur in an NHPP with ROCOF $w(t)$, and let $W(t)$ denote the mean number of failures in the interval $(0, t]$. On page 281 we showed that the time-transformed process with occurrence times $W(S_1), W(S_2), \ldots$ is an HPP with rate 1. In the transformed process, the mean time between failures (and renewals) will then be 1. Lindqvist (1993, 1998) generalized this model, by replacing the HPP with rate 1 with a renewal process with

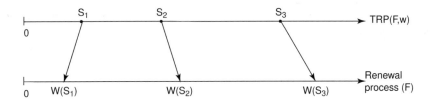

Fig. 7.23 Illustration of the transformation of a TRP(F, w) to a renewal process. (Adapted from Lindqvist 1999).

underlying distribution $F(\cdot)$ with mean 1. He called the resulting process a trend-renewal process, TRP(F, w). To specify the process we need to specify the rate $w(t)$ of the initial NHPP and the distribution $F(t)$.

If we have a TRP(F, w) with failure times $S_1, S_2, \ldots$, the time-transformed process with occurrence times $W(S_1), W(S_2), \ldots$ will be a renewal process with underlying distribution $F(t)$. The transformation is illustrated in Fig. 7.23. The requirement that $F(t)$ has mean value 1 is made for convenience. The scale is then taken care of by the rate $w(t)$. Lindqvist (1998) shows that the conditional ROCOF of the TRP(F, w) is

$$w_C^{\text{TRP}}(t) = z\left(W(t) - W(S_{N(t-)})\right) w(t) \qquad (7.151)$$

where $z(t)$ is the failure rate (FOM) function of the distribution $F(t)$. The conditional ROCOF of the TRP(F, w) is hence a product of a factor, $w(t)$, that depends on the age t of the system, and a factor that depends on the (transformed) time from the previous failure. When both the failure rate (FOM) function $z(t)$ and the initial ROCOF $w(t)$ are increasing functions, then the conditional ROCOF (7.151) at time t after a failure at time s_0 is

$$z\left(W(t + s_0) - W(s_0)\right) w(t + s_0)$$

To check the properties of the TRP we may look at some special cases:

- If $z(t) = \lambda$ and $w(t) = \beta$ are both constant, the conditional ROCOF is also constant, $w_C(t) = \lambda \cdot \beta$. Hence the HPP is a special case of the TRP.
- If $z(t) = \lambda$ is constant, the conditional ROCOF is $w_C(t) = \lambda \cdot w(t)$, and the NHPP is hence a special case of the TRP.
- If $z(0) = 0$, the conditional ROCOF is equal to 0 just after each failure, that is, $w_C(S_{N(t_+)}) = 0$.
- If $w(t) = \beta$ is constant, we have an ordinary renewal process, $w_C(t) = z(t - S_{N(t_-)})$.
- If $z(0) > 0$, the conditional ROCOF just after a failure is $z(0) \cdot w(S_{N(t_+)})$ and is increasing with t when $w(t)$ is an increasing function

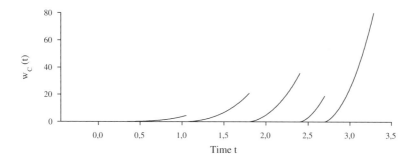

Fig. 7.24 Illustration of the conditional ROCOF $w_C(t)$ in Example 7.17 for some possible failure times.

- If $z(t)$ is the failure rate (FOM) function of a Weibull distribution with shape parameter α and $w(t)$ is a power law (Weibull) process with shape parameter β, the conditional ROCOF will have a Weibull form with shape parameter $\alpha\beta - 1$.

Example 7.17
Consider a trend renewal process with initial ROCOF $w(t) = 2\theta^2 t$, that is, a linearly increasing ROCOF, and a distribution $F(t)$ with failure rate (FOM) function $z(t) = 2.5\lambda^{2.5} \cdot t^{1.5}$, that is, a Weibull distribution with shape parameter $\alpha = 2.5$ and scale parameter λ. For the mean value of $F(t)$ to be equal to 1, the scale parameter must be $\lambda \approx 0.88725$. The conditional ROCOF in the interval until the first failure is from (7.151)

$$w_C(t) = 5\lambda^{2.5}\theta^5 \cdot t^4 \quad \text{for } 0 \le t < S_1$$

Just after the first failure, $w_C(S_{1+}) = 0$. Generally, we can find $w_C(t)$ from (7.151). Between failure n and failure $n+1$, the conditional ROCOF is

$$w_C(t) = 5\lambda^{2.5}\theta^5 \cdot (t^2 - S_n^2)^{1.5} \cdot t \quad \text{for } S_n \le t < S_{n+1}$$

The conditional ROCOF $w_C(t)$ is illustrated for some possible failure times $S_1, S_2, \ldots$ in Fig. 7.24. □

The trend renewal process is further studied by Lindqvist (1993, 1998) and Elvebakk (1999) who also provides estimates for the parameters of the model.

7.6 MODEL SELECTION

A simple framework for model selection for a repairable system is shown in Fig. 7.25. The figure is inspired by a figure in Ascher and Feingold (1984), but new aspects have been added.

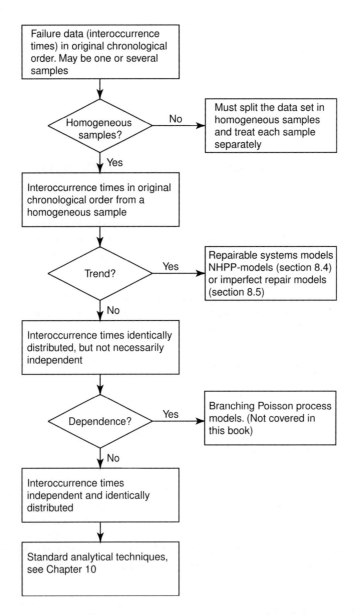

Fig. 7.25 Model selection framework.

We will illustrate the model selection framework by a simple example. In OREDA (2002), failure data from 449 pumps were collected from 61 different installations. A total of 524 critical failures were recorded, that is, on the average 1.17 failures per pump. To get adequate results we have to merge failure data from several valves. It is important that the data that are merged are homogeneous, meaning that the valves are of the same type and that the operational and environmental stresses are comparable. Since there are very few data from each valve, this analysis will have to be qualitative. The total data set should be split into homogeneous subsets and each subset has to be analyzed separately. A very simple problem related to inhomogeneous samples is illustrated in Section 2.9.

We now continue with a subset of the data that is deemed to be homogeneous. The next step is to check whether or not there is a trend in the ROCOF. This may be done by establishing a Nelson-Aalen plot as described in Section 7.4.3 on page 282. If the plot is approximately linear we conclude that the ROCOF is close to constant. If the plot is convex (concave) we conclude that the ROCOF is increasing (decreasing). The ROCOF may also be increasing in one part of the lifelength and decreasing in another part.

If we conclude that the ROCOF is increasing or decreasing, we may use either a NHPP or one of the imperfect repair models described in Section 7.5. Which model to use must (usually) be decided by a qualitative analysis of the repair actions, whether it is a minimal repair or and age, or failure rate, reduction repair. In some cases we may have close to minimal repairs during a period followed by a major overhaul. In the Norwegian offshore sector, such overhauls are often carried our during annual revision stops. When we have decided a model, we may use the methods described in this chapter to analyze the data. More detailed analyses are described, for example, in Crowder et al. (1991).

If no trend in the ROCOF is detected, we conclude that the intervals between failures are identically distributed, but not necessarily independent. The next step is then to check whether or not the data may be considered as independent. Several plotting techniques and formal tests are available. These methods are, however, not covered in this book. An introduction to such methods may, for example, be found in Crowder et al. (1991).

If we can conclude that the intervals between failures are independent and identically distributed, we have a renewal process, and we can use the methods described in Chapter 11 to analyze the data.

If the intervals are dependent, we have to use methods that are not described in this book. Please consult, for example, Crowder et al. (1991) for relevant approaches.

PROBLEMS

7.1 Consider a homogeneous Poisson process (HPP) $\{N(t), t \geq 0\}$ and let $t, s \geq 0$. Determine

$$E(N(t) \cdot N(t+s))$$

7.2 Consider an HPP $\{N(t), t \geq 0\}$ with rate $\lambda > 0$. Verify that

$$\Pr(N(t) = k \mid N(s) = n) = \binom{n}{k}\left(\frac{t}{s}\right)^k \left(1 - \frac{t}{s}\right)^{n-k} \quad \text{for } 0 < t < s \text{ and } 0 \leq k \leq n$$

7.3 Let T_1 denote the time to the first occurrence of an HPP $\{N(t), t \geq 0\}$ with rate λ. Show that

$$\Pr(T_1 \leq s \mid N(t) = 1) = \frac{s}{t} \quad \text{for } s \leq t$$

7.4 Let $\{N(t), t \geq 0\}$ be a counting process, with possible values $0, 1, 2, 3, \ldots$. Show that the mean value of $N(t)$ can be written

$$E(N(t)) = \sum_{n=1}^{\infty} \Pr(N(t) \geq n) = \sum_{n=0}^{\infty} \Pr(N(t) > n) \tag{7.152}$$

7.5 Let $S_1, S_2, \ldots$ be the occurrence times of an HPP $\{N(t), t \geq 0\}$ with rate λ. Assume that $N(t) = n$. Show that the random variables $S_1, S_2, \ldots, S_n$ have the joint probability density function

$$f_{S_1,\ldots,S_n \mid N(t)=n}(s_1, \ldots, s_n) = \frac{n!}{t^n} \quad \text{for } 0 < s_1 < \cdots < s_n \leq t$$

7.6 Consider a renewal process $\{N(t), t \geq 0\}$. Is it true that

(a) $N(t) < r$ if and only if $S_r > t$?

(b) $N(t) \leq r$ if and only if $S_r \geq t$?

(c) $N(t) > r$ if and only if $S_r < t$?

7.7 Consider a nonhomogeneous Poisson process (NHPP) with rate

$$w(t) = \lambda \cdot \frac{t+1}{t} \quad \text{for } t \geq 0$$

(a) Make a sketch of $w(t)$ as a function of t.

(b) Make a sketch of the cumulative ROCOF, $W(t)$, as a function of t.

7.8 Consider an NHPP $\{N(t), t \geq 0\}$ with rate:

$$w(t) = \begin{cases} 6 - 2t & \text{for } 0 \leq t \leq 2 \\ 2 & \text{for } 2 < t \leq 20 \\ -18 + t & \text{for } t > 20 \end{cases}$$

(a) Make a sketch of $w(t)$ as a function of t.

(b) Make a sketch of the corresponding cumulative ROCOF, $W(t)$, as a function of t.

(c) Estimate the number of failures/events in the interval (0, 12)

7.9 In Section 7.3.8 it is claimed that the superposition of independent renewal processes is generally *not* a renewal process. Explain why the superposition of independent homogeneous Poisson processes (HPP) is a renewal process. What is the renewal density of this superimposed process?

7.10 Table 7.2 shows the intervals in operating hours between successive failures of air-conditioning equipment in a Boeing 720 aircraft. The data are from Proschan (1963).

Table 7.2 Time Between Failures in Operating Hours of Air-conditioning Equipment.

413	14	58	37	100	65	9	169
447	184	36	201	118	34	31	18
18	67	57	62	7	22	34	

First interval is 413, the second is 14, and so on. Source: Proschan (1963).

(a) Establish the Nelson-Aalen plot ($N(t)$ plot) of the data set. Describe (with words) the shape of the ROCOF.

7.11 Atwood (1992) uses the following parametrization for the power law model, the linear model and the log-linear model:

$$\begin{aligned} w(t) &= \lambda_0 \, (t/t_0)^\beta & \text{(power law model)} \\ w(t) &= \lambda_0 [1 + \beta(t - t_0)] & \text{(linear model)} \\ w(t) &= \lambda_0 \, e^{\beta(t - t_0)} & \text{(log-linear model)} \end{aligned}$$

(a) Discuss the meaning of t_0 item[(b)] Show that Atwood's parameterization is compatible with the parameterization used in Section 7.4.4.

(c) Show that $w(t) = \lambda_0$ when $t = t_0$ for all the three models.

(d) Show that $w(t)$ is increasing if $\beta > 0$, is constant if $\beta = 0$, and decreasing if $\beta < 0$, for all the three models.

7.12 Use the MIL-HDBK test described in Section 7.4.5 to check if the "increasing trend" of the data in Example 7.1 is significant (5% - level).

7.13 Table 7.3 shows the intervals in days between successive failures of a piece of software developed as part of a large data system. The data are from Jelinski and Moranda (1972).

Table 7.3 Intervals in Days Between Successive Failures of a Piece of Software.

9	12	11	4	7	2	5	8	5	7
1	6	1	9	4	1	3	3	6	1
11	33	7	91	2	1	87	47	12	9
135	258	16	35						

First interval is 9, the second is 12, and so on. Source: Jelinski and Moranda (1972).

(a) Establish the Nelson-Aalen plot ($N(t)$ plot) of the data set. Is the ROCOF increasing or decreasing?

(b) Assume that the ROCOF follows a log-linear model, and find the maximum likelihood estimates (MLE) for the parameters of this model.

(c) Draw the estimated cumulative ROCOF in the same diagram as the Nelson-Aalen plot. Is the fit acceptable?

(d) Use the Laplace test to determine whether the ROCOF is decreasing or not (use a 5% level of significance).

8
Markov Processes

8.1 INTRODUCTION

The models in the first six chapters are all based on the assumption that the components and the systems can be in one out of two possible states: a *functioning state* or a *failed state*. We have also seen that the models are rather static and not well suited for analysis of repairable systems.

Stochastic processes were introduced in Chapter 7. In this chapter we will introduce a special type of stochastic processes, called Markov[1] chains, to model systems with several states and the *transitions* between the states. A Markov chain is a stochastic process $\{X(t), t \geq 0\}$ that possesses the Markov property. (We will define the Markov property clearly later.) The random variable $X(t)$ denotes the *state* of the process at *time t*. The collection of all possible states is called the *state space*, and we will denote it by $\mathcal{X}$. The state space $\mathcal{X}$ is either finite or countable infinite. In most of our applications the state space will be *finite* and the states will correspond to real states of a system (see Example 8.1). Unless stated otherwise, we take $\mathcal{X}$ to be $\{0, 1, 2, \ldots, r\}$, such that $\mathcal{X}$ contains $r+1$ different states. The time may be discrete, taking values in $\{0, 1, 2, \ldots\}$, or continuous. When the time is discrete, we have a *discrete-time Markov chain*; and when the time is continuous, we have a *continuous-time Markov chain*. A continuous-time Markov chain is also called a *Markov process*. When the time is discrete, we denote the time by n and the discrete-time Markov chain by $\{X_n, n = 0, 1, 2, \ldots\}$.

---

[1] Named after the Russian mathematician Andrei A. Markov (1856–1922).

Table 8.1 Possible States of a System of Two Components.

State	Component 1	Component 2
3	Functioning	Functioning
2	Functioning	Failed
1	Failed	Functioning
0	Failed	Failed

The presentation of the theoretical basis of the Markov chains in this book is rather brief and limited. The reader should consult a textbook on stochastic processes for more details. An excellent introduction to Markov chains may be found in, for example, Ross (1996). A very good description of continuous-time Markov chains and their application in reliability engineering is given by Cocozza-Thivent (1997).

The main focus in this book is on continuous-time Markov chains and how these chains can be used to model the reliability and availability of a system. In the following, a continuous-time Markov chain will be called a *Markov process*. In this chapter, we start by defining the Markov property and Markov processes. A set of linear, first order differential equations, called the *Kolmogorov equations*, are established to determine the probability distribution $P(t) = [P_0(t), P_1(t), \ldots, P_r(t)]$ of the Markov process at time t, where $P_i(t)$ is the probability that the process (the system) is in state i at time t. We then show that $P(t)$, under specific conditions, will approach a limit P when $t \to \infty$. This limit is called the *steady-state* distribution of the process (the system). Several system performance measures – like state visit frequency, system availability, and mean time to first system failure – are introduced. The steady-state distribution and the system performance measures are then determined for some simple systems like series and parallel systems, systems with dependent components, and various types of standby systems. Some approaches to analysis of complex systems are discussed. The time-dependent solution of the Kolmogorov equations is briefly discussed. The chapter ends by a brief discussion of semi-Markov processes, a generalization of the Markov processes.

Example 8.1
Consider a parallel structure of two components. Each component is assumed to have two states, a functioning state and a failed state. Since each of the components has two possible states, the parallel structure has $2^2 = 4$ possible states. These states are listed in Table 8.1. The state space is therefore $\mathcal{X} = \{0, 1, 2, 3\}$. The system is fully functioning when the state is 3 and failed when the state is 0. In states 1 and 2 the system is operating with only one component functioning. □

When the system has n components, and each component has two states (functioning and failed), the system will have at most 2^n different states. In some applications, we will introduce more than two states for each component. A pump may, for example, have three states: operating, standby, or failed. A producing unit may operate with 100% capacity, 80% capacity, and so on. In other applications it is important to distinguish the various failure modes of an item, and we may define the failure modes as states. For a complex system, the number of states may hence be overwhelming, and we may need to simplify the system model, and separately consider modules of the system.

8.1.1 Markov Property

Consider a stochastic process $\{X(t), t \geq 0\}$ with continuous time and state space $\mathcal{X} = \{0, 1, 2, \ldots, r\}$. Assume that the state of the process at time s is $X(s) = i$. The conditional probability that the process will be in state j at time $t + s$ is

$$\Pr(X(t+s) = j \mid X(s) = i, X(u) = x(u), 0 \leq u < s)$$

where $\{x(u), 0 \leq u < s\}$ denotes the "history" of the process up to, but not including, time s.

The process is said to have the Markov property if

$$\Pr(X(t+s) = j \mid X(t) = i, X(u) = x(u), 0 \leq u < s)$$
$$= \Pr(X(t+s) = j \mid X(s) = i) \qquad (8.1)$$
$$\text{for all possible } x(u), \ 0 \leq u < s$$

In other words, when the *present* state of the process is known, the future development of the process is independent of anything that has happened in the past.

A stochastic process satisfying the Markov property (8.1) is called a *Markov process* (or a continuous-time Markov chain).

We will further assume that the Markov process for all i, j in $\mathcal{X}$ fulfills

$$\Pr(X(t+s) = j \mid X(s) = i) = \Pr(X(t) = j \mid X(0) = i) \quad \text{for all } s, t \geq 0$$

which says that the probability of a transition from state i to state j does not depend on the global time and only depends on the time interval available for the transition. A process with this property is known as a process with *stationary transition probabilities*, or as a *time-homogeneous* process.

From now on we will only consider Markov processes (i.e., processes fulfilling the Markov property) that have stationary transition probabilities. A consequence of this assumption is that a Markov process cannot be used to model a system where the transition probabilities are influenced by long-term trends and/or seasonal variations. To use a Markov process, we have to assume that the environmental and operational conditions for the system are relatively stable as a function of time.

8.2 MARKOV PROCESSES

Consider a Markov process $\{X(t), t \geq 0\}$ with state space $\mathcal{X} = \{0, 1, 2, \ldots, r\}$ and stationary transition probabilities. The transition probabilities of the Markov process

$$P_{ij}(t) = \Pr(X(t) = j \mid X(0) = i) \quad \text{for all } i, j \in \mathcal{X}$$

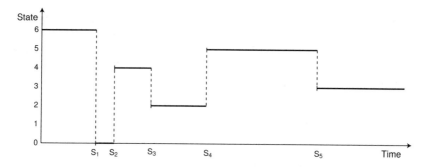

Fig. 8.1 Trajectory of a Markov process.

may be arranged as a matrix

$$\mathbb{P}(t) = \begin{pmatrix} P_{00}(t) & P_{01}(t) & \cdots & P_{0r}(t) \\ P_{10}(t) & P_{11}(t) & \cdots & P_{1r}(t) \\ \vdots & \vdots & \ddots & \vdots \\ P_{r0}(t) & P_{r1}(t) & \cdots & P_{rr}(t) \end{pmatrix} \qquad (8.2)$$

Since all entries in $\mathbb{P}(t)$ are probabilities, we have that

$$0 \leq P_{ij}(t) \leq 1 \quad \text{for all } t \geq 0, \ i, j \in \mathcal{X}$$

When a process is in state i at time 0, it must either be in state i at time t or have made a transition to a different state. We must therefore have

$$\sum_{j=0}^{r} P_{ij}(t) = 1 \quad \text{for all } i \in \mathcal{X} \qquad (8.3)$$

The sum of each row in the matrix $\mathbb{P}$ is therefore equal to 1. Note that the entries in row i represent the transitions out of state i (for $j \neq i$), and that the entries in column j represent the transition into state j (for $i \neq j$).

Let $0 = S_0 \leq S_1 \leq S_2 \leq \cdots$ be the times at which transitions occur, and let $T_i = S_{i+1} - S_i$ be the ith interoccurrence time, or *sojourn time*, for $i = 1, 2, \ldots$. A possible "path" of a Markov process is illustrated in Fig. 8.1. The path is sometimes called the *trajectory* of the process. We define S_i such that transition i takes place immediately before S_i, in which case the trajectory of the process is continuous from the right. The Markov process in Fig. 8.1 starts out at time $t = 0$ in state 6, and stays in this state a time T_1. At time $S_1 = T_1$ the process has a transition to state 0 where it stays a time T_2. At time $S_2 = T_1 + T_2$ the process has a transition to state 4, and so on.

Consider a Markov process that enters state i at time 0, such that $X(0) = i$. Let $\tilde{T}_i$ be the sojourn time in state i. [Note that T_i denotes the ith interoccurrence time, while $\widetilde{T}_i$ is the time spent during a visit to state i.] We want to find the probability $\Pr(\widetilde{T}_i > t)$. Let us now assume that we observe that the process is still in state i at time s, that is, $\widetilde{T}_i > s$, and that we are interested in finding the probability that it will remain in state i for t time units more. We hence want to find $\Pr(\widetilde{T}_i > t+s \mid \widetilde{T}_i > s)$. Since the process has the Markov property, the probability for the process to stay for t more time units is determined only by the current state i. The fact that the process has been staying there for s time units is therefore irrelevant. Thus

$$\Pr(\widetilde{T}_i > t+s \mid \widetilde{T}_i > s) = \Pr(\widetilde{T}_i > t) \quad \text{for } s, t \geq 0$$

Hence the random variable $\widetilde{T}_i$ is memoryless and must be exponentially distributed.

The sojourn times $T_1, T_2, \ldots$ must therefore also be independent and exponentially distributed. The independence follows from the Markov property. See Ross (1996, p. 232) for a more detailed discussion.

We will now disregard the time spent in the various states and only consider the transitions that take place at times $S_1, S_2, \ldots$. Let $X_n = X(S_n)$ denote the state immediately after transition n. The process $\{X_n, n = 1, 2, \ldots\}$ is called the *skeleton* of the Markov process. Transitions of the skeleton may be considered to take place at discrete times $n = 1, 2, \ldots$. The skeleton may be imagined as a chain where all the sojourn times are deterministic and of equal length. It is straightforward to show that the skeleton of a Markov process is a discrete-time Markov chain; see Ross (1996). The skeleton is also called the *embedded* Markov chain.

We may now (Ross, 1996, p. 232) construct a Markov process as a stochastic process having the properties that each time it enters a state i:

1. The amount of time $\widetilde{T}_i$ the process spends in state i before making a transition into a different state is exponentially distributed with rate, say α_i.

2. When the process leaves state i, it will next enter state j with some probability P_{ij}, where $\sum_{\substack{j=0 \\ j \neq i}}^{r} P_{ij} = 1$.

The mean sojourn time in state i is therefore

$$E(\widetilde{T}_i) = \frac{1}{\alpha_i}$$

If $\alpha_i = \infty$, state i is called an *instantaneous* state, since the mean sojourn time in such a state is zero. When the Markov process enters such a state, the state is instantaneously left. In this book, we will assume that the Markov process has no instantaneous states, and that $0 \leq \alpha_i < \infty$ for all i. If $\alpha_i = 0$, then state i is called *absorbing* since once entered it is never left. In Sections 8.2 and 8.3 we will assume that there are no absorbing states. Absorbing states are further discussed in Section 8.5.

We may therefore consider a Markov process as a stochastic process that moves from state to state in accordance with a discrete-time Markov chain. The amount of time it spends in each state, before going to the next state, is exponentially distributed.

The amount of time the process spends in state i, and the next state visited, must be independent random variables.

Let a_{ij} be defined by

$$a_{ij} = \alpha_i \cdot P_{ij} \quad \text{for all } i \neq j \tag{8.4}$$

Since α_i is the rate at which the process leaves state i and P_{ij} is the probability that it goes to state j, it follows that a_{ij} is the rate when in state i that the process makes a transition into state j. We call a_{ij} the *transition rate* from i to j.

Since $\sum_{j \neq i} P_{ij} = 1$, it follows from (8.4) that

$$\alpha_i = \sum_{\substack{j=0 \\ j \neq i}}^{r} a_{ij} \tag{8.5}$$

Let T_{ij} be the time the process spends in state i before entering into state $j (\neq i)$. The time T_{ij} is exponentially distributed with rate a_{ij}.

Consider a short time interval Δt. Since T_{ij} and $\widetilde{T}_i$ are exponentially distributed, we have that

$$P_{ii}(\Delta t) = \Pr(\widetilde{T}_i > \Delta t) = e^{-\alpha_i \Delta t} \approx 1 - \alpha_i \Delta t$$
$$P_{ij}(\Delta t) = \Pr(T_{ij} \leq \Delta t) = 1 - e^{-a_{ij} \Delta t} \approx a_{ij} \Delta t$$

when Δt is "small". We therefore have that

$$\lim_{\Delta t \to 0} \frac{1 - P_{ii}(\Delta t)}{\Delta t} = \lim_{\Delta t \to 0} \frac{\Pr(\widetilde{T}_i < \Delta t)}{\Delta t} = \alpha_i \tag{8.6}$$

$$\lim_{\Delta t \to 0} \frac{P_{ij}(\Delta t)}{\Delta t} = \lim_{\Delta t \to 0} \frac{\Pr(T_{ij} < \Delta t)}{\Delta t} = a_{ij} \quad \text{for } i \neq j \tag{8.7}$$

For a formal proof, see Ross (1996, p. 239).

Since we, from (8.4) and (8.5), can deduce α_i and P_{ij} when we know a_{ij} for all i, j in $\mathcal{X}$, we may equally well define a Markov process by specifying (i) the state space $\mathcal{X}$ and (ii) the transition rates a_{ij} for all $i \neq j$ in $\mathcal{X}$. The second definition is often more natural and will be our main approach in the following.

We may arrange the transition rates a_{ij} as a matrix:

$$\mathbb{A} = \begin{pmatrix} a_{00} & a_{01} & \cdots & a_{0r} \\ a_{10} & a_{11} & \cdots & a_{1r} \\ \vdots & \vdots & \ddots & \vdots \\ a_{r0} & a_{r1} & \cdots & a_{rr} \end{pmatrix} \tag{8.8}$$

where we have introduced the following notation for the diagonal elements:

$$a_{ii} = -\alpha_i = -\sum_{\substack{j=0 \\ j \neq i}}^{r} a_{ij} \tag{8.9}$$

We will call $\mathbb{A}$ the *transition rate matrix* of the Markov process. Some authors refer to the matrix $\mathbb{A}$ as the *infinitesimal generator* of the process.

Observe that the entries of row i are the transition rates out of state i (for $j \neq i$). We will call them *departure rates* from state i. According to (8.5) $-a_{ii} = \alpha_i$ is the sum of the departure rates from state i, and hence the *total departure rate* from state i. The entries of column i are transition rates into state i (for $j \neq i$). Notice that the sum of the entries in row i is equal to 0, for all $i \in \mathcal{X}$.

Procedure to Establish the Transition Rate Matrix To establish the transition rate matrix $\mathbb{A}$, we have to:

1. List and describe all relevant system states. Non-relevant states should be removed, and identical states should be merged (e.g, see Example 8.3). Each of the remaining states must be given a unique identification. In this book we use the integers from 0 up to r. We let r denote the best functioning state of the system and 0 denote the worst state. The state space of the system is thus $\mathcal{X} = \{0, 1, \ldots, r\}$. Any other sequence of numbers, or letters may, however, also be used.

2. Specify the transition rates a_{ij} for all $i \neq j$ and $i, j \in \mathcal{X}$. Each transition will usually involve a failure or a repair. The transition rates will therefore be failure rates and repair rates, and combinations of these.

3. Arrange the transition rates a_{ij} for $i \neq j$ as a matrix, similar to the matrix (8.8). (Leave the diagonal entries a_{ii} open.)

4. Fill in the diagonal elements a_{ii} such that the sum of all entries in each *row* is equal to zero, or by using (8.9).

A Markov process may be represented graphically by a *state transition diagram* that records the a_{ij} of the possible transitions of the Markov process. The state transition diagram is also known as a Markov diagram. In the state transition diagram, circles are used to represent states, and directed arcs are used to represent transitions between the states. An example of a state transition diagram is given in Fig. 8.2.

Example 8.2
Reconsider the parallel system of two independent components in Example 8.1. It is assumed that the following corrective maintenance strategy is adopted: When a component fails, a repair action is initiated to bring this component back to its initial functioning state. After the repair is completed, the component is assumed to be *as good as new*. Each component is assumed to have its own dedicated repair crew.

We assume that the components have constant failure rates λ_i and constant repair rates μ_i, for $i = 1, 2$. The transitions between the four system states in Table 8.1 are illustrated in the state transition diagram in Fig. 8.2.

Assume that the system is in state 3 at time 0. The first transition may either be to state 2 (failure of component 2) or to state 1 (failure of component 1). The

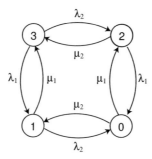

Fig. 8.2 State transition diagram of the parallel structure in Example 8.2.

transition rate to state 2 is $a_{32} = \lambda_2$, and the transition rate to state 1 is $a_{31} = \lambda_1$. The sojourn time in state 3 is therefore $\widetilde{T}_3 = \min\{T_{31}, T_{32}\}$, where T_{ij} is the time to the first transition from state i to state j. $\widetilde{T}_3$ has an exponential distribution with rate $a_{31} + a_{32} = \lambda_1 + \lambda_2$, and the mean sojourn time in state 3 is $1/(\lambda_1 + \lambda_2)$.

When the system is in state 2, the next transition may either be to state 3 (with rate $a_{23} = \mu_2$) or to state 0 (with rate $a_{20} = \lambda_1$). The probability that the transition is to state 3 is $\mu_2/(\mu_2 + \lambda_1)$, and the probability that it goes to state 0 is $\lambda_1/(\mu_2 + \lambda_1)$. The memoryless property of the exponential distribution ensures that component 1 is as good as new when the system enters state 2. In this example we assume that component 1 has the same failure rate λ_1 in state 3, where both components are functioning, as it has in state 2, where only component 1 is functioning. The failure rate a_{20} of component 1 in state 2 may, however, easily be changed to a failure rate λ_1' that is different from (e.g., higher than) λ_1.

When the system is in state 0, both components are in a failed state and two independent repair crews are working to bring the components back to a functioning state. The repair times T_{01} and T_{02} are independent and exponentially distributed with repair rates μ_1 and μ_2, respectively. The sojourn time $\widetilde{T}_0$ in state 0, $\min\{T_{01}, T_{02}\}$ is exponentially distributed with rate $(\mu_1 + \mu_2)$, and the mean downtime (MDT) of the system is therefore $1/(\mu_1 + \mu_2)$. When the system enters state 0, one of the components will already have failed and be under repair when the other component fails. The memoryless property of the exponential distribution ensures, however, that the time to complete the repair is independent of how long the component has been under repair.

The transition rate matrix of the system is thus

$$\mathbb{A} = \begin{pmatrix} -(\mu_1+\mu_2) & \mu_2 & \mu_1 & 0 \\ \lambda_2 & -(\lambda_2+\mu_1) & 0 & \mu_1 \\ \lambda_1 & 0 & -(\lambda_1+\mu_2) & \mu_2 \\ 0 & \lambda_1 & \lambda_2 & -(\lambda_1+\lambda_2) \end{pmatrix} \quad (8.10)$$

In this model we disregard the possibility of common cause failures. Thus a transition between state 3 and state 0 is assumed to be impossible during a time interval of length Δt.

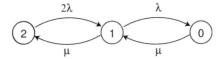

Fig. 8.3 State transition diagram of the parallel structure in Example 8.3.

Notice that when drawing the state transition diagram we consider a very short time interval, such that the transition diagram only records events of single transitions. Analogous with the Poisson process, the probability of having two or more events in a short time Δt is $o(\Delta t)$, and hence events of multiple transitions are not included in the state transition diagram. It is therefore not possible to have a transition from state 1 to state 2 in Fig. 8.2, since this would involve failure of component 2 and at the same time completed repair of component 1. A common cause failure can be modeled as a transition from state 3 to state 0 in Fig. 8.2. Such a transition will involve the failure of two components but may be considered as a single event. □

Example 8.3

Again consider the parallel system in Example 8.1, but assume that the two components are independent and identical with the same failure rate λ. In this case it is not necessary to distinguish between the states 1 and 2 in Table 8.1, and we may reduce the state space to the three states:

2   Both components are functioning
1   One component is functioning, and one is failed
0   Both components are in a failed state

Assume that the system is taken care of by a single repair crew that has adopted a first-fail-first-repair policy. The repair time of a component is assumed to be exponentially distributed with repair rate μ. The mean repair time (downtime) is then $1/\mu$. The transitions between the three system states are illustrated in Fig. 8.3.

A transition from state 2 to state 1 will take place as soon as one of the two independent components fails. The transition rate is therefore $a_{21} = 2\lambda$. When the system is in state 1, it will either go to state 2 [with probability $\mu/(\mu + \lambda)$], or to state 0 [with probability $\lambda/(\mu + \lambda)$].

The transition rate matrix of the system is

$$\mathbb{A} = \begin{pmatrix} -\mu & \mu & 0 \\ \lambda & -(\mu+\lambda) & \mu \\ 0 & 2\lambda & -2\lambda \end{pmatrix}$$

The mean sojourn times in the three states are

$$E(\tilde{T}_0) = \frac{1}{\mu}, \quad E(\tilde{T}_1) = \frac{1}{\mu + \lambda}, \quad E(\tilde{T}_2) = \frac{1}{2\lambda}$$

that is, the inverse of the absolute value of the corresponding diagonal entry in $\mathbb{A}$.

Fig. 8.4 State transition diagram for a homogeneous Poisson process (HPP).

An alternative repair strategy for state 0 would be to repair both components at the same time and only start up the system when both components are functioning again. If the repair time for this common repair action has rate μ_C, we have to modify the state transition diagram in Fig. 8.2 and introduce $a_{01} = 0$ and $a_{02} = \mu_C$ (a_{12} is still μ). □

Example 8.4
Consider a homogeneous Poisson process (HPP) $\{X(t), t \geq 0\}$ with rate λ. The HPP is a Markov process with countable *infinite* state space $\mathcal{X} = \{0, 1, 2, \ldots\}$. In this case we have $\alpha_i = \lambda$ for $i = 0, 1, 2, \ldots$, and $a_{ij} = \lambda$ for $j = i + 1$, and 0 for $j \neq i + 1$. The transition rate matrix for the HPP is thus

$$\mathbb{A} = \begin{pmatrix} -\lambda & \lambda & 0 & \cdots \\ 0 & -\lambda & \lambda & \cdots \\ 0 & 0 & -\lambda & \cdots \\ \vdots & \vdots & \vdots & \ddots \end{pmatrix}$$

The state transition diagram for the HPP is illustrated in Fig. 8.4. □

Chapman-Kolmogorov Equations By using the Markov property and the law of total probability, we realize that

$$P_{ij}(t+s) = \sum_{k=0}^{r} P_{ik}(t) P_{kj}(s) \quad \text{for all } i, j \in \mathcal{X}, t, s > 0 \tag{8.11}$$

Equations (8.11) are known as the *Chapman-Kolmogorov equations*. The equations may, by using (8.2), be written in matrix terms as

$$\mathbb{P}(t+s) = \mathbb{P}(t) \cdot \mathbb{P}(s)$$

Notice that $\mathbb{P}(0) = \mathbb{I}$ is the identity matrix. Notice also that if t is an integer, it follows that $\mathbb{P}(t) = [\mathbb{P}(1)]^t$. It can be shown that this also holds when t is not an integer.

8.2.1 Kolmogorov Differential Equations

We will try to establish a set of differential equations that may be used to find $P_{ij}(t)$, and therefore start by considering the Chapman-Kolmogorov equations

$$P_{ij}(t + \Delta t) = \sum_{k=0}^{r} P_{ik}(\Delta t) P_{kj}(t)$$

Note that we here split the interval $(0, t + \Delta t)$ in two parts. First, we consider a transition from state i to state k in the small interval $(0, \Delta t)$, and thereafter a transition from state k to state j in the rest of the interval. We now consider

$$P_{ij}(t + \Delta t) - P_{ij}(t) = \sum_{\substack{k=0 \\ k \neq i}}^{r} P_{ik}(\Delta t) P_{kj}(t) - [1 - P_{ii}(\Delta t)] P_{ij}(t)$$

By dividing by Δt and then taking the limit as $\Delta t \to 0$, we obtain

$$\lim_{\Delta t \to 0} \frac{P_{ij}(t + \Delta t) - P_{ij}(t)}{\Delta t} = \lim_{\Delta t \to 0} \sum_{\substack{k=0 \\ k \neq i}}^{r} \frac{P_{ik}(\Delta t)}{\Delta t} P_{kj}(t) - \alpha_i P_{ij}(t) \qquad (8.12)$$

Since the summing index is finite, we may interchange the limit and summation on the right-hand side of (8.12) and obtain, using (8.6) and (8.7),

$$\dot{P}_{ij}(t) = \sum_{\substack{k=0 \\ k \neq i}}^{r} a_{ik} P_{kj}(t) - \alpha_i P_{ij}(t) = \sum_{k=0}^{r} a_{ik} P_{kj}(t) \qquad (8.13)$$

where $a_{ii} = -\alpha_i$, and the following notation for the time derivative is introduced:

$$\dot{P}_{ij}(t) = \frac{d}{dt} P_{ij}(t)$$

The differential equations (8.13) are known as the Kolmogorov *backward equations*.[2] They are called backward equations because we start with a transition back by the start of the interval.

The Kolmogorov backward equations may also be written in matrix format as

$$\dot{\mathbb{P}}(t) = \mathbb{A} \cdot \mathbb{P}(t) \qquad (8.14)$$

We may also start with the following equation:

$$P_{ij}(t + \Delta t) = \sum_{k=0}^{r} P_{ik}(t) P_{kj}(\Delta t)$$

Here we split the time interval $(0, t + \Delta t)$ into two parts. We consider a transition from i to k in the interval $(0, t)$, and then a transition from k to j in the small interval $(t, t + \Delta t)$. We consider

$$P_{ij}(t + \Delta t) - P_{ij}(t) = \sum_{\substack{k=0 \\ k \neq j}}^{r} P_{ik}(t) P_{kj}(\Delta t) - [1 - P_{jj}(\Delta t)] P_{ij}(t)$$

---

[2] Named after the Russian mathematician Andrey N. Kolmogorov (1903–1987).

By dividing by Δt and then taking the limit as $\Delta t \to 0$, we obtain

$$\lim_{\Delta t \to 0} \frac{P_{ij}(t+\Delta t) - P_{ij}(t)}{\Delta t} = \lim_{\Delta t \to 0} \left[\sum_{\substack{k=0 \\ k \neq j}}^{r} P_{ik}(t) \frac{P_{kj}(\Delta t)}{\Delta t} - \frac{1 - P_{jj}(\Delta t)}{\Delta t} P_{ij}(t) \right]$$

Since the summation index is finite, we may interchange limit with summation and obtain

$$\dot{P}_{ij}(t) = \sum_{\substack{k=0 \\ k \neq j}}^{r} a_{kj} P_{ik}(t) - \alpha_j P_{ij}(t) = \sum_{k=0}^{r} a_{kj} P_{ik}(t) \qquad (8.15)$$

where, as before, $a_{jj} = -\alpha_j$. The differential equations (8.15) are known as the Kolmogorov *forward equations*. The interchange of the limit and the sum above does not hold in all cases but is always valid when the state space is finite.

The Kolmogorov forward equations may be written in matrix terms as

$$\dot{\mathbb{P}}(t) = \mathbb{P}(t) \cdot \mathbb{A} \qquad (8.16)$$

For the Markov processes we are studying in this book the backward and the forward equations have the same unique solution $\mathbb{P}(t)$, where $\sum_{j=0}^{r} P_{ij}(t) = 1$ for all i in $\mathcal{X}$. In the following, we will mainly use the forward equations.

8.2.2 State Equations

Let us assume that we know that the Markov process has state i at time 0, that is, $X(0) = i$. This can be expressed as

$$\begin{aligned} P_i(0) &= \Pr(X(0) = i) = 1 \\ P_k(0) &= \Pr(X(0) = k) = 0 \quad \text{for } k \neq i \end{aligned}$$

Since we know the state at time 0, we may simplify the notation by writing $P_{ij}(t)$ as $P_j(t)$. The vector $\boldsymbol{P}(t) = [P_0(t), P_1(t), \ldots, P_r(t)]$ then denotes the distribution of the Markov process at time t, when we *know* that the process started in state i at time 0. As in (8.3) we know that $\sum_{j=1}^{r} P_j(t) = 1$.

The distribution $\boldsymbol{P}(t)$ may be found from the Kolmogorov forward equations (8.15)

$$\dot{P}_j(t) = \sum_{k=0}^{r} a_{kj} P_k(t) \qquad (8.17)$$

where, as before, $a_{jj} = -\alpha_j$. In matrix terms, this may be written

$$[P_0(t), \ldots, P_r(t)] \cdot \begin{pmatrix} a_{00} & a_{01} & \cdots & a_{0r} \\ a_{10} & a_{11} & \cdots & a_{1r} \\ \vdots & \vdots & \ddots & \vdots \\ a_{r0} & a_{r1} & \cdots & a_{rr} \end{pmatrix} = [\dot{P}_0(t), \ldots, \dot{P}_r(t)] \qquad (8.18)$$

or in a more compact form as

$$P(t) \cdot \mathbb{A} = \dot{P}(t) \qquad (8.19)$$

Equations (8.19) are called the *state equations* for the Markov process.

Remark: Some authors prefer to present the state equations as the transpose of (8.19), that is $\mathbb{A}^T \cdot P(t)^T = \dot{P}(t)^T$. In this case the vectors will be column vectors, and equations (8.18) can be written in a more compact form as

$$\begin{pmatrix} a_{00} & a_{10} & \cdots & a_{r0} \\ a_{01} & a_{11} & \cdots & a_{r1} \\ \vdots & \vdots & \ddots & \vdots \\ a_{0r} & a_{1r} & \cdots & a_{rr} \end{pmatrix} \cdot \begin{bmatrix} P_0(t) \\ P_1(t) \\ \vdots \\ P_r(t) \end{bmatrix} = \begin{bmatrix} \dot{P}_0(t) \\ \dot{P}_1(t) \\ \vdots \\ \dot{P}_r(t) \end{bmatrix}$$

In this format the indexes do not follow standard matrix notation. The entries in *column i* represent the departure rates from state i, and the sum of all the entries in a *column* will be 0. The reader may choose in which format he wants to present the state equations. Both formats will give the same result. In this book, we will, however, present the state equations in the format of (8.18) and (8.19). □

Since the sum of the entries in each row in $\mathbb{A}$ is equal to 0, the determinant of $\mathbb{A}$ is 0 and the matrix is singular. Consequently, equations (8.19) do not have a unique solution. However, by using that

$$\sum_{j=0}^{r} P_j(t) = 1$$

and the known initial state $[P_i(0) = 1]$, we are often able to compute the probabilities $P_j(t)$ for $j = 0, 1, 2, \ldots, r$. [Conditions for existence and uniqueness of the solutions are discussed, for example, by Cox and Miller (1965).]

Example 8.5
Consider a single component. The component has two possible states:

1   The component is functioning
0   The component is in a failed state

Transition from state 1 to state 0 means that the component fails, and transition from state 0 to state 1 means that the component is repaired. The transition rate a_{10} is thus the failure rate of the component, and the transition rate a_{01} is the repair rate of the component. In this example we will use the following notation:

$a_{10} = \lambda$ The failure rate of the component
$a_{01} = \mu$ The repair rate of the component

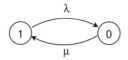

Fig. 8.5 State transition diagram for a single component (function-repair cycle).

The mean sojourn time in state 1 is the mean time to failure, MTTF $= 1/\lambda$, and the mean sojourn time in state 0 is the mean downtime, MDT $= 1/\mu$. The mean downtime is sometimes called the mean time to repair (MTTR).

The state transition diagram for the single component is illustrated in Fig. 8.5. The state equations are

$$[P_0(t), P_1(t)] \cdot \begin{pmatrix} -\mu & \mu \\ \lambda & -\lambda \end{pmatrix} = [\dot{P}_0(t), \dot{P}_1(t)] \tag{8.20}$$

The component is assumed to be functioning at time $t = 0$,

$$P_1(0) = 1, \quad P_0(0) = 0$$

Since the two equations we get from (8.20) are linearly dependent, we use only one of them, for example,

$$-\mu P_0(t) + \lambda P_1(t) = \dot{P}(t)$$

and combine this equation with $P_0(t) + P_1(t) = 1$. The solution is

$$P_1(t) = \frac{\mu}{\mu + \lambda} + \frac{\lambda}{\mu + \lambda} e^{-(\lambda + \mu)t} \tag{8.21}$$

$$P_0(t) = \frac{\lambda}{\mu + \lambda} - \frac{\lambda}{\mu + \lambda} e^{-(\lambda + \mu)t} \tag{8.22}$$

For a detailed solution of the differential equation, see Ross (1996, p. 243).

$P_1(t)$ denotes the probability that the component is functioning at time t, that is, the *availability* of the component (see Section 9.4).

The limiting availability $P_1 = \lim_{t \to \infty} P_1(t)$ is from (8.21),

$$P_1 = \lim_{t \to \infty} P_1(t) = \frac{\mu}{\lambda + \mu} \tag{8.23}$$

The limiting availability may therefore be written as the well-known formula

$$P_1 = \frac{\text{MTTF}}{\text{MTTF+MDT}} \tag{8.24}$$

When there is no repair ($\mu = 0$), the availability is $P_1(t) = e^{-\lambda t}$ which coincides with the survivor function of the component. The availability $P_1(t)$ is illustrated in Fig. 8.6. □

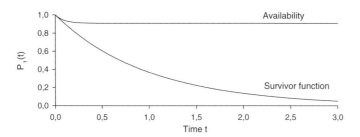

Fig. 8.6 The availability and the survivor function of a single component ($\lambda = 1$, $\mu = 10$).

8.3 ASYMPTOTIC SOLUTION

In many applications only the long-run (steady-state) probabilities are of interest, that is, the values of $P_j(t)$ when $t \to \infty$. In Example 8.5 the state probabilities $P_j(t)$ ($j = 0, 1$) approached a steady-state P_j when $t \to \infty$. The same steady-state value would have been found irrespective of whether the system started in the operating state or in the failed state.

Convergence toward steady-state probabilities is assumed of the Markov processes we are studying in this chapter. The process is said to be *irreducible* if every state is reachable from every other state (see Ross 1996).

For an irreducible Markov process, it can be shown that the limits

$$\lim_{t \to \infty} P_j(t) = P_j \quad \text{for } j = 0, 1, 2, \ldots, r$$

always exist and are independent of the initial state of the process (at time $t = 0$). For a proof, see Ross (1996, p. 251). Hence a process that has been running for a long time has lost its dependency of its initial state $X(0)$. The process will converge to a process where the probability of being in state j is

$$P_j = P_j(\infty) = \lim_{t \to \infty} P_j(t) \quad \text{for } j = 0, 1, \ldots, r$$

These asymptotic probabilities are often called the *steady-state probabilities* for the Markov process.

If $P_j(t)$ tends to a constant value when $t \to \infty$, then

$$\lim_{t \to \infty} \dot{P}_j(t) = 0 \quad \text{for } j = 0, 1, \ldots, r$$

The steady-state probabilities $\mathbf{P} = [P_0, P_1, \ldots, P_r]$ must therefore satisfy the matrix equation:

$$[P_0, P_1, \ldots, P_r] \cdot \begin{pmatrix} a_{00} & a_{01} & \cdots & a_{0r} \\ a_{10} & a_{11} & \cdots & a_{1r} \\ \vdots & \vdots & \ddots & \vdots \\ a_{r0} & a_{r1} & \cdots & a_{rr} \end{pmatrix} = [0, 0, \ldots, 0] \quad (8.25)$$

which may be abbreviated to

$$P \cdot \mathbb{A} = \mathbf{0} \tag{8.26}$$

where as before

$$\sum_{j=0}^{r} P_j = 1$$

To calculate the steady-state probabilities, $P_0, P_1, \ldots, P_r$, of such a process, we use r of the $r+1$ linear algebraic equation from the matrix equation (8.25) and in addition the fact that the sum of the state probabilities is always equal to 1. The initial state of the process has no influence on the steady-state probabilities. Note that P_j also may be interpreted as the average, long-run proportion of time the system spends in state j.

Example 8.6
Consider a power station with two generators, 1 and 2. Each generator can have two states: a functioning state (1) and a failed state (0). A generator is considered to be in the failed state (0) also during repair. Generator 1 is supplying 100 MW when it is functioning and 0 MW when it is not functioning. Generator 2 is supplying 50 MW when it is functioning and 0 MW when it is not functioning.

The possible states of the system are:

System state	State of Generator 1	State of Generator 2	System output
3	1	1	150 MW
2	1	0	100 MW
1	0	1	50 MW
0	0	0	0 MW

We assume that the generators fail independent of each other and that they are operated on a continuous basis. The failure rates of the generators are

λ_1 Failure rate of generator 1
λ_2 Failure rate of generator 2

When a generator fails, a repair action is started to bring the generator back into operation. The two generators are assumed to be repaired independent of each other, by two independent repair crews. The repair rates of the generators are

μ_1 Repair rate of generator 1
μ_2 Repair rate of generator 2

The corresponding state transition diagram is shown in Fig. 8.7. The transition

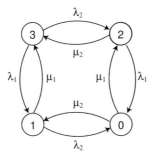

Fig. 8.7 State transition diagram of the generators in Example 8.6.

matrix is

$$\mathbb{A} = \begin{pmatrix} -(\mu_1 + \mu_2) & \mu_2 & \mu_1 & 0 \\ \lambda_2 & -(\lambda_2 + \mu_1) & 0 & \mu_1 \\ \lambda_1 & 0 & -(\lambda_1 + \mu_2) & \mu_2 \\ 0 & \lambda_1 & \lambda_2 & -(\lambda_1 + \lambda_2) \end{pmatrix}$$

We can use (8.26) to find the steady-state probabilities P_j for $j = 0, 1, 2, 3$, and we get the following equations:

$$-(\mu_1 + \mu_2)P_0 + \lambda_2 P_1 + \lambda_1 P_2 = 0$$
$$\mu_2 P_0 - (\lambda_2 + \mu_1)P_1 + \lambda_1 P_3 = 0$$
$$\mu_1 P_0 - (\lambda_1 + \mu_2)P_2 + \lambda_2 P_3 = 0$$
$$P_0 + P_1 + P_2 + P_3 = 1$$

Note that we use three of the steady-state equations from (8.26) and in addition the fact that $P_0 + P_1 + P_2 + P_3 = 1$. Note also that we may choose any three of the four steady-state equations, and get the same solution.

The solution is

$$P_0 = \frac{\lambda_1 \lambda_2}{(\lambda_1 + \mu_1)(\lambda_2 + \mu_2)}$$

$$P_1 = \frac{\lambda_1 \mu_2}{(\lambda_1 + \mu_1)(\lambda_2 + \mu_2)}$$

$$P_2 = \frac{\mu_1 \lambda_2}{(\lambda_1 + \mu_1)(\lambda_2 + \mu_2)}$$

$$P_3 = \frac{\mu_1 \mu_2}{(\lambda_1 + \mu_1)(\lambda_2 + \mu_2)} \quad (8.27)$$

Now for $i = 1, 2$ let

$$q_i = \frac{\lambda_i}{\lambda_i + \mu_i} = \frac{\text{MDT}_i}{\text{MTTF}_i + \text{MDT}_i}$$

MARKOV PROCESSES

$$p_i = \frac{\mu_i}{\lambda_i + \mu_i} = \frac{\text{MTTF}_i}{\text{MTTF}_i + \text{MDT}_i}$$

where $\text{MDT}_i = 1/\mu_i$ is the mean downtime required to repair component i, and $\text{MTTF}_i = 1/\lambda_i$ is the mean time to failure of component i ($i = 1, 2$). Thus q_i denotes the average, or limiting, unavailability of component i, while p_i denotes the average (limiting) availability of component i ($i = 1, 2$). The steady-state probabilities may thus be written as

$$\begin{aligned} P_0 &= q_1 q_2 \\ P_1 &= q_1 p_2 \\ P_2 &= p_1 q_2 \\ P_3 &= p_1 p_2 \end{aligned} \quad (8.28)$$

In this simple example, the components fail and are repaired independently of each other. We may therefore use direct reasoning to obtain the results in (8.28):

$P_0 = $ Pr(component 1 is failed) · Pr(component 2 is failed) $= q_1 q_2$
$P_1 = $ Pr(component 1 is failed) · Pr(component 2 is functioning) $= q_1 p_2$
$P_2 = $ Pr(component 1 is functioning) · Pr(component 2 is failed) $= p_1 q_2$
$P_3 = $ Pr(component 1 is functioning) · Pr(component 2 is functioning) $= p_1 p_2$

Note: In this simple example, where all failures and repairs are independent events, we do not need to use Markov methods to find the steady-state probabilities. The steady-state probabilities may easily be found by using standard probability rules for independent events. Please notice that this only applies for systems with independent failures and repairs.

Assume now that we have the following data:

	Generator 1	Generator 2
MTTF_i	6 months ≈ 4380 hours	8 months ≈ 5840 hours
Failure rate, λ_i	$2.3 \cdot 10^{-4}$ hours^{-1}	$1.7 \cdot 10^{-4}$ hours^{-1}
MDT_i	12 hours	24 hours
Repair rate, μ_i	$8.3 \cdot 10^{-2}$ hours^{-1}	$4.2 \cdot 10^{-2}$ hours^{-1}

Note that the steady-state probabilities can be interpreted as the mean proportion of time the system stays in the state concerned. The steady-state probability of state 1 is, for example, equal to

$$P_1 = \frac{\lambda_1 \mu_2}{(\lambda_1 + \mu_1)(\lambda_2 + \mu_2)} = q_1 p_2 \approx 2.72 \cdot 10^{-3}$$

Hence

$$P_1 = 0.00272 \left[\frac{\text{year}}{\text{year}}\right] = 0.00272 \cdot 8760 \left[\frac{\text{hours}}{\text{year}}\right] \approx 23.8 \left[\frac{\text{hours}}{\text{year}}\right]$$

In the long run the system will stay in state 1 approximately 23.8 hours per year. This does *not* mean that state 1 occurs on average once per year and lasts for 23.8 hours each time.

With the given data, we obtain

System State	System Output	Steady-State Probability	Average Hours in State per Year
3	150 MW	0.9932	8700.3
2	100 MW	$4.08 \cdot 10^{-3}$	35.8
1	50 MW	$2.72 \cdot 10^{-3}$	23.8
0	0 MW	$1.12 \cdot 10^{-5}$	0.1

□

8.3.1 System Performance Characteristics

Several system performance measures that may be used in the *steady-state* situation are introduced in this section. Examples are provided in Sections 8.5 to 8.7.

Visit Frequency The Kolmogorov forward equation (8.15) was

$$\dot{P}_{ij}(t) = \sum_{\substack{k=0 \\ k \neq j}}^{r} a_{kj} P_{ik}(t) - \alpha_j P_{ij}(t)$$

When we let $t \to \infty$, then $P_{ij}(t) \to P_j$, and $\dot{P}_{ij}(t) \to 0$. Since the summation index in (8.15) is finite, we may interchange the limit and the sum and get, as $t \to \infty$,

$$0 = \sum_{\substack{k=0 \\ k \neq j}}^{r} a_{kj} P_k - \alpha_j P_j$$

that can be written as

$$P_j \alpha_j = \sum_{\substack{k=0 \\ k \neq j}}^{r} P_k a_{kj} \qquad (8.29)$$

The (unconditional) probability of a departure from state j in the time interval $(t, t + \Delta t]$ is

$$\sum_{\substack{k=0 \\ k \neq j}}^{r} \Pr((X(t + \Delta t) = k) \cap (X(t) = j))$$

$$= \sum_{\substack{k=0 \\ k \neq j}}^{r} \Pr(X(t + \Delta t) = k \mid X(t) = j) \cdot \Pr(X(t) = j) = \sum_{\substack{k=0 \\ k \neq j}}^{r} P_{jk}(\Delta t) \cdot P_j(t)$$

When $t \to \infty$, this probability tends to $\sum_{\substack{k=0 \\ k \neq j}}^{r} P_{jk}(\Delta t) \cdot P_j$, and the steady-state frequency of departures from state j is, with the same argument as we used to derive equation (8.5),

$$\nu_j^{\text{dep}} = \lim_{\Delta t \to 0} \frac{\sum_{\substack{k=0 \\ k \neq j}}^{r} P_{jk}(\Delta t) \cdot P_j}{\Delta t} = P_j \alpha_j$$

The left-hand side of (8.29) is hence the steady-state frequency of departures from state j. The frequency of departures from state j is seen to be the proportion of time P_j spent in state j times the transition rate α_j out of state j.

Similarly, the frequency of transitions from state k into state j is $P_k a_{kj}$. The total frequency of arrivals into state j is therefore

$$\nu_j^{\text{arr}} = \sum_{\substack{k=0 \\ k \neq j}}^{r} P_k a_{kj}$$

Equation (8.29) says that the frequency of departures from state j is equal to the frequency of arrivals into state j, for $j = 0, 1, \ldots, r$, and is therefore sometimes referred to as the *balance equations*. In the steady-state situation, we define the *visit frequency* to state j as

$$\nu_j = P_j \alpha_j = \sum_{\substack{k=0 \\ k \neq j}}^{r} P_k a_{kj} \qquad (8.30)$$

and the mean time between visits to state j is $1/\nu_j$.

Mean Duration of a Visit When the process arrives at state j, the system will stay in this state a time $\widetilde{T}_j$ until the process departs from that state, $j = 0, 1, \ldots, r$. We have called $\widetilde{T}_j$ the sojourn time in state j and shown that $\widetilde{T}_j$ is exponentially distributed with rate α_j. The mean sojourn time, or mean duration of a visit, is hence

$$\theta_j = E(\widetilde{T}_j) = \frac{1}{\alpha_j} \quad \text{for } j = 0, 1, \ldots, r \qquad (8.31)$$

By combining (8.30) and (8.31) we obtain

$$\begin{aligned} \nu_j &= P_j \alpha_j = \frac{P_j}{\theta_j} \\ P_j &= \nu_j \theta_j \end{aligned} \qquad (8.32)$$

The mean proportion of time, P_j, the system is spending in state j is thus equal to the visit frequency to state j multiplied by the mean duration of a visit in state j for $j = 0, 1, \ldots, r$.

System Availability Let $\mathcal{X} = \{0, 1, \ldots, r\}$ be the set of all possible states of a system. Some of these states represent system functioning according to some specified criteria. Let B denote the subset of states in which the system is functioning, and let $F = \mathcal{X} - B$ denote the states in which the system is failed.

The average, or long-term *availability* of the system is the mean proportion of time when the system is functioning; that is, its state is a member of B. The average system availability A_s is thus defined as

$$A_s = \sum_{j \in B} P_j \qquad (8.33)$$

In the following we will omit the term average and call A_s the system availability.

The system unavailability $(1 - A_s)$ is then

$$1 - A_s = \sum_{j \in F} P_j \qquad (8.34)$$

The unavailability $(1 - A_s)$ of the system is the mean proportion of time when the system is in a failed state.

Frequency of System Failures The frequency ω_F of system failures is the steady-state frequency of transitions from a functioning state (in B) to a failed state (in F):

$$\omega_F = \sum_{j \in B} \sum_{k \in F} P_j \cdot a_{jk} \qquad (8.35)$$

Mean Duration of a System Failure The mean duration θ_F of a system failure is defined as the mean time from when the system enters into a failed state (F) until it is repaired/restored and brought back into a functioning state (B).

Analogous with (8.32) it is obvious that the system unavailability $(1 - A_s)$ is equal to the frequency of system failures multiplied by the mean duration of a system failure. Hence

$$1 - A_s = \omega_F \cdot \theta_F \qquad (8.36)$$

Mean Time between System Failures The mean time between system failures, MTBF_s, is the mean time between consecutive transitions from a functioning state (B) into a failed state (F). The MTBF_s may be computed from the frequency of system failures by

$$\text{MTBF}_s = \frac{1}{\omega_F} \qquad (8.37)$$

Mean Functioning Time until System Failure The mean functioning time ("up-time") until system failure, $E(U)_s$, is the mean time from a transition from a failed state (F) into a functioning state (B) until the first transition back to a failed state (F). It is obvious that

$$\mathrm{MTBF}_s = E(U)_s + \theta_F \qquad (8.38)$$

Note the difference between the mean functioning time ("up-time") and the mean time to system failure MTTF$_S$. The MTTF$_S$ is normally calculated as the mean time until system failure when the system initially is in a *specified* functioning state.

8.4 PARALLEL AND SERIES STRUCTURES

In this section we study the steady-state properties of parallel and series structures of independent components.

8.4.1 Parallel Structures of Independent Components

Reconsider the parallel structure of two independent components in Example 8.6. For this system we get

Mean Duration of the Visits From (8.31), we get

$$\begin{aligned}
\theta_0 &= 1/(\mu_1 + \mu_2) \\
\theta_1 &= 1/(\lambda_1 + \mu_2) \\
\theta_2 &= 1/(\lambda_2 + \mu_1) \\
\theta_3 &= 1/(\lambda_1 + \lambda_2)
\end{aligned} \qquad (8.39)$$

Visit Frequency From (8.31) and (8.39), we get

$$\begin{aligned}
\nu_0 &= P_0(\mu_1 + \mu_2) \\
\nu_1 &= P_1(\lambda_1 + \mu_2) \\
\nu_2 &= P_2(\lambda_2 + \mu_1) \\
\nu_3 &= P_3(\lambda_1 + \lambda_2)
\end{aligned} \qquad (8.40)$$

The parallel structure is functioning when at least one of its two components is functioning. When the system is in state 1, 2, or 3 the system is functioning, while state 0 corresponds to system failure.

The average system unavailability is

$$1 - A_s = P_0 = q_1 q_2 \qquad (8.41)$$

and the average system availability is

$$A_s = P_1 + P_2 + P_3 = 1 - q_1 q_2$$

The frequency of system failures ω_F is equal to the visit frequency to state 0, which is

$$\omega_F = \nu_0 = P_0(\mu_1 + \mu_2) = (1 - A_s) \cdot (\mu_1 + \mu_2) \tag{8.42}$$

The mean duration of a system failure θ_F is in this case equal to the mean duration of a stay in state 0. Thus

$$\theta_F = \theta_0 = \frac{1}{\mu_1 + \mu_2} = \frac{1 - A_s}{\omega_F} \tag{8.43}$$

For a parallel structure of n independent components, the above results may be generalized as follows: For system unavailability,

$$1 - A_s = \prod_{i=1}^{n} q_i = \prod_{i=1}^{n} \frac{\lambda_i}{\lambda_i + \mu_i} \tag{8.44}$$

For frequency of system failures,

$$\omega_F = (1 - A_s) \cdot \sum_{i=1}^{n} \mu_i \tag{8.45}$$

For mean duration of a system failure,

$$\theta_F = \frac{1}{\sum_{i=1}^{n} \mu_i} \tag{8.46}$$

The mean functioning time (up-time) $E(U)_P$ of the parallel structure can be determined from

$$1 - A_s = \frac{\theta_F}{\theta_F + E(U)_P}$$

Hence

$$E(U)_P = \frac{\theta_F A_s}{1 - A_s} = \frac{1 - \prod_{i=1}^{n} \lambda_i/(\lambda_i + \mu_i)}{\prod_{i=1}^{n} \lambda_i/(\lambda_i + \mu_i) \cdot \sum_{j=1}^{n} \mu_j} \tag{8.47}$$

When the component availabilities are very high (i.e., $\lambda_i \ll \mu_i$ for all $i = 1, 2, \ldots, n$), then

$$\frac{\lambda_i}{\lambda_i + \mu_i} = \frac{\lambda_i \, \mathrm{MDT}_i}{1 + \lambda_i \, \mathrm{MDT}_i} \approx \lambda_i \, \mathrm{MDT}_i$$

The frequency ω_F of system failures can now be approximated as

$$\omega_F = (1 - A_s) \cdot \sum_{i=1}^{n} \mu_i = \prod_{i=1}^{n} \frac{\lambda_i}{\lambda_i + \mu_i} \cdot \sum_{j=1}^{n} \mu_j$$

$$\approx \prod_{i=1}^{n} \lambda_i \, \mathrm{MDT}_i \cdot \sum_{j=1}^{n} \frac{1}{\mathrm{MDT}_j} \tag{8.48}$$

324 MARKOV PROCESSES

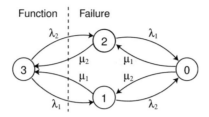

Fig. 8.8 Partitioning the state transition diagram of a series structure of two independent components.

For two components (8.48) reduces to

$$\omega_F \approx \lambda_1 \lambda_2 \cdot (MDT_1 + MDT_2) \qquad (8.49)$$

For three components (8.48) reduces to

$$\omega_F \approx \lambda_1 \lambda_2 \lambda_3 \cdot (MDT_1 \cdot MDT_2 + MDT_1 \cdot MDT_3 + MDT_2 \cdot MDT_3)$$

8.4.2 Series Structures of Independent Components

Consider a series structure of two independent components. The states of the system and the transition rates are as defined in Example 8.6. The state transition diagram of the series structure is shown in Fig. 8.8. The corresponding steady-state equations are equal to those found for the parallel structure in Example 8.6.

The average availability of the structure, A_s, is equal to P_3 which was found in (8.28) to be

$$A_s = P_3 = \frac{\mu_1 \mu_2}{(\lambda_1 + \mu_1)(\lambda_2 + \mu_2)} = p_1 \cdot p_2 \qquad (8.50)$$

where

$$p_i = \frac{\mu_i}{\lambda_i + \mu_i} \quad \text{for } i = 1, 2$$

The frequency of system failures, ω_F, is the same as the frequency of visits to state 3. Thus

$$\omega_F = \nu_3 = P_3 \cdot (\lambda_1 + \lambda_2) = A_s \cdot (\lambda_1 + \lambda_2) \qquad (8.51)$$

The mean duration of a system failure θ_F is equal to

$$\theta_F = \frac{1 - A_s}{\omega_F} \qquad (8.52)$$

For a series structure of n independent components the above results can be generalized as follows: For system availability

$$A_s = \prod_{i=1}^{n} p_i = \prod_{i=1}^{n} \frac{\mu_i}{\lambda_i + \mu_i} \qquad (8.53)$$

For frequency of system failures

$$\omega_F = A_s \sum_{i=1}^{n} \lambda_i \qquad (8.54)$$

For mean duration of a system failure

$$\begin{aligned}
\theta_F &= \frac{1 - A_s}{\omega_F} \\
&= \frac{1}{\sum_{i=1}^{n} \lambda_i} \frac{1 - A_s}{A_s} \\
&= \frac{1 - \prod_{i=1}^{n} \mu_i/(\lambda_i + \mu_i)}{\prod_{i=1}^{n} \mu_i/(\lambda_i + \mu_i) \cdot \sum_{j=1}^{n} \lambda_j}
\end{aligned} \qquad (8.55)$$

When all the component availabilities are very high such that $\lambda_i \ll \mu_i$ for all i, then $A_s \approx 1$ and the frequency of system failures is approximately

$$\omega_F \approx \sum_{i=1}^{n} \lambda_i \qquad (8.56)$$

which is the same as the failure rate of a nonrepairable series structure of n independent components.

The mean duration of a system failure θ_F may be approximated as

$$\begin{aligned}
\theta_F &= \frac{1}{\sum_{i=1}^{n} \lambda_i} \frac{1 - A_s}{A_s} = \frac{1}{\sum_{i=1}^{n} \lambda_i} \left(\frac{1}{A_s} - 1 \right) = \frac{1}{\sum_{i=1}^{n} \lambda_i} \left(\prod_{i=1}^{n} \frac{1}{p_i} - 1 \right) \\
&= \frac{1}{\sum_{i=1}^{n} \lambda_i} \left(\prod_{i=1}^{n} \left(1 + \frac{\lambda_i}{\mu_i} \right) - 1 \right) \approx \frac{1}{\sum_{i=1}^{n} \lambda_i} \left(1 + \sum_{i=1}^{n} \frac{\lambda_i}{\mu_i} - 1 \right) \\
&= \frac{\sum_{i=1}^{n} \lambda_i/\mu_i}{\sum_{i=1}^{n} \lambda_i} = \frac{\sum_{i=1}^{n} \lambda_i \cdot \mathrm{MDT}_i}{\sum_{i=1}^{n} \lambda_i}
\end{aligned} \qquad (8.57)$$

where $\mathrm{MDT}_i = 1/\mu_i$ as before is the mean downtime required to repair component i, $i = 1, 2, \ldots, n$. Equation (8.57) is a commonly used approximation for the mean duration of a failure in series structures of high reliability.

8.4.3 Series Structure of Components Where Failure of One Component Prevents Failure of the Other

Consider a series structure of two components. When one of the components fails, the other component is immediately taken out of operation until the failed component is repaired.[3] After a component is taken out of operation, it is not exposed to any stress,

---
[3] The same model is discussed by Barlow and Proschan (1975, pp. 194–201) in a more general context that does not assume constant failure and repair rates.

326 MARKOV PROCESSES

Table 8.2 Possible States of a Series Structure of Two Components Where Failure of One Component Prevents Failure of the Other.

State	Component 1	Component 2
2	Functioning	Functioning
1	Taken out of operation	Functioning
0	Functioning	Taken out of operation

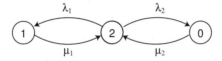

Fig. 8.9 State transition diagram of a series structure of two components where failure of one component prevents failure of the other component.

and we therefore assume that it will not fail. This dependence between the failures prevents a simple solution by direct reasoning as was possible in Example 8.6. This system has three possible states as described in Table 8.2.

The following transition rates are assumed:

$$a_{21} = \lambda_1 \quad \text{Failure rate of component 1}$$
$$a_{20} = \lambda_2 \quad \text{Failure rate of component 2}$$
$$a_{12} = \mu_1 \quad \text{Repair rate of component 1}$$
$$a_{02} = \mu_2 \quad \text{Repair rate of component 2}$$

The state transition diagram of the series structure is illustrated in Fig. 8.9. The steady-state equations for this system are

$$[P_0, P_1, P_2] \cdot \begin{pmatrix} -\mu_2 & 0 & \mu_2 \\ 0 & -\mu_1 & \mu_1 \\ \lambda_2 & \lambda_1 & -(\lambda_1 + \lambda_2) \end{pmatrix} = [0, 0, 0] \quad (8.58)$$

The steady-state probabilities may be found from the equations

$$-\mu_2 P_0 + \lambda_2 P_2 = 0$$
$$-\mu_1 P_1 + \lambda_1 P_2 = 0$$
$$P_0 + P_1 + P_2 = 1$$

PARALLEL AND SERIES STRUCTURES

The solution is

$$P_2 = \frac{\mu_1 \mu_2}{\lambda_1 \mu_2 + \lambda_2 \mu_1 + \mu_1 \mu_2} = \frac{1}{1 + (\lambda_1/\mu_1) + (\lambda_2/\mu_2)} \qquad (8.59)$$

$$P_1 = \frac{\lambda_1}{\mu_1} P_2 \qquad (8.60)$$

$$P_0 = \frac{\lambda_2}{\mu_2} P_2 \qquad (8.61)$$

Since the series structure is only functioning when both the components are functioning (state 2), the average system availability is

$$A_s = P_2 = \frac{\mu_1 \mu_2}{\lambda_1 \mu_2 + \lambda_2 \mu_1 + \mu_1 \mu_2} = \frac{1}{1 + (\lambda_1/\mu_1) + (\lambda_2/\mu_2)}$$

Observe that in this case the availability of the series structure is *not* equal to the product of the component availabilities.

The mean durations of the stays in each state are

$$\theta_2 = \frac{1}{\lambda_1 + \lambda_2}$$

$$\theta_1 = \frac{1}{\mu_1}$$

$$\theta_0 = \frac{1}{\mu_2}$$

The frequency of system failures ω_F is the same as the frequency of visits to state 2.

$$\omega_F = \nu_2 = P_2(\lambda_1 + \lambda_2) = A_s(\lambda_1 + \lambda_2) \qquad (8.62)$$

The mean duration of a system failure θ_F is

$$\theta_F = \frac{1 - A_s}{\omega_F} = \frac{1}{\lambda_1 + \lambda_2} \cdot \frac{1 - A_s}{A_s}$$

$$= \frac{1}{\mu_1} \cdot \frac{\lambda_1}{\lambda_1 + \lambda_2} + \frac{1}{\mu_2} \cdot \frac{\lambda_2}{\lambda_1 + \lambda_2} \qquad (8.63)$$

Equation (8.63) may also be written

$$\theta_F = \text{MDT}_1 \cdot \Pr(\text{Component 1 fails} \mid \text{system failure})$$
$$+ \text{MDT}_2 \cdot \Pr(\text{Component 2 fails} \mid \text{system failure})$$

This formula is obvious since the duration of a system failure will be equal to the repair time of component 1 when component 1 fails and equal to the repair time of component 2 when component 2 fails.

The mean time between system failures, MTBF_s, is

$$\text{MTBF}_s = \text{MTTF}_S + \theta_F = \frac{1}{\lambda_1 + \lambda_2} + \frac{1}{\mu_1} \frac{\lambda_1}{\lambda_1 + \lambda_2} + \frac{1}{\mu_2} \frac{\lambda_2}{\lambda_1 + \lambda_2}$$

$$= \frac{1 + (\lambda_1/\mu_1) + (\lambda_2/\mu_2)}{\lambda_1 + \lambda_2} \qquad (8.64)$$

The frequency of system failures may also be expressed as

$$\omega_F = \frac{1}{\text{MTBF}_s} = (\lambda_1 + \lambda_2) \cdot \frac{1}{1 + (\lambda_1/\mu_1) + (\lambda_2/\mu_2)} = A_s \cdot (\lambda_1 + \lambda_2)$$

For a series structure of n components the above results can be generalized as follows:
For system availability

$$A_s = \frac{1}{1 + \sum_{i=1}^{n}(\lambda_i/\mu_i)} \qquad (8.65)$$

For mean time to system failure

$$\text{MTTF} = \frac{1}{\sum_{i=1}^{n} \lambda_i} \qquad (8.66)$$

For mean duration of a system failure

$$\theta_F = \sum_{i=1}^{n} \frac{1}{\mu_i} \cdot \frac{\lambda_i}{\sum_{j=1}^{n} \lambda_j} = \frac{1}{\sum_{j=1}^{n} \lambda_j} \cdot \sum_{i=1}^{n} \frac{\lambda_i}{\mu_i} \qquad (8.67)$$

For frequency of system failures

$$\omega_F = A_s \cdot \sum_{i=1}^{n} \lambda_i = \frac{\sum_{i=1}^{n} \lambda_i}{1 + \sum_{i=1}^{n}(\lambda_i/\mu_i)} \qquad (8.68)$$

8.5 MEAN TIME TO FIRST SYSTEM FAILURE

8.5.1 Absorbing states

All the processes we have studied so far in this chapter have been irreducible, which means that every state is reachable from every other state.

We will now introduce Markov processes with *absorbing states*. An absorbing state is a state that, once entered, cannot be left until the system starts a new mission. The popular saying is that the system is *trapped* in an absorbing state.

Example 8.7
Reconsider the parallel system in Example 8.3 with two independent and identical components with failure rate λ. When one of the components fails, it is repaired. The repair time is assumed to be exponentially distributed with repair rate μ. When both components have failed, the system is considered to have failed and no recovery is possible. Let the number of functioning components denote the state of the system. The state space is thus $\mathcal{X} = \{0, 1, 2\}$, and state 0 is an absorbing state. The state transition diagram of the system is given in Fig. 8.10.

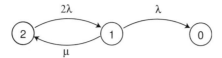

Fig. 8.10 State transition diagram for a parallel system with two identical components.

We assume that both components are functioning (state 2) at time 0. That is $P_2(0) = 1$. The transition rate matrix of this system is thus

$$\mathbb{A} = \begin{pmatrix} 0 & 0 & 0 \\ \lambda & -(\lambda + \mu) & \mu \\ 0 & 2\lambda & -2\lambda \end{pmatrix} \qquad (8.69)$$

Since state 0 is an absorbing state, all the transition rates from this state are equal to zero. Thus the entries of the row corresponding to the absorbing state are all equal to zero.

Since the matrix $\mathbb{A}$ does not have full rank, we may remove one of the three equations without loosing any information about $P_0(t)$, $P_1(t)$, and $P_2(t)$. In this case we remove the first of the three equations. This is accomplished by removing the first column of the matrix. Hence we get the state equations

$$[P_0(t), P_1(t), P_2(t)] \cdot \begin{pmatrix} 0 & 0 \\ -(\lambda + \mu) & \mu \\ 2\lambda & -2\lambda \end{pmatrix} = [\dot{P}_1(t), \dot{P}_2(t)]$$

Since all the elements of the first column of the matrix are equal to zero, $P_0(t)$ will "disappear" in the solution of the equations. We may therefore reduce the matrix equations to

$$[P_1(t), P_2(t)] \cdot \begin{pmatrix} -(\lambda + \mu) & \mu \\ 2\lambda & -2\lambda \end{pmatrix} = [\dot{P}_1(t), \dot{P}_2(t)] \qquad (8.70)$$

The matrix

$$\begin{pmatrix} -(\lambda + \mu) & \mu \\ 2\lambda & -2\lambda \end{pmatrix}$$

has full rank if $\lambda > 0$. Therefore (8.70) determines $P_1(t)$ and $P_2(t)$. $P_0(t)$ may thereafter be found from $P_0(t) = 1 - P_1(t) - P_2(t)$. This solution of the reduced matrix equations (8.70) is identical to the solution of the initial matrix equations. The reduced matrix is seen to be obtained by deleting the row and the column corresponding to the absorbing state.

Since state 0 is absorbing and reachable from the other states, it is obvious that

$$\lim_{t \to \infty} P_0(t) = 1$$

The Laplace transforms of the reduced matrix equations (8.70) are

$$(P_1^*(s), P_2^*(s)) = \begin{bmatrix} -(\lambda + \mu) & \mu \\ 2\lambda & -2\lambda \end{bmatrix} = (sP_1^*(s), sP_2^*(s) - 1)$$

when the system is assumed to be in state 2 at time $t = 0$. Thus

$$-(\lambda + \mu)P_1^*(s) + 2\lambda P_2^*(s) = sP_1^*(s)$$
$$\mu P_1^*(s) - 2\lambda P_2^*(s) = sP_2^*(s) - 1$$

Solving for $P_1^*(s)$ and $P_2^*(s)$, we get (see Appendix B)

$$P_1^*(s) = \frac{2\lambda}{s^2 + (3\lambda + \mu)s + 2\lambda^2}$$

$$P_2^*(s) = \frac{\lambda + \mu + s}{s^2 + (3\lambda + \mu)s + 2\lambda^2}$$

Let $R(t)$ denote the survivor function of the system. Since the system is functioning as long as the system is either in state 2 or in state 1, the survivor function is equal to

$$R(t) = P_1(t) + P_2(t) = 1 - P_0(t)$$

The Laplace transform of $R(t)$ is thus

$$R^*(s) = P_1^*(s) + P_2^*(s) = \frac{3\lambda + \mu + s}{s^2 + (3\lambda + \mu)s + 2\lambda^2} \tag{8.71}$$

The survivor function $R(t)$ may now be determined by inverting the Laplace transform, or we may consider $P_0(t) = 1 - R(t)$ which denotes the distribution function of the time T_s to system failure. The Laplace transform of $P_0(t)$ is

$$P_0^*(s) = \frac{1}{s} - P_1^*(s) - P_2^*(s) = \frac{2\lambda^2}{s[s^2 + (3\lambda + \mu)s + 2\lambda^2]}$$

Let $f_s(t)$ denote the probability density function of the time T_s to system failure, that is, $f_s(t) = dP_0(t)/dt$. The Laplace transform of $f_s(t)$ is thus

$$f_s^*(s) = sP_0^*(s) - P_0(0) = \frac{2\lambda^2}{s^2 + (3\lambda + \mu)s + 2\lambda^2} \tag{8.72}$$

The denominator of (8.72) can be written

$$s^2 + (3\lambda + \mu)s + 2\lambda^2 = (s - k_1)(s - k_2)$$

where

$$k_1 = \frac{-(3\lambda + \mu) + \sqrt{\lambda^2 + 6\lambda\mu + \mu^2}}{2}$$

$$k_2 = \frac{-(3\lambda + \mu) - \sqrt{\lambda^2 + 6\lambda\mu + \mu^2}}{2}$$

The expression for $f_s^*(s)$ can be rearranged so that

$$f_s^*(s) = \frac{2\lambda^2}{k_1 - k_2}\left(\frac{1}{s+k_2} - \frac{1}{s+k_1}\right)$$

By inverting this transform, we get

$$f_s(t) = \frac{2\lambda^2}{k_1 - k_2}\left(e^{-k_2 t} - e^{-k_1 t}\right)$$

The mean time to system failure, MTTF$_S$, is now given by (the integration is left to the reader as an exercise)

$$\text{MTTF}_S = \int_0^\infty t f_s(t)\, dt = \frac{3}{2\lambda} + \frac{\mu}{2\lambda^2} \tag{8.73}$$

Note that the MTTF$_S$ of a two-component parallel system, without any repair (i.e., $\mu = 0$) is equal to $3/2\lambda$. The repair facility thus increases the MTTF$_S$ by $\mu/2\lambda^2$. □

8.5.2 Survivor Function

As discussed on page 321, the set of states $\mathcal{X}$ of a system may be grouped in a set B of functioning states and a set $F = \mathcal{X} - B$ of failed states. In the present section we will assume that the failed states are absorbing states.

Consider a system that is in a specified functioning state at time $t = 0$. The survivor function $R(t)$ determines the probability that a system does not leave the set B of functioning states during the time interval $(0, t]$. The survivor function is thus

$$R(t) = \sum_{j \in B} P_j(t) \tag{8.74}$$

The Laplace transform of the survivor function is

$$R^*(s) = \sum_{j \in B} P_j^*(s)$$

8.5.3 Mean Time to System Failure

The mean time to system failure, MTTF$_S$, may according to Section 2.6 be determined by

$$\text{MTTF}_S = \int_0^\infty R(t)\, dt \tag{8.75}$$

The Laplace transform of $R(t)$ is given by

$$R^*(s) = \int_0^\infty R(t) e^{-st}\, dt \tag{8.76}$$

The MTTF$_S$ of the system may thus be determined from (8.76) by inserting $s = 0$. Thus

$$R^*(0) = \int_0^\infty R(t)\,dt = \text{MTTF}_S \qquad (8.77)$$

Example 8.7 (Cont.)
The Laplace transform of the survivor function for the two-component parallel system was in (8.71) found to be

$$R^*(s) = \frac{3\lambda + \mu + s}{s^2 + (3\lambda + \mu)s + 2\lambda^2}$$

By introducing $s = 0$, we get

$$\text{MTTF}_S = R^*(0) = \frac{3\lambda + \mu}{2\lambda^2} = \frac{3}{2\lambda} + \frac{\mu}{2\lambda^2}$$

which is in accordance with (8.73). □

Procedure for Finding the MTTF As indicated in Example 8.7, the following procedure may be used to find the mean time to first failure, MTTF, of a system with state space $\mathcal{X} = \{0, 1, \ldots, r\}$. See Billinton and Allen (1983, p. 217) and Pagès and Gondran (1980, p. 133) for details and justification.

1. Establish the transition rate matrix $\mathbb{A}$, and let $\boldsymbol{P}(t) = [P_0(t), P_1(t), \ldots, P_r(t)]$ denote the distribution of the process at time t. Observe that $\mathbb{A}$ is a $(r+1) \times (r+1)$ matrix.

2. Define the initial distribution $\boldsymbol{P}(0) = [P_0(0), P_1(0), \ldots, P_r(0)]$ of the process, and verify that $\boldsymbol{P}(0)$ means that the system has a functioning state.

3. Identify the failed states of the system, and define these states as absorbing states. Assume that there are k absorbing states.

4. Delete the rows and columns of $\mathbb{A}$ corresponding to the absorbing states, that is, if j is an absorbing state, remove the entries a_{ji} and a_{ij} for all i from $\mathbb{A}$. Let $\mathbb{A}_R$ denote the reduced transition rate matrix. The dimension of $\mathbb{A}_R$ is $(r+1-k) \times (r+1-k)$.

5. Let $\boldsymbol{P}^*(s) = [P_0^*(s), P_1^*(s), \ldots, P_r^*(s)]$ denote the Laplace transform of $\boldsymbol{P}(t)$ and remove the entries of $\boldsymbol{P}^*(s)$ corresponding to absorbing states. Let $\boldsymbol{P}_R^*(s)$ denote the reduced vector. Notice that $\boldsymbol{P}_R^*(s)$ has dimension $(r+1-k)$.

6. Remove the entries of $s\boldsymbol{P}^*(s) - \boldsymbol{P}(0)$ corresponding to absorbing states. Let $[s\boldsymbol{P}^*(s) - \boldsymbol{P}(0)]_R$ denote the reduced vector.

7. Establish the equation

$$\boldsymbol{P}_R^*(s) \cdot \mathbb{A}_R = [s\boldsymbol{P}^*(s) - \boldsymbol{P}(0)]_R$$

set $s = 0$ and determine $P_R^*(0)$

8. The MTTF is determined by

$$\text{MTTF} = \sum P_j^*(0)$$

where the sum is taken over all j representing the $(r + 1 - k)$ nonabsorbing states.

Example 8.8
Reconsider the parallel structure of two independent components in Example 8.2, where the components have failure rates λ_1 and λ_2 and repair rates μ_1 and μ_2, respectively. The states of the system are defined in Table 8.1. The system is assumed to start out at time 0 in state 3 with both components functioning. The system is functioning as long as at least one of the components is functioning. The set B of functioning states is thus $\{1, 2, 3\}$. The system fails when both components are in a failed state, state 0.

In this example we are primarily interested in determining the MTTF_S. We therefore define state 0 to be an absorbing state and set all departure rates from state 0 equal to zero. The transition rate matrix is then

$$\begin{pmatrix} 0 & 0 & 0 & 0 \\ \lambda_2 & -(\lambda_2 + \mu_1) & 0 & \mu_1 \\ \lambda_1 & 0 & -(\lambda_1 + \mu_2) & \mu_2 \\ 0 & \lambda_1 & \lambda_2 & -(\lambda_1 + \lambda_2) \end{pmatrix}$$

and the survivor function is

$$R(t) = P_1(t) + P_2(t) + P_3(t)$$

We now reduce the matrix equations by removing the row and the column corresponding to the absorbing state (state 0) and take Laplace transforms:

$$[P_1^*(0), P_2^*(0), P_3^*(0)] \cdot \begin{pmatrix} -(\lambda_2 + \mu_1) & 0 & \mu_1 \\ 0 & -(\lambda_1 + \mu_2) & \mu_2 \\ \lambda_1 & \lambda_2 & -(\lambda_1 + \lambda_2) \end{pmatrix} = [0, 0, -1]$$

This means that

$$P_1^*(0) = \frac{\lambda_1}{\lambda_2 + \mu_1} P_3^*(0) \tag{8.78}$$

$$P_2^*(0) = \frac{\lambda_2}{\lambda_1 + \mu_2} P_3^*(0) \tag{8.79}$$

$$\left(\frac{\lambda_1 \mu_1}{\lambda_2 + \mu_1} + \frac{\lambda_2 \mu_2}{\lambda_1 + \mu_2} - (\lambda_1 + \lambda_2) \right) P_3^*(0) = -1 \tag{8.80}$$

The last equation leads to

$$P_3^*(0) = \frac{1}{\lambda_1 \lambda_2 [1/(\lambda_1 + \mu_2) + 1/(\lambda_2 + \mu_1)]} \qquad (8.81)$$

Finally

$$\begin{aligned}
\text{MTTF}_S &= R^*(0) = P_1^*(0) + P_2^*(0) + P_3^*(0) \\
&= \frac{\lambda_1/(\lambda_2 + \mu_1) + \lambda_2/(\lambda_1 + \mu_2) + 1}{\lambda_1 \lambda_2 [1/(\lambda_1 + \mu_2) + 1/(\lambda_2 + \mu_1)]}
\end{aligned} \qquad (8.82)$$

where $P_1^*(0)$ and $P_2^*(0)$ are determined by inserting (8.81) in (8.78) and (8.79), respectively.

Some Special Cases:

1. Nonrepairable system ($\mu_1 = \mu_2 = 0$)

$$\text{MTTF}_S = \frac{(\lambda_2/\lambda_1) + (\lambda_1/\lambda_2) + 1}{\lambda_1 + \lambda_2}$$

When the two components have identical failure rates, $\lambda_1 = \lambda_2 = \lambda$, this expression is reduced to

$$\text{MTTF}_S = \frac{3}{2} \frac{1}{\lambda} \qquad (8.83)$$

2. The two components have identical failure rates and identical repair rates ($\lambda_1 = \lambda_2 = \lambda$ and $\mu_1 = \mu_2 = \mu$). Then

$$\text{MTTF}_S = \frac{3}{2\lambda} + \frac{\mu}{2\lambda^2}$$

□

8.6 SYSTEMS WITH DEPENDENT COMPONENTS

In this section we illustrate how a Markov model can be used to model dependent failures. Two simple situations are described: systems exposed to *common cause failures* and load-sharing systems that are exposed to *cascading failures*. Dependent failures were discussed in Chapter 6.

Common Cause Failures Consider a parallel structure of two identical components. The components may fail due to aging or other inherent defects. Such failures occur independent of each other with failure rate λ_I. The components are repaired independent of each other with repair rate μ.

An external event may occur that causes all functioning components to fail at the same time. Failures caused by the external event are called *common cause failures*.

SYSTEMS WITH DEPENDENT COMPONENTS

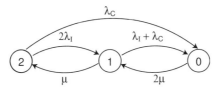

Fig. 8.11 State transition diagram for a parallel system with two components which is exposed to common cause failures.

The external events occur with rate λ_C which is denoted the common cause failure rate.

The states of the system are named according to the number of components functioning. Thus the state space is $\{0, 1, 2\}$. The state transition diagram of the parallel system with common cause failures is shown in Fig. 8.11.

The corresponding transition rate matrix is

$$\mathbb{A} = \begin{pmatrix} -2\mu & 2\mu & 0 \\ \lambda_C + \lambda_I & -(\lambda_I + \lambda_C + \mu) & \mu \\ \lambda_F & 2\lambda_I & -(2\lambda_I + \lambda_C) \end{pmatrix}$$

Assume that we are interested in determining the MTTF$_S$. Since the system fails as soon as it enters state 0, we define state 0 as an absorbing state and remove the row and the column from the transition rate matrix corresponding to state 0.

As before, we assume that the system is in state 2 (both components are functioning) at time $t = 0$. By introducing Laplace transforms, we get the following matrix equations

$$[P_1^*(0), P_2^*(0)] \cdot \begin{pmatrix} -(\lambda_I + \lambda_C + \mu) & \mu \\ 2\lambda_I & -(2\lambda_I + \lambda_C) \end{pmatrix} = [0, -1]$$

The solutions are

$$P_1^*(0) = \frac{2\lambda_I}{(2\lambda_I + \lambda_C)(\lambda_I + \lambda_C) + \lambda_C\mu}$$

$$P_2^*(0) = \frac{\lambda_I + \lambda_C + \mu}{(2\lambda_I + \lambda_C)(\lambda_I + \lambda_C) + \lambda_C\mu}$$

and the mean time to system failure is

$$\text{MTTF}_S = P_2^*(0) + P_1^*(0) = \frac{3\lambda_I + \lambda_C + \mu}{(2\lambda_I + \lambda_C)(\lambda_I + \lambda_C) + \lambda_C\mu} \tag{8.84}$$

Define a common cause factor β by

$$\beta = \frac{\lambda_C}{\lambda_C + \lambda_I} = \frac{\lambda_C}{\lambda}$$

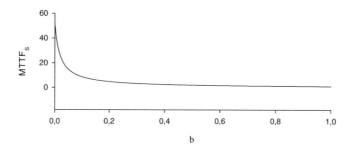

Fig. 8.12 The MTTF$_S$ of a parallel system as a function of the common cause factor β ($\lambda = 1$, and $\mu = 100$).

Here $\lambda = \lambda_C + \lambda_I$ is the total failure rate of a component, and the factor β denotes the fraction of common cause failures among all failures of a component. Some sources of reliability data present the *total* failure rate λ, while other sources present the *independent* failure rate λ_I. See Chapter 14 for details about reliability data sources. To investigate how the common cause factor β affects the MTTF$_S$, we insert β and λ into (8.84) and get

$$\begin{aligned} \text{MTTF}_S &= \frac{3(1-\beta)\lambda + \beta\lambda + \mu}{(2(1-\beta)\lambda + \beta\lambda)\lambda + \beta\lambda\mu} \\ &= \frac{3 - 2\beta\lambda + \mu}{(2-\beta)\lambda^2 + \beta\lambda\mu} = \frac{1}{\lambda}\frac{\lambda(3-2\beta) + \mu}{(2-\beta)\lambda + \beta\mu} \end{aligned} \quad (8.85)$$

Fig. 8.12 illustrates how the MTTF$_S$ of a parallel system depends on the common cause factor β.

Let us consider two simple cases.

1. $\beta = 0$ (i.e., only *independent failures*, $\lambda = \lambda_I$):

$$\text{MTTF}_S = \frac{3}{2\lambda_I} + \frac{\mu}{2\lambda_I^2}$$

which is what we obtained in Example 8.8.

2. $\beta = 1$ (i.e., all failures are common cause failures, $\lambda = \lambda_C$):

$$\text{MTTF}_S = \frac{1}{\lambda_C}\frac{\lambda_C + \mu}{\lambda_C + \mu} = \frac{1}{\lambda_C}$$

The last result is evident. Only common cause failure are occurring and they affect both components simultaneously with failure rate λ_C. This β-factor model is further discussed in Chapter 6.

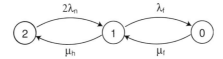

Fig. 8.13 Parallel system with two components sharing a common load.

Load-Sharing Systems Consider a parallel system with two identical components. The components share a common load. If one component fails, the other component has to carry the whole load and the failure rate of this component is assumed to increase immediately when the load is increased. Thus the failures of the two components are dependent. In Chapter 6, this type of dependency was referred to as *cascading failures*. The components may, for example, be pumps, compressors, or power generators. The following failure rates are assumed:

λ_n = failure rate at normal load (i.e., when both components are functioning)
λ_f = failure rate at full load (i.e., when one of the components is failed)

Let μ_h denote the repair rate of a component when only one component has failed, and let μ_f denote the repair rate of a component when both components have failed. Let the number of components that are functioning denote the state of the system. The state space is thus {0, 1, 2}. When the system has failed (state 0), all available repair resources are used to repair one of the components (usually the component that failed first). The system is stated up again (in state 1) as soon as this component is repaired. The state transition diagram of the system is given in Fig. 8.13.

The transition rate matrix is

$$\begin{pmatrix} -\mu_f & \mu_f & 0 \\ \lambda_f & -(\mu_h + \lambda_f) & \mu_h \\ 0 & 2\lambda_n & -2\lambda_n \end{pmatrix}$$

The system fails when both components fail (i.e., in state 0). To determine the MTTF$_S$ we define state 0 as an absorbing state and remove the row and the column corresponding to this state from the transition rate matrix. If we assume that the system starts out at time $t = 0$ with both components functioning (state 2), and take Laplace transforms with $s = 0$, we get

$$[P_1^*(0), P_2^*(0)] \cdot \begin{pmatrix} -(\mu_h + \lambda_f) & \mu_h \\ 2\lambda_n & -2\lambda_n \end{pmatrix} = [0, -1]$$

The solution is

$$P_1^*(0) = \frac{1}{\lambda_f}$$

$$P_2^*(0) = \frac{\lambda_f + \mu_h}{2\lambda_n \lambda_f}$$

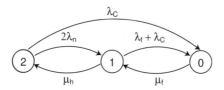

Fig. 8.14 State transition diagram for the generator system with load-sharing and common cause failures.

The survivor function is $R(t) = P_1(t) + P_2(t)$, and the mean time to system failure is thus

$$\text{MTTF}_S = R^*(0) = P_1^*(0) + P_2^*(0) = \frac{1}{\lambda_f} + \frac{1}{2\lambda_n} + \frac{\mu_h}{2\lambda_n \lambda_f} \tag{8.86}$$

Note that when no repair is carried out ($\mu_h = 0$)

$$\text{MTTF}_S = \frac{1}{\lambda_f} + \frac{1}{2\lambda_n} \tag{8.87}$$

When the load on the remaining component is not increased, such that $\lambda_f = \lambda_n$, we get $\text{MTTF}_S = 3/(2\lambda_n)$ in accordance with equation (8.83).

Example 8.9

Consider a power station with two generators of the same type. During normal operation, the generators are sharing the load and each generator has failure rate $\lambda_n = 1.6 \cdot 10^{-4}$ (hours)$^{-1}$. When one of the generators fails, the load on the remaining generator will be increased, and the failure rate will increase to $\lambda_f = 8.0 \cdot 10^{-4}$ (hours)$^{-1}$ (five times as high as the normal failure rate). In addition, the system is exposed to common cause failures. All generators in operation will fail at the same time when common cause events occur. The common cause failure rate is $\lambda_C = 2.0 \cdot 10^{-5}$ hours^{-1}. When one generator fails, it is repaired. The mean downtime to repair, MDT$_h$ is 12 hours, and the repair rate is therefore $\mu_h \approx 8.3 \cdot 10^{-2}$ (hours)$^{-1}$. When the system fails, the mean downtime to repair one generator is MDT$_f = 8$ hours, and the repair rate is $\mu_f = 1.25 \cdot 10^{-1}$ (hours)$^{-1}$. The state transition diagram of the generator system with load-sharing and common cause failures is shown in Fig. 8.14.

The steady-state probabilities can be found by the same approach as we have shown several times (e.g., see Example 8.6).

$$P_2 = \frac{\mu_n \mu_f}{(\lambda_f + \lambda_C + \mu_f)(\lambda_C + 2\lambda_n) + \lambda_C \mu_n + \mu_n \mu_f} \approx 0.99575$$

$$P_1 = \frac{(\lambda_C + 2\lambda_n)\mu_f}{(\lambda_f + \lambda_C + \mu_f)(\lambda_C + 2\lambda_n) + \lambda_C \mu_n + \mu_n \mu_f} \approx 0.00406$$

$$P_0 = \frac{(\lambda_f + \lambda_C)(\lambda_C + 2\lambda_n) + \lambda_C \mu_n}{(\lambda_f + \lambda_C + \mu_f)(\lambda_C + 2\lambda_n) + \lambda_C \mu_n + \mu_n \mu_f} \approx 0.00019$$

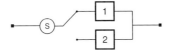

Fig. 8.15 Two-item standby system.

The mean time to system failure is found from the Laplace transforms:

$$[P_1^*(0), P_2^*(0)] \cdot \begin{pmatrix} -(\lambda_f + \lambda_C + \mu_n) & \mu_h \\ 2\lambda_n & -(\lambda_C + 2\lambda_n) \end{pmatrix} = [0, -1]$$

We find that

$$\text{MTTF}_S = P_1^* + P_2^* = \frac{2\lambda_n + \lambda_f + \lambda_C + \mu_n}{(\lambda_C + 2\lambda_n)(\lambda_f + \lambda_C + \mu_n) - 2\lambda_n \mu_n}$$
$$\approx 43\,421 \text{ hours} \approx 4.96 \text{ years}$$

□

8.7 STANDBY SYSTEMS

Standby systems were introduced in Section 4.6 where the survivor function $R(t)$ and the mean time to failure MTTF_S were determined for some simple non-repairable standby systems. In the present section we will discuss some simple two-item repairable standby systems in the light of Markov models. The system considered is illustrated in Fig. 8.15. Item A is initially (at time $t = 0$) the operating item and S is the sensing and changeover device.

A standby system may be operated and repaired in a number of different ways:

- The standby item may be cold or partly loaded.
- The changeover device may have several failure modes, like "fail to switch," "spurious switching," and "disconnect".
- Failure of the standby item may be hidden (nondetectable) or detectable.

In the present section a few operation and repair modes of a standby system are illustrated. Generalizations to more complex systems and operational modes are often straightforward, at least in theory. The computations may, however, require a computer.

8.7.1 Parallel System with Cold Standby and Perfect Switching

Since the standby item is passive, it is assumed not to fail in the standby state. The switching is assumed to be perfect. Failure of the active item is detected immediately, and the standby item is activated with probability 1. The failure rate of item i in

Table 8.3 Possible States of a Two-Item Parallel System with Cold Standby and Perfect Switching.

System State	State of Item A	State of Item B
4	O	S
3	F	O
2	S	O
1	O	F
0	F	F

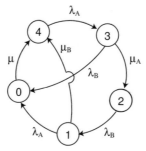

Fig. 8.16 State transition diagram of a two-item parallel system with cold standby and perfect switching.

operating state is denoted λ_i for $i = A, B$. When the active item has failed, a repair action is initiated immediately. The time to repair is exponentially distributed with repair rate μ_i for $i = A, B$. When a repair action is completed, the item is placed in standby state.

The possible states of the system are listed in Table 8.3 where O denotes operating state, S denotes standby state, and F denotes failed state. System failure occurs when the operating item fails before repair of the other item is completed. The failed state of the system is thus state 0 in Table 8.3. When both items have failed, they are repaired simultaneously and the system is thus brought back to state 4. The repair rate in this case is denoted μ. The state transition diagram of the standby system is illustrated in Fig. 8.16.

The transition rate matrix is

$$\mathbb{A} = \begin{pmatrix} -\mu & 0 & 0 & 0 & \mu \\ \lambda_A & -(\lambda_A + \mu_B) & 0 & 0 & \mu_B \\ 0 & \lambda_B & -\lambda_B & 0 & 0 \\ \lambda_B & 0 & \mu_A & -(\lambda_B + \mu_A) & 0 \\ 0 & 0 & 0 & \lambda_A & -\lambda_A \end{pmatrix} \quad (8.88)$$

The steady-state probabilities may be determined according to the procedures described in Section 8.3. The survivor function $R(t)$ and the MTTF$_S$ of the system can be determined by considering the failed state of the system (state 0) to be an absorbing state. Suppose that the initial state at $t = 0$ is state 4. By deleting the row and the column of the transition rate matrix corresponding to the absorbing state 0, we get the reduced matrix $\mathbb{A}_R$:

$$\mathbb{A}_R = \begin{pmatrix} -(\lambda_A + \mu_B) & 0 & 0 & \mu_B \\ \lambda_B & -\lambda_B & 0 & 0 \\ 0 & \mu_A & -(\lambda_B + \mu_A) & 0 \\ 0 & 0 & \lambda_A & -\lambda_A \end{pmatrix}$$

By taking Laplace transforms (with $s = 0$), we get the equations

$$[P_1^*(0), P_2^*(0), P_3^*(0), P_4^*(0)] \cdot \mathbb{A}_R = [0, 0, 0, -1]$$

The solution is

$$P_2^*(0) = \frac{\lambda_A + \mu_B}{\lambda_B} P_1^*(0)$$

$$P_3^*(0) = \frac{\lambda_A + \mu_B}{\mu_A} P_1^*(0)$$

$$P_4^*(0) = \frac{\lambda_B + \mu_A}{\lambda_A} P_3^*(0)$$

$$= \frac{(\lambda_A + \mu_B)(\lambda_B + \mu_A)}{\lambda_A \mu_A} P_1^*(0)$$

$$= \frac{1 + \mu_B P_1^*(0)}{\lambda_A}$$

Thus

$$P_1^*(0) = \frac{\mu_A}{\lambda_A \lambda_B + \lambda_A \mu_A + \lambda_B \mu_B}$$

The mean time to failure of the system is now, according to (8.77),

$$\begin{aligned} \text{MTTF}_S &= R^*(0) = P_1^*(0) + P_2^*(0) + P_3^*(0) + P_4^*(0) \\ &= \frac{1}{\lambda_A} + \frac{1}{\lambda_B} + \frac{\mu_A}{\lambda_B} \left(\frac{1}{\lambda_B} - \frac{1}{\lambda_B + \mu_A + \frac{\lambda_B}{\lambda_A}\mu_B} \right) \end{aligned} \quad (8.89)$$

For a nonrepairable system, $\mu_A = \mu_B = 0$. Then

$$\text{MTTF}_S = \frac{1}{\lambda_A} + \frac{1}{\lambda_B}$$

which is an obvious result.

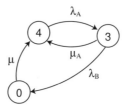

Fig. 8.17 State transition diagram of a two-item parallel system with cold standby and perfect switching (item A is the main operating item).

8.7.2 Parallel System with Cold Standby and Perfect Switching (Item A Is the Main Operating Item)

Reconsider the standby system in Fig. 8.15 but assume that item A is the main operating item. This means that item B is only used when A is in a failed state and under repair. Item A will thus be put into operation again as soon as the repair action is completed. System failure occurs when the operating item B fails before repair of item A is completed. The failed state of the system is thus state 0 in Table 8.3. When both items have failed, they are repaired simultaneously and brought back to state 4. The repair rate in this case is denoted μ. State 1 and state 2 in Table 8.3 are therefore irrelevant states for this system. The state transition diagram of this system is illustrated in Fig. 8.17.

The transition rate matrix is

$$\begin{pmatrix} -\mu & 0 & \mu \\ \lambda_B & -(\lambda_B + \mu_A) & \mu_A \\ 0 & \lambda_A & -\lambda_A \end{pmatrix} \quad (8.90)$$

The steady-state probabilities are determined by

$$[P_0, P_3, P_4] \cdot \begin{pmatrix} -\mu & 0 & \mu \\ \lambda_B & -(\lambda_B + \mu_A) & \mu_A \\ 0 & \lambda_A & -\lambda_A \end{pmatrix} = [0, 0, 0]$$

and

$$P_0 + P_3 + P_4 = 1$$

The solution is

$$P_0 = \frac{\lambda_A \lambda_B}{\lambda_A \lambda_B + \lambda_A \mu + \lambda_B \mu + \mu \mu_A}$$

$$P_3 = \frac{\lambda_A \mu}{\lambda_A \lambda_B + \lambda_A \mu + \lambda_B \mu + \mu \mu_A}$$

$$P_4 = \frac{\lambda_B \mu + \mu \mu_A}{\lambda_A \lambda_B + \lambda_A \mu + \lambda_B \mu + \mu \mu_A}$$

where P_j is the mean proportion of time the system is spending in state j for $j = 0, 3, 4$.

The frequency of system failures, ω_F, is in this case equal to the visit frequency to state 0, that is,

$$\omega_F = \nu_0 = \frac{P_0}{\mu}$$

The MTTF$_S$ of the system is determined as on page 332. By deleting the row and the column of the transition rate matrix in (8.90) and taking Laplace transforms (with $s = 0$), we obtain

$$[P_3^*(0), P_4^*(0)] \cdot \begin{pmatrix} -(\lambda_B + \mu_A) & \mu_A \\ \lambda_A & -\lambda_A \end{pmatrix} = [0, -1]$$

The solution is

$$P_3^*(0) = \frac{1}{\lambda_B}$$

$$P_4^*(0) = \frac{1}{\lambda_A} + \frac{\mu_A}{\lambda_A \lambda_B}$$

The mean time to failure of the system is thus

$$\text{MTTF}_S = R^*(0) = P_3^*(0) + P_4^*(0) = \frac{1}{\lambda_A} + \frac{1}{\lambda_B} + \frac{\mu_A}{\lambda_A \cdot \lambda_B} \quad (8.91)$$

The mean downtime required to repair the system is

$$\text{MDT}_S = \frac{1}{\mu}$$

The average availability A of the system is thus

$$A = \frac{\text{MTTF}_S}{\text{MTTF}_S + \text{MDT}_S} = \frac{1/\lambda_A + 1/\lambda_B + \mu_A/(\lambda_A \lambda_B)}{1/\lambda_A + 1/\lambda_B + \mu_A/(\lambda_A \lambda_B) + 1/\mu}$$

8.7.3 Parallel System with Cold Standby and Imperfect Switching (Item A Is the Main Operating Item)

Reconsider the standby system in Fig. 8.15, but assume that the switching is no longer perfect. When the active item A fails, the standby item B will be activated properly with probability $(1 - p)$. The probability p may also include a "fail to start" probability of the standby item. The state transition diagram of the system is illustrated in Fig. 8.18. From state 4 the system may show a transition to state 3 with rate $(1 - p)\lambda_A$ and to state 0 with rate $p\lambda_A$.

The steady-state probabilities are determined by

$$[P_0, P_3, P_4] \cdot \begin{pmatrix} -\mu & 0 & \mu \\ \lambda_B & -(\lambda_B + \mu_A) & \mu_A \\ p\lambda_A & (1-p)\lambda_A & -\lambda_A \end{pmatrix} = [0, 0, 0] \quad (8.92)$$

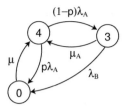

Fig. 8.18 State transition diagram of a two-item parallel system with cold standby and imperfect switching (item A is the main operating item).

and

$$P_0 + P_3 + P_4 = 1$$

The solution is

$$P_0 = \frac{\lambda_A \lambda_B + p\lambda_A \mu_A}{\lambda_A \lambda_B + p\lambda_A \mu_A + (1-p)\lambda_A \mu + \lambda_B \mu + \mu\mu_A}$$

$$P_3 = \frac{\lambda_A \mu (1-p)}{\lambda_A \lambda_B + p\lambda_A \mu_A + (1-p)\lambda_A \mu + \lambda_B \mu + \mu\mu_A}$$

$$P_4 = \frac{\lambda_B \mu + \mu\mu_A}{\lambda_A \lambda_B + p\lambda_A \mu_A + (1-p)\lambda_A \mu + \lambda_B \mu + \mu\mu_A}$$

The MTTF$_S$ can be determined from

$$[P_3^*(0), P_4^*(0)] \cdot \begin{pmatrix} -(\lambda_B + \mu_A) & \mu_A \\ (1-p)\lambda_A & -\lambda_A \end{pmatrix} = [0, -1]$$

which leads to

$$P_3^*(0) = \frac{1-p}{\lambda_B + p\mu_A}$$

$$P_4^*(0) = \frac{\lambda_B + \mu_A}{\lambda_A(\lambda_B + p\mu_A)}$$

Thus

$$\text{MTTF}_S = R^*(0) = P_3^*(0) + P_4^*(0) = \frac{(1-p)\lambda_A + \lambda_B + \mu_A}{\lambda_A(\lambda_B + p\mu_A)} \quad (8.93)$$

8.7.4 Parallel System with Partly Loaded Standby and Perfect Switching (Item A is the Main Operating Item)

Reconsider the standby system in Fig. 8.15 but assume that the standby item B may fail in standby mode and have a hidden failure when activated. The failure rate of item B in standby mode is denoted λ_B^s and is normally less than the corresponding

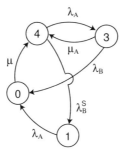

Fig. 8.19 State transition diagram of a two-item parallel system with partly loaded standby and perfect switching (item A is the main operating item).

failure rate during operation. In addition to the transition in Fig. 8.17, this system may also have transitions from state 4 to state 1 (in Table 8.3) and from state 1 to state 0. The state transition diagram is illustrated in Fig. 8.19.

The steady-state probabilities are determined by

$$[P_0, P_1, P_3, 4] \cdot \begin{pmatrix} -\mu & 0 & 0 & \mu \\ \lambda_A & -\lambda_A & 0 & 0 \\ \lambda_B & 0 & -(\lambda_B + \mu_A) & \mu_A \\ 0 & \lambda_B^s & \lambda_A & -(\lambda_A + \lambda_B^s) \end{pmatrix} = [0, 0, 0, 0]$$

and

$$P_0 + P_1 + P_3 + P_4 = 1$$

The MTTF$_S$ can be determined from

$$[P_1^*(0), P_3^*(0), P_4^*(0)] \cdot \begin{pmatrix} -\lambda_A & 0 & 0 \\ 0 & -(\lambda_B + \mu_A) & \mu_A \\ \lambda_B^s & \lambda_A & -(\lambda_A + \lambda_B^s) \end{pmatrix} = [0, 0, -1]$$

$$P_1^*(0) = \frac{\frac{\lambda_B^s}{\lambda_A}(\lambda_B + \mu_A)}{\lambda_A \lambda_B + \lambda_B \lambda_B^s + \lambda_B^s \mu_A}$$

$$P_3^*(0) = \frac{\lambda_A}{\lambda_A \lambda_B + \lambda_B \lambda_B^s + \lambda_B^s \mu_A}$$

$$P_4^*(0) = \frac{\lambda_B + \mu_A}{\lambda_A \lambda_B + \lambda_B \lambda_B^s + \lambda_B^s \mu_A}$$

Thus

$$\begin{aligned} \text{MTTF}_S &= R^*(0) = P_1^*(0) + P_3^*(0) + P_4^*(0) \\ &= \frac{(\frac{\lambda_B^s}{\lambda_A} + 1)(\lambda_B + \mu_A) + \lambda_A}{\lambda_A \lambda_B + \lambda_B \lambda_B^s + \lambda_B^s \mu_A} \end{aligned} \quad (8.94)$$

346   MARKOV PROCESSES

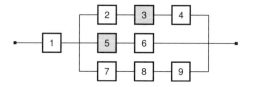

Fig. 8.20 Reliability block diagram of a system where components 3 and 5 are supercomponents that have to be modeled by Markov methods.

Let us now assume that we have two items of the same type and no repair is carried out. Let $\lambda_a = \lambda_B = \lambda$, and $\lambda_A^S = \lambda_B^S = \lambda^S$. In this case, the mean time to failure is

$$\mathrm{MTTF}_S = \frac{1}{\lambda + \lambda^S}\left(2 + \frac{\lambda^S}{\lambda}\right) \qquad (8.95)$$

Note that when $\lambda = \lambda^S$, equation (8.95) reduces to the mean time to failure of an active parallel system.

8.8 COMPLEX SYSTEMS

In principle, we can establish Markov models of systems with a large number of components. In practice, however, this will soon become unmanageable. Several approaches have been suggested for complex systems. We will briefly discuss a few of these approaches.

8.8.1 Markov "Modules" in Complex Systems

In Chapters 3 and 4 we discussed how to model complex systems by reliability block diagrams and by fault trees. We found that these approaches were suitable for rather static systems but were not able to account for dynamic features like complex maintenance and complex switching systems. Most systems will, however, have some modules that are rather static and other modules that have dynamic features. A possible approach is then to isolate the dynamic effects in as small modules as possible and treat these modules by Markov analysis. As far as possible, these modules should be defined in such a way that they are independent of each other. Thereafter, we may introduce these modules as supercomponents into a reliability block diagram (or the fault tree), as illustrated in the reliability block diagram in Fig. 8.20, and do the system calculations according to the approach we presented in Chapter 4. Dependencies between the various items in the reliability block diagram, that we are not able to model explicitly, may be analyzed by the methods described in Chapter 6.

The supercomponents 3 and 5 in Fig. 8.20 will normally comprise several components. When we establish a Markov model for the supercomponents, we will usually define several states for each of them and find the steady-state probability for each state. To find the system reliability by the methods we presented in Chapter 4, we have to define two states for each item in the reliability block diagram. The states, resulting from the Markov analysis, must therefore be merged into a functioning state (1) and a failed state (0). By this merging we will loose a lot of information that might be useful.

8.8.2 Independent Modules

In some cases we may be able to split a complex system into manageable modules that may be regarded as independent. If we are able to establish Markov models for each module and calculate the steady-state probabilities, we may use standard probability rules to find the system steady-state probabilities. This approach is analogous to what we did in Example 8.6 for a simple parallel system.

When we have a large number of independent modules with several states for each module, the total number of possible system states may be overwhelming. An alternative approach is Kronecker[4] sums and products (see Appendix C). We will illustrate this approach by a simple example.

Example 8.10
Reconsider the parallel system in Example 8.6 with two independent items 1 and 2, with two independent repair crews. The failure rates and repair rates are λ_i and μ_i for $i = 1, 2$, respectively. We may split the system into two independent modules, where each module consists of one single item. The transition rate matrix for item i is from Example 8.5

$$\mathbb{A}_i = \begin{pmatrix} -\mu_i & \mu_i \\ \lambda_i & -\lambda_i \end{pmatrix}$$

The Kronecker sum of the two transition rate matrices is

$$\mathbb{A}_1 \oplus \mathbb{A}_2 = \mathbb{A}_1 \otimes \mathbb{I} + \mathbb{I} \otimes \mathbb{A}_2$$

$$= \begin{pmatrix} -\mu_1 & \mu_1 \\ \lambda_1 & -\lambda_1 \end{pmatrix} \otimes \begin{pmatrix} 1 & 0 \\ 0 & 1 \end{pmatrix} + \begin{pmatrix} 1 & 0 \\ 0 & 1 \end{pmatrix} \otimes \begin{pmatrix} -\mu_2 & \mu_2 \\ \lambda_2 & -\lambda_2 \end{pmatrix}$$

$$= \begin{pmatrix} -(\mu_1+\mu_2) & \mu_1 & \mu_2 & 0 \\ \lambda_1 & -(\lambda_1+\mu_2) & 0 & \mu_2 \\ \lambda_2 & 0 & -(\lambda_2+\mu_1) & \mu_1 \\ 0 & \lambda_2 & \lambda_1 & -(\lambda_1+\lambda_2) \end{pmatrix}$$

which we recognize as the transition rate matrix for the parallel system in Example 8.6. The Kronecker sum of the transition rate matrices for the two independent modules is therefore equal to the transition rate matrix for the whole system. □

---

[4]Named after the German mathematician Leopold Kronecker (1823–1891).

It has been shown that the result in Example 8.10 also is valid in the general case. If a system comprises n independent modules with transition rate matrices $\mathbb{A}_1, \mathbb{A}_2, \ldots, \mathbb{A}_n$, then the transition rate matrix $\mathbb{A}$ of the system may be written as

$$\mathbb{A} = \mathbb{A}_1 \oplus \mathbb{A}_2 \oplus \cdots \oplus \mathbb{A}_n = \bigoplus_{i=1}^{n} \mathbb{A}_i \tag{8.96}$$

For details, see Amoia and Santomauro (1977).

If we are able to split the system into manageable and independent modules and establish transition rate matrices for the various modules, the Kronecker sum may then be used to establish the total system transition rate matrix. Several of the most popular mathematical programs have specific subroutines that may be used to find the system transition rate matrix.

The Kronecker product is very efficient when it comes to solving linear equations. Let $\mathbb{A}_i$ denote the transition rate matrix of module i, and let $\boldsymbol{P}^{(i)} = [P_0^{(i)}, \ldots, P_{r_i}^{(i)}]$ denote the steady-state probabilities of module i. We know from (8.26) that $\boldsymbol{P}^{(i)} \cdot \mathbb{A} = \boldsymbol{0}$. Let us now assume that we have a system with two independent modules with transition rate matrices $\mathbb{A}_1$ and $\mathbb{A}_2$ and steady-state probabilities $\boldsymbol{P}^{(1)}$ and $\boldsymbol{P}^{(2)}$, respectively. We know from (8.96) that the transition rate matrix for the system is given by $\mathbb{A} = \mathbb{A}_1 \oplus \mathbb{A}_2$. The system steady-state probabilities $\boldsymbol{P}$ must fulfill

$$\boldsymbol{P} \cdot \mathbb{A} = \boldsymbol{P} \cdot (\mathbb{A}_1 \oplus \mathbb{A}_2) = \boldsymbol{0}$$

The question is then: Will it be possible to find $\boldsymbol{P}$ from $\boldsymbol{P}^{(1)}$ and $\boldsymbol{P}^{(2)}$? Before we answer this question, we look at an example.

Example 8.11
Reconsider the parallel system of two independent items in Example 8.10. The failure rates and repair rates are λ_i and μ_i for $i = 1, 2$, respectively. From Example 8.5 we know that the steady-state probabilities of item i are $\boldsymbol{P}^{(i)} = [P_0^{(i)}, P_1^{(i)}]$ for $i = 1, 2$ where

$$P_0^{(i)} = \frac{\lambda_i}{\lambda_i + \mu_i} \quad \text{and} \quad P_1^{(i)} = \frac{\mu_i}{\lambda_i + \mu_i}$$

In Example 8.6 we found that the steady-state probabilities for the parallel system were given by

$$\boldsymbol{P} = [P_0^{(1)} P_0^{(2)}, P_1^{(1)} P_0^{(2)}, P_0^{(1)} P_1^{(2)}, P_1^{(1)} P_1^{(2)}]$$

This is seen to be the Kronecker product of $\boldsymbol{P}^{(1)}$ and $\boldsymbol{P}^{(2)}$. For the parallel system of two independent items, we have therefore shown that the steady-state probability $\boldsymbol{P}$ is equal to the Kronecker product of the steady-state probabilities of the modules, that is, $\boldsymbol{P} = \boldsymbol{P}^{(1)} \otimes \boldsymbol{P}^{(2)}$. □

It is rather straightforward to show (e.g., see Graham 1981) that the result in Example 8.11 also holds in a more general case. We may therefore use the following approach to find the steady-state probabilities $\boldsymbol{P}$ for a complex system.

COMPLEX SYSTEMS

1. Split the complex system into a set of n manageable and coherent modules. The various modules must be independent. Components within a module may, however, be dependent. Dependent components must belong to the same module.

2. Find the transition rate matrix $\mathbb{A}_i$ and the corresponding steady-state probabilities $\boldsymbol{P}^{(i)}$ for each module $i = 1, 2, \ldots, n$. (It might be wise to build a "library" of standard modules)

3. If of interest, the transition rate matrix for the system may be determined by $\mathbb{A}_1 \oplus \mathbb{A}_2 \oplus \cdots \oplus \mathbb{A}_n$.

4. Determine the steady-state probabilities for the system by $\boldsymbol{P} = \boldsymbol{P}^{(1)} \otimes \boldsymbol{P}^{(2)} \otimes \cdots \otimes \boldsymbol{P}^{(n)}$.

In practice, it might be a problem to keep track of the indexes in $\boldsymbol{P}$, that is, to realize which system state corresponds to a specific index. It is therefore important to be very systematic when defining the indexes for each module.

The Kronecker product approach has been applied to protective relays in transformer stations by Svendsen (2002). Application of the Kronecker product to dependent modules was discussed by Lesanovský (1988).

8.8.3 Markov Analysis in Fault Tree Analysis

We will now illustrate how results from Markov analysis can be used in fault tree analysis. Assume that a fault tree has been established with respect to a TOP event (a system failure or accident) in a specific system. The fault tree has n basic events (components) and k minimal cut sets $K_1, K_2, \ldots, K_k$.

The probability of the fault tree TOP event may be approximated by the upper bound approximation (4.50)

$$Q_0(t) \approx 1 - \prod_{j=1}^{k} (1 - \check{Q}_j(t)) \qquad (8.97)$$

Let us assume that the TOP event is a system failure, such that $Q_0(t)$ is the system unavailability. The average (limiting) system unavailability is thus approximately

$$Q_0 \approx 1 - \prod_{j=1}^{k} (1 - \check{Q}_j) \qquad (8.98)$$

where $\check{Q}_j$ denotes the average unavailability of the minimal cut parallel structure corresponding to the minimal cut set K_j, $j = 1, 2, \ldots, k$.

In the rest of this section we will assume that component i has constant failure rate λ_i, mean downtime to repair MDT_i, and constant repair rate $\mu_i = 1/\text{MDT}_i$ for $i = 1, 2, \ldots, n$. Furthermore, we assume that $\lambda_i \ll \mu_i$ for all $i = 1, 2, \ldots, n$.

The average unavailability q_i of component i is $\mu_i/(\mu_i + \lambda_i)$, which may be approximated by $\lambda_i \cdot \text{MDT}_i$, such that

$$\check{Q}_j = \prod_{i \in K_j} \frac{\mu_i}{\mu_i + \lambda_i} \approx \prod_{i \in K_j} \lambda_i \cdot \text{MDT}_i \tag{8.99}$$

The TOP event probability (system unavailability) is thus approximately

$$Q_0 \approx 1 - \prod_{j=1}^{k} \left(1 - \prod_{i \in K_j} \lambda_i \cdot \text{MDT}_i \right) \tag{8.100}$$

or

$$Q_0 \approx \sum_{j=1}^{k} \prod_{i \in K_j} \lambda_i \cdot \text{MDT}_i \tag{8.101}$$

Cut Set Information Consider a specific minimal cut parallel structure K_j, for $j = 1, 2, \ldots, k$. As before we assume that the components fail and are repaired independent of each other.

When all the components of the cut set K_j are in a failed state, we have a *cut set failure*. The mean duration of a failure of cut set K_j is from (8.46)

$$\text{MDT}_j = \frac{1}{\sum_{i \in K_j} \mu_i} \tag{8.102}$$

The expected frequency of cut set failures ω_j is from (8.48)

$$\omega_j \approx \left(\prod_{i \in K_j} \frac{\lambda_i}{\mu_i} \right) \cdot \left(\sum_{i \in K_j} \mu_i \right) \tag{8.103}$$

and, the mean time between failures (MTBF) of cut set K_j is

$$\text{MTBF}_j = \frac{1}{\omega_K}$$

Note that MTBF_j also includes the mean downtime of the cut parallel structure. The downtime is, however, usually negligible compared to the uptime.

System Information The system may be considered as a series structure of its k minimal cut parallel structures. If the cut parallel structures were independent and the downtimes were negligible, the frequency ω_S of system failures would be

$$\omega_S = \sum_{j=1}^{k} \omega_j \tag{8.104}$$

In general, this formula is not correct because (1) the minimal cut parallel structures are usually not independent, and (2) the downtimes of the minimal cut parallel structures are often not negligible.

For a system with very high availability, (8.104) is a good approximation for the expected frequency ω_S of system failures.

The mean time between system failures, MTBF$_S$, in the steady-state situation is approximately

$$\text{MTBF}_S \approx \frac{1}{\omega_S}$$

The mean system downtime per system failure is from (8.57) approximately

$$\text{MDT}_S \approx \frac{\sum_{j=1}^{k} \omega_j \text{MDT}_j}{\sum_{j=1}^{k} \omega_j}$$

The average system availability may now be approximated by

$$A_S = \frac{\text{MTBF}_S}{\text{MTBF}_S + \text{MDT}_S}$$

The formulas in this section are used in some of the computer programs for fault tree analysis, for example, CARA Fault Tree.

8.9 TIME-DEPENDENT SOLUTION

Reconsider the Kolmogorov forward equations (8.19)

$$\boldsymbol{P}(t) \cdot \mathbb{A} = \dot{\boldsymbol{P}}(t)$$

where $\boldsymbol{P}(t) = [P_0(t), P_1(t), \ldots, P_r(t)]$ is the distribution of the process at time t. Assume that we know the distribution of the system state at time 0, $\boldsymbol{P}(0)$. Usually, we know that the system is in a specific state i at time 0, but sometimes we only know that it has a specific distribution.

It is, in principle, possible to solve the Kolmogorov equations and find $\boldsymbol{P}(t)$ by

$$\boldsymbol{P}(t) = \boldsymbol{P}(0) \cdot e^{t\mathbb{A}} = \boldsymbol{P}(0) \cdot \sum_{k=0}^{\infty} \frac{t^k \mathbb{A}^k}{k!} \qquad (8.105)$$

where $\mathbb{A}^0$ is the identity matrix $\mathbb{I}$. To determine $\boldsymbol{P}(t)$ from (8.105) is sometimes time-consuming and inefficient.

When we study a system with absorbing states, like the parallel system in Example 8.7, we may define a column vector $\boldsymbol{C}$ with entries 1 and 0, where 1 corresponds to a functioning state, and 0 corresponds to a failed state. In Example 8.7, the states

1 and 2 are functioning, and state 0 is failed. The (column) vector is therefore $C = [0, 1, 1]^T$. The survival probability of the system is then given by

$$R(t) = \boldsymbol{P}(0) \cdot \sum_{k=0}^{\infty} \frac{t^k \mathbb{A}^k}{k!} \cdot \boldsymbol{C} \qquad (8.106)$$

It is also possible to use that

$$e^{t\mathbb{A}} = \lim_{k \to \infty} (\mathbb{I} + t \cdot \mathbb{A}/k)^k$$

and approximate $\boldsymbol{P}(t)$ by

$$\boldsymbol{P}(t) \approx \boldsymbol{P}(0) \cdot (\mathbb{I} + t \cdot \mathbb{A}/n)^n \qquad (8.107)$$

for a "sufficiently" large n. See Bon (1995, pp. 176–182) for further approximations and discussions.

Laplace Transforms An alternative approach is to use Laplace transforms. An introduction to Laplace transforms is given in Appendix B.

Again, assume that we know $\boldsymbol{P}(0)$, the distribution of the Markov process at time 0. The state equations (8.19) for the Markov process at time t are seen to be a set of linear, first order differential equations. The easiest and most widely used method to solve such equations is by Laplace transforms.

The Laplace transform of the state probability $P_j(t)$ is denoted by $P_j^*(s)$, and the Laplace transform of the time derivative of $P_j(t)$ is, according to Appendix B,

$$\mathcal{L}[\dot{P}_j(t)] = s P_j^*(s) - P_j(0) \quad \text{for } j = 0, 1, 2, \ldots, r$$

The Laplace transform of the state equations (8.19) is thus in matrix terms

$$\boldsymbol{P}^*(s) \cdot \mathbb{A} = s \boldsymbol{P}^*(s) - \boldsymbol{P}(0) \qquad (8.108)$$

By introducing the Laplace transforms, we have reduced the differential equations to a set of linear equations. The Laplace transforms $P_j^*(s)$ may now be computed from (8.108). Afterwards the state probabilities $P_j(t)$ may be determined from the inverse Laplace transforms.

Example 8.12
Reconsider the single component in Example 8.5, with transition rate matrix

$$\mathbb{A} = \begin{pmatrix} -\mu & \mu \\ \lambda & -\lambda \end{pmatrix}$$

We assume that the component is functioning at time $t = 0$, such that $\boldsymbol{P}(0) = (P_0(0), P_1(0)) = (0, 1)$. The Laplace transform of the state equation is then from (8.108)

$$(P_0^*(s), P_1^*(s)) \cdot \begin{bmatrix} -\mu & \mu \\ \lambda & -\lambda \end{bmatrix} = (s P_0^*(s) - 0, s P_1^*(s) - 1)$$

Thus

$$-\mu P_0^*(s) + \lambda P_1^*(s) = s P_0^*(s)$$
$$\mu P_0^*(s) - \lambda P_1^*(s) = s P_1^*(s) - 1 \quad (8.109)$$

By adding these two equations, we get

$$s P_0^*(s) + s P_1^*(s) = 1$$

Thus

$$P_0^*(s) = \frac{1}{s} - P_1^*(s)$$

By inserting this $P_0^*(s)$ into (8.109), we obtain

$$\frac{\mu}{s} - \mu P_1^*(s) - \lambda P_1^*(s) = s P_1^*(s) - 1$$

$$P_1^*(s) = \frac{1}{\lambda + \mu + s} + \frac{\mu}{s} \cdot \frac{1}{\lambda + \mu + s}$$

To find the inverse Laplace transform, we rewrite this expression as

$$P_1^*(s) = \frac{\lambda}{\lambda + \mu} \cdot \frac{1}{\lambda + \mu + s} + \frac{\mu}{\lambda + \mu} \cdot \frac{1}{s} \quad (8.110)$$

From Appendix B, the inverse Laplace transform of (8.110) is

$$P_1(t) = \frac{\mu}{\mu + \lambda} + \frac{\lambda}{\mu + \lambda} e^{-(\lambda + \mu)t}$$

which is the same result we gave in Example 8.5. □

To find the time-dependent state probabilities for a complex system is usually a difficult task and will not be discussed any further in this book. In most practical applications we are primarily interested in the steady-state probabilities and do not need to find the time-dependent probabilities.

8.10 SEMI-MARKOV PROCESSES

In Section 8.2 we defined a Markov process as a stochastic process having the properties that each time it enters a state i:

1. The amount of time the process spends in state i before making a transition into a different state is exponentially distributed with rate, say α_i.

2. When the process leaves state i, it will next enter state j with some probability P_{ij}, where $\sum_{\substack{j=0 \\ j \neq i}}^{r} P_{ij} = 1$.

An obvious extension to this definition is to allow the time the process spends in state i (the sojourn time in state i) to have a general "life" distribution, and also to let this distribution be dependent on the state to which the process will go. Ross (1996, p. 213) therefore defines a *semi-Markov process* as a stochastic process $\{X(t), t \geq 0\}$ with state space $\mathcal{X} = \{0, 1, 2, \ldots, r\}$ such that whenever the process enters state i:

1. The next state it will enter is state j with probability P_{ij}, for i, j in $\mathcal{X}$.

2. Given that the next state to be entered is state j, the time until the transition from i to j occurs has distribution F_{ij}.

The skeleton of the semi-Markov process is defined in the same way as for the Markov process (see Section 8.2), and will be a discrete-time Markov chain. The semi-Markov process is said to be irreducible if the skeleton is irreducible.

The distribution of the sojourn time $\widetilde{T}_i$ in state i is

$$F_i(t) = \sum_{\substack{j=0 \\ j \neq i}}^{r} P_{ij} \, F_{ij}(t)$$

The mean sojourn time in state i is

$$\mu_i = E(\widetilde{T}_i) = \int_0^\infty t \, dF_i(t)$$

We notice that if $F_{ij}(t) = 1 - e^{\alpha_i t}$, the semi-Markov process is an ordinary Markov process.

Let T_{ii} denote the time between successive transitions into state i, and let $\mu_{ii} = E(T_{ii})$. The visits to state i will now be a renewal process, and we may use the theory of renewal processes described in Chapter 7.

If we let $N_i(t)$ denote the number of times in $[0, t]$ that the process is in state i, the family of vectors

$$[N_0(t), N_1(t), \ldots, N_r(t)], \quad \text{for } t \geq 0$$

is called a *Markov renewal process*.

If the semi-Markov process is irreducible and if T_{ii} has a nonlattice distribution with finite mean, then

$$\lim_{t \to \infty} \Pr(X(t) = i \mid X(0) = j) = P_i$$

exists and is independent of the initial state. Furthermore

$$P_i = \frac{\mu_i}{\mu_{ii}}$$

For proof, see Ross (1996, p. 214). P_i is the proportion of transitions into state i and is also equal to the long-run proportion of time the process is in state i.

When the skeleton (the embedded process) is irreducible and positive recurrent, we may find the stationary distribution of the skeleton $\boldsymbol{\pi} = [\pi_0, \pi_1, \ldots, \pi_r]$ as the unique solution of

$$\pi_j = \sum_{i=0}^{r} \pi_i P_{ij}$$

where $\sum_i \pi_i = 1$ and $\pi_j = \lim_{n \to \infty} \Pr(X_n = j)$ [since we assume that the Markov process is aperiodic]. Since the π_j is the proportion of transitions that are into state j, and μ_j is the mean time spent in state j per transition, it seems intuitive that the limiting probabilities should be proportional to $\pi_j \mu_j$. In fact

$$P_j = \frac{\pi_j \mu_j}{\sum_i p_i \mu_i}$$

For a proof, see Ross (1996, p. 215).

Semi-Markov processes are not discussed any further in this book. Details about semi-Markov processes may be found in Ross (1996), Cocozza-Thivent (1997), and Limnios and Oprisan (2001).

PROBLEMS

8.1 A fail-safe valve has two main failure modes: premature/spurious closure (PC) and fail to close (FTC), with constant failure rates:

$$\lambda_{PC} = 10^{-3} \text{ PC failures per hour}$$
$$\lambda_{FTC} = 2 \cdot 10^{-4} \text{ FTC failures per hour}$$

The mean time to repair a PC failure is assumed to be 1 hour, while the mean time to repair an FTC failure is 24 hours. The repair times are assumed to be exponentially distributed.

(a) Explain why the operation of the valve may be described by a Markov process with three states. Establish the state transition diagram and the state equations for this process.

(b) Calculate the average availability of the valve and the mean time between failures.

8.2 Two identical pumps are operated as a parallel system. During normal operation, both pumps are functioning. When the first pump fails, the other pump has to do the whole job alone with a higher load than when both pumps are in operation. The pumps are assumed to have constant failure rates:

$\lambda_H = 1.5 \cdot 10^{-4}$ failure per hour
(the failure rate when the pumps are sharing the load, i.e., "half-load")
$\lambda_F = 3.5 \cdot 10^{-4}$ failure per hour
(the failure rate at "full load," when one of the pumps is in a failed state).

Both pumps may fail at the same time due to some external stresses (*common cause failure*, see Chapter 6). The failure rate with respect to common cause failures has been estimated to be $\lambda_C = 3.0 \cdot 10^{-5}$ common cause failures per hour. This type of external stresses affects the system at a rate λ_C irrespective of how many of its items that are functioning. The common cause failure rate must therefore be added to the "individual" failure rate also when only one of the pumps is functioning.

Repair is initiated as soon as one of the pumps fails. The mean downtime of a pump has been estimated to be 15 hours. When both pumps are in a failed state at the same time, the whole process system has to be shut down. In this case the system will not be put into operation again until both pumps have been repaired. The mean downtime when both pumps have failed has been estimated to be 25 hours.

(a) Establish a state transition diagram for the system consisting of the two pumps.

(b) Write down the state equations for the system in matrix format.

(c) Explain what is meant by the steady-state probabilities, and determine the steady-state probabilities for each of the states of the pump system.

(d) Determine the percentage of time when:

   (i) Both the pumps are functioning.

   (ii) Only one of the pumps is functioning.

   (iii) Both pumps are in a failed state.

(e) Determine the mean number of pump repairs that are necessary during a period of 5 years.

(f) How many times must we expect to have a total pump failure (i.e., both pumps in a failed state at the same time) during a period of 5 years.

8.3 The water chlorination system of a small town has two separate pipelines, each with a pump which supplies chlorine to the water at prescribed rates. The two pumps are denoted A and B, respectively. During normal operation both pumps are functioning and thus are sharing the load. In this case each pump is operated on approximately 60% of its capacity (cap% = 0.60). When one of the pumps fails, the corresponding pipeline is closed down, and the other pump has to supply chlorine at a higher rate. In this case the single pump is operated at full capacity (cap% = 1.00). We assume that the pumps have the following constant failure rates:

$$\lambda_{\text{cap}\%} = \text{cap}\% \cdot 6.3 \text{ failures per year}$$

Assume that the probability of common cause failures is negligible. Repair is initiated as soon as one of the pumps fails. The mean time to repair a pump has been estimated to be eight hours, and the pump is put into operation again as soon as the repair is completed. Repairs are carried out independent of each other (i.e., maintenance crew is thus not a limiting factor). If both pumps are in a failed state at the same time, unchlorinated water will be supplied to the customers.

Both pumps are assumed to be functioning at time $t = 0$.

(a) Define the possible system states and establish a state transition diagram for the system.

(b) Write down the corresponding state equations on matrix format.

(c) Determine the steady-state probabilities for each of the system states.

(d) Determine the mean number of pump repairs during a period of 3 years.

(e) Determine the percentage of time exactly one of the pumps is in a failed state.

(f) Determine the mean time to the first system failure, that is, the mean time until unchlorinated water is supplied to the customers for the first time after time $t = 0$.

(g) Determine the percentage of time unchlorinated water is supplied to the customers.

8.4 Consider a parallel structure of three independent and identical components with failure rate λ and repair rate μ. The components are repaired independently. All the three components are assumed to be functioning at time $t = 0$.

(a) Establish the state transition diagram and the state equations for the parallel structure.

(b) Show that the mean time to the first system failure is given by

$$\text{MTTF} = \frac{11}{6\lambda} + \frac{7\mu}{6\lambda^2} + \frac{\mu^2}{3\lambda^3} \qquad (8.111)$$

8.5 Consider a parallel structure of four independent and identical components with failure rate λ and repair rate μ. The components are repaired independently. All the four components are assumed to be functioning at time $t = 0$.

(a) Establish the state transition diagram and the state equations for the parallel structure.

(b) Determine the mean time to the first system failure.

(c) Is it possible to find a general formula for a parallel structure of n components? (Compare with the preceding problem.)

8.6 Fig. 8.21 illustrates a standby system of two compressors and a switching item. When the active compressor A fails, the standby compressor B is to be put into

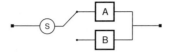

Fig. 8.21 Standby system of two compressors.

operation. Compressor B is assumed not to fail while in passive state (i.e., cold standby).

The probability of successful changeover to the standby compressor B is estimated to be $(1 - p)$. The probability p of unsuccessful changeover also includes the fail-to-start probability for the standby compressor. Compressor A is the main compressor. When we have a system failure (i.e., when both compressors are in a failed state), both compressors are repaired simultaneously, and the system is not started until both compressors have been repaired. The system is always started up again with compressor A as the active compressor. Common cause failures are considered to be negligible.

(a) Define the possible system states and establish a state transition diagram.

(b) The following input data are assumed:

   Failure rate for compressor A: $\lambda_A = 5.0 \cdot 10^{-4}$ failures per hour.
   Failure rate for compressor B: $\lambda_B = 2.0 \cdot 10^{-3}$ failures per hour.
   Mean time to repair compressor A: $\frac{1}{\mu_A} = 25$ hours.
   Mean time to repair both the compressors: $\frac{1}{\mu} = 35$ hours
   Probability of unsuccessful changeover: $p = 0.03$.

   Establish the state equations for the system and determine the steady-state probabilities.

(c) Determine the average availability A_{av} for the system.

(d) How many compressor repairs may we anticipate over a period of 5 years?

(e) How many system failures may we anticipate over a period of 5 years?

(f) Determine the mean time to the first system failure, when both compressors are functioning at time $t = 0$ with compressor A as the operating compressor.

(g) Assume next that we are considering an alternative system where both compressors are operated as an ordinary parallel system (i.e., with active redundancy). All the input data are as above, except that when the two compressors share the load, their failure rates are reduced by 20%. The mean time to repair compressor B is assumed to be 20 hours. Establish the state transition diagram for the alternative system, and compute the availability A_{av}. Discuss the result.

(h) Also determine the mean time to the first system failure for the alternative system in (g).

8.7 Consider a system which is subject to two types of repair. Initially the system has a constant failure rate λ_1. When the system fails for the first time, a partial repair is performed to restore the system to the functioning state. This partial repair is not perfect, and the failure rate λ_2 after this partial repair is therefore larger than λ_1. After the system fails the second time, a thorough repair is performed that restores the system to an "as good as new" condition. The third repair will be a partial repair, and so on. Let μ_1 denote the constant repair rate of a partial repair and μ_2 be the constant repair rate of a complete repair ($\mu_1 > \mu_2$). Assume that the system is put into operation at time $t = 0$ in an "as good as new" condition.

(a) Establish the state transition diagram and the state equations for this process.

(b) Determine the steady-state probabilities of the various states.

9
Reliability of Maintained Systems

9.1 INTRODUCTION

Aspects related to reliability of maintained systems were discussed in Chapters 7 and 8. In Chapter 7 we studied four types of repair processes: homogeneous Poisson processes (HPP), renewal processes, nonhomogeneous Poisson processes (NHPP), and imperfect repair processes. The presentation was restricted to single items, and the main reliability measure was the rate of occurrence of failures, ROCOF. In most of Chapter 7 we only studied corrective maintenance, that is, maintenance (repair) that is carried out after a failure has occurred to bring the item back to a functioning state. In Chapter 8 we discussed reliability assessment of repairable systems by Markov methods. The models were restricted to corrective maintenance, and all functioning times and repair times had to be exponentially distributed.

In this chapter we give a brief general introduction to maintenance concepts that are relevant for system reliability assessment. We discuss how the reliability of a maintained system can be analyzed and give results for some specific maintenance policies, like age and block replacement. We also give an introduction to condition-based replacement policies. We further introduce two approaches to maintenance planning and optimization, reliability centered maintenance (RCM) and total productive maintenance (TPM). The presentation in this chapter is rather brief, and the reader is therefore advised to consult some of the references cited for more details.

Maintenance is defined as "the combinations of all technical and corresponding administrative actions, including supervision actions, intended to retain an entity in, or restore it to, a state in which it can perform its required function" [IEC 50(191)]. The main reliability measure for a maintained item is the *availability* $A(t)$. The avail-

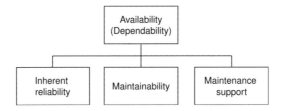

Fig. 9.1 The availability (dependability) as a function of the inherent reliability, the maintainability and the maintenance support.

ability was defined in Chapter 1 as "the ability of an item (under combined aspects of its reliability, maintainability and maintenance support) to perform its required function at a stated instant of time or over a stated period of time" (BS 4778). Some authors and standards use the term *dependability* instead of *availability*. The availability is a function of (i) the (inherent) reliability of the item, (ii) the maintainability of the item, and (iii) the maintenance support, as illustrated in Fig. 9.1. The *maintainability* of an item is "the ability of the item, under stated conditions of use, to be retained in, or restored to, a state in which it can perform its required functions, when maintenance is performed under stated conditions and using prescribed procedures and resources" (BS 4778). The maintainability of an item depends on design factors like ease of access to the item, ease of dismantling, ease of reinstallation, and so on. The maintenance support depends on the maintenance personnel, their availability, skills, and tools and on the availability and quality of spare parts.

Maintenance management has traditionally been a reverse engineering activity, where the decision process has been highly correlated with the technical and mechanical education of the maintenance staff and their own practical experience. Even if accumulated technical experience is essential, it should not be the only basis for maintenance related decisions (Christer 1999; Péres 1996). Maintenance decisions have to take into account a large number of decision criteria that may sometimes be contradictory. To choose the "best" maintenance task at the "best" possible time is a complex task that does not depend only on the current state of the item, but also on future factors like the consequences of this choice for the long term exploitation of the item.

It is often recommended (Christer 1999, Scarf 1997) to establish mathematical models that can be used to assess the impacts of maintenance decisions. This approach seems to give promising results but has not yet been sufficiently developed in an industrial context. By using mathematical/stochastic models it may be possible to "simulate" maintenance strategies and to reveal the associated effects and maintenance costs and operational performance. The simulation may, in some cases, be used to determine the best maintenance strategy to implement.

Some main building blocks of such a mathematical model are presented and discussed in this chapter.

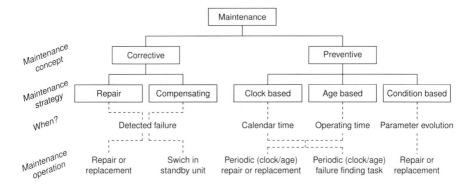

Fig. 9.2 Classification of maintenance types.

9.2 TYPES OF MAINTENANCE

Maintenance tasks may be classified in many different ways. Some of the most common designations are illustrated in Fig. 9.2 and described below:

1. *Preventive maintenance (PM)* is planned maintenance performed when an item is functioning properly to prevent future failures. PM seeks to reduce the probability of failure of the item. It may involve inspection, adjustments, lubrication, parts replacement, calibration, and repair of items that are beginning to wear out. PM is generally performed on a regular basis, regardless of whether or not functionality or performance is degraded. PM tasks can be classified into the following categories:

    (a) *Age-based maintenance*. In this case PM tasks are carried out at a specified age of the item. The age may be measured as time in operation or by other time concepts, like number of kilometers for an automobile or number of take-offs/landings for an aircraft. The age replacement policy discussed in Section 9.6 is an example of age-based maintenance.

    (b) *Clock-based maintenance*. In this case PM tasks are carried out at specified calendar times. The block replacement policy discussed in Section 9.6 is an example of clock-based maintenance. A clock-based maintenance policy is generally easier to administer than an age-based maintenance policy, since the maintenance tasks can be scheduled to predefined times.

    (c) *Condition-based maintenance*. In this case PM tasks are based on measurements of one or more condition variables of the item. Maintenance is initiated when a condition variable approaches or passes a threshold value. Examples of condition variables include vibration, temperature, and number of particles in the lube oil. The condition variables may be monitored continuously or at regular intervals. Condition-based maintenance is also called *predictive maintenance*.

(d) *Opportunity maintenance.* This is applicable for multi-item systems, where maintenance tasks on other items or a system shutdown/intervention provides an opportunity for carrying out maintenance on items that were not the cause of the opportunity.

2. *Corrective maintenance* (CM). This type of maintenance is often called *repair* and is carried out after an item has failed. The purpose of corrective maintenance is to bring the item back to a functioning state as soon as possible, either by repairing or replacing the failed item or by switching in a redundant item. Corrective maintenance is also called *breakdown maintenance* or *run-to-failure maintenance*.

3. *Failure-finding maintenance.* This is a special type of preventive maintenance that involves functional and operational checks or tests to verify proper operation of off-line functions, like protective devices or backup systems. Failure-finding tasks are carried out to reveal *hidden* failures that have already occurred. Failure-finding maintenance is generally carried out on a regular basis, with a specified interval between failure-finding tasks. Failure-finding maintenance is further discussed in Chapter 10.

9.3 DOWNTIME AND DOWNTIME DISTRIBUTIONS

The *downtime* of an item (component or system) is the time in a specified *mission period* where the item is not able to perform one or more of its intended functions.

9.3.1 Planned vs. Unplanned Downtime

The downtime can be split in two types:

1. *Unplanned downtime* is the downtime caused by item failures and internal and external (random) events, for example, human errors, environmental impacts, loss of utility functions, labor conflicts (strikes), and sabotage. In some applications (e.g., electro-power generation) the unplanned downtime is called the *forced outage* time.

2. *Planned downtime* is the downtime caused by planned preventive maintenance, planned operations (e.g., change of tools), and planned breaks, holidays, and the like. What is to be included as planned downtime will depend on how the mission period is defined. We may, for example, define the mission period as one year (8760 hours) or the net planned time in operation during one year, excluding all holidays and breaks, and all planned operational stops. In some applications it is common to split the unplanned in two types: (i) *scheduled downtime* that is planned a long time in advance (e.g., planned preventive maintenance, breaks, and holidays) and (ii) *unscheduled planned downtime* initiated by condition monitoring, detection of incipient failures, and events

that may require a preventive task to improve or maintain the quality of the system functions or to reduce the probability of a future failure. The associated remedial tasks can usually be postponed (within some limits) and carried out when it is suitable from an operational point of view.

The scheduled downtime can often be regarded as deterministic, and can be estimated from the operational plans. The unscheduled planned downtime may be subject to random variations, but it will usually be rather straightforward to estimate a mean value.

The unplanned downtime is generally strongly dependent of the cause of the downtime. Assume that we have identified n independent causes of unplanned downtime, and let D_i be the random downtime associated to cause i for $i = 1, 2, \ldots, n$. Let $F_{D_i}(d)$ denote the distribution function of D_i, and let p_i be the probability that a specific downtime has cause i. The distribution of the downtime D is then $F_D(d) = \sum_{i=1}^{n} p_i \cdot F_{D_i}(d)$, and the mean downtime is

$$\text{MDT} \approx \sum_{i=1}^{n} p_i \cdot \text{MDT}_i$$

where $\text{MDT}_i = E(D_i)$ denotes the mean downtime associated with cause i for $i = 1, 2, \ldots, n$.

9.3.2 Downtime Caused by Failures

In the following we will confine ourselves to discussing the downtime caused by item failures and assume that the planned downtime and the unplanned downtime from other causes are treated separately. When we use the term *downtime* in the following, we tacitly assume that the downtime is caused by item failures.

The downtime of an item can usually be regarded as a sum of elements like access time, diagnosis time, active repair time, checkout time, and so on. The elements are further discussed by Smith (1997). The length of the various elements are influenced by a number of system-specific factors, like ease of access, maintainability, and availability of maintenance personnel, tools, and spare parts. The downtime associated to a specific failure therefore has to be estimated based on knowledge of all these factors.

For detailed reliability assessments, it is important to choose an adequate downtime distribution as basis for the estimation. Three distributions are commonly used: the exponential, the normal, and the lognormal distribution (Ebeling 1997). We will briefly discuss the adequacy of these distributions.

Exponential Distribution The exponential distribution is the most simple downtime distribution we can choose, since it has only one parameter, the *repair rate* μ. The exponential distribution was discussed in detail in Section 2.9. We will briefly mention some of the main features of the exponential distribution.

The mean downtime is MDT = $1/\mu$, and the probability that a downtime D is longer than a value d is $\Pr(D > d) = e^{-\mu d}$. The exponential distribution has no memory. This implies that if a downtime has lasted a time d, the mean *residual* downtime is $1/\mu$ regardless of the value of d. This feature is not realistic for most downtimes, except for situations where the main part of the downtime is spent on search for failures, and where failures are found more or less at random.

In many applications the exponential distribution is chosen as a downtime distribution, not because it is realistic, but because it is easy to use.

Example 9.1
Consider a repairable item with downtime D related to a specific type of failure. The downtime is assumed to be exponentially distributed with repair rate μ. The MDT for this specific type of failure has been estimated to be 5 hours. The repair rate is then $\mu = 1/\text{MDT} = 0.20$ hours^{-1}. The probability that the downtime, D, is longer than 7 hours is $\Pr(D > 7) = e^{-7\mu} \approx 0.247 = 24.7\%$. □

Normal Distribution The rationale for choosing a normal (Gaussian) downtime distribution is motivated by the fact that the downtime may be considered as a sum of many independent elements. The normal distribution was discussed in Section 2.13. Estimation of the MDT and the standard deviation is straightforward in the normal model. When using the normal distribution, the repair rate function $\mu(d)$ may be approximated by a straight line as a function of the elapsed downtime d. The probability of being able to complete an ongoing repair task within a short interval will therefore increase almost linearly with time.

Lognormal Distribution The lognormal distribution is often used as a model for the repair time distribution. The lognormal distribution was discussed in Section 2.14. When using the lognormal distribution, the repair rate $\mu(d)$ increases up to a maximum, and thereafter decreases asymptotically down to zero as a function of the elapsed downtime d. When an item has been down for a very long time, this indicates serious problems, for example, that there are no spare parts available on the site or that the maintenance crew is not able to get access to or correct the failure. It is therefore natural to believe that the repair rate is decreasing after a certain period of time.

9.3.3 Mean System Downtime

Consider a series structure of n *independent* components. Component i has constant failure rate λ_i. When component i fails, the system will have a downtime MDT$_i$, for $i = 1, 2, \ldots, n$. The probability that the system failure is caused by component i is $\lambda_i / \sum_{j=1}^{n} \lambda_j$, and the mean system downtime for an unspecified failure is

$$\text{MDT} \approx \frac{\sum_{i=1}^{n} \lambda_i \cdot \text{MDT}_i}{\sum_{j=1}^{n} \lambda_j} \tag{9.1}$$

The MDT is equal to the right-hand side of (9.1) only when the failure of a component prevents the other components from failing, in which case the components are not independent. The exact formula is given in (8.56). In most applications, however, equation (9.1) gives a very good approximation.

Example 9.2
Consider an item with n independent failure modes. Failure mode i occurs with constant failure rate λ_i, and the mean downtime required to restore the item from failure mode i is MDT_i for $i = 1, 2, \ldots, n$. The item may be considered as a series system of n independent virtual components, where component i only can fail with failure mode i. The mean downtime of the item is therefore given by (9.1). □

Equation (9.1) may also be used as an approximation for the mean downtime caused by an unspecified component failure of a nonseries structure of independent components. In this case it is important to realize that MDT_i denotes the *system* downtime caused by failure of component i for $i = 1, 2, \ldots, n$.

9.4 AVAILABILITY

Consider a repairable item that is put into operation at time $t = 0$. When the item fails, a repair action is initiated to restore the function of the item. The state of the item at time t is given by the state variable:

$$X(t) = \begin{cases} 1 & \text{if the item is functioning at time } t \\ 0 & \text{otherwise} \end{cases}$$

In this section we only consider the unplanned downtime caused by failures. The mean time to repair the item is denoted MTTR. The total mean downtime, MDT, or mean forced outage time, is the mean time the item is in a nonfunctioning state. The MDT is usually significantly longer than the MTTR, and will include time to detect and diagnose the failure, logistic time, and time to test and startup of the item. When the item is put into operation again it is considered to be "as good as new." The mean up-time, MUT, of the item is equal to what we in Chapter 2 called mean time to failure, MTTF. Both concepts may be used, but MUT is more commonly used in maintenance applications. The mean time between consecutive occurrences of failures is denoted MTBF. The state variable and the various time concepts are illustrated in Fig. 9.3.

The reliability of a repairable item may be measured by the *availability* of the item at time t.

Definition 9.1 The availability $A(t)$ at time t of a repairable item is the probability that the item is functioning at time t:

$$A(t) = \Pr(X(t) = 1) \tag{9.2}$$

□

368 RELIABILITY OF MAINTAINED SYSTEMS

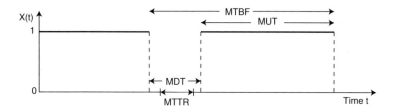

Fig. 9.3 Average "behavior" of a repairable item and main time concepts.

$A(t)$ is sometimes referred to as the point availability. Note that if the item is not repaired, then $A(t) = R(t)$, the *survivor* function.

Definition 9.2 The unavailability $\bar{A}(t)$ at time t of a repairable item is the probability that the item is not in a functioning state at time t:

$$\bar{A}(t) = 1 - A(t) = \Pr(X(t) = 0) \tag{9.3}$$

□

Sometimes we are interested in the *interval* or *mission availability* in the time interval (t_1, t_2), defined by:

Definition 9.3 The (average) interval or mission availability $A_{\text{av}}(t_1, t_2)$ in the time interval (t_1, t_2) is

$$A_{\text{av}}(t_1, t_2) = \frac{1}{t_2 - t_1} \int_{t_1}^{t_2} A(t)\, dt \tag{9.4}$$

□

$A_{\text{av}}(t_1, t_2)$ is just the average value of the point availability $A(t)$ over a specified interval (t_1, t_2).

In some applications we are interested in the interval or mission availability from startup, that is, in an interval $(0, \tau)$. This is defined as

$$A_{\text{av}}(\tau) = \frac{1}{\tau} \int_0^\tau A(t)\, dt \tag{9.5}$$

The average availability $[A_{\text{av}}(t_1, t_2)$ and $A_{\text{av}}(\tau)]$ may be interpreted as the mean proportion of time in the interval where the item is able to function.

When $\tau \to \infty$ the average interval availability (9.5) will approach a limit called the long run *average availability* of the item.

Definition 9.4 The long run average availability is

$$A_{\text{av}} = \lim_{\tau \to \infty} A_{\text{av}}(\tau) = \lim_{\tau \to \infty} \frac{1}{\tau} \int_0^\tau A(t)\, dt \tag{9.6}$$

The long run average availability A_{av} may be interpreted as the average proportion of a long period of time where the item is able to function.

The long-term average unavailability $\bar{A}_{\text{av}} = 1 - A_{\text{av}}$ is in some application areas (e.g., electro-power generation) called the *forced outage rate*.

Example 9.3
Consider a time interval of 1000 hours and let the steady state (average) availability of a certain item in this interval be 0.95. We would then expect the item to be able to function 950 of those 1000 hours. Note that the availability does not tell anything about how many times the item will fail in this interval. □

When the up-times and downtimes are nonlattice (see Definition 7.2), the point availability $A(t)$ will approach a limit A when $t \to \infty$. The limit A is called the *limiting availability* of the item.

Definition 9.5 The limiting availability is

$$A = \lim_{t \to \infty} A(t) \tag{9.7}$$

when the limit exists. □

The limiting availability is sometimes called the *steady-state availability*. When the limiting availability exists, it is equal to the long run average availability, that is, $A_{\text{av}} = A$.

9.4.1 Availability with Perfect Repair

Consider a repairable item that is put into operation and is functioning at time $t = 0$. Whenever the item fails, it is replaced by a new item of the same type or repaired to an "as good as new" condition. We then get a sequence of lifetimes or *up-times* $T_1, T_2, \ldots$ for the items. We assume that $T_1, T_2, \ldots$ are independent and identically distributed, with distribution function $F_T(t) = \Pr(T_i \leq t)$, $i = 1, 2, \ldots$, and mean time to failure MTTF.

We further assume that the downtimes $D_1, D_2, \ldots$ are independent and identically distributed with distribution function $F_D(t) = \Pr(D_i \leq t)$ for $i = 1, 2, \ldots$, and mean downtime MDT. Finally we assume that $T_i + D_i$ for $i = 1, 2, \ldots$ are independent. The state variable $X(t)$ of the item is illustrated in Fig. 9.4.

Suppose that we have observed an item until repair n is completed. Then we have obtained the lifetimes $T_1, T_2, \ldots, T_n$ and the downtimes $D_1, D_2, \ldots, D_n$. According to the law of large numbers, then under relatively general assumptions (e.g., see Dudewicz and Mishra 1988, p. 302), with probability one

$$\frac{1}{n} \sum_{i=1}^{n} T_i \to E(T) = \text{MTTF} \quad \text{when } n \to \infty$$

$$\frac{1}{n} \sum_{i=1}^{n} D_i \to E(D) = \text{MDT} \quad \text{when } n \to \infty$$

370 RELIABILITY OF MAINTAINED SYSTEMS

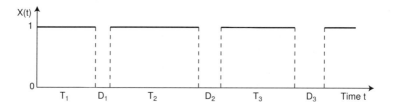

Fig. 9.4 States of a repairable item.

The proportion of the time in which the item has been functioning is

$$\frac{\sum_{i=1}^{n} T_i}{\sum_{i=1}^{n} T_i + \sum_{i=1}^{n} D_i} = \frac{(1/n) \sum_{i=1}^{n} T_i}{(1/n) \sum_{i=1}^{n} T_i + (1/n) \sum_{i=1}^{n} D_i} \qquad (9.8)$$

Heuristically we may expect that the right-hand side of (9.8) will tend to

$$\frac{E(T)}{E(T) + E(D)} \quad \text{as } n \to \infty \qquad (9.9)$$

which is the average proportion of time where the item has been functioning, when we consider a long period of time. We have therefore found the *long run average availability* (see, Definition 9.4) of the item

$$A_{\text{av}} = \frac{E(T)}{E(T) + E(D)} = \frac{\text{MTTF}}{\text{MTTF} + \text{MDT}} \qquad (9.10)$$

Example 9.4
A machine with MTTF = 1000 hours and MDT = 5 hours, has average availability

$$A_{\text{av}} = \frac{\text{MTTF}}{\text{MTTF} + \text{MDT}} = \frac{1000}{1000 + 5} = 0.995$$

On the average the machine will function 99.5% of the time. The average unavailability is thus 0.5% which corresponds to approximately 44 hours of downtime per year, when the machine is supposed to run continuously. □

Example 9.5
Consider a repairable item where the up-times are independent and exponentially distributed with failure rate λ. The downtimes are assumed to be independent and exponentially distributed with parameter μ. The mean downtime is thus

$$\text{MDT} = 1/\mu$$

The parameter μ is called the *repair rate*, even if this term is in conflict with the time concepts we introduced in Section 9.3.

In Chapter 8 we showed (8.21) that the availability $A(t)$ of the item is

$$A(t) = \frac{\mu}{\lambda + \mu} + \frac{\lambda}{\lambda + \mu} e^{-(\lambda + \mu)t} \qquad (9.11)$$

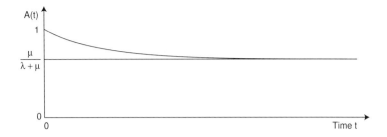

Fig. 9.5 The availability $A(t)$ of a item in Example 9.5, with failure rate λ and repair rate μ.

The availability $A(t)$ is illustrated in Fig. 9.5.

For these up-time and downtime distributions the availability $A(t)$ is seen to approach a constant A when $t \to \infty$.

$$A = \lim_{t \to \infty} A(t) = \frac{\mu}{\lambda + \mu} = \frac{1/\lambda}{1/\lambda + 1/\mu} = \frac{\text{MTTF}}{\text{MTTF} + \text{MDT}} \quad (9.12)$$

The limiting availability A (see Definition 9.5) therefore exists and is equal to the average availability A_{av}.

When the item is not repaired, that is, when $\mu = 0$, then the availability is equal to the survivor function $R(t)$.

$$A(t) = R(t) = e^{-\lambda t} \quad \text{when} \quad \mu = 0$$

□

Example 9.6
Consider an item with independent up-times with constant failure rate λ. The downtimes are independent and identically distributed with mean MDT. Since usually MDT $\ll$ MTTF the average unavailability of the item is approximately

$$\bar{A}_{\text{av}} = \frac{\text{MDT}}{\text{MTTF} + \text{MDT}} = \frac{\lambda \cdot \text{MDT}}{1 + \lambda \cdot \text{MDT}} \approx \lambda \cdot \text{MDT} \quad (9.13)$$

This approximation is often used in hand calculations. □

When planning supplies of spare parts, it is of interest to know how many failures that may be expected in a given time interval. Let $W(t)$ denote the mean number of repairs carried out in the time interval $(0, t)$. Obviously $W(t)$ will then depend on the distributions of the up-times and the downtimes. It is often difficult to find an exact expression for $W(t)$ (see Chapter 7). When t is relatively large, however, the following approximation may be used:

$$W(t) \approx \frac{t}{\text{MTTF} + \text{MDT}} \quad (9.14)$$

9.4.2 Operational Availability

The *operational availability* A_{OP} of a item is defined as the mean proportion of a mission period the item is able to perform its intended functions. To determine A_{OP} we have to specify the mission period and estimate the mean total planned downtime and the mean total unplanned downtime in the mission period. These concepts were discussed in Section 9.3. See also Ebeling (1997). The *operational unavailability* $\bar{A}_{OP} = 1 - A_{OP}$ may be determined from

$$\bar{A}_{OP} = \frac{\text{Mean total planned downtime} + \text{Mean total unplanned downtime}}{\text{Mission period}}$$

When using the concepts availability and operational availability we only consider two states: a functioning state and a failed state. The output from a production system may vary a lot, and the availability is therefore not a fully adequate measure of the system's performance. Several alternative measures have been proposed. In NORSOK Z-016 the concept production *regularity* is introduced as a measure of the operational performance of an oil/gas production system. Regularity is a term used to describe the capability of a production system of meeting demands for deliveries or performance. The NORSOK standard was developed for the oil and gas industry, but most of the concepts may also be used in other industries. Production regularity is further discussed by Kawauchi and Rausand (2002).

Alternative Metrics Several metrics have been proposed for operational performance. Among these are:

Deliverability is defined by NORSOK Z-016 as the ratio between the actual deliveries and the planned/agreed deliveries over a specified period of time, when the effect of compensating elements such as substitution from other producers and downstream buffer storage are included:

$$\text{Deliverability} = \frac{\text{Actual deliveries}}{\text{Planned or agreed deliveries}}$$

The deliverability is a measure of the system's ability to meet demands agreed with a customer. Failures and other problems in the production system may be compensated using products from a storage or by purchasing products from other suppliers. The North Sea operators supply gas to Europe through subsea pipelines. The deliverability is measured at the interface between the subsea pipeline and the national gas network (e.g., in Germany). A relatively short downtime of a production unit will have no effect on the outlet of the pipeline due to the large volume of gas and the high pressure in the pipeline. A longer downtime may be compensated by increasing the gas production from other production units, connected to the same pipeline.

The *on-stream availability*, OA, is defined as the mean proportion of time, in a specified time period, in which the production (delivery) is greater than zero. In this case,

$1 - OA$ denotes the mean proportion of time the system is not producing at all.

The *100% production availability*, A_{100}, in a time interval (t_1, t_2) is defined as the mean proportion of the time in this interval the system is producing with full production (time is measured in hours):

$$A_{100} = \frac{\text{No. of hours in the interval } (t_1, t_2) \text{ with full production}}{t_2 - t_1}$$

In this definition we are only concerned with full production. We do not distinguish between 90% production and no production. We may also define the production availability at a reduced capacity, for example, 80%:

$$A_{80} = \frac{\text{No. of hours in } (t_1, t_2) \text{ the system is producing with } \geq 80\% \text{ capacity}}{t_2 - t_1}$$

9.5 SYSTEM AVAILABILITY ASSESSMENT

9.5.1 Introduction

The availability of a system may be analyzed by several different approaches. The most commonly used approaches are:

1. *Reliability block diagrams.* The reliability block diagram approach may be used to find an approximative value for the average availability of a system of independent components that are restored to an "as good as new" condition after each failure. The approach is easy to use and may give adequate average results for some systems. A brief introduction to the reliability block diagram approach is given in Section 9.5.2.

   The reliability block diagram approach requires each component to have only two possible states, a functioning state and a failed state. The availability is found from the structure function by using the formulas described in Chapter 4.

   A reliability block diagram is a static "picture" of a system's ability to perform a specified function, and is therefore not suitable as a model for a repairable system with a complex maintenance strategy.

2. *Fault tree.* Availability assessment based on a fault tree model is similar to the approach based on reliability block diagrams. Most computer programs for fault tree analysis can be used for repairable systems, with the same limitations as described for reliability block diagrams.

3. *Markov methods.* Markov methods can be used to analyze repairable systems with rather complex repair strategies. All failure rates and repair rates are assumed to be constant, and it is not possible to take into account any long-term trends or seasonal effects. The number of system states increases rapidly with the number of components, and the workload may therefore be overwhelming

even for systems of moderate complexity. The average proportion of time the system spends in the various states (the steady-state probabilities) can be calculated. If we know the production (system output) in the various system states, this may be combined with the steady-state probabilities as input to a production regularity assessment. Markov methods were discussed in Chapter 8.

4. *Flow networks.* A flow network is a graph describing all possible paths from one or more input nodes to one or more output nodes. The various paths may have different capacities. An example of a flow network is a water distribution network, from one or more sources to one or more customers. The water pipes may have different diameters and different pressures. A flow network may be analyzed by graph theory methods (see Aven 1992).

   A simplified single source – single terminal flow network may be drawn as a reliability block diagram where the components have different capacities. The approach described in Chapter 4 based on structure functions can no longer be used. Such networks are usually analyzed by Monte Carlo simulation (see Section 9.5.3). Several computer programs have been developed for this purpose. This approach represents an extension to the reliability block diagram approach but do not have the same dynamic properties as Markov methods. The approach is not further discussed in this book.

5. *Petri nets* Petri nets are rather flexible and can be used to analyze systems with rather complex repair strategies. Petri nets for availability assessment are discussed by Signoret (1986) but are not further discussed in this book.

6. *Monte Carlo next event simulation.* Monte Carlo next event simulation is the most flexible approach to availability assessment of repairable systems, and can be used to analyze almost any type of systems. The simulation must be carried out by a computer simulation program, and the program may set certain limitations. The approach is further discussed on in Section 9.5.3.

A listing of computer programs for the various approaches may be found on the book's web page.

9.5.2 Reliability Block Diagrams

Consider a system with n components and structure function $\phi(X(t))$. If the state variables $X_1(t), X_2(t), \ldots, X_n(t)$ are independent random variables, the system availability, $A_S(t)$, can be determined by the procedure described in Section 4.2:

$$A_S(t) = E(\phi(X(t))) \tag{9.15}$$

Example 9.7
The system in Fig. 9.6 has structure function

$$\phi(X(t)) = X_1(t)(X_2(t) + X_3(t) - X_2(t)X_3(t)) \tag{9.16}$$

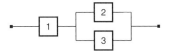

Fig. 9.6 Reliability block diagram (Example 9.7).

The three components are assumed to fail and be repaired independent of each other. We want to determine the average availability of the system. The MTTFs and MDTs of the three components are listed below, together with the average component availabilities, calculated by

$$A_{\text{av},i} = \frac{\text{MTTF}_i}{\text{MTTF}_i + \text{MDT}_i} \quad \text{for } i = 1, 2, 3$$

To simplify the notation we often omit the explicit reference to average (av), and write A_i instead of $A_{\text{av},i}$.

i	MTTF$_i$ (hours)	MDT$_i$ (hours)	A_i
1	1000	10	0.990
2	500	10	0.980
3	500	10	0.980

The average availability of the system is

$$A_S = A_1(A_2 + A_3 - A_2 A_3) \approx 0.9896$$

The average system unavailability is thus $\bar{A}_S \approx 0.0104$ which corresponds to approximately 91 hours of downtime per year (when the system is supposed to be operated on a continuous basis). □

To use the approach in Example 9.7 we have to assume that the system components fail and are repaired independent of each other. This means that when a component is down for repair, all the other components continue to operate as if nothing had happened. This assumption is often not realistic. Some types of dependencies may be modeled by using the methods in Chapter 6, but generally this approach will not give fully realistic results.

The approach outlined in Example 9.7 is often used in practical analyses because it is easy to use. Most of the computer programs for fault tree analysis analyze repairable systems by the fault tree equivalent of the approach in Example 9.7. When using this approach we should be aware of the inherent limitations in the approach: All components are assumed to fail and be repaired independent of each other, meaning that:

- The operation and maintenance of a component are not influenced by the status of the other components.

- It is not possible to model load-sharing systems (where the load on a component increases if one or more redundant components fail).
- It is not possible to merge repair actions and start up again only when a certain number of components are able to function.
- There is no limitation with respect to repair resources.

Some of these limitations may partly be overcome by careful modeling of the fault tree (or the reliability block diagram), but this is not a straightforward task and is usually not done in practical analyses.

A fault tree (reliability block diagram) only provides a *static* picture of the causes of system failure (requirements for system function) and is therefore not made for analyses of systems with dynamic features, like systems subject to maintenance/repair.

9.5.3 Monte Carlo Next Event Simulation

Monte Carlo next event simulation is carried out by simulating "typical" lifetime scenarios for a system on a computer. We start with a model of the system, usually in the form of flow diagrams and reliability block diagrams. Random events (i.e., events associated to item failures) are generated in the computer model, and scheduled events (e.g., preventive maintenance actions), and conditional events (i.e., events initiated by the occurrence of other events) are included to create a simulated lifetime scenario that is as close to a real lifetime scenario as possible. Several types of input data have to be available:

- A description of the system based on flow diagrams, control schematics, and component information
- Knowledge of component failure modes, failure effects, and failure consequences, usually in the form of a failure modes, effects, and criticality analysis (FMECA)
- Component failure and repair data (failure mode specific life and downtime distributions and estimates of the required parameters)
- Intervention and repair strategies and durations for the various failure modes
- Frequency and duration of inspections and planned maintenance actions
- Opportunity maintenance strategies
- Resource data (e.g., availability of spare parts and maintenance resources)
- Throughput data and system/component capacities

When a "typical" lifetime scenario has been simulated on the computer, this scenario is treated as a real experiment, and performance measures are calculated. We may, for example, calculate

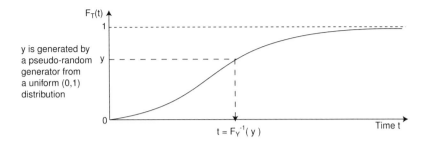

Fig. 9.7 Generation of a random variable with distribution $F_T(t)$.

- The observed availability of the system in the simulated time period (e.g., the observed uptime divided by the length of the simulated period)
- The number of system failures
- The number of failures for each component
- The contribution to system unavailability from each component
- The use of maintenance resources
- The system throughput (production) as a function of time

The simulation can be repeated to generate a number of "independent" lifetime scenarios. From these scenarios we may deduct estimates of the performance measures of interest.

Generation of Random Variables with a Specified Distribution Let T denote a random variable, not necessarily a time to failure, with distribution function $F_T(t)$ which is strictly increasing for all t, such that $F_T^{-1}(y)$ is uniquely determined for all $y \in (0, 1)$. Further let $Y = F_T(T)$. Then the distribution function $F_Y(y)$ of Y is

$$\begin{aligned} F_Y(y) &= \Pr(Y \leq y) = \Pr(F_T(T) \leq y) \\ &= \Pr(T \leq F_T^{-1}(y)) = F_T(F_T^{-1}(y)) = y \quad \text{for } 0 < y < 1 \end{aligned}$$

Hence $Y = F_T(T)$ has a uniform distribution over $(0, 1)$. This implies that if a random variable Y has a uniform distribution over $(0, 1)$, then $T = F_T^{-1}(Y)$ has the distribution function $F_T(t)$.

This result can be used to generate random variables $T_1, T_2, \ldots$ with a specified distribution function $F_T(t)$ on a computer. Variables $Y_1, Y_2, \ldots$, which are uniformly distributed over $(0, 1)$, may be generated by a pseudo-random number generator. The variables $T_i = F_T^{-1}(Y_i)$ for $i = 1, 2, \ldots$, will then have distribution function $F_Y(t)$. The generation of random variable is illustrated in Fig. 9.7. Alternative methods to generate random variables from specific distribution classes are discussed, for example, by Ripley (1987). A wide range of pseudo-random number generators are available (e.g., as part of computer programs for statistical analysis). Most of these pseudo-random number generators are able to generate variables $Y_1, Y_2, \ldots$ that are approximately independent with a uniform distribution over (0, 1).

Next Event Simulation We will illustrate the next event simulation by a very simple example, a single repairable item with only one failure mode. A lifetime scenario for the item may be simulated as follows:

1. The simulation is started at time $t = 0$ (the simulator clock is set to 0 that may correspond to a specified date). The item is assumed to be functioning at time $t = 0$.

2. The time t_1 to the first failure is generated from the life distribution $F_{T_1}(t)$. The life distribution $F_{T_1}(t)$ has to be specified by the analyst. The simulator clock is now set to t_1.

3. The repair or restoration time d_1 is generated from a specified repair time distribution $F_{D_1}(d)$. The repair time distribution $F_{D_1}(d)$ has to be specified by the analyst, and may, for example, depend on the season (the date) and the time of the day of the failure. The repair time may, for example, be longer for a failure that occurs during the night than for the same failure occurring during ordinary working hours. The simulator clock is now set to $t_1 + d_1$

4. The time t_2 to the second failure is generated from a the life distribution $F_{T_2}(t)$. The item may not be "as good as new" after the repair action and the life distribution $F_{T_2}(t)$ may therefore be different from $F_{T_1}(t)$. The simulator clock is set to $t_1 + d_1 + t_2$.

5. The repair or restoration time d_2 is generated from a specified repair time distribution $F_{D_2}(d)$.

The simulation is continued until the simulator clock reaches a predefined time, for example, 10 years. The computer creates a chronological log file where all events (failures, repairs) and the (simulator clock) time for each event is recorded. From this log file we are able to calculate the number of failures in the simulated period, the accumulated use of repair resources and utilities, the observed availability, and so on for this specific life scenario. The observed availability A_1 is, for example, calculated as the accumulated time the item has been functioning divided by the length of the simulated period.

The simulation described above is repeated n times (with different *seed* values), and the parameters of interest are calculated for each simulation. Let A_i be the observed availability in simulation i for $i = 1, 2, \ldots, n$. The average availability A of the item is then calculated as the sample mean $\sum_{i=1}^{n} A_i/n$. The sample standard deviation may be used as a measure of the uncertainty of A. It is also possible to split the simulation period into a number of intervals and calculate the average availability within each interval. The availability may, for example, be reported per year. A variety of approaches to reduce the variation in the estimates are available. Variance reduction methods are discussed, for example, by Mitrani (1982) and Ripley (1987).

The simulation on a computer can theoretically take into account virtually any aspects and contingencies of an item:

- Seasonal and daily variations
- Variations in loading and output
- Periodic testing and interventions into the item
- Phased mission schemes
- Planned shutdown periods
- Interactions with other components and systems
- Dependencies between functioning times and downtimes

Simulation of a life scenario for a complex system requires a lot of input data to the computer. In addition, we have to establish a set of decision rules for the various events and combinations of events. These rules must state which actions should be a consequence of each event. Examples of such decision rules are

- Setting priorities between repair actions of simultaneous failures when we have limited repair resources
- Switching policies between standby items
- Deciding to replace or refurbish some additional components of the same subsystem when a component fails
- Deciding to shut down the whole subsystem after a failure of a component, until repair action of the component is completed

To obtain estimates of satisfactory accuracy we have to simulate a rather high number of life histories of the system. The number of replicated simulations will depend on the system complexity and the reliabilities of the various system components. Systems with a high reliability will in general require more replications than systems with low reliability. The simulation time will be especially long when the model involves extremely rare events with extreme consequences. For complex systems, we may need several thousands of replications. The simulation time will often be excessive even on a fast computer, and the log file may be very large.

A number of simulation programs have been developed for availability—and production regularity—assessment of specific systems. A listing of relevant programs may be found on the book's web page.

Example 9.8
Consider a system of two production items as illustrated in Fig. 9.8. When both items are functioning, 60% of the system output comes from item 1 and 40% from item 2. The system is started up on a specific date (e.g., 1 January 2010). The times to failure are assumed to be independent and Weibull distributed with known parameters

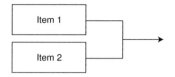

Fig. 9.8 System of two production items.

(α_i, λ_i), for $i = 1, 2$. The simulation is started by generating two Weibull distributed times to failure t_1 and t_2. Let us assume that $t_1 < t_2$. From time t_1, item 1 is out of operation during a random downtime that has a lognormal distribution with known parameters (ν_1, τ_1) that depend on the date at which item 1 failed. The production from item 2 is increased to 60% to partly compensate for the outage of item 1. The time to failure of item 2 with 60% production is Weibull distributed with parameters (α_2, λ_2^1). (A conditional Weibull distribution might be selected.) The next step of the simulation is to generate the repair time d_1 of item 1, and the time to failure t_2^1 of item 2 with increased production. Let us assume that $d_1 < t_2^1$. At time d_1 item 1 is put into operation again, with 60% production and the load on item 2 is reduced to 40%. Time to failure distributions are allocated to the two items. Conditional distributions, given the time in operation, may be used. New times to failure are generated according to the same procedure as described above. Periodic stops with adjustments, cleaning, and lubrication may easily be included in the simulation. If item 2 fails, the load on item 1 is increased to 80%. The simulation is illustrated in Fig. 9.9 together with the resulting simulated production. Several other metrics may be recorded, such as total item downtime, use of repair resources and spare parts. The simulation of times to failure may further be split into different failure modes. The simulation is repeated a large number of times to give average values. □

9.6 PREVENTIVE MAINTENANCE POLICIES

In this section we present and discuss some preventive maintenance policies.[1] We start with the classical time-based age and block replacement policies and present some examples and extensions to these policies. We then present a condition-based replacement policy where the item (component, system) is inspected at regular intervals, and replacement is decided based on measurement of a deterioration variable. The presentation in this section is rather brief. A thorough introduction to preventive maintenance modeling is given, for example, by Barlow and Proschan (1965), Pierskalla and Voelker (1979), Valdez-Flores and Feldman (1989), Cho and Parlar (1991), Gertsbakh (2000), and Wang (2002). A description of condition-based maintenance modeling is given, for example, by Castanier (2001).

---

[1] Section 9.6 is coauthored by Dr. **Bruno Castanier**, Ecole des Mines de Nantes.

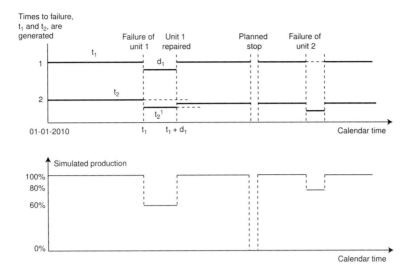

Fig. 9.9 Simulation of the performance of the production system in Fig. 9.8.

We also present a specific inspection and replacement model; the *PF interval* model. The PF interval model is commonly used in reliability centered maintenance (RCM) which is discussed in Section 9.7.

Throughout this section it is important to specify clearly which time concept we are using. The time may be measured as calendar time, operating time, and other time concepts like kilometers driven by a car. It may also be relevant to combine different time criteria, and, for example, say that a part of a car should be replaced every 80000 kilometers and/or every 4 years, whichever come first.

9.6.1 Age Replacement

Under an *age replacement* policy an item (component, system) is replaced upon failure or at a specified operational age t_0, whichever comes first. This policy makes sense if the failure replacement cost is higher than the cost of a planned replacement, and if the failure rate of the item is increasing.

Consider a process where the item is subject to age replacement at age t_0, which is nonrandom. Let T denote the (potential) time to failure of the item. T is assumed to be continuous with distribution function $F(t)$, density $f(t)$, and mean time to failure (MTTF). The time required to replace the failed item is considered to be negligible, and after replacement, the item is assumed to be "as good as new." The time between two consecutive replacements is called a *replacement period*. The mean time between replacements/renewals (MTBR) with replacement age t_0 is

$$\text{MTBR}(t_0) = \int_0^{t_0} t\, f(t)\, dt + t_0 \cdot \Pr(T \geq t_0) = \int_0^{t_0} (1 - F(t))\, dt \qquad (9.17)$$

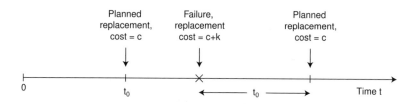

Fig. 9.10 Age replacement policy and costs.

Notice that MTBR(t_0) is always less than t_0, and that $\lim_{t_0 \to \infty}$ MTBR(t_0) = MTTF. The mean number of replacements, $E_{t_0}(N(t))$, in a long time interval of length t is therefore approximately

$$E_{t_0}(N(t)) \approx \frac{t}{\text{MTBR}(t_0)} = \frac{t}{\int_0^{t_0}(1 - F(t))\,dt} \tag{9.18}$$

When the item has reached the age t_0, the cost of a preventive replacement is c, and the cost of replacing a failed item (before age t_0) is $c + k$. The cost c covers the hardware and man-hour costs, while k is the extra cost incurred by the unplanned replacement, such as production loss. The costs are illustrated in Fig. 9.10. By the age replacement policy the replacement times cannot be fully scheduled, and the policy may therefore be complex to administer when we have a high number of items. The age of each item has to be monitored, and the replacement actions will be spread out in time.

The total cost per replacement period is equal to the replacement cost c plus the extra cost k whenever a failure occurs. The mean total cost per replacement period is thus

$$c + k \cdot \Pr(\text{"failure"}) = c + k \cdot \Pr(T < t_0) = c + k \cdot F(t_0)$$

The total mean cost per time unit $C_A(t_0)$ with replacement age t_0 is determined by

$$C_A(t_0) \cdot \text{MTBR}(t_0) = c + k \cdot F(t_0)$$

Hence

$$C_A(t_0) = \frac{c + k \cdot F(t_0)}{\int_0^{t_0}(1 - F(t))\,dt} \tag{9.19}$$

The objective is now to determine the age t_0 that minimizes $C(t_0)$. A graphical approach to finding the optimal t_0 is shown in Example 11.12.

If we let $t_0 \to \infty$ in (9.19), we get

$$C_A(\infty) = \lim_{t_0 \to \infty} C_A(t_0) = \frac{c + k}{\int_0^\infty (1 - F(t))\,dt} = \frac{c + k}{\text{MTTF}} \tag{9.20}$$

When $t_0 \to \infty$, this means that no age replacement takes place. All replacements are corrective replacements and the cost of each replacement is $c + k$. The time between replacements is MTTF, and (9.20) is therefore an obvious result. The ratio

$$\frac{C_A(t_0)}{C_A(\infty)} = \frac{c + k \cdot F(t_0)}{\int_0^{t_0}(1 - F(t))\,dt} \cdot \frac{\text{MTTF}}{c + k}$$
$$= \frac{1 + r \cdot F(t_0)}{\int_0^{t_0}(1 - F(t))\,dt} \cdot \frac{\text{MTTF}}{1 + r} \qquad (9.21)$$

where $r = k/c$, can therefore be used as a measure of the cost efficiency of the age replacement policy with replacement interval t_0. A low value of $C_A(t_0)/C_A(\infty)$ indicates a high cost efficiency.

Example 9.9
Consider an item with Weibull life distribution $F(t)$ with scale parameter λ and shape parameter α. To find the optimal replacement age t_0, we have to find the t_0 that minimizes (9.19), or alternatively (9.21). By using (9.21), we get

$$\frac{C_A(t_0)}{C_A(\infty)} = \frac{1 + r\left(1 - e^{-(\lambda t_0)^\alpha}\right)}{\int_0^{t_0} e^{-(\lambda t)^\alpha}\,dt} \cdot \frac{\Gamma(1/\alpha + 1)/\lambda}{1 + r} \qquad (9.22)$$

By introducing $x_0 = \lambda t_0$, (9.22) may be written as

$$\frac{\tilde{C}_A(x_0)}{C_A(\infty)} = \frac{1 + r(1 - e^{-x_0^\alpha})}{\int_0^{x_0} e^{-x^\alpha}\,dx} \cdot \frac{\Gamma(1/\alpha + 1)}{1 + r} \qquad (9.23)$$

To find the x_0 for which (9.23) attains its minimum by analytical methods is not straightforward. The optimal x_0 may be found graphically by plotting $\tilde{C}_A(x_0)/C_A(\infty)$ as a function of x_0. An example is shown in Fig. 9.11 where $\tilde{C}_A(x_0)/C_A(\infty)$ is plotted for $\alpha = 3$, and some selected values of $r = k/c$. The optimal x_0, and thereby the optimal replacement age $t_0 = x_0/\lambda$, can be found from Fig. 9.11 as the value minimizing the ratio $\tilde{C}_A(x_0)/C_A(\infty)$. Notice that when $\tilde{C}_A(x_0)/C_A(\infty) > 1$ no age replacement should take place. The cost efficiency of the age replacement policy is seen to decrease when t_0 increases. □

Time Between Failures Let $Y_1, Y_2, \ldots$ denote the times between the actual consecutive failures. This may be represented as a renewal process where the renewals are the actual failures. The renewal periods, Y_i, are composed of a random number, N_i of time periods of length t_0 (corresponding to replacements without failure), plus a last time period in which the item fails at an age Z_i, less than t_0.
Thus

$$Y_i = N_i \cdot t_0 + Z_i \quad \text{for } i = 1, 2, \ldots$$

The random variable N_i has a *geometric distribution* (see Section 2.8)

$$\Pr(N_i = n) = (1 - F(t_0))^n F(t_0) \quad \text{for } n = 0, 1, \ldots$$

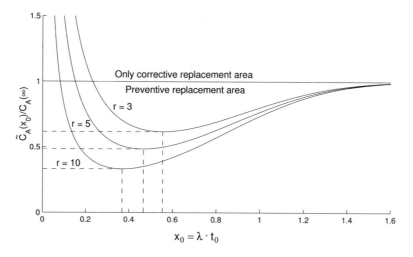

Fig. 9.11 The ratio $\tilde{C}_A(x_0)/C_A(\infty)$ as a function of x_0 for the Weibull distribution with shape parameter $\alpha = 3$, and $r = 3, 5,$ and 10.

The mean number of replacements without failure is thus

$$E(N_i) = \sum_{n=0}^{\infty} n \cdot \Pr(N_i = n) = \frac{1 - F(t_0)}{F(t_0)}$$

The distribution of Z_i is

$$\Pr(Z_i \leq t) = \Pr(T \leq t \mid T \leq t_0) = \frac{F(t)}{F(t_0)} \quad \text{for } 0 \leq t \leq t_0$$

Hence

$$E(Z_i) = \int_0^{t_0} \left(1 - \frac{F(t)}{F(t_0)}\right) dt = \frac{1}{F(t_0)} \int_0^{t_0} (F(t_0) - F(t)) \, dt$$

The mean time between actual failures is thus

$$\begin{aligned} E(Y_i) &= t_0 \cdot E(N_i) + E(Z_i) \\ &= \frac{1}{F(t_0)} \left(t_0(1 - F(t_0)) + \int_0^{t_0} (F(t_0) - F(t)) \, dt \right) \\ &= \frac{1}{F(t_0)} \int_0^{t_0} (1 - F(t)) \, dt \end{aligned} \qquad (9.24)$$

Age Replacement – Availability Criterion In some applications the unavailability of the item is more important than the cost of replacement/repair, and we may be interested in finding a replacement age t_0 that minimizes the average unavailability of the item. Let MDT_P be the mean downtime for a planned replacement, and MDT_F

be the mean downtime needed to restore the function after a failure. The total mean downtime in a replacement interval of length t_0 is

$$\begin{aligned} \text{MDT}(t_0) &= \text{MDT}_F \cdot F(t_0) + \text{MDT}_P \cdot (1 - F(t_0)) \\ &= [\text{MDT}_F - \text{MDT}_P] + \text{MDT}_P \cdot F(t_0) \end{aligned}$$

The mean time between replacements is

$$\begin{aligned} \text{MTBR}(t_0) &= \int_0^{t_0} (1 - F(t))\,dt + \text{MDT}_F \cdot F(t_0) + \text{MFD}_P \cdot (1 - F(t_0)) \\ &= \int_0^{t_0} (1 - F(t))\,dt + \text{MDT}_P + [\text{MDT}_F - \text{MDT}_P] \cdot F(t_0) \end{aligned}$$

The average unavailability with the age replacement policy for age t_0 is therefore

$$\begin{aligned} \bar{A}_{\text{av}}(t_0) &= \frac{\text{MDT}(t_0)}{\text{MTBR}(t_0)} \\ &= \frac{[\text{MDT}_F - \text{MDT}_P] + \text{MDT}_P \cdot F(t_0)}{\int_0^{t_0} (1 - F(t))\,dt + \text{MDT}_P + [\text{MDT}_F - \text{MDT}_P] \cdot F(t_0)} \end{aligned} \qquad (9.25)$$

The optimal t_0 is the value of t_0 that minimizes $\bar{A}_{\text{av}}(t_0)$ in (9.26). This value may be found by the same approach we used for the cost criterion.

9.6.2 Block Replacement

An item (component, system) that is maintained under a block replacement policy is preventively replaced at regular time intervals ($t_0, 2t_0, \ldots$) regardless of age. The block replacement policy is easier to administer than an age replacement policy since only the elapsed (calendar) time since last replacement must be monitored, rather than the operational time since last replacement. The block replacement policy is therefore commonly used when there is a large number of similar items in service. The main drawback of the block replacement policy is that it is rather wasteful, since almost new items may be replaced at planned replacement times.

Consider an item that is put into operation at time $t = 0$. The time to failure T of the item has distribution function $F(t) = \Pr(T \le t)$. The item is operated under a block replacement policy where it is preventively replaced at times $t_0, 2t_0, \ldots$. The preventive replacement cost is c. If the item fails in an interval, it is immediately repaired or replaced. The cost of the unplanned repair is k. Let $N(t_0)$ denote the number of failures/replacements in an interval of length t_0, and let $W(t_0) = E(N(t_0))$ denote the mean number of failures/repairs in the interval.

The mean cost in an interval is $c + k \cdot W(t_0)$. The average costs $C_B(t_0)$ per time unit when using a block interval of length t_0 is equal to

$$C_B(t_0) = \frac{c + k \cdot W(t_0)}{t_0} \qquad (9.26)$$

In the original block replacement model, the items are replaced with a new item of the same type after each failure. We therefore have a *renewal process* within each interval of length t_0 (see Section 7.3). In this case $W(t_0)$ is the renewal function, and we may use the formulas developed in Section 7.3.

Example 9.10
Consider a block replacement model where the replacement interval t_0 is considered to be so short that the probability of having more than one failure in a block interval is negligible. In this case we may use the approximation

$$W(t_0) = E(N(t_0)) = \sum_{n=0}^{\infty} n \cdot \Pr(N(t_0) = n)$$
$$\approx \Pr(N(t_0) = 1) = \Pr(T \le t_0) = F(t_0)$$

The average costs $C_B(t_0)$ per time unit is

$$C_B(t_0) \approx \frac{c + k \cdot F(t_0)}{t_0}$$

The minimum of $C_B(t_0)$ may be found by solving $d\,C_B(t_0)/dt_0 = 0$ which gives

$$\frac{c}{k} + F(t_0) = t\,F'(t_0)$$

Let us now assume that $F(t)$ is a Weibull distribution with shape parameter $\alpha > 1$ and shape parameter λ. We can then find the optimal replacement interval by solving

$$\frac{c}{k} + 1 - e^{-(\lambda t_0)^\alpha} = t_0 \cdot \alpha \lambda^\alpha t_0^{\alpha-1} e^{-(\lambda t_0)^\alpha}$$

which can be written as

$$\frac{c}{k} + 1 = \left(1 + \alpha(\lambda t_0)^\alpha\right) \cdot e^{-(\lambda t_0)^\alpha} \qquad (9.27)$$

For this model to be realistic the preventive replacement cost c must be small compared to the corrective replacement cost k. By introducing $x = (\lambda t_0)^\alpha$, and using the approximation $e^x \approx 1 + x + x^2/2$, we can solve (9.27) and get the approximative solution (when remembering that t_0 is small):

$$x \approx \frac{\alpha}{1+c/k} - 1 - \sqrt{\left(\frac{\alpha}{1+c/k} - 1\right)^2 - 2\left(1 - \frac{1}{1+c/k}\right)}$$

If we assume that $c/k = 0.1$ and $\alpha = 2$, we get the optimal value $t_0 = 1/\lambda \cdot x^{1/\alpha} \approx 0.35 \cdot 1/\lambda \approx 0.39 \cdot \text{MTTF}$. With the same value of c/k and $\alpha = 3$, we get the optimal value $t_0 \approx 0.39 \cdot 1/\lambda \approx 0.44 \cdot \text{MTTF}$. In Fig. 9.12 the optimal replacement interval t_0 is plotted as a function of α. The optimal value t_0 is equal to $h \cdot \text{MTTF}$, where MTTF is the mean of the Weibull distribution with parameters α and λ. □

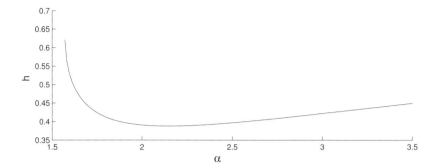

Fig. 9.12 The optimal replacement interval t_0 in Example 9.10 as a function of the shape parameter α of the Weibull distribution. The optimal value t_0 is equal to $h \cdot$MTTF.

Block Replacement with Minimal Repair The block replacement policy may be modified by only carrying out *minimal repair* when items fail in the block interval. The assumption is that a minimal repair is often adequate until the next planned replacement. In this case we will have a nonhomogeneous Poisson process (NHPP) within the block interval of length t_0, and we may use the formulas developed in Section 7.4 to determine $W(t_0)$. This modified block replacement model was proposed and studied by Barlow and Hunter (1961).

Another approach would be to assume that we carry out normal (imperfect) repairs in the block interval. In that case we may use the theory described in Section 7.5 to determine $W(t_0)$.

Block Replacement with Limited Number of Spares Consider an item that is operated under a block replacement policy. We will now assume that the number m of spares that may be used in a replacement interval is limited. In this case we may run out of spares and the item's function may therefore be unavailable during a part of the replacement interval. The times to failure $T_1, T_2, \ldots$ of the items are assumed to be independent and identically distributed with distribution function $F(t)$.

Let k_u denote the cost per time unit when the item function is not available, and let $\tilde{T}_u(t_0)$ be the time the item remains unavailable in a replacement interval of length t_0. Hence, we have $\tilde{T}_u(t_0; m) = t_0 - \sum_{i=1}^{m+1} T_i$ if the initial item and the m spares fail in the replacement interval, and $\tilde{T}_u(t_0; m) = 0$ if less than $m + 1$ failures occur.

The same number m of spares are assumed to be made available for each replacement interval. All intervals will therefore have the same stochastic properties, and we may therefore confine ourselves to studying the first interval $(0, t_0)$.

The mean cost in a replacement interval is $c + k \cdot E(N(t_0)) + k_u \cdot E(\tilde{T}_u(t_0; m))$, and the average cost $C_{\tilde{B}}(t_0; m)$ per time unit when using a block interval of length t_0 is

$$C_{\tilde{B}}(t_0; m) = \frac{c + k \cdot E(N(t_0)) + k_u \cdot E(\tilde{T}_u(t_0; m))}{t_0} \qquad (9.28)$$

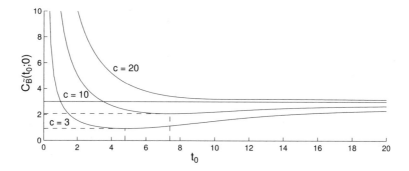

Fig. 9.13 The average cost per time unit for a block replacement policy with no spares when the time to failure distribution is a Weibull distribution with $\alpha = 3$ and $\lambda = 0.1$, $k = 10$, $k_u = 3$, and $c = 3, 10, 20$.

where $N(t_0)$ is the number of replacements in $(0, t_0)$.

Example 9.11
Consider an item that is operated under a block replacement policy without any spare item ($m = 0$) in each block interval. In this case (9.28) can be written

$$C_{\tilde{B}}(t_0; 0) = \frac{c + k \cdot F(t_0) + k_u \cdot \int_0^{t_0} F(t)\, dt}{t_0}$$

Let $F(t)$ be a Weibull distribution with shape parameter $\alpha = 3$ and scale parameter $\lambda = 0.1$. In Fig. 9.13 $C_{\tilde{B}}(t_0; 0)$ is plotted as a function of t_0 for some selected cost values c, k, and k_u that gives three different shapes.

When the replacement period t_0 tends toward infinity, the block replacement policy is equivalent to leave the item as it is, and not replace it. Then the average cost per time unit will tend to k_u. When $c = 3$, the shape of the curve is quite similar to the corresponding curve for the classical age replacement policy with optimal replacement period. When $c = 10$, the optimal block replacement cost $C_{\tilde{B}}(t_0; 0)$ is close to k_u. When t_0 increases, $C_{\tilde{B}}(t_0; 0)$ remains close to the replacement cost until the influence of $\tilde{T}_u(t_0)$ becomes sufficiently large. When $c = 20$, the curve does not have a very distinctive minimum, and we may as well choose a very long replacement interval.
□

Example 9.12
Consider an item that is operated under a block replacement policy with m spare items in each block interval. The time to failure T is assumed to be gamma distributed with parameters λ and α. The probability density function of T is

$$f_T(t) = \frac{\lambda}{\Gamma(\alpha)} (\lambda t)^{\alpha-1} e^{-\lambda t}$$

The item function will be unavailable when the initial item and the m spares have failed. The time to system failure is therefore $T_s = \sum_{i=1}^{m+1} T_i$ where the times to

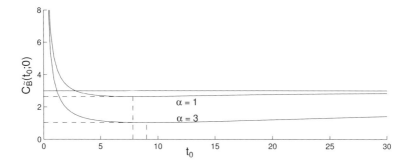

Fig. 9.14 The average cost per time unit for a block replacement policy with $m = 5$ spares when the time to failure distribution is a gamma distribution with parameters α and $\lambda = 1$, for $\alpha = 1$ and 3, and $k = 10$, $k_u = 3$, and $c = 3$.

individual failure $T_1, T_2, \ldots, T_{m+1}$ are assumed to be independent and identically distributed with probability density function $f_T(t)$. Let $F^{(m+1)}(t)$ denote the distribution function of T_s. The distribution $F^{(m+1)}(t)$ can be found by taking the $(m+1)$-fold convolution of $F(t)$ (see Section 7.3). Since the gamma distribution is "closed under addition," T_s is gamma distributed with parameters λ and $(m+1)\alpha$. In this case (9.28) can be written

$$C_{\tilde{B}}(t_0; m) = \frac{c + k \cdot F^{(m+1)}(t_0) + k_u \cdot \int_0^{t_0} F^{(m+1)}(t)\, dt}{t_0}$$

In Fig. 9.14 $C_{\tilde{B}}(t_0; m)$ is plotted as a function of t_0 for some selected values of the parameter α, and cost values c, k, and k_u. □

The cost k of a repair/replacement in the block interval may also be extended to be time dependent and also to include other types of costs, for example, if the item deteriorates during the interval and will require increasing operating costs.

9.6.3 Condition-Based Maintenance

Condition-based maintenance (CBM) is a maintenance policy where the maintenance action is decided based on measurement of one, or more, variables that are correlated to a degradation, or a loss of performance, of the system. The variables may be *physical* variables (e.g., thickness of material, erosion percentage, temperature, or pressure), system *performance* variables (e.g., quality of produced items or number of discarded items), or variables related to the *residual life* of the system. In the latter case, the expression *predictive maintenance* is often used instead of condition-based maintenance.

The CBM policy requires a monitoring system that can provide measurements of selected variables, and a mathematical model that can predict the behavior of the system deterioration process. The type of maintenance action, and the date of the action are decided based on an analysis of measured values. A decision is often taken

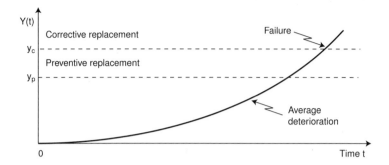

Fig. 9.15 Average deterioration with preventive and corrective replacement thresholds.

when a measurement (of a variable) passes a predefined threshold value. The threshold values make it possible to divide the system state space into different decision areas, where each area represents a specific maintenance decision. This type of maintenance policy is often called a *control limit policy* and is obviously only relevant for systems with an increasing failure rate.

Condition-Based Replacements We will now consider a special type of CBM with preventive *replacements*. Let $Y(t)$ be a random variable describing the deterioration of the item at time t, and assume that $Y(t)$ is measured on a continuous scale. The item is supposed to be deteriorating in such a way that $Y(t)$ is nondecreasing as a function of t. The item is inspected and the deterioration $Y(t)$ is measured at specific points of time $t_1, t_2, \ldots$. The variable $Y(t)$ is only measured at the inspections at times $t_1, t_2, \ldots$, and not between these points of time. When a measurement $Y(t) \geq y_p$, the item should be preventively replaced. If a measurement $Y(t) \geq y_c$ ($> y_p$), the item is in a failed state and has to be correctively replaced. A failure is not detected immediately when $Y(t)$ passes the failure limit y_c. The failure will be detected at the first inspection after $Y(t)$ has passed y_c. The corrective replacement cost will be significantly higher than the preventive replacement cost. After a replacement (preventive or corrective) the item is assumed to be as good as new. The performance behavior is illustrated in Fig. 9.15.

Example 9.13
Consider the deterioration (wear) of the brake pads on the front wheels a car, and let $Y(t)$ be the wear (the reduction of the thickness of the brake pads) at time t, where t is the number of kilometers driven since the brake pads were new. The wear $Y(t)$ is measured (controlled) when the car is at the garage for service at regular intervals of length τ (e.g., 15000 km). The brake pads should be preventively replaced when the wear is greater than y_p. If the wear is greater than y_c, the brake effect is reduced, the pad holders will make scratches in the brake discs, and the discs will have to be replaced. The cost of this replacement will be significantly higher than the cost of only replacing the brake pads. In addition the risk cost due to reduced braking efficiency should be considered. □

In some applications it has been found to be realistic to model the deterioration as a *gamma stochastic process* $\{Y(t), t \geq 0\}$, with the following characteristics:

1. $Y(0) = 0$.

2. The process $\{Y(t), t \geq 0\}$ has independent increments (see Section 7.1).

3. For all $0 \leq s < t$, the random variable $Y(t) - Y(s)$ has a gamma distribution with parameters $(\alpha(t-s), \beta)$, with probability density function (see Section 2.11)

$$f_{(s,t)}(y) = \frac{\beta}{\Gamma(\alpha(t-s))} (\beta y)^{\alpha(t-s)-1} e^{-\beta y} \quad \text{for } y \geq 0 \qquad (9.29)$$

The mean deterioration in the interval (s, t) is from (2.43):

$$E[Y(t) - Y(s)] = \frac{\alpha}{\beta}(t-s)$$

When using the gamma process, the mean deterioration is therefore a linear function of time with deterioration speed (slope) α/β, and not a convex function like the one illustrated in Fig. 9.15. The application of gamma processes to model deterioration is further discussed, for example, by Grall et al. (2002).

Example 9.14

Consider a condition-based replacement policy where the deterioration $Y(t)$ is modeled by a gamma process. The mean deterioration is therefore a linear function of time. The system is inspected after regular intervals of length τ. The deterioration $Y(t)$ is measured at the inspections, and never between two inspections. Let ΔY_i be the deterioration within inspection interval i, for $i = 1, 2, \ldots$. Since we have assumed a gamma process, $\Delta Y_1, \Delta Y_2, \ldots$, will be independent and identically gamma distributed with distribution function $F(y)$. Let $\mu \tau$ denote the mean deterioration in an inspection interval. The parameter μ is hence the deterioration speed.

Let c denote the cost of a preventive replacement, let k be the additional cost of a corrective replacement, and let k_i be the cost of an inspection. In this example we will use heuristic arguments to determine the optimal interval τ between inspections.

Let n_p denote the mean number of inspections until the deterioration process reaches the preventive replacement threshold y_p. We then have that $n_p \cdot \mu \tau \approx y_p$. Since the crossing of the threshold y_p is only detected in the first inspection after it has been passed, the crossing will be detected in inspection $\tilde{n}_p$ where

$$\tilde{n}_p = \frac{y_p}{\mu \tau} + 1$$

The mean time between replacements with this replacement policy will therefore be

$$\text{MTBR}(\tau) = \left(\frac{y_p}{\mu \tau} + 1\right) \cdot \tau$$

The average cost per replacement cycle is

$$\begin{aligned} c + k_i \cdot n_p + k \cdot \Pr(\text{failure}) &= c + k_i \cdot n_p + k \cdot \Pr(\Delta Y > (y_c - y_p)) \\ &= c + k_i \cdot n_p + k \cdot [1 - F(y_c - y_p)] \end{aligned}$$

The average cost $C_{\text{CB}}(\tau)$ per time unit using this policy is

$$\begin{aligned} C_{\text{CB}}(\tau) &= \frac{c + k_i \cdot n_p + k \cdot [1 - F(y_p - y_c)]}{(y_p/\mu\tau + 1) \cdot \tau} \\ &= \frac{c + k_i \cdot y_p/\mu\tau + k \cdot [1 - F(y_c - y_p)]}{(y_p/\mu\tau + 1) \cdot \tau} \end{aligned} \quad (9.30)$$

If we know the distribution function $F(y)$ and the costs, (9.30) may be used to determine the optimal inspection interval τ. $\square$

Consider the condition-based replacement situation in Example 9.14, and assume that the deterioration process $\{Y(t), t \geq 0\}$ can be modeled as a gamma process. The item is inspected at regular intervals of length τ. The deteriorations $\Delta Y_1, \Delta Y_2, \ldots$ are assumed to be independent and gamma distributed with probability density function $f(y)$ given by (9.29) with parameters $\alpha\tau$ and β. (The time interval is $t - s = \tau$.)

Let $N(y_p)$ be the (smallest) number of inspection intervals until the accumulated deterioration has passed the preventive replacement threshold y_p, that is, until $\sum_{i=1}^{k} \Delta Y_i \geq y_p$. Notice that $N(y_p) = n$ is equivalent to $\sum_{i=1}^{n-1} \Delta Y_i < y_p \cap \sum_{i=1}^{n-1} \Delta Y_i + \Delta Y_n \geq y_p$. The probability density function $f^{(n-1)}(y)$ of $\sum_{i=1}^{n-1} \Delta Y_i$ is the $(n-1)$ convolution of the probability density function $f(y)$ for the deterioration ΔY in one interval. Since the gamma distribution is "closed under addition," the density $f^{(n-1)}(y)$ is a gamma distribution with parameters $(n-1)\alpha\tau$ and β.

The mean number of inspection intervals until the deterioration process has crossed the threshold y_p is

$$\begin{aligned} E(N(y_p)) &= \sum_{n=1}^{\infty} n \cdot \Pr(N(y_p) = n) \\ &= \sum_{n=1}^{\infty} n \int_{y_p}^{\infty} \int_{0}^{y_p} f^{(n-1)}(u)\, f(v-u)\, du\, dv \end{aligned} \quad (9.31)$$

The mean time between replacements with replacement threshold y_p is

$$\text{MTBR}(y_p) = E(N(y_p)) \cdot \tau \quad (9.32)$$

where $E(N(y_p))$ is given by (9.31). Let the cost of a preventive replacement be c, and k be the extra cost if a failure occurs. The mean total cost $C_{\text{CB}}(y_p)$ in a replacement period with preventive replacement threshold y_p is $c + k \cdot (1 - F(y_c - y_p))$ where $F(y)$ is the distribution function corresponding to the probability density function $f(y)$. In this case we have assumed that the inspection cost k_i is so small that it may be disregarded. The inspection cost may, however, be included in the same way as in Example 9.14. The mean cost per time unit of this condition-based replacement policy is thus

$$C_{CB}(y_p) = \frac{c + k \cdot (1 - F(y_c - y_p))}{\text{MTBR}(y_p)} \qquad (9.33)$$

Let $C_{CB}(y_c)$ denote the mean cost per time unit when we do not use a preventive replacement threshold y_p, but replace the item only when the accumulated deterioration has crossed the corrective replacement threshold y_c. The cost efficiency of the preventive replacement policy may now be evaluated based on the ratio

$$\frac{C_{CB}(y_p)}{C_{CB}(y_c)} = \frac{c + k \cdot F(y_c - y_p)}{c + k} \cdot \frac{\text{MTBR}(y_c)}{\text{MTBR}(y_p)} \qquad (9.34)$$

The ratio (9.34) can be used to find the most cost-efficient preventive replacement threshold y_p for a given inspection interval τ. The most cost-efficient inspection interval τ for a specified threshold y_p may be determined by minimizing (9.33).

Example 9.15
Consider the same situation as described above, but assume that the deterioration ΔY_i within inspection interval i is exponentially distributed with rate β, for $i = 1, 2, \ldots$. Remember that the exponential distribution with rate β is a gamma distribution with parameters 1 and β, such that this is a special case of the situation described above with $\alpha = 1/\tau$.

The accumulated deterioration in the first k intervals is $Y(k \cdot \tau) = \sum_{i=1}^{k} \Delta Y_i$, that has a gamma (Erlangian) distribution with parameters k and β. The mean deterioration in the interval $(0, k\tau)$ is $E(Y(k\tau)) = \alpha(k\tau)/\beta = k/\beta$.

The mean time between replacements is from (9.32)

$$\text{MTBR}(y_p) = (\theta \cdot y_p + 1) \cdot \tau \qquad (9.35)$$

The ratio in (9.34) is in this case

$$\frac{C_{CB}(y_p)}{C_{CB}(y_c)} = \frac{1 + r \cdot e^{-\beta(y_c - y_p)}}{1 + r} \cdot \frac{\beta \cdot y_c + 1}{\beta \cdot y_p + 1}$$

where $r = k/c$. The ratio $C_{CB}(y_p)/C_{CB}(y_c)$ is plotted in Fig. 9.16 as a function of y_p, for $\beta = 3$, $y_c = 2$ and for $r = 3, 5, 10$. From Fig. 9.16 we are able to determine the most cost-efficient preventive replacement threshold y_p for the various values of $r = k/c$. □

Remark: The situation in Examples 9.14 and 9.15 is not always realistic, since the same inspection interval is applied throughout. In most practical applications, an observed deterioration close to the preventive replacement threshold y_p would imply a replacement action within a specified interval that is determined by the value of $Y(t)$. If, for example, the wear of the brake pads in Example 9.13 is observed to be

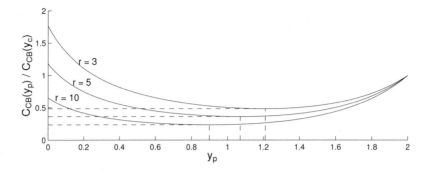

Fig. 9.16 The ratio $C_{CB}(y_p)/C_{CB}(y_c)$ plotted as a function of the preventive replacement threshold y_p for $\beta = 3$, $y_c = 2$ and for $r = 3, 5, 10$.

so close to y_p in the inspection at "time" $t = 90,000$ km that the repairman fears that they will pass y_c within the next inspection at "time" $t = 105,000$ km, the owner of the car may be advised to replace the brake pads before "time" $t = 97,500$ km, at the same time when he is due to change the motor oil.

A simultaneous optimization of the preventive replacement threshold and the time to the next inspection (and replacement action) is discussed by Grall et al. (2002). More complex condition-based maintenance policies are discussed, for example, by Castanier (2001) and Bérenguer et al. (2003). □

9.6.4 PF Intervals

We will now study an inspection and replacement policy known as the PF interval approach. The PF interval approach is briefly discussed in most of the main RCM references.

Consider an item that is exposed to random shocks (events). We assume that the shocks occur as a homogeneous Poisson process (HPP) with rate λ. The time between two consecutive shocks is then, according to Section 7.2, exponentially distributed with rate λ and mean $1/\lambda$. When a shock occurs, it produces a weakness (potential failure) in the item that, in time, will develop/deteriorate into a critical failure. We are not able to observe the shocks but may be able to reveal potential failures some time after the shock has occurred. Let P be the point of time (after a shock) when an indication of a potential failure can be first detected, and let F be the point of time where the item has functionally failed. The time interval from P to F is called the PF interval and is generally a random variable. If a potential failure is detected between P and F in Fig. 9.17, this is the time interval in which it is possible to take action to prevent the failure and to avoid its consequences. The cost of a *preventive replacement* (or repair) is C_P, and the cost of a *corrective replacement* after a critical failure has occurred is C_C.

The item is inspected at regular intervals of length τ, and the cost of each inspection is C_I. The inspections may be observations using human senses (sight, smell, sound),

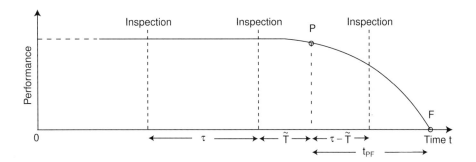

Fig. 9.17 Average behavior and concepts used in PF interval models.

or we may use some monitoring equipment. In the most simple setup we assume that the inspection procedure is perfect, such that all potential failures are detected by the inspection. In many cases this is not a realistic assumption, and the probability of successful detection may be a function of the time since P, the time of the year, and so on. Our main objective in this section is to find the optimal inspection interval τ, that is, the value of τ that gives the lowest mean average cost.

The length of the PF interval will generally depend on the materials and characteristics of the item, the failure mode, the failure mechanisms, and the environmental and operational conditions. Estimates of PF intervals are not available in reliability data sources and must be estimated by expert judgment by operators, specialists on deteriorating mechanisms, and equipment designers. The length of the PF interval may be regarded as a random variable T_{PF} with a subjective distribution function (see Chapter 13).

Example 9.16
Vatn and Svee (2002) have studied crack occurrences and crack detection in (railroad) rails. In their model cracks are initiated at random. The frequency λ of initiated cracks may be measured as the number of initiated cracks per unit length of rails and per time unit. The frequency will generally depend on the traffic load, the material and geometry of the rail, and various environmental factors but may also be caused by particles on the rails or "shocks" from trains with noncircular wheels. In the first phase the cracks are very small and very difficult to detect. A special rail-car equipped with ultrasonic inspection equipment is used to inspect the rails. When a crack has grown to a specific size, it should be detectable by ultrasonic inspection. This crack size corresponds to the potential failure P described above. The PF interval is the time interval from an observable crack P is present until a critical failure F occurs. The critical failure F will, in this case, be breakage of a rail and possible derailment of a train. Ultrasonic inspection is carried out at regular intervals, at a rather high cost. It is therefore of interest to find an optimal inspection interval that balances the inspection cost and the costs related to replacements and potential accidents. □

Our objective is to find the inspection interval τ that minimizes the mean total cost. In the general setup this is a rather difficult task. We therefore start by solving the problem in the most simple situation, with known (deterministic) PF interval and known repair time. Thereafter we present some ideas on how to solve the problem in a more realistic setup.

Deterministic PF Interval and Repair Time and Perfect Inspection To simplify the problem we assume that the length of the PF interval t_{PF} is known (deterministic). The time from when a potential failure P is detected (during the first inspection after P) until the failure has been corrected, t_R, is also assumed to be known (deterministic). We further assume that the inspections are perfect such that all potential failures are detected during the inspections. From Fig. 9.17 is is easy to see that we will have a preventive replacement when $\tau - t + t_R < t_{\text{PF}}$, and a corrective replacement if $\tau - t + t_R > t_{\text{PF}}$. If $\tau + t_R < t_{\text{PF}}$ all the replacements will be preventive, and there is no problem to optimize. We therefore assume that $\tau + t_R > t_{\text{PF}}$ (see the remark on page 397).

Assume that we start observing the item at time $t = 0$, and that the potential failure P is observable a short time after the shock occurs. The time T from startup to P is exponentially distributed with failure rate λ. Let $N(\tau)$ denote the number of inspection intervals before a shock occurs. The event $N(\tau) = n$ hence means that we observe n inspection intervals without any shock, and the shock occurs in inspection interval $n + 1$. The random variable $N(\tau)$ has a geometric distribution (see Section 2.8) with point probability

$$\Pr(N(\tau) = n) = \left(e^{-\lambda \tau}\right)^n \left(1 - e^{-\lambda \tau}\right) \quad \text{for } n = 0, 1, \ldots$$

and mean value

$$E(N(\tau)) = \frac{e^{-\lambda \tau}}{1 - e^{-\lambda \tau}}$$

Let us assume that a shock and an observable potential failure P has occurred in inspection interval $n + 1$. Let $\tilde{T}$ denote the time from inspection n till P. The probability distribution of $\tilde{T}$ is

$$\Pr(\tilde{T} \leq t) = \Pr(T \leq t \mid T \leq \tau) = \frac{1 - e^{-\lambda t}}{1 - e^{-\lambda \tau}} \quad \text{for } 0 < t \leq \tau$$

A preventive replacement will therefore take place with probability

$$P_P(\tau) = \Pr(\tilde{T} > \tau + t_R - t_{\text{PF}}) = 1 - \frac{1 - e^{-\lambda(\tau + t_R - t_{\text{PF}})}}{1 - e^{-\lambda \tau}}$$

A corrective replacement will take place with probability

$$P_C(\tau) = \Pr(\tilde{T} < \tau + t_R - t_{\text{PF}}) = \frac{1 - e^{-\lambda(\tau + t_R - t_{\text{PF}})}}{1 - e^{-\lambda \tau}}$$

If we know that the potential failure will result in a critical failure (corrective maintenance), the mean time to this failure is $1/\lambda + t_{\text{PF}}$. On the other hand, if we know that the potential failure will result in a preventive replacement, the mean time to this replacement is $E(N(\tau) + 1) \cdot \tau + t_R$. The mean time between replacements is therefore

$$\text{MTBR}(\tau) = \left(\frac{1}{\lambda} + t_{\text{PF}}\right) \cdot P_C(\tau) + (E(N(\tau)) + 1) \cdot \tau + t_R) \cdot P_P(\tau)$$

$$= \left(\frac{1}{\lambda} + t_{\text{PF}}\right) \cdot P_C(\tau) + \left(\frac{\tau}{1 - e^{-\lambda\tau}} + t_R\right) \cdot P_P(\tau) \quad (9.36)$$

The mean total cost $C_T(\tau)$ in a replacement interval is

$$C_T(\tau) = C_P \cdot P_P(\tau) + C_C \cdot P_C(\tau) + C_I \cdot \left(E(N(\tau)) + \Pr(\tilde{T} > \tau - t_{\text{PF}})\right)$$

where $\Pr(\tilde{T} > \tau - t_{\text{PF}})$ is the probability that the item will not fail within the inspection interval where the potential failure occurred, and consequently that the next inspection will be carried out. When $\tau - t_{\text{PF}} > 0$ this probability is

$$\Pr(\tilde{T} > \tau - t_{\text{PF}}) = \Pr(T > \tau - t_{\text{PF}} \mid T \leq \tau) = \frac{e^{-\lambda(\tau - t_{\text{PF}})} - e^{-\lambda\tau}}{1 - e^{-\lambda\tau}}$$

We therefore have that

$$\Pr(\tilde{T} > \tau - t_{\text{PF}}) = \begin{cases} \dfrac{e^{-\lambda(\tau - t_{PF})} - e^{-\lambda\tau}}{1 - e^{-\lambda\tau}} & \text{for } \tau - t_{\text{PF}} > 0 \\ 1 & \text{for } \tau - t_{\text{PF}} < 0 \end{cases}$$

The mean total cost $C_T(\tau)$ in a replacement interval is therefore

$$C_T(\tau) = \begin{cases} C_P \cdot P_P(\tau) + C_C \cdot P_C(\tau) + C_I \cdot \dfrac{e^{-\lambda(\tau - t_{PF})}}{1 - e^{-\lambda\tau}} & \text{for } \tau - t_{\text{PF}} > 0 \\ C_P \cdot P_P(\tau) + C_C \cdot P_C(\tau) + C_I \cdot \left(\dfrac{e^{-\lambda\tau}}{1 - e^{-\lambda\tau}} + 1\right) & \text{for } \tau - t_{\text{PF}} < 0 \end{cases}$$

The mean total cost per time unit with inspection interval τ is

$$C(\tau) = \frac{C_T(\tau)}{\text{MTBR}(\tau)} \quad (9.37)$$

To find the value of τ for which (9.37) attains its minimum is not a straightforward task. The optimal τ may be found graphically by plotting $C(\tau)$ as a function of τ. An example is shown in Fig. 9.18.

Remark: In the case when $\tau + t_R < t_{\text{PF}}$ all the replacements will be preventive, and the mean time between replacements is $\text{MTBR}(\tau) = (E(N(\tau)) + 1) \cdot \tau + t_R$. The total cost in a replacement period is $C_T(\tau) = C_P + C_I \cdot (E(N(\tau)) + 1)$. The optimal replacement interval (with the restriction that $\tau + t_R < t_{\text{PF}}$) can therefore be found by minimizing

$$C(\tau) = \frac{C_T(\tau)}{\text{MTBR}(\tau)} = \frac{C_I/(1 - e^{-\lambda\tau}) + C_P}{\tau/(1 - e^{-\lambda\tau}) + t_R}$$

□

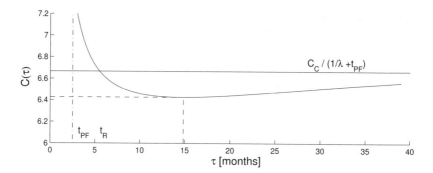

Fig. 9.18 The mean cost $C(\tau)$ per time unit as a function of τ for $\lambda = 1/12$ months^{-1}, $t_{\text{PF}} = 3$ months, $t_R = 0.5$ month, $C_C = 100$, $C_P = 20$, and $C_I = 15$.

Stochastic PF Interval, Deterministic Repair Time and Nonperfect Inspection

We now consider the same situation as described above but assume that the inspection is not perfect. In general, the probability of detecting a potential failure will depend on the time since the potential failure became observable. When a crack in the rail in Example 9.16 has been initiated, it will grow with time. The probability of detecting the crack is assumed to increase with the size of the crack. A model where the probability of successful detection is a function of the crack size will be rather complex. We therefore simplify the situation and introduce $\theta_i(\tau)$ to be the probability that the potential failure is *not* detected in inspection i after an observable potential failure P has occurred, for $i = 1, 2, \ldots$. The probability is assumed to be a function of the inspection interval τ. We assume that $1 > \theta_1(\tau) \geq \theta_2(\tau) \geq \cdots$.

The PF interval, T_{PF} is assumed to be a random variable with distribution function $F_{\text{PF}}(t)$. The repair time t_R is assumed to be known (deterministic). Let $T_F = \tilde{T} + T_{\text{PF}}$. The variable T_F is hence the time from the last inspection before P until a (possible) critical failure. The distribution of T_F can be found by the convolution of the distribution of $\tilde{T}$ and $F_{\text{PF}}(t)$.

$$\begin{aligned} F_F(t) &= \Pr(T_F \leq t) = \int_0^\tau F_{\text{PF}}(t-u)\, dF_{\tilde{T}}(u) \\ &= \frac{\lambda}{1 - e^{-\lambda \tau}} \int_0^\tau F_{\text{PF}}(t-u) \cdot e^{-\lambda u}\, du \end{aligned} \quad (9.38)$$

Let $\bar{F}_F(t) = 1 - F_F(t)$, and let $Z(\tau)$ denote the number of inspections carried out after a potential failure P has occurred. We want to find the probabilities $\Pr(Z(\tau) \geq k)$ for $k = 0, 1, \ldots$. It is obvious that $\Pr(Z(\tau) \geq 0) = 1$. At least one inspection will be carried out if $T_F = \tilde{T} + T_{\text{PF}} > \tau$, that is,

$$\Pr(Z(\tau) \geq 1) = \Pr(T_F > \tau) = \bar{F}_F(\tau)$$

At least two inspections will be carried out if $T_F > \tau$, the failure is not detected in the first inspection, and if $T_F > 2\tau$. Since $\Pr(T_F > \tau \cap T_F > 2\tau) = \Pr(T_F > 2\tau)$, we get

$$\Pr(Z(\tau) \geq 2) = \theta_1(\tau) \cdot \bar{F}_F(2\tau)$$

By continuing this argument we get in the general case that [we define $\theta_0(\tau) = 0$]

$$\Pr(Z(\tau) \geq k) = \left(\prod_{j=0}^{k-1} \theta_j(\tau)\right) \bar{F}_F(k\tau) \quad \text{for } k = 1, 2, \ldots \quad (9.39)$$

The mean number of inspections is therefore (see Problem 2.7)

$$E(Z(\tau)) = \sum_{k=1}^{\infty} \Pr(Z(\tau) \geq k) = \sum_{k=1}^{\infty} \left(\prod_{j=0}^{k-1} \theta_j(\tau)\right) \bar{F}_F(k\tau)$$

A preventive replacement will take place with probability

$$\begin{aligned} P_P(\tau) &= (1 - \theta_1(\tau)) \cdot \Pr(T_F > \tau + t_R) \\ &\quad + \theta_1(\tau) \cdot (1 - \theta_2(\tau)) \cdot \Pr(T_F > 2\tau + t_R) + \cdots \end{aligned}$$

which can be written as

$$\begin{aligned} P_P(\tau) &= \sum_{k=1}^{\infty} (1 - \theta_k(\tau)) \prod_{j=0}^{k-1} \theta_j(\tau) \cdot \Pr(T_F > k\tau + t_R) \\ &= \sum_{k=1}^{\infty} (1 - \theta_k(\tau)) \prod_{j=0}^{k-1} \theta_j(\tau) \cdot \bar{F}_F(k\tau + t_R) \quad (9.40) \end{aligned}$$

A corrective replacement will take place with probability $P_C(\tau) = 1 - P_P(\tau)$. Let $Z_P(\tau)$ be the number of inspections that are carried out after a potential failure P has occurred, when we know that the item will be preventively replaced. By using the same argument as we used to find (9.39), we get

$$\Pr(Z_P(\tau) \geq k) = \prod_{j=1}^{k-1} \theta_j(\tau)$$

and the mean value is

$$E(Z_P(\tau)) = \sum_{k=1}^{\infty} \prod_{j=1}^{k-1} \theta_j(\tau)$$

The mean time between replacements is therefore

$$\begin{aligned} \text{MTBR}(\tau) &= \left(\frac{1}{\lambda} + E(T_{\text{PF}})\right) \cdot P_C(\tau) \\ &\quad + [(E(N(\tau)) + E(Z_P(\tau))) \cdot \tau + t_R] \cdot P_P(\tau) \quad (9.41) \end{aligned}$$

The mean total cost $C_T(\tau)$ in a replacement interval is

$$\begin{aligned} C_T(\tau) &= C_P \cdot P_P(\tau) + C_C \cdot P_C(\tau) \\ &\quad + C_I \cdot (E(N(\tau)) + E(Z(\tau))) \end{aligned} \quad (9.42)$$

The optimal inspection interval τ may in this case be determined as the value of τ that minimizes $C(\tau) = C_T(\tau)/\text{MTBR}(\tau)$.

The models described in this section may be extended in many different ways. An obvious extension is to let the repair time be a random variable T_R. Another extension is to let the time to potential failure P have an increasing failure rate function.

Delay-Time Models Few references are available discussing quantitative assessment related to the PF interval approach. Some further developments have, however, be made based on the *delay-time* concept that was introduced in maintenance applications by Christer (1992). The delay-time model assumes that a failure is dependent on the occurrence of a defect (an incipient or potential failure). The time to failure T of an item can therefore be divided in two parts: (i) the time T_P from startup until a defect occurs and (ii) the delay-time T_{PF} from when the defect occurred until the item fails.

Several inspection models have been developed based on the delay-time principle covering, for example, imperfect inspections, nonconstant defect rate, and nonstationary inspection rules (Christer 2002). Some of these models have been applied in industry [e.g., see Dekker (1996) and Scarf (1997), for a review of industrial applications of delay-time models].

9.7 MAINTENANCE OPTIMIZATION

Maintenance tasks and resources have traditionally been allocated based on (i) requirements in legislation, (ii) company standards, (iii) recommendations from manufacturers and vendors of the equipment, and (iv) in-house maintenance experience. The maintenance strategy development is illustrated in Fig. 9.19. Maintenance theory and maintenance models have, so far, very seldom been used to develop system-specific maintenance strategies (Dekker 1996).

Many companies are faced with laws and regulations related to personnel safety and environmental protection that set requirements to their maintenance strategies. The oil companies operating in the Norwegian sector of the North Sea are, for example, required to test well barriers and safety functions according to regulations by the Norwegian Petroleum Directorate.

Recommendations from manufacturers are not always based on real experience data. Many manufacturers get very little feedback from the users of their equipment after the guarantee period is over. It is also sometimes claimed that manufacturers' recommendations may be more slanted toward maximizing the sales of consumable spares rather than minimizing the downtime for the user. Fear of product liability claims may also influence the manufacturers' recommendations.

MAINTENANCE OPTIMIZATION

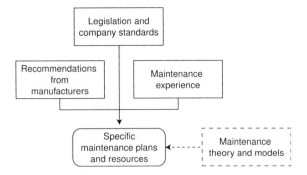

Fig. 9.19 Maintenance strategy development.

Maintenance aspects should preferably be considered during system design from the early concept phase. However, all too often the maintainability considerations are postponed until it is too late to make any significant system changes. Detailed maintenance strategies should also be established before the system is put into operation. Very often these strategies are only rudimentary and made on an ad hoc basis as problems occur. A promising approach to maintenance optimization has been proposed by Vatn et al. (1996).

9.7.1 Reliability Centered Maintenance

As many modern maintenance practices, the reliability centered maintenance (RCM) concept originated within the aircraft industry. RCM has now been applied with considerable success for more than 25 years - first within the aircraft industry, and later within the military forces, the nuclear power industry, the offshore oil and gas industry, and many other industries. Experiences from these industries show significant reductions in preventive maintenance (PM) costs while maintaining, or even improving, the availability of the systems.[2]

Definition 9.6 Reliability centered maintenance is a systematic consideration of system functions, the way functions can fail, and a priority-based consideration of safety and economics that identifies applicable and effective PM tasks (EPRI). □

The main focus of RCM is on the system *functions* and not on the system hardware. The main objective of RCM is to reduce the maintenance cost by focusing on the most important functions of the system and avoiding or removing maintenance tasks that are not strictly necessary. If a maintenance program already exists, the result of an RCM analysis will often be to eliminate inefficient PM tasks.

The RCM concept is described in several standards, reports, and textbooks. Among these are Nowlan and Heap (1978), IEC 60300-3-11, SAE JA1012, NASA (2000),

---

[2]This section is adapted from M. Rausand. 1998. Reliability centered maintenance, *Reliability Engineering and System Safety* **60**:121–132. Copyright 1998, with permission from Elsevier.

DEF-STD 02-45 (2000), MIL-STD 2173 (AS), and Moubray (1997). The main ideas presented in the various sources are more or less the same, but the detailed procedures may be rather different.

The maintenance tasks considered in the RCM approach are related to failures and functional degradation. Maintenance carried out, for example, to preserve or improve the aesthetic appearance of a system by cleaning and painting is outside the scope of RCM, at least when such maintenance has no effect on the system functions. However, planning of such tasks should be integrated with the planning of RCM relevant tasks.

What is RCM? RCM is a technique for developing a PM program. It is based on the assumption that the inherent reliability of the equipment is a function of the design and the build quality. An effective PM program will ensure that the inherent reliability is maintained. It should be realized that RCM will never be a substitute for poor design, inadequate build quality, or bad maintenance practices. RCM cannot improve the inherent reliability of the system. This is only possible through redesign or modification.

The application of PM is often misunderstood. It is easy to erroneously believe that the more an item is routinely maintained, the more reliable it will be. Often the opposite is the case, due to maintenance-induced failures. RCM was designed to balance the costs and benefits to obtain the most cost-effective PM program. To achieve this, the desired system performance standards have to be specified. PM will not prevent all failures, and therefore the potential consequences of each failure must be identified and the likelihood of failure must be known. PM tasks are chosen to address each failure by using a set of applicability and effectiveness criteria. To be effective, a PM task must provide a reduced expected loss related to personnel injuries, environmental damage, production loss, and/or material damage.

An RCM analysis basically provides answers to the following seven questions.

1. What are the functions and associated performance standards of the equipment in its present operating context?

2. In what ways can it fail to fulfill its functions?

3. What is the cause of each functional failure?

4. What happens when each failure occurs?

5. In what way does each failure matter?

6. What can be done to prevent each failure?

7. What should be done if a suitable preventive task cannot be found?

Experience has shown that approximately 30% of the efforts of an RCM analysis is involved in defining functions and performance standards, that is, answering question 1.

Main Steps of an RCM Analysis The RCM analysis may be carried out as a sequence of activities or steps, some of which are overlapping in time.

1. Study preparation
2. System selection and definition
3. Functional failure analysis (FFA)
4. Critical item selection
5. Data collection and analysis
6. FMECA
7. Selection of maintenance actions
8. Determination of maintenance intervals
9. Preventive maintenance comparison analysis
10. Treatment of noncritical items
11. Implementation
12. In-service data collection and updating

The various steps are discussed in the following.

Step 1: Study Preparation In step 1 an RCM project group is established. The project group must define and clarify the objectives and the scope of the analysis. Requirements, policies, and acceptance criteria with respect to safety and environmental protection should be made visible as boundary conditions for the RCM analysis.

Overall drawings and process diagrams, like piping and instrumentation diagrams, must be made available. Possible discrepancies between the as-built documentation and the real plant must be identified. The resources that are available for the analysis are usually limited. The RCM group should therefore be sober with respect to what to look into, realizing that analysis cost should not dominate potential benefits.

Step 2: System Selection and Definition Before a decision to perform an RCM analysis at a plant is taken, two questions should be considered.

1. To which systems are an RCM analysis beneficial compared with more traditional maintenance planning?
2. At what level of assembly (plant, system, subsystem) should the analysis be conducted?

All systems may in principle benefit from an RCM analysis. With limited resources, we must, however, make priorities, at least when introducing RCM in a new plant. We should start with the systems we assume will benefit most from the analysis. Most operating plants have developed some sort of assembly hierarchy. In the offshore oil and gas industry this hierarchy is referred to as a tag number system. The following terms will be used for the levels of the assembly hierarchy:

Plant is a set of systems that function together to provide some sort of output. An offshore gas production platform is, for example, considered to be a plant.

System is a set of subsystems that perform a main function in the plant (e.g., generate electro-power, supply steam). The gas compression system on an offshore gas production platform may, for example, be considered as a system. Note that the compression system may consist of several compressors with a high degree of redundancy. Redundant items performing the same main function should be included in the same system.

The system level is recommended as the starting point for the RCM analysis. This means that on an offshore oil/gas platform the starting point of the analysis should, for example, be the gas compression system, and not the whole platform.

The systems may be broken down into subsystems, and subsubsystems, and so on. For the purpose of the RCM analysis, the lowest level of the hierarchy is called *maintainable items*.

Maintainable item is an item that is able to perform at least one significant function as a stand-alone item (e.g., pumps, valves, and electric motors). By this definition a shutdown valve is, for example, a maintainable item, while the valve actuator is not. The actuator is a supporting equipment to the shutdown valve and only has a function as part of the valve. The importance of distinguishing the maintainable items from their supporting equipment is clearly seen in the FMECA in step 6. If a maintainable item is found to have no significant failure modes, then none of the failure modes or causes of the supporting equipment are important, and therefore do not need to be addressed. Similarly, if a maintainable item has only one significant failure mode, then the supporting equipment only needs to be analyzed to determine if there are failure causes that may affect that particular failure mode. Therefore only the failure modes and effects of the maintainable items need to be analyzed in the FMECA in step 6.

By the RCM approach all maintenance tasks and maintenance intervals are decided for the maintainable items. When it comes to the execution of a particular maintenance task on a maintainable item, this will usually involve repair, replacement, or testing of an item or part of the maintainable item. These components/parts are identified in the FMECA in step 6. The RCM analyst should always try to keep the analysis at the highest practical indenture level. The lower the level, the more difficult it is to define performance standards.

It is important that the maintainable items are selected and defined in a clear and unambiguous way in this initial phase of the RCM analysis, since the following steps of the analysis will be based on these items.

Step 3: Functional Failure Analysis A specific system was selected in step 2. The objectives of this step are to:

(i) Identify and describe the system's required functions and performance criteria

System:	Performed by:							Page: of
Drawing no.:	Date:							
Operational mode	System function	Functional requirements	Functional failure	Criticality				Frequency
				S	E	A	C	

Fig. 9.20 Functional failure analysis (FFA) worksheet.

(ii) Describe input interfaces required for the system to operate

(iii) Identify the ways in which the system might fail to function

Step 3(i): Identification of System Functions The system will usually have a high number of different functions. It is essential for the RCM analysis that all the important system functions are identified. The analyst may benefit from using the approach outlined in Chapter 3.

Step 3(ii): Identification of Interfaces The various system functions may be represented by functional block diagrams to illustrate the input interfaces to a function. In some cases we may want to split system functions into subfunctions on an increasing level of detail, down to functions of maintainable items. This may be accomplished by functional block diagrams or reliability block diagrams.

Step 3(iii): Functional Failures The next step is a functional failure analysis (FFA) to identify and describe the potential system failure modes. In most of the RCM references the system failure modes are denoted *functional failures*. Classification schemes for failure modes were discussed in Chapter 3. Such schemes may be used to secure that all relevant functional failures are identified.

The functional failures are recorded on a specific FFA worksheet, that is rather similar to a standard FMECA worksheet. An example of an FFA worksheet is shown in Fig. 9.20. In the first column of the worksheet the various operational modes of the system are recorded. For each operational mode, all the relevant system functions are recorded in column 2. The performance requirements to each function, like target values and acceptable deviations, are listed in column 3. For each system function (in column 2) all the relevant functional failures are listed in column 4. In columns 5 to 8 a criticality ranking of each functional failure in that particular operational mode is given. The reason for including the criticality ranking is to be able to limit the extent of the further analysis by not wasting time on insignificant functional failures. For complex systems such a screening is often important in order not to waste time and money.

The criticality must be judged on the plant level and should be ranked in the four consequence classes:

S: Safety of personnel

E: Environmental impact

A: Production availability

M: Material loss

For each of these consequence classes the criticality may be ranked as high (H), medium (M), low (L), or negligible (N), where the definition of the categories will depend on the specific application. If at least one of the four entries is medium (M) or high (H), the criticality of the functional failure should be classified as significant and be subject to further analysis.

The frequency of the functional failure may also be classified in four categories. The frequency classes may be used to prioritize between the significant functional failures. If all the four criticality entries of a functional failure are low or negligible, and the frequency is also low, the failure is classified as insignificant and disregarded in the further analysis.

Step 4: Critical Item Selection The objective of this step is to identify the maintainable items that are potentially critical with respect to the functional failures identified in step 3 (iii). These maintainable items are denoted *functional significant items* (FSI). Note that some of the less critical functional failures are disregarded at this stage of the analysis.

For simple systems the FSIs may be identified without any formal analysis. In many cases it is obvious which maintainable items have influence on the system functions.

For complex systems with an ample degree of redundancy or with buffers, we may need a formal approach to identify the FSIs. Depending on the complexity of the system, importance ranking based on techniques like fault tree analysis, reliability block diagrams, or Monte Carlo simulation may be suitable. In a petroleum production plant there is often a variety of buffers and rerouting possibilities. For such systems, Monte Carlo next event simulation may often be the only feasible approach.

In addition to the FSIs, we should identify items with high failure rate, high repair costs, low maintainability, long lead time for spare parts, and items requiring external maintenance personnel. These maintainable items are denoted *maintenance cost significant items* (MCSI). The combination of the FSIs and the MCSIs are denoted *maintenance significant items* (MSI).

In the FMECA in step 6, each of the MSIs will be analyzed to identify potential failure modes and effects.

Step 5: Data Collection and Analysis The various steps of the RCM analysis require a variety of input data, like design data, operational data, and reliability data. Reliability data sources are discussed in Chapter 14. Reliability data is necessary to decide the criticality, to mathematically describe the failure process, and to optimize the time between PM tasks.

In some situations there is a complete lack of reliability data. This is the case when developing a maintenance program for new systems. The maintenance program development starts long before the equipment enters service. Helpful sources of information may then be experience data from similar equipment, recommendations from manufacturers, and expert judgments. The RCM method will even in this situation provide useful information.

Step 6: Failure Modes, Effects, and Criticality Analysis The objective of this step is to identify the dominant failure modes of the MSIs identified in step 4. A variety of different FMECA worksheets are proposed in the main RCM references. The FMECA worksheet used in our approach is presented in Fig. 9.21 and is more detailed than most of the FMECA worksheets in the main RCM references. The various columns in our FMECA worksheet are as follows:

- *MSI*. In this column we record the maintainable item number in the assembly hierarchy (tag number), optionally with a descriptive text.

- *Operational mode*. The MSI may have various operational modes, for example, running and standby. The operational modes are listed, one by one.

- *Function*. The various functions for each operational mode of the MSI are listed.

- *Failure mode*. The failure modes for each function are listed.

- *Effect of failure/severity class*. The effect of a failure is described in terms of the "worst-case" outcome for the S, E, A, and C categories introduced in step 3 (iii). The criticality may be specified by the same four classes as in step 3 (iii), or by some numerical severity measure. A failure of an MSI will not necessarily give a worst-case outcome resulting from redundancy, buffer capacities, and the like. Conditional likelihood columns are therefore introduced.

- *Worst-case probability*. The worst-case probability is defined as the probability that an equipment failure will give the worst-case outcome. To obtain a numerical probability measure, a system model is sometimes required. This will often be inappropriate at this stage of the analysis, and a descriptive measure may be used.

- *MTTF*. Mean time to failure for each failure mode is recorded. Either a numerical measure or likelihood classes may be used.

The information described so far should be entered for all failure modes. A screening may now be appropriate, giving only dominant failure modes, that is, items with high criticality.

System:					Performed by:												
Drawing no.:					Date:							Page: of					

	Description of item		Failure mode	Effect of failure			MTTF	Criti-cality	Failure cause	Failure mecha-nism	%MTTF	Failure charac-teristic	Mainte-nance action	Failure charac-teristic measure	Recom-mended interval
MSI	Operational mode	Function		Consequence class		Worst case probability									
				S E A C		S E A C									

Fig. 9.21 RCM-FMECA worksheet.

- *Criticality.* The criticality field is used to tag off the dominant failure modes according to some criticality measure. A criticality measure should take failure effect, worst-case probability, and MTTF into account. "Yes" is used to tag off the dominant failure modes.

For the dominant failure modes the following fields are required:

- *Failure cause.* For each failure mode there may be several failure causes. An MSI failure mode will typically be caused by one or more component failures. Note that supporting equipment to the MSIs entered in the FMECA worksheet is for the first time considered in this step. In this context a failure cause may therefore be a failure mode of a supporting equipment. A "fail to close" failure of a safety valve may, for example, be caused by a broken spring in the failsafe actuator.

- *Failure mechanism.* For each failure cause, there is one or several failure mechanisms. Examples of failure mechanisms are fatigue, corrosion, and wear.

- *%MTTF.* The MTTF was entered on an MSI failure mode level. It is also interesting to know the (marginal) MTTF for each failure mechanism. To simplify, a percent is given, and the (marginal) MTTF may be estimated for each failure mechanism. The %MTTF will obviously only be an approximation since the effects of the various failure mechanisms usually are strongly interdependent.

- *Failure characteristic.* Failure propagation may be divided into three classes.

    1. The failure propagation may be measured by one or several (condition monitoring) indicators. The failure is referred to as a gradual failure.

    2. The failure probability is age dependent, that is, there is a predictable wear-out limit. The failure is referred to as an *aging failure*.

    3. Complete randomness. The failure cannot be predicted by either condition monitoring indicators or by measuring the age of the item. The time to failure can only be described by an exponential distribution, and the failure is referred to as a sudden failure.

- *Maintenance action.* For each failure mechanism, an appropriate maintenance action may hopefully be found by the decision logic in step 7. This field can therefore not be completed until step 7 is performed.

- *Failure characteristic measure.* For gradual failures, the condition monitoring indicators are listed by name. Aging failures are described by an aging parameter, that is, the shape parameter (α) in the Weibull distribution is recorded.

- *Recommended maintenance interval.* In this column the interval between consecutive maintenance tasks is given. The length of the interval is determined in step 8.

Step 7: Selection of Maintenance Actions This step is the most novel compared to other maintenance planning techniques. A decision logic is used to guide the analyst through a question and answer process. The input to the RCM decision logic is the dominant failure modes from the FMECA in step 6. The main idea is for each dominant failure mode to decide whether a PM task is applicable and effective or it will be best to let the item deliberately run to failure and afterwards carry out a corrective maintenance task. There are generally three main reasons for doing a PM task:

1. To prevent a failure

2. To detect the onset of a failure

3. To discover a hidden failure

The following basic maintenance tasks are considered:

1. Scheduled on-condition task

2. Scheduled overhaul

3. Scheduled replacement

4. Scheduled function test

5. Run to failure

Scheduled on-condition task is a task to determine the condition of an item, for example, by condition monitoring. There are three criteria that must be met for an on-condition task to be applicable.

1. It must be possible to detect reduced failure resistance for a specific failure mode.

2. It must be possible to define a potential failure condition that can be detected by an explicit task.

3. There must be a reasonable consistent age interval between the time of potential failure (P) is detected and the time of functional failure (F).

The time interval from which it is possible to reveal a potential failure (P) by the currently used monitoring technique until a functional failure (F) occurs is called the *PF interval*. The PF interval can be regarded as the potential warning time in advance of a functional failure. The longer the PF interval, the more time one has to make a good decision and plan actions. PF intervals were discussed on page 394.

Scheduled overhaul of an item may be performed at or before some specified age limit and is often called hard time maintenance. An overhaul task is considered applicable to an item only if the following criteria are met:

1. There must be an identifiable age at which there is a rapid increase in the item's failure rate function.

2. A large proportion of the items must survive to that age.

3. It must be possible to restore the original failure resistance of the item by reworking it.

Scheduled replacement is replacement of an item (or one of its parts) at or before some specified age or time limit. A scheduled replacement task is applicable only under the following circumstances:

1. The item must be subject to a critical failure.

2. The item must be subject to a failure that has major potential consequences.

3. There must be an identifiable age at which the item shows a rapid increase in the failure rate function.

4. A large proportion of the items must survive to that age.

Scheduled function test is a scheduled failure-finding task or inspection of a hidden function to identify failures. Failure-finding tasks are preventive only in the sense that they prevent surprises by revealing failures of hidden functions. A scheduled function test task is applicable to an item under the following conditions:

1. The item must be subject to a functional failure that is not evident to the operating crew during the performance of normal duties: tasks that have to be based on information about the failure rate function, the likely consequences and costs of the failure the PM task is supposed to prevent, the cost and risk of the PM task, and so on.

2. The item must be one for which no other type of task is applicable and effective.

Run to failure is a deliberate decision to run to failure because the other tasks are not possible or the economics are less favorable.

Preventive maintenance will not prevent all failures. Consequently, if there is a clear identifiable failure mode that cannot be adequately addressed by an applicable and effective PM task that will reduce the probability of failure to an acceptable level, then there is need to redesign or modify the item. If the consequences of failures are related to safety or the environment, redesign will normally be mandatory. For operational and economic consequences of failure this may be desirable, but a cost-benefit assessment has to be performed. The criteria given for using the various tasks should only be considered as guidelines for selecting an appropriate task. A task might be found appropriate even if some of the criteria are not fulfilled.

A variety of different RCM decision logic diagrams are used in the main RCM references. Some of these are rather complex. The decision logic diagram shown in Fig. 9.22 is much simpler than those found in other RCM references. The resulting maintenance tasks will, however, in many cases be the same. It should be emphasized that such a logic can never cover all situations. In the case of a hidden function with aging failures, a combination of scheduled replacements and function tests is required.

412 RELIABILITY OF MAINTAINED SYSTEMS

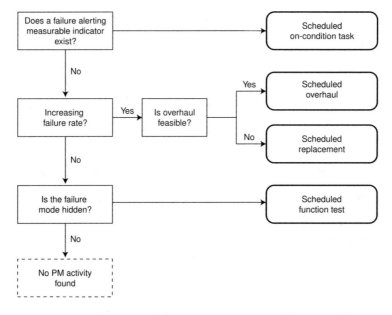

Fig. 9.22 Maintenance task assignment/decision logic.

Step 8: Determination of Maintenance Intervals Some of the PM tasks are to be performed at regular intervals. To determine the optimal interval is a very difficult task that has to be based on information about the failure rate function, the likely consequences and costs of the failure the PM task is supposed to prevent, the cost and risk of the PM task, and so on. Some models were discussed in Section 9.6.

In practice the various maintenance tasks have to be grouped into maintenance packages that are carried out at the same time, or in a specific sequence. The maintenance intervals can therefore not be optimized for each single item. The whole maintenance package has, at least to some degree, to be treated as an entity.

Step 9: Preventive Maintenance Comparison Analysis Two overriding criteria for selecting maintenance tasks are used in RCM. Each task selected must meet two requirements:

1. It must be applicable.
2. It must be effective.

Applicability means that the task is applicable in relation to our reliability knowledge and in relation to the consequences of failure. If a task is found based on the preceding analysis, it should satisfy the applicability criterion. A PM task is applicable if it can eliminate a failure, or at least reduce the probability of the occurrence of failure to an acceptable level, or reduce the impact of the failure.

Cost-effectiveness means that the task does not cost more than the failure(s) it is going to prevent.

A PM task's effectiveness is a measure of how well it accomplishes its purpose and if it is worth doing. Clearly, when evaluating the effectiveness of a task, we are balancing the cost of performing the maintenance with the cost of not performing it. The cost of a PM task may include:

1. The risk/cost related to maintenance induced failures
2. The risk the maintenance personnel is exposed to during the task
3. The risk of increasing the likelihood of failure of another item while the one is out of service
4. The use and cost of physical resources
5. The unavailability of physical resources elsewhere while in use on this task
6. Production unavailability during maintenance
7. Unavailability of protective functions during maintenance

In contrast, the cost of a failure may include:

1. The consequences of the failure should it occur (loss of production, possible violation of laws or regulations, reduction in plant or personnel safety, or damage to other equipment)
2. The consequences of not performing the PM task even if a failure does not occur (e.g., loss of warranty)
3. Increased premiums for emergency repairs (such as overtime, expediting costs, or high replacement power cost)

Step 10: Treatment of non-MSIs In step 4 critical items (MSIs) were selected for further analysis. A remaining question is what to do with the items which are not analyzed. For plants already having a maintenance program, a brief cost evaluation should be carried out. If the existing maintenance cost related to the non-MSIs is insignificant, it is reasonable to continue this program. See Paglia et al. (1991) for further discussion.

Step 11: Implementation A necessary basis for implementing the result of the RCM analysis is that the organizational and technical maintenance support functions are available. A main issue is therefore to ensure that these support functions are available. Experience shows that many accidents occur either during maintenance or because of inadequate maintenance. When implementing a maintenance program it is therefore of vital importance to consider the risk associated with the various maintenance tasks. For complex maintenance operations it may be relevant to perform a job safety analysis combined with a human HAZOP to reveal possible hazards and human errors related to the maintenance task. See also Hoch (1990) for a further discussion on implementing the RCM analysis results.

Step 12: In-service Data Collection and Updating The reliability data we have access to at the outset of the analysis may be scarce, or even second to none. In our opinion, one of the most significant advantages of RCM is that we systematically analyze and document the basis for our initial decisions and, hence, can better utilize operating experience to adjust that decision as operating experience data become available. The full benefit of RCM is therefore only obtained when operation and maintenance experience is fed back into the analysis process.

The updating process should be concentrated on three major time perspectives:

1. Short-term interval adjustments

2. Medium-term task evaluation

3. Long-term revision of the initial strategy

For each significant failure that occurs in the system, the failure characteristics should be compared with the FMECA. If the failure was not covered adequately in the FMECA, the relevant part of the RCM analysis should, if necessary, be revised.

The short-term update may be considered as a revision of previous analysis results. The input to such an analysis is updated failure information and reliability estimates. This analysis should not require much resources, as the framework for the analysis is already established. Only steps 5 to 8 in the RCM process will be affected by short-term updates.

The medium-term update should carefully review the basis for the selection of maintenance tasks in step 7. Analysis of maintenance experience may identify significant failure causes not considered in the initial analysis, requiring an updated FMECA in step 6.

The long-term revision should consider all steps in the analysis. It is not sufficient to consider only the system being analyzed; it is required to consider the entire plant with its relations to the outside world, for example, contractual considerations, new laws regulating environmental protection, and so on.

9.7.2 Total Productive Maintenance

Total productive maintenance (TPM) is an approach to maintenance management that was developed in Japan (Nakajima 1988) to support the implementation of just-in-time manufacturing and associated efforts to improve product quality. TPM activities focus on eliminating the *six major losses*:

- Availability losses

    1. *Equipment failure (breakdown) losses.* Associated costs include downtime, labor, and spare part cost.

    2. *Setup and adjustment losses* that occur during product changeovers, shift change, or other changes in operating conditions

- Performance (speed) losses

3. *Idling and minor stoppages* that typically last up to 10 minutes. These include machine jams and other brief stoppages that are difficult to record and consequently usually are hidden from efficiency reports. When combined, they can represent substantial equipment downtime.

4. *Reduced speed losses* that occur when equipment must be slowed down to prevent quality defects or minor stoppages. In most cases, this loss is not recorded because the equipment continues to operate, albeit at a lower speed. Speed losses obviously have a negative effect on productivity and asset utilization.

- Quality losses

5. *Defects in process and reworking losses* that are caused by manufacture of defective or substandard products that must be reworked or scrapped. These losses include the labor and material costs (if scrapped) associated with off-specification production.

6. *Yield losses* reflect the wasted raw materials associated with the quantity of rejects and scrap that result from startups, changeovers, equipment limitations, poor product design, and so on. It excludes the category 5 defect losses that result during normal production.

The six major losses determine the *overall equipment effectiveness* (OEE), which is a multiplicative combination of equipment availability losses (1 and 2), equipment performance losses (3 and 4), and quality losses (5 and 6). The time concepts used in TPM are illustrated in Fig. 9.23. The factors used to determine the OEE are:

$$
\begin{aligned}
&\text{Operational availability} && A_O = t_F/t_R \\
&\text{Performance rate} && R_P = t_N/t_F \\
&\text{Quality rate} && R_Q = t_U/t_F
\end{aligned}
$$

The quality rate may alternatively be measured as

$$\text{Quality rate} = R_Q = \frac{\text{No. of processed products} - \text{No. of rejected products}}{\text{No. of processed products}}$$

The OEE is defined as

$$\text{OEE} = A_O \cdot R_P \cdot R_Q \qquad (9.43)$$

The OEE is used as an indicator of how well machines, production lines, and processes are performing in terms of availability, performance, and quality. An OEE $\geq 85\%$ is considered to be "world class."

Total productive maintenance has been described as a partnership approach to maintenance. Under TPM, small groups or teams create a cooperative relationship between maintenance and production. Production workers become involved in performing maintenance work allowing them to play a role in equipment monitoring and upkeep. This raises the skill of production workers and allows them to be more

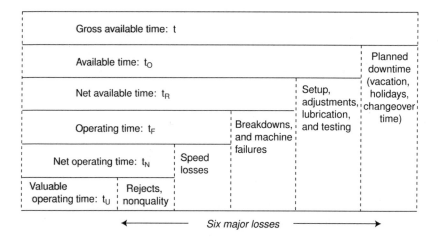

Fig. 9.23 Time concepts used in total productive maintenance.

effective in maintaining the equipment in good condition. Team-based activities play an important role in TPM. Team-based activities involve groups from maintenance, production, and engineering. The technical skill of engineers and the experience of maintenance workers and equipment operators are communicated through these teams. The objective of the team-based activities is to improve equipment performance through better communication of current and potential equipment problems. Maintainability improvement and maintenance prevention are two team-based TPM activities. TPM has several benefits. The efforts of maintenance improvement teams should result in improved equipment availability and reduced maintenance costs. Maintainability improvement should result in increased maintenance efficiency and reduced repair time. TPM resembles total quality management (TQM) in several aspects, such as (i) total commitment to the program from upper level management is required, (ii) employees must be empowered to initiate corrective actions, and (iii) a long range outlook must be accepted as TPM may take a year or more to implement and is an ongoing process.

PROBLEMS

9.1 Consider an item that is replaced with a new item of the same type after regular intervals of length τ. If the item fails within a replacement interval, it is repaired to an "as good as new" condition. Show that the limiting availability A of the item does not exist.

9.2 An item has constant failure rate $\lambda = 5 \cdot 10^{-4}$ hours^{-1}. When the item fails, it is repaired to an "as good as new" condition. The associated mean downtime is 6 hours. The item is supposed to be in continuous operation.

(a) Find the average availability A_{av} of the item.

(b) How many hours per year will the item on average be out of operation?

9.3 Consider an item with time to failure T that has a Weibull distribution with shape parameter 2.25 and scale parameter $\lambda = 4 \cdot 10^{-4}$ hours^{-1}. When the item fails it is repaired to an as good as new condition. The repairtime (downtime) D has a lognormal distribution with median equal to 4.5 hours, and error factor 2 (see Section 2.14). The item is supposed to be in continuous operation.

(a) Find the average availability of the item.

(b) A preventive maintenance task that takes 5 hours is performed every 300 hours. Find the operational availability of the item.

9.4 An machine with constant failure rate $\lambda = 2 \cdot 10^{-3}$ hours^{-1} is operated 8 hours per day, 230 days per year. The mean downtime required to repair the machine and bring it back into operation is MDT $= 5$ hours. The machine can only fail during active operation. If a repair action cannot be completed within normal working hours, overtime will be used to complete the repair such that the machine is available next morning.

(a) Determine the average availability of the machine (during the planned working hours).

(b) Determine the average availability of the machine if the use of overtime were not allowed.

9.5 An item is exposed to wear and has failure rate function $z_1(t) = \beta t$.

(a) Determine the survival probability $R(t)$ of the item at time $t = 2000$ hours, when $\beta = 5 \cdot 10^{-8}$ hours^{-2}.

The item will be overhauled after regular intervals of length τ. We assume that the overhaul will reduce the failure rate and that we may use the following model:

$$z(t) = \beta t - \alpha k \tau \quad \text{for } k\tau < t \leq (k+1)\tau$$

where k is the number of overhauls after time $t = 0$.

(b) Draw a sketch of $z(t)$. Explain what is meant by the term $\alpha k \tau$. Do you consider this model to be realistic?

(c) Determine the survival probability $R(t)$ for time $t = k\tau$, that is, just before overhaul k. Draw a sketch of $R(t)$ as a function of t.

(d) Find the conditional probability that the item is functioning just before overhaul $k + 1$, when you know that it was functioning just before overhaul k.

418 RELIABILITY OF MAINTAINED SYSTEMS

9.6 Consider the age replacement policy, and find the mean time between actual item failures, $E(Y_i)$ by equation (9.24) when the distribution of the time to failure T of the item has

  (a) An exponential distribution with failure rate λ. Give a "physical" interpretation of the result you get.

  (b) A gamma distribution with parameters $(2, \lambda)$.

9.7 Show that the mean time between replacements in Example 9.15 is equal to $(\beta \cdot y_p + 1) \cdot \tau$.

9.8 Consider the block replacement policy that is described on page 387, and find the optimal number of spares when the cost of a spare, c_s, per spare item and per time unit is included.

  (a) Determine the optimal average maintenance cost including the average spare cost as a function of the block replacement time t_0 and the number m of sparse units.

  (b) Plot the curve of the optimal average maintenance cost as a function of m. Assume that the time to failure T has a gamma distribution with parameters (α, β). Select realistic values for the necessary input parameters and generate the plot.

9.9 Use a spreadsheet program (like Excel) to generate 20 pseudo-random numbers in the interval $(0, 1)$. Use the transformation illustrated in Fig. 9.7 to generate corresponding times to failure that are Weibull distributed with scale parameter $\lambda = 5 \cdot 10^{-5}$ hours^{-1} and shape parameter $\alpha = 2.7$. Find the sample mean and the sample standard deviation, and compare with the mean and standard deviation of the Weibull distribution.

10
Reliability of Safety Systems

10.1 INTRODUCTION

In this chapter we discuss reliability aspects of safety systems that are designed to be activated upon hazardous process deviations (*process demands*) to protect people, the environment, and material assets. In Example 3.2 we discussed the safety systems of a gas/oil separator. The safety system had three *protection layers:*

1. An inlet shutdown system comprising pressure sensors, a logic solver, and shutdown valves

2. A pressure relief system comprising two pressure relief valves

3. A rupture disc

In Example 3.2 the process demand was a blockage of the gas outlet line. The process demand will cause a rapid increase of the pressure in the separator, and the separator might rupture if safety systems were not available. The system the protection layers are installed to protect is often referred to as the *equipment under control* (EUC). In this example, the EUC is the separator. An EUC may have several hazardous process demands that require their own safety systems. In the process industry, the potential process demands are usually identified by a hazards and operability (HAZOP) study (e.g., see IEC 61882).

Process demands may be classified according to their frequency of occurrence. Some process demands occur so frequently that the safety system is operated almost continuously. An example of such a safety system is the brakes of a car. "Process" demands for the brakes will occur several times each time we drive the car, and brake

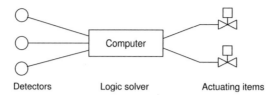

Fig. 10.1 Sketch of a simple safety instrumented system (SIS).

failures and malfunctions may therefore be detected almost immediately. The brakes are said to be a safety system with a *high demand mode of operation*.

Other process demands occur very infrequently and the safety system is therefore in a passive state for long periods of time. An example of such a system is the airbag system in a car. The airbag system remains passive until a "process" demand occurs, and is said to be a safety system with a *low demand mode of operation*. Such a safety system may fail in passive state, and the failure may remain *hidden* until a process demand occurs or until the system is tested. To reveal hidden failures, safety systems with low demand mode of operation are normally function tested at regular intervals.

A safety system composed of sensors, logic solvers, and actuating items is called a *safety instrumented system* (SIS). A brief introduction to SISs is given in Section 10.2. Several standards have been issued setting requirements to safety instrumented systems. The most important of these standards is IEC 61508 "Functional safety of electrical/electronic/programmable electronic safety-related systems" that is briefly introduced in Section 10.6.

In Section 10.3 we introduce the main reliability models for the elements of safety systems and discuss various issues related to the analysis of such systems. The discussion is mainly limited to systems with a low demand of operation, that are periodically tested. Problems related to common cause failures and spurious activation of the systems are discussed. A Markov approach to analyzing safety systems is introduced in Section 10.8.

10.2 SAFETY INSTRUMENTED SYSTEMS

A safety instrumented system (SIS) is an independent protection layer that is installed to mitigate the risk associated with the operation of a specified hazardous system, which is referred to as the *equipment under control* (EUC). The EUC may be various types of equipment, machinery, apparatus, or plant used for manufacturing, process, transportation, medical, or other activities. An SIS is composed of *sensors*, *logic solvers*, and *actuating items*.[1] The actuating items may, for example, be shutdown valves or brakes. A sketch of a simple SIS is shown in Fig. 10.1. SISs are used in many sectors of our society, for example, as emergency shutdown systems in haz-

---

[1] Actuating items are called *final elements* in some standards.

ardous chemical plants, fire and gas detection and alarm systems, pressure protection systems, dynamic positioning systems for ships and offshore platforms, automatic train stop (ATS) systems, fly-by-wire operation of aircraft flight control surfaces, antilock brakes, and airbag systems in automobiles, and systems for interlocking and controlling the exposure dose of medical radiotherapy machines. Recent developments include network-based safety-related systems, often facilitated by Internet technology.

A *safety instrumented function* (SIF) is a function that is implemented by an SIS and that is intended to achieve or maintain a safe state for the EUC with respect to a specific process demand. An SIS may consist of one or more SIFs.

In addition to the elements that are illustrated in Fig. 10.1 (detectors, logic solver, and actuating items) an SIS will usually comprise electric power supply, user interface, pneumatic and/or hydraulic system, electrical connections, and various process connections.

In the standard IEC 61508 an SIS is referred to as an "electrical/electronic/programmable electronic (E/E/PE) safety-related system."

Main Functions An SIS has two main system functions:

1. When a predefined process demand (deviation) occurs in the EUC, the deviation shall be detected by the SIS *sensors*, and the required *actuating items* shall be activated and fulfill their intended functions.

2. The SIS shall not be activated spuriously, that is, without the presence of a predefined process demand (deviation) in the EUC.

A failure to perform the first system function is called a *fail to function* (FTF), and a failure of the second function is called a *spurious trip* (ST).

Example 10.1 Safety Systems on Offshore Oil and Gas Platforms
The safety systems on an offshore oil and gas platform are usually grouped into three categories:

1. Process control (PC) system

2. Process Shutdown (PSD) system

3. Fire and gas detection (FGD) and emergency shutdown (ESD) system

The objective of the process control system is to keep an EUC process within preset limits. Various process control valves and regulators are used to control the process, based on signals from temperature, pressure, level, and other types of transmitters. When the process deviates from normal values, the process shutdown system is activated and will close down the EUC. The required actions for each type of deviation/demand is programmed into the logic solver. The actions may involve activation of alarms, closure of shutdown valves, and opening of relief valves. The process control and process shutdown systems are *local* systems that are related to a specific EUC. For some types of process demands that have a potential for a major accident the ESD system is activated. Relevant process demands include fires, gas leaks, and loss of main power. The required ESD actions are usually grouped into several levels, depending on the type of deviation/demand that is detected and where it is detected. The top ESD level will usually involve shutdown of the whole platform and evacuation of the personnel. □

10.2.1 Testing of SIS Functions

Many SISs are passive systems that are only activated when a specified process demand occurs in the EUC. A fire detection and extinguishing system should, for example, only be activated when a fire occurs. Such a system may fail in the passive position and the failure may remain undetected (hidden) until the system is activated or tested.

Diagnostic Self-Testing In modern SISs the logic solver is often programmable and may carry out *diagnostic self-testing* during on-line operation. The logic solver may send frequent signals to the detectors and to the actuating items and compare the responses with predefined values. The diagnostic testing can reveal failures of input and output devices, and to an increasing degree, also failures of detectors and actuating items. In many cases the logic solver consists of two or more redundant computers that can carry out diagnostic self-testing of each other. The fraction of failures that can be revealed by diagnostic self-testing is called the *diagnostic coverage*. The self-testing may be carried out so often that failures are detected almost immediately.

Function Testing The diagnostic self-testing cannot reveal all failure modes and failure causes, and the various parts of the SIS are therefore often function tested at regular intervals. The objective of a function test is to reveal hidden failures, and to verify that the system is (still) able to perform the required functions if a process demand should occur. It is sometimes not feasible to carry out a fully realistic function test, because it may not be technically feasible or be very time consuming. Another reason may be that the test itself leads to unacceptable hazards. It is, for example, not realistic to fill a room with toxic gases to test a gas detector. The gas detector is rather tested with a nontoxic test gas that is directly input to the gas detector through a test pipe.

Consider a safety valve that is installed in a pipeline. During normal operation the valve is kept in open position. If a specified process demand occurs, the valve should close and stop the flow in the pipeline. A realistic test of the safety valve would imply to close the valve and apply a pressure to the upstream side of the valve that is equal to the maximum expected shut-in pressure in a demand situation. This may not be possible, and we may have to suffice with only checking that the valve is able to close on demand, and perhaps to check the valve for leakage with normal shut-in pressure. In some cases it may be possible to pressure test the valve from the downstream side. In this case, we may be able to test the valve to maximum shut-in pressure, but the wrong side of the seals is tested. In some situations it may be hazardous to shut down a flow, and closure of the valve should therefore be avoided.

Some valve functions may be tested by partly closing the valve (a gate valve may be moved some few millimeters, and a ball valve may be rotated some degrees). This type of testing is called *partial stroke testing*.

Some actuating items employ an actuating principle that is not possible to function test without destroying the item. This is, for example, the case for the pyrotechnic seat belt tensioners in automobiles.

10.2.2 Failure Classification

A general introduction to failures and failure classification was given in Chapter 3. For an SIS and the SIS subsystems we may use the following failure mode classification (see IEC 61508):

1. *Dangerous (D)*. The SIS does not fulfill its required safety-related functions upon demand. These failures may further be split into:

    (a) *Dangerous undetected (DU)*. Dangerous failures are preventing activation on demand and are revealed only by testing or when a demand occurs. DU failures are sometimes called *dormant* failures.

    (b) *Dangerous detected (DD)*. Dangerous failures that are detected immediately when they occur, for example, by an automatic, built-in self-test. The average period of unavailability due to a DD failure is equal to the mean downtime, MDT, that is, the mean time elapsing from the failure is detected by the built-in self-test until the function is restored.

2. *Safe failures (S)*. The SIS has a nondangerous failure. These failures may further be split into:

    (a) *Safe undetected (SU)*. Nondangerous failures that are not detected by automatic self-testing.

    (b) *Safe detected (SD)*. Non-dangerous failures that are detected by automatic self-testing. In some configurations early detection of failures may prevent an actual spurious trip of the system.

The failure mode classification is illustrated in Fig. 10.2.

The failure modes may also be classified according to the cause of the failure, as (see IEC 61508 and Corneliussen and Hokstad 2003):

1. *Random hardware failures.* These are physical failures where the supplied service deviates from the specified service due to physical degradation of the item. Random hardware failures can further be split into

    (a) *Aging failures*. These failures occur under conditions within the design envelope of the item. Aging failures are also called *primary failures*

    (b) *Stress failures*. These failures occur due to excessive stresses on the item. The excessive stresses may be caused by external causes or by human

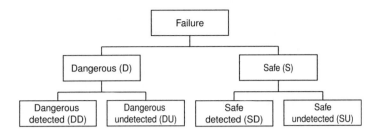

Fig. 10.2 Failure mode classification.

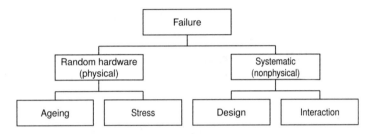

Fig. 10.3 Failure classification by cause of failure. (Adapted from Corneliussen and Hokstad 2003).

errors during operation and maintenance. Stress failures are also called *secondary failures*.

2. *Systematic failures*. These failures are nonphysical failures where the supplied service deviates from the specified service without any physical degradation of the item. The failures can only be eliminated by a modification of the design or of the manufacturing process, operational procedures, or documentation. The systematic failures can further be split into

   (a) *Design failures*. These failures are initiated during engineering, manufacturing, or installation and may be latent from the first day of operation. Examples include software failures, sensors that do not discriminate between true and false demands, and fire/gas detectors that are installed in a wrong place, where they are prohibited from detecting the demand.

   (b) *Interaction failures*. These failures are initiated by human errors during operation or maintenance/testing. Examples are loops left in the override position after completion of maintenance and miscalibration of sensors during testing. Scaffolding that cover up a sensor making it impossible to detect an actual demand is another example of an interaction failure.

The failure mode classification by the cause of failure is shown in Fig. 10.3.

Example 10.2 Safety Shutdown Valve
A safety shutdown valve is installed in a gas pipeline feeding a production system.

If an emergency occurs in the production system, the valve should close and stop the gas flow. The valve is a hydraulically operated gate valve. The actual open/close function is performed by sliding a rectangular gate, having a bore equal to the bore of the conduct. The gate is moved by a hydraulic piston connected to the gate by a stem. The gate valve has a *fail-safe* actuator. The valve is opened and kept open by hydraulic control pressure on the piston. The fail-safe function is achieved by a steel spring that is compressed by the hydraulic pressure. The valve is automatically closed by the spring force when the hydraulic pressure is bled off.

The valve is connected to an ESD system. When an emergency situation is detected in the production system, an electric signal is sent to the valve control system and the pressure is bled off.

In this example, we will only consider the valve but will come back to the rest of the ESD system later in the chapter.

The main failure modes of the valve are:

- *Fail to close on command* (FTC). This failure mode may be caused by a broken spring, blocked return line for the hydraulic fluid, too high friction between the stem and the stem seal, too high friction between the gate and the seats, or by sand, debris, or hydrates in the valve cavity

- *Leakage (through the valve) in closed position* (LCP). This failure mode is mainly caused by corrosion and/or erosion on the gate or the seat. It may also be caused by misalignment between the gate and the seat.

- *Spurious trip* (ST). This failure mode occurs when the valve closes without a signal from the ESD system. It is caused by a failure in the hydraulic system or a leakage in the supply line from the control system to the valve.

- *Fail to open on command* (FTO). When the valve is closed, it may fail to reopen. Possible causes may be leakage in the control line, too high friction between the stem seals and the stem, too high friction between the gate and the seats, and sand, debris, or hydrates in the valve cavity.

The valve has been installed to close the flow (and keep tight) following a demand. The failure modes FTC and LCP prevent this function and are *dangerous* failure modes with respect to safety. ST and FTO failures are normally not dangerous with respect to safety but will cause production shutdown and lost income.

Since the valve is normally in open position, we are not able to detect the dangerous failure modes, FTC and LCP, unless we try to close the valve. These dangerous failure modes are *hidden* during normal operation and are therefore called *dangerous undetected* (DU) failure modes. To reveal and repair DU failures the valve is tested periodically with test interval τ. This means that the valve is tested at times $0, \tau, 2\tau, \ldots$. A typical test interval may be 3 to 12 months. During a standard test, the valve is closed and tested for leakage. The cause of a DU failure may occur at a random point of time within a test interval, but will not be manifested (revealed) until the valve is tested or attempted closed due to operational reasons. The safety unavailability (SU) of the valve will obviously be lower with a short test interval than with a long test interval. The gas flow has to be closed down during the test, and the test will usually lead to a production loss. In some situations the shutdown and startup procedure may also have safety implications. The length of the test interval τ must therefore be a compromise between safety and economic considerations.

In some situations, it may be impractical and even dangerous to close the valve, and we have to suffice with *partial stroke* testing. In this case we move the gate slightly and monitor the movement of the valve stem. The test will reveal some of the DU failure causes, but not all. A hidden LCP failure will, for example, not be revealed.

The ST failure will stop the flow and will usually be detected immediately. An ST failure is therefore called an *evident* failure. In some systems, an ST failure may also have significant safety implications.

The FTO failure may occur after a test and is an evident failure. The FTO failure will cause a repair intervention but will have no extra safety implications, since the gas flow is shut down when the failure occurs. The FTO failure is therefore called a *noncritical* or *safe* failure. □

10.3 PROBABILITY OF FAILURE ON DEMAND

In this section we consider a safety item (component or system) that is tested periodically, in the same way as the safety valve in Example 10.2. We assume that no diagnostic self-testing is carried out, and that that *all* hidden failures are revealed by the function testing. Some of the main concepts that we use in this section are introduced in Example 10.2. The reader should therefore study the example carefully before reading this section.

Consider a safety item that is put into operation at time $t = 0$. The item may be a safety valve (e.g., shutdown valve or relief valve), a sensor (e.g., fire/gas detector, pressure sensor, or level sensor), or a logic solver. The item is tested and, if necessary, repaired or replaced after regular time intervals of length τ. The time required to test and repair the item is considered to be negligible. After a test (repair) the item is considered to be "as good as new." We say that the item is functioning as a safety *barrier* if a DU failure mode is not present.

The state variable $X(t)$ of an item with respect to DU failures is

$$X(t) = \begin{cases} 1 & \text{if the item is able to function as a safety barrier,} \\ & \text{(i.e., no DU failure is present)} \\ 0 & \text{if the item is not able to function as a safety barrier,} \\ & \text{(i.e., a DU failure is present)} \end{cases}$$

The state variable $X(t)$ is illustrated in Fig. 10.4.

Probability of Failure on Demand Let T denote the time to DU failure of the item, with distribution function $F(t)$.

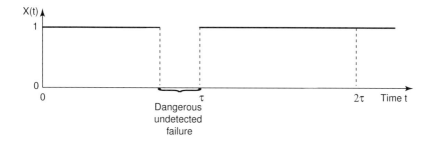

Fig. 10.4 The state $X(t)$ of a periodically tested item with respect to DU failures.

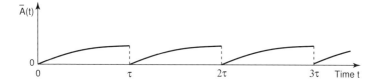

Fig. 10.5 The safety unavailability $\bar{A}(t)$ of a periodically tested item.

The safety *unavailability* $\bar{A}(t)$ of the item in the *first* test interval $(0, \tau]$ is

$$\begin{aligned} \bar{A}(t) &= \Pr(\text{a DU failure has occurred at, or before, time } t) \\ &= \Pr(T \leq t) = F(t) \end{aligned} \quad (10.1)$$

Since the item is assumed to be "as good as new" after each test, the test intervals $(0, \tau], (\tau, 2\tau], \ldots$, are all equal from a stochastic point of view. Hence the safety unavailability $\bar{A}(t)$ of the item is as illustrated in Fig. 10.5. Note that $\bar{A}(t)$ is discontinuous for $t = n\tau$, for $n = 1, 2, \ldots$. If a demand for the safety item occurs at time t, the safety unavailability $\bar{A}(t)$ denotes the probability that the item will fail to respond adequately to the demand. The safety unavailability $\bar{A}(t)$ is therefore often called the *probability of failure on demand (PFD)* at time t.

In most applications we are not interested in the PFD as a function of time. It is sufficient to know the long run average value of PFD. The average value is denoted PFD, without reference to the time t. Because of the periodicity of $\bar{A}(t)$, the long run average PFD is equal to the average value of $\bar{A}(t)$ in the first test interval $(0, \tau]$,

$$\text{PFD} = \frac{1}{\tau} \int_0^\tau \bar{A}(t)\, dt = \frac{1}{\tau} \int_0^\tau F(t)\, dt \quad (10.2)$$

Let $R(t)$ denote the survivor function of the item with respect to DU failure. Since $R(t) = 1 - F(t)$, (10.2) may alternatively be written

$$\text{PFD} = 1 - \frac{1}{\tau} \int_0^\tau R(t)\, dt \quad (10.3)$$

Consider a test interval, and let T_1 be the part of this test interval where the item is able to function as a safety barrier. Let D_1 be the part of the interval where the item is in a failed state (i.e., a DU failure is present but has not been detected), such that $T_1 + D_1 = \tau$.

The PFD in (10.2) is the average safety unavailability in a test interval. Since the average safety unavailability is the mean proportion of time the item is not functioning as a safety barrier, the PFD may be written as

$$\text{PFD} = \frac{E(D_1)}{\tau} \qquad (10.4)$$

The mean downtime in a test interval is therefore

$$E(D_1) = \int_0^\tau F(t)\,dt \qquad (10.5)$$

and the mean uptime in a test interval is

$$E(T_1) = \tau - \int_0^\tau F(t)\,dt = \int_0^\tau R(t)\,dt \qquad (10.6)$$

The PFD may from (10.4) also be interpreted as the mean proportion of time the item is not functioning as a safety barrier upon demand. The PFD is therefore sometimes called the *mean fractional deadtime (MFDT)* of the item.

Example 10.3 Single Item
A fire detector, that is tested at regular intervals of length τ, has constant failure rate λ_{DU} with respect to DU failures. The survivor function of the fire detector is $R(t) = e^{-\lambda_{\text{DU}} t}$ and the PFD is from (10.3)

$$\begin{aligned}\text{PFD} &= 1 - \frac{1}{\tau} \int_0^\tau R(t)\,dt = 1 - \frac{1}{\tau} \int_0^\tau e^{-\lambda_{\text{DU}} t}\,dt \\ &= 1 - \frac{1}{\lambda_{\text{DU}} \tau}\left(1 - e^{-\lambda_{\text{DU}} \tau}\right)\end{aligned} \qquad (10.7)$$

If we replace $e^{-\lambda_{\text{DU}} \tau}$ in (10.7) by its Maclaurins series, we get

$$\begin{aligned}\text{PFD} &= 1 - \frac{1}{\lambda_{\text{DU}} \tau}\left(\lambda_{\text{DU}} \tau - \frac{(\lambda_{\text{DU}} \tau)^2}{2} + \frac{(\lambda_{\text{DU}} \tau)^3}{3!} - \frac{(\lambda_{\text{DU}} \tau)^4}{4!} + \cdots\right) \\ &= 1 - \left(1 - \frac{\lambda_{\text{DU}} \tau}{2} + \frac{(\lambda_{\text{DU}} \tau)^2}{3!} - \frac{(\lambda_{\text{DU}} \tau)^3}{4!} + \cdots\right)\end{aligned}$$

When $\lambda_{\text{DU}} \tau$ is small, then

$$\text{PFD} \approx \frac{\lambda_{\text{DU}} \tau}{2} \qquad (10.8)$$

This approximation is often used in practical calculation. The approximation is always conservative, meaning that the approximated value in (10.8) is greater that the correct value in (10.7).

According to OREDA (2002) the failure rate of a specific type of fire detectors is $\lambda_{DU} = 0.21 \cdot 10^{-6}$ DU failures per hour. If we use a test interval $\tau = 3$ months ≈ 2190 hours, the PFD is

$$\text{PFD} \approx \frac{\lambda_{DU}\tau}{2} = \frac{0.21 \cdot 10^{-6} \cdot 2190}{2} \approx 0.00023 = 2.30 \cdot 10^{-4}$$

If a demand for the fire detector occurs, the (average) probability that the detector will not be able to detect the fire is: PFD ≈ 0.00023. This means that approximately one out of 4350 fires will not be detected by the fire detector.

The mean proportion of time the detector is not able to detect a fire is MFDT ≈ 0.00023. This means that the fire detector is not able to detect a fire in 0.023% of the time, or approximately 2 hours per year, when we assume that the detector is in continuous operation, and that a year is 8760 hours. We also say that we are *unprotected* by the fire detector in 0.023% of the time. □

Example 10.4 Parallel System

Assume that we have two independent fire detectors of the same type with failure rate λ_{DU} with respect to DU failures, that are tested at the same time with test interval τ. The fire detectors are operated as a 1-out-of-2 (1oo2) system, where it is sufficient that one detector is functioning for the system to function. The survivor function for the system is

$$R(t) = 2e^{-\lambda_{DU}t} - e^{-2\lambda_{DU}t}$$

and the PFD is from (10.3)

$$\begin{aligned}\text{PFD} &= 1 - \frac{1}{\tau}\int_0^\tau 2e^{-\lambda_{DU}t} - e^{-2\lambda_{DU}t}\,dt \\ &= 1 - \frac{2}{\lambda_{DU}\tau}\left(1 - e^{-\lambda_{DU}\tau}\right) + \frac{1}{2\lambda_{DU}\tau}\left(1 - e^{-2\lambda_{DU}\tau}\right) \quad (10.9)\end{aligned}$$

If we replace $e^{-\lambda_{DU}\tau}$ by its Maclaurin series, we may use the following approximation:

$$\text{PFD} \approx \frac{1}{3}(\lambda_{DU}\tau)^2 \quad (10.10)$$

when $\lambda_{DU}\tau$ is small.

Let us now introduce the same data as we used for one single fire detector in Example 10.3, $\lambda_{DU} = 0.21 \cdot 10^{-6}$ hours^{-1} and $\tau = 3$ months. The average unavailability of the parallel system is then

$$\bar{A}_{av} \approx \frac{1}{3}(\lambda_{DU}\tau)^2 = \frac{1}{3}(0.21 \cdot 10^{-6} \cdot 2190)^2 \approx 7.1 \cdot 10^{-8}$$

If a demand for the fire detector system occurs, the (average) probability that the system will not be able to detect the fire is PFD $\approx 7.1 \cdot 10^{-8}$, that is, a very high reliability. □

Remark: Since the parallel system will only fail when both of its components fail, the probability, $Q_S(t)$, that the system is in a failed state at time t is equal to $q_1(t) \cdot q_2(t)$, where $q_i(t)$ is the probability that component i is in a failed state at time t, for $i = 1, 2$. Since the (average) probability that component i is in a failed state is $\text{PFD}_i \approx \lambda_{\text{DU}}\tau/2$, we should expect that the average unavailability (PFD) of the system would be approximately $(\lambda_{\text{DU}}\tau/2)^2 = (\lambda_{\text{DU}}\tau)^2/4$ instead of $(\lambda_{\text{DU}}\tau)^2/3$ as we found in (10.10). The result in (10.10) is the correct result. The reason for this difference is the fact that the average of a product is not the same as the product of averages. Several computer programs for fault tree analysis do this failure. A bad effect is that the wrong approach produces a nonconservative result. □

Example 10.5 2-out-of-3 System

Assume that we have three independent fire detectors of the same type with failure rate λ_{DU} with respect to DU failures, that are tested at the same time with test interval τ. The fire detectors are operated as a 2-out-of-3 (2oo3) system, where two detectors have to function for the system to function. The survivor function for the system is

$$R(t) = 3 e^{-2\lambda_{\text{DU}}t} - 2 e^{-3\lambda_{\text{DU}}t}$$

and the PFD is from (10.3)

$$\begin{aligned}
\text{PFD} &= 1 - \frac{1}{\tau} \int_0^\tau 3 e^{-2\lambda_{\text{DU}}t} - 2 e^{-3\lambda_{\text{DU}}t} \, dt \\
&= 1 - \frac{3}{2\lambda_{\text{DU}}\tau} \left(1 - e^{-2\lambda_{\text{DU}}\tau}\right) + \frac{2}{3\lambda_{\text{DU}}\tau} \left(1 - e^{-3\lambda_{\text{DU}}\tau}\right) \quad (10.11)
\end{aligned}$$

If we replace $e^{-\lambda_{\text{DU}}\tau}$ by its Maclaurin series, we may use the following approximation:

$$\text{PFD} \approx (\lambda_{\text{DU}}\tau)^2 \quad (10.12)$$

when $\lambda_{\text{DU}}\tau$ is small.

Let us now introduce the same data as we used for one single fire detector in Example 10.3, $\lambda_{\text{DU}} = 0.21 \cdot 10^{-6}$ hours^{-1} and $\tau = 3$ months. The average unavailability of the parallel system is then

$$\text{PFD} \approx (\lambda_{\text{DU}}\tau)^2 = (0.21 \cdot 10^{-6} \cdot 2190)^2 \approx 2.1 \cdot 10^{-7}$$

If a demand for the fire detector system occurs, the (average) probability that the system will not be able to detect the fire is $\text{PFD} \approx 2.1 \cdot 10^{-7}$. □

The PFD of a 2oo3 system is seen to be approximately three times as high as for a parallel system. In Chapter 4 we saw that a 2oo3 system may be represented as a series system of three 1oo2, parallel systems. Each of these parallel systems will have an average unavailability of $(\lambda_{\text{DU}}\tau)^2/3$. When $\lambda_{\text{DU}}\tau$ is small, the probability of two parallel systems being in a failed state at the same time will be negligible, and the average unavailability of the 2oo3 system should be approximately the sum of the average availabilities of the three parallel systems, which is the case.

Example 10.6 Series System
Assume that we have two independent items with failure rate $\lambda_{DU,1}$ and $\lambda_{DU,2}$ respectively, with respect to DU failures. The items are tested at the same time with test interval τ. The items are operated as a 2-out-of-2 (2oo2) system, where both items have to function for the system to function. The survivor function for the system is

$$R(t) = e^{-(\lambda_{DU,1}+\lambda_{DU,2})t}$$

and the PFD is from (10.3)

$$\begin{aligned} \text{PFD} &= 1 - \frac{1}{\tau} \int_0^\tau e^{-(\lambda_{DU,1}+\lambda_{DU,2})t} \, dt \\ &\approx \frac{(\lambda_{DU,1}+\lambda_{DU,2})\tau}{2} = \frac{\lambda_{DU,1}\tau}{2} + \frac{\lambda_{DU,2}\tau}{2} \end{aligned} \quad (10.13)$$

when $\lambda_{DU,i}\tau$ is small, for $i = 1, 2$. When we have a series system, the PFD of the system is hence approximately the sum of the PFDs of the individual items. □

10.3.1 Approximation Formulas

Assume that we have a system of n independent components with constant failure rates $\lambda_{DU,i}$, for $i = 1, 2, \ldots, n$. The distribution function $F_{T_i}(t)$ of item i is approximated by

$$F_{T_i}(t) = 1 - e^{-\lambda_{DU,i}t} \approx \lambda_{DU,i}t$$

By using fault tree terminology the unavailability of component i in the first test interval is

$$\begin{aligned} q_i(t) &= \Pr(\text{Component } i \text{ is in a failed state at time } t) \\ &= F_{T_i}(t) \approx \lambda_{DU,i}t \end{aligned}$$

Let $K_1, K_2, \ldots, K_k$ be the k minimal cut sets of the system. The probability that the minimal cut parallel structure corresponding to the minimal cut set K_j is failed at time t is

$$\check{Q}_j(t) = \prod_{i \in K_j} q_i(t) \approx \prod_{i \in K_j} \lambda_{DU,i}t \quad \text{for } j = 1, 2, \ldots, k$$

The probability that the system is failed (has a hidden failure) at time t is

$$\begin{aligned} Q_0(t) = F_S(t) &\approx \sum_{j=1}^k \check{Q}_j(t) \approx \sum_{j=1}^k \prod_{i \in K_j} \lambda_{DU,i}t \\ &= \sum_{j=1}^k \left(\prod_{i \in K_j} \lambda_{DU,i} \right) t^{|K_j|} \end{aligned} \quad (10.14)$$

Table 10.1 PFD of Some *koon* Systems of Identical and Independent Components with Failure Rate λ_{DU} and Test Interval τ.

$k \backslash n$	1	2	3	4
1	$\dfrac{\lambda_{DU}\tau}{2}$	$\dfrac{(\lambda_{DU}\tau)^2}{3}$	$\dfrac{(\lambda_{DU}\tau)^3}{4}$	$\dfrac{(\lambda_{DU}\tau)^4}{5}$
2	–	$\lambda_{DU}\tau$	$(\lambda_{DU}\tau)^2$	$(\lambda_{DU}\tau)^3$
3	–	–	$\dfrac{3\lambda_{DU}\tau}{2}$	$2(\lambda_{DU}\tau)^2$
4	–	–	–	$2\lambda_{DU}\tau$

where $|K_j|$ denotes the *order* of the minimal cut set K_j, $j = 1, 2, \ldots, k$.

The PFD of the system that is tested periodically with test interval τ is, by combining (10.2) and (10.14), approximately

$$\text{PFD} = \frac{1}{\tau}\int_0^\tau F_s(t)\,dt \approx \sum_{j=1}^k \prod_{i \in K_j} \lambda_{DU,i} \frac{1}{\tau}\int_0^\tau t^{|K_j|}\,dt \qquad (10.15)$$

Hence

$$\text{PFD} \approx \sum_{j=1}^k \frac{1}{|K_j|+1} \prod_{i \in K_j} \lambda_{DU,i}\tau \qquad (10.16)$$

Assume now that we have a *k*-out-of-*n* (*koon*) system of identical and independent components with failure rate λ_{DU}. A *koon* system has $\binom{n}{n-k+1}$ minimal cut sets of order $(n-k+1)$. The PFD of the *koon* system is thus

$$\begin{aligned}
\text{PFD} &\approx \int_0^\tau \binom{n}{n-k+1} \cdot (\lambda_{DU}t)^{n-k+1}\,dt \\
&= \binom{n}{n-k+1} \frac{(\lambda_{DU}\tau)^{n-k+1}}{n-k+2}
\end{aligned} \qquad (10.17)$$

The PFD of some simple *koon* systems are listed in Table 10.1.

10.3.2 Mean Downtime in a Test Interval

The mean downtime $E(D_1)$ in a test interval was found in (10.5) to be

$$E(D_1) = \int_0^\tau F(t)\,dt$$

Suppose that we test an item at time τ and find that the item is in a failed state [i.e., $X(\tau) = 0$]. What is the (conditional) mean downtime in the interval $(0, \tau]$ when the item is found in a failed state at time τ?

By using double expectation, the mean downtime $E(D_1)$ may be written

$$\begin{aligned} E(D_1) &= E[E(D_1 \mid X(\tau))] \\ &= E(D_1 \mid X(\tau) = 0) \cdot \Pr(X(\tau) = 0) \\ &\quad + E(D_1 \mid X(\tau) = 1) \cdot \Pr(X(\tau) = 1) \end{aligned}$$

If the component is functioning at time τ, the downtime D_1 is equal to 0. Therefore $E(D_1 \mid X(\tau) = 1) = 0$. Furthermore

$$\Pr(X(\tau) = 0) = \Pr(T \leq \tau) = F(\tau)$$

Hence

$$E(D_1) = E(D_1 \mid X(\tau) = 0) \cdot F(\tau)$$

By using (10.6) and (10.3)

$$\begin{aligned} E(D_1 \mid X(\tau) = 0) &= \frac{E(D_1)}{F(\tau)} = \frac{1}{F(\tau)} \int_0^\tau F(t)\,dt \\ &= \frac{\tau}{F(\tau)} \cdot \frac{1}{\tau} \int_0^\tau F(t)\,dt = \frac{\tau}{F(\tau)} \cdot \text{PFD} \quad (10.18) \end{aligned}$$

Example 10.3 (Cont.)
With a single item, the conditional mean downtime in (10.18) is approximately

$$E(D_1 \mid X(\tau) = 0) = \frac{\tau}{F(\tau)} \cdot \text{PFD} \approx \frac{\tau}{1 - e^{-\lambda_{DU}\tau}} \cdot \frac{\lambda_{DU} \cdot \tau}{2} \approx \frac{\tau}{2}$$

which is an intuitive result. □

Example 10.4 (Cont.)
With a parallel system of two independent, and identical items, the conditional mean downtime in (10.18) is

$$E(D_1 \mid X(\tau) = 0) = \frac{\tau}{F(\tau)} \cdot \text{PFD} \approx \frac{\tau}{1 - 2e^{-\lambda_{DU}\tau} + e^{-2\lambda_{DU}\tau}} \cdot \frac{(\lambda_{DU} \cdot \tau)^2}{3} \approx \frac{\tau}{3}$$

The last approximation follows since the distribution function of the parallel structure $1 - 2e^{-\lambda_{DU}\tau} + e^{-2\lambda_{DU}\tau}$ can be approximated by $(\lambda_{DU} \cdot \tau)^2$ by using Maclaurin series. □

10.3.3 Mean Number of Test Intervals Until First Failure

Let us next determine the mean number of test intervals *until* the first failure occurs. Let C_i denote the event that the component does *not* fail in test interval i for $i = 1, 2, \ldots$. Then

$$\Pr(C_i) = \Pr(T > \tau) = R(\tau)$$

Since the events $C_1, C_2, \ldots$ are independent with the same probability $p = R(\tau)$, the number of test intervals, Z, *until* the component fails for the first time, has a geometric distribution with point probability

$$\Pr(Z = z) = \Pr(C_1 \cap C_2 \cap \ldots \cap C_z \cap C_{z+1}^c) = p^z(1-p)$$

for $z = 0, 1, \ldots$

The mean number of test intervals *until* the components fails is then

$$E(Z) = \sum_{z=0}^{\infty} z \Pr(Z=z) = \frac{p}{1-p} = \frac{R(\tau)}{F(\tau)} \qquad (10.19)$$

Let T' denote the time the component is put into operation until its first failure. Then

$$\begin{aligned}
E(T') &= \tau E(Z) + (\tau - E(D_1 \mid X(\tau) = 0)) \\
&= \tau \frac{R(\tau)}{F(\tau)} + \tau - \frac{1}{F(\tau)} \left(\tau - \int_0^\tau R(t)\,dt \right) \\
&= \frac{1}{F(\tau)} \int_0^\tau R(t)\,dt \qquad (10.20)
\end{aligned}$$

Example 10.7
If in particular the component has constant failure rate λ_{DU}, then

$$E(T') = \frac{1}{F(\tau)} \int_0^\tau R(t)\,dt = \frac{1}{1 - e^{-\lambda_{DU}\tau}} \int_0^\tau e^{-\lambda_{DU} t}\,dt = \frac{1}{\lambda_{DU}}$$

This result also follows directly from the properties of the exponential distribution. □

10.3.4 Staggered Testing

When we have two items in parallel, we may reduce the system PFD by testing the items at different times. Let us now assume that we have two independent items with constant failure rates $\lambda_{DU,1}$ and $\lambda_{DU,2}$, respectively, with respect to dangerous undetected failures. Item 1 is tested at times $0, \tau, 2\tau, \ldots$, while item 2 is tested at times $t_0, \tau + t_0, 2\tau + t_0, \ldots$. This testing is called *staggered testing* with interval t_0. We assume that the time necessary for testing and repair is so short that it can be neglected. Let us further assume that the process has been running some time and that time 0 is the time for a test of item 1.

PROBABILITY OF FAILURE ON DEMAND   435

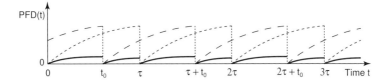

Fig. 10.6 Probability of failure on demand PFD(t) of a parallel system of two items with staggered testing. Item 1 (short dash) is tested at times $0, \tau, 2\tau, \ldots$, while item 2 (long dash) is tested at times $t_0, \tau + t_0, 2\tau + t_0, \ldots$. The system PFD($t$) is the fully drawn curve.

The PFD of the two items as a function of time is illustrated in Fig. 10.6. In the first test interval $(0, \tau]$ the items have the following unavailabilities:

$$\begin{aligned} q_i(t) &= 1 - e^{-\lambda_{DU,1} t} & \text{for} \quad 0 < t \le \tau \\ q_2(t) &= 1 - e^{-\lambda_{DU,2}(t + \tau - t_0)} & \text{for} \quad 0 \le t \le t_0 \\ &= 1 - e^{-\lambda_{DU,2}(t - t_0)} & \text{for} \quad t_0 < t \le \tau \end{aligned}$$

The unavailability of item 1, $q_1(t)$, is illustrated by a short-dashed line in Fig. 10.6, while the unavailability of item 2, $q_2(t)$, is illustrated by a long-dashed. The system unavailability $q_s(t) = q_1(t) \cdot q_2(t)$ is illustrated by a fully drawn line in Fig. 10.6.

$$\begin{aligned} q_s(t) &= \left(1 - e^{-\lambda_{DU,1} t}\right)\left(1 - e^{-\lambda_{DU,2}(t + \tau - t_0)}\right) & \text{for} \quad 0 < t \le t_0 \\ &= \left(1 - e^{-\lambda_{DU,2} t}\right)\left(1 - e^{-\lambda_{DU,2}(t - t_0)}\right) & \text{for} \quad t_0 < t \le \tau \end{aligned}$$

The average unavailability in $(0, \tau]$ is equal to the PFD and is a function of t_0

$$\begin{aligned} \text{PFD}(t_0) &= \frac{1}{\tau} \int_0^\tau q_s(t)\, dt \\ &= 1 - \frac{1}{\lambda_{DU,1}\tau}\left(1 - e^{-\lambda_{DU,1} t_0}\right) - \frac{e^{-\lambda_{DU,2}(\tau - t_0)}}{(\lambda_{DU,1} + \lambda_{DU,2})\tau}\left(1 - e^{-\lambda_{DU,2} t_0}\right) \\ &\quad + \frac{e^{-\lambda_{DU,2}(\tau - t_0)}}{(\lambda_{DU,1} + \lambda_{DU,2})\tau}\left(1 - e^{-(\lambda_{DU,1} + \lambda_{DU,2}) t_0}\right) \\ &\quad - \frac{1}{\lambda_{DU,1}\tau}\left(e^{-\lambda_{DU,1} t_0} - e^{-\lambda_{DU,1}\tau}\right) \\ &\quad - \frac{e^{\lambda_{DU,2} t_0}}{\lambda_{DU,2}\tau}\left(e^{-\lambda_{DU,2} t_0} - e^{\lambda_{DU,2}\tau}\right) \\ &\quad + \frac{e^{\lambda_{DU,2} t_0}}{(\lambda_{DU,1} + \lambda_{DU,2})\tau}\left(e^{-(\lambda_{DU,1} + \lambda_{DU,2}) t_0} - e^{-(\lambda_{DU,1} + \lambda_{DU,2})\tau}\right) \end{aligned}$$

PFD(t_0) will attain its minimum for

$$t_0 = \frac{1}{\lambda_{DU,1} + \lambda_{DU,2}} \ln\left(\frac{\lambda_{DU,1}\left(1 - e^{\lambda_{DU,2}\tau}\right)}{\lambda_{DU,2}\left(e^{\lambda_{DU,1}\tau} - 1\right)}\right) + \tau$$

When the two items have the same failure rate, $\lambda_{DU,1} = \lambda_{DU,2}$, we get $t_0 = \tau/2$, which is an intuitive result.

10.3.5 Nonnegligible Repair Time

In some situations the repair time after a failure is so long that it cannot be neglected. This is, for example, illustrated by the following example.

Example 10.8
A downhole safety valve (DHSV) is located in the oil/gas production tubing in subsea production wells. The DHSV is an integral part of the tubing approximately 100 meters below the sea bottom. The valve has a spring-loaded hydraulic fail-safe actuator and is held open by hydraulic pressure. The operation of the DHSV is comparable to the gate valve described in Example 10.2, and the DHSV has the same failure modes as the gate valve. The DHSV is tested periodically, with a test interval of 3 to 6 months. To repair a failed valve is a long, hazardous, and extremely costly operation. A semi-submersible intervention rig has to be moved from its permanent location out to the offshore field. The tubing string has to be pulled and the well pressure has to be controlled during the intervention. The operation may last several weeks, depending on the system and the weather conditions. In addition, we may have to wait months before an intervention rig becomes available. In this case the repair time is far from negligible. □

As illustrated in Example 10.8, the item may sometimes be unavailable as a safety barrier during the repair action and while waiting for repair. This unavailability may, however, be different from the unavailability in the test interval, since we now know that the item is in a failed state, and may take precautions to reduce the risk. The time from when a failure is detected until the function is restored is sometimes called the *restoration time*. The risk associated to the restoration time may depend on:

- The *failure mode*. The various failure modes of the item may require different repair actions, and the risk during waiting for repair may also be different.

- The various *phases* of the restoration time may have different risk levels. The risk during waiting for repair may, for example, be different from the risk during actual repair.

It may therefore be necessary to find the unavailability for each failure mode and for the various phases of the restoration time.

10.4 SAFETY UNAVAILABILITY

The *safety unavailability*, SU, of a safety system is the probability that the system is *not* able to perform its required function upon a demand. The safety unavailability may be split in four categories, as illustrated in Fig. 10.7. The categories of the safety

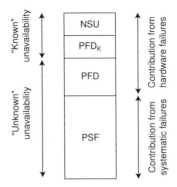

Fig. 10.7 Contributions to safety unavailability.

unavailability are further discussed by Corneliussen and Hokstad (2003), who also define more detailed categories.

NSU = Noncritical safety unavailability of the item, mainly caused by functional testing. In this case it is known that the item is unavailable, and other preventive actions may be taken.

PFD = The (unknown) safety unavailability due to dangerous undetected (DU) failures during the test interval when it is not known that the function is unavailable.

PFD_K = Safety unavailability of the item due to restoration actions after a failure has been revealed. In this case we know that the item is unavailable. The various phases of the restoration actions may give rise to different levels of risk, as discussed on page 436.

PSF = The probability that a systematic failure will prevent the item from performing its intended function. Systematic failures (see page 424) are not revealed by periodic testing. The PSF is approximately equal to the probability that an item that has just been functionally tested will fail on demand. Unavailability due to imperfect testing, like partial stoke testing of valves, may adequately be included in the PSF.

10.4.1 Probability of Critical Situation

Consider a safety system that has been installed as a barrier against a specific type of accidental events. We may, for example, assume that the safety system is a fire detector system, and that the accidental events are fires (in an early phase). Assume that fires occur randomly according to a homogeneous Poisson process (HPP) with intensity β. The parameter β denotes the mean number of fires per time unit and is sometimes called the *process demand rate*.

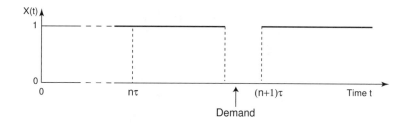

Fig. 10.8 Critical situation – fire detector system. $X(t)$ is the state of the fire detector system.

A *critical situation* occurs if a fire occurs while the fire detector system is in a failed state. This situation is illustrated in Fig. 10.8.

Each time a fire occurs, there is a probability SU that the fire detector system is not able to detect the fire. In Section 7.2 we showed how to combine an HPP with Bernoulli trials, such that critical situations will occur as an HPP with intensity $\beta \cdot \text{SU}$.

Let $N_C(t)$ denote the number of critical situations in the interval $(0, t)$. The probability of having n critical situations in the interval is

$$\Pr(N_C(t) = n) = \frac{(\beta \cdot \text{SU} \cdot t)^n}{n!} e^{-\beta \cdot \text{SU} \cdot t} \quad \text{for } n = 0, 1, \ldots \quad (10.21)$$

The mean number of critical situations in the time interval $(0, t)$ is

$$E(N_C(t)) = \beta \cdot \text{SU} \cdot t \quad (10.22)$$

10.4.2 Spurious Trips

For many safety items, the rate of spurious trips (ST) may be comparable, and even higher, than the rate of dangerous undetected (DU) failures. Spurious trips will usually imply significant costs and also reduce the confidence in the system.

Consider a safety system comprising m independent subsystems. The system may, for example, comprise a flame detector subsystem, a heat detector subsystem, a smoke detector subsystem, a logic solver subsystem, and safety shutdown valves. Each subsystem may comprise several items. The system is considered to be a series structure of the subsystems with respect to ST failures. A subsystem ST failure will therefore give a system ST failure. Let $\lambda_{ST}^{(j)}$ denote the rate of spurious trips of the safety subsystem j, and let $\text{MDT}_{ST}^{(j)}$ denote the mean system downtime associated with the spurious trip, for $j = 1, 2, \ldots, m$. The safety unavailability of the system caused by spurious trips is approximately

$$\bar{A}_{ST} \approx \sum_{j=1}^{m} \lambda_{ST}^{(j)} \cdot \text{MDT}_{ST}^{(j)} \quad (10.23)$$

Example 10.9 Parallel System
Consider a fire detector subsystem of n independent detectors. Detector i has constant

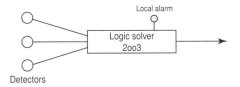

Fig. 10.9 A 2oo3 detector system.

failure rate $\lambda_{ST,i}$ with respect to spurious trips, for $i = 1, 2, \ldots, n$. The subsystem is a parallel structure with respect to safety, meaning that if one of the detectors is activated, the subsystem will raise an alarm. The subsystem is therefore a 1oo$n$ system with respect to safety. With this configuration, a spurious signal (a false alarm) from any of the detectors will raise an alarm. The subsystem is therefore a series ($n$oo$n$) system with respect to spurious trips, and the spurious trip rate from the subsystem is

$$\lambda_{ST}^{1oon} = \sum_{i=1}^{n} \lambda_{ST,i} \quad (10.24)$$

A high degree of redundancy may therefore lead to many spurious trips. □

Example 10.10 2-out-of-3 System
Consider a subsystem of three independent fire detectors of the same type, and let λ_{ST} denote the constant failure rate with respect to spurious trips from one detector. The detectors are connected to a logic solver with a 2oo3 voting logic. The system is illustrated in Fig. 10.9. Two detectors have to send a signal to the logic solver to raise an alarm. We assume that the logic solver is so reliable that failures may be neglected. Since the fire detectors are independent, spurious trips (false alarms) will occur as single failures. When a detector gives a false alarm, the system will only give a false alarm if a second detector gives a false alarm before the first false alarm is detected and repaired. Let us assume that when the logic solver receives a signal from a detector, a local alarm is raised. The operators may therefore check the status and repair the detector that has given the false alarm. Assume that the restoration time is t_r. If a second alarm is not received by the logic solver before the first failure is repaired, there will be no system false alarm. The spurious trip (false alarm) rate from the 2oo3 subsystem is therefore

$$\begin{aligned}\lambda_{ST}^{2oo3} &= 3\lambda_{ST} \cdot \int_0^{t_r} \left(1 - e^{-2\lambda_{ST}t}\right) dt \\ &= 3\lambda_{ST} \left(t_r - \frac{1}{2\lambda_{ST}} \left(1 - e^{-2\lambda_{ST}t_r}\right)\right) \quad (10.25)\end{aligned}$$

Let $\lambda_{ST} = 5 \cdot 10^{-5}$ ST failures per hour, and $t_r = 2$ hours. In this case we get $\lambda_{ST}^{2oo3} \approx 3 \cdot 10^{-8}$ hours^{-1}, that is, a very low spurious trip rate. □

Table 10.2 gives a brief comparison of three simple systems with independent items of the same type, with constant failure rate λ_{DU} with respect to DU failures and

440 RELIABILITY OF SAFETY SYSTEMS

Table 10.2 Probability of Failure on Demand (PFD) and Spurious Trip Rate for Three Simple Systems.

System	PFD	Rank	Spurious trip rate	Rank
Single item				
1oo1	$\dfrac{\lambda_{DU}\tau}{2}$	(3)	λ_{ST}	(2)
Parallel system				
1oo2	$\dfrac{(\lambda_{ST}\tau)^2}{3}$	(1)	$2\lambda_{ST}$	(3)
2-out-of-3 system				
2oo3	$(\lambda_{ST}\tau)^2$	(2)	≈ 0	(1)

constant failure rate λ_{ST} with respect to ST failures. The test interval is τ. The 2oo3 system is often chosen as the best configuration for detector systems, because it has a PFD in the same order of magnitude as a parallel system, and because it can be made much more reliable than a parallel system when it comes to spurious trips.

10.4.3 Failures Detected by Diagnostic Self-Testing

Many failures of a modern safety instrumented system may be revealed by diagnostic self-testing. This applies both for dangerous failures and safe failures as defined on page 423. The diagnostic testing is assumed to be carried out so frequently that the failures are revealed immediately. In subsystems with redundant items a failure may sometimes be repaired while the subsystem is on-line and is able to perform its safety function. In other cases the subsystem has to be taken off-line to repair the failure. Let $\lambda_{DT,i}^{(j)}$ denote the rate of failures of item i in subsystem j that are revealed by diagnostic self-testing, for $i = 1, 2, \ldots, n_j$ and $j = 1, 2, \ldots, m$. If we assume that all items are independent, the rate of failures of subsystem j that are revealed by diagnostic self-testing is

$$\lambda_{DT}^{(j)} = \sum_{i=1}^{n_j} \lambda_{DT,i}^{(j)}$$

Let $\text{MDT}_{DT}^{(j)}$ denote the mean downtime of subsystem j to repair a failure of an item in subsystem j that has been revealed by diagnostic self-testing. (For some configurations the mean downtime may be zero). The system unavailability caused by failures that are revealed by diagnostic self-testing is therefore

$$\bar{A}_{DT} \approx \sum_{j=1}^{m} \lambda_{DT}^{(j)} \cdot \text{MDT}_{DT}^{(j)} \qquad (10.26)$$

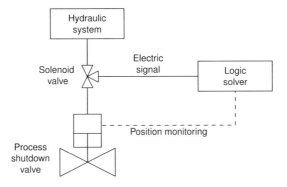

Fig. 10.10 A process shutdown valve with fail-safe hydraulic actuator.

In (10.26) the mean downtime is given for each subsystem. For subsystems with different types of items, it may be more appropriate to give the mean downtime associated to repair of each type of items.

The *diagnostic coverage* of the diagnostic self-test of item i is defined by

$$c_{DT,i} = \frac{\lambda_{DT,i}}{\lambda_i}$$

where λ_i denotes the total failure rate (of a specified category) of item i, for $i = 1, 2, \ldots, n$. A diagnostic self-testing with test coverage 70% will hence reveal 70% of all the failures of the item. The term *diagnostic coverage* is mainly used for dangerous failures and is then the percentage of dangerous failures that can be detected by self-testing. The term may, however, also be used for safe failures.

Example 10.11
Consider a process shutdown valve, as illustrated by the sketch in Fig. 10.10. The valve has a fail-safe actuator and is held open by hydraulic pressure. When a process demand occurs, the logic solver will send an electric signal to the solenoid valve to open and bleed off the hydraulic pressure. Diagnostic self-testing may be carried out by sending on/off electric signals to the solenoid valve. The solenoid valve will start to open and bleed off hydraulic pressure, and the shutdown valve will start to close. The movement of the valve actuator may be monitored by the logic solver. When the valve actuator has moved some few millimeters, full hydraulic pressure is again applied to the actuator and the valve will fully open. By this testing we can reveal failures of the electrical cables, the solenoid valve, and the process shutdown valve. The test coverage for the electrical cables will be 100%. The test coverage of the solenoid valve and the hydraulic flow will depend on the design of the system, and may be made close to 100%. This type of testing of the shutdown valve is called *partial stroke testing* and will only reveal some failure causes of the valve. The partial stroke testing will reveal some main causes of fail to close (FTC) failures but will not reveal leakage in closed position (LCP) failures.

In most applications, only the electrical cables will be tested by very frequent diagnostic testing. To avoid excessive wear of the valve seals, the diagnostic testing of the solenoid valve, and the shutdown valve will be less frequent. □

10.5 COMMON CAUSE FAILURES

So far in this chapter, we have assumed that all items are independent. This is, however, not always the case in practice. Safety systems will often have a high degree of redundancy, and the system reliability will therefore be strongly influenced by potential common cause failures. It is therefore important to identify potential common cause failures and take the necessary precautions to prevent such failures.

Checklists that may be used to identify common cause failure problems of an SIS during its life cycle have been developed, for example, by Summers and Raney (1999).

When we are able to identify the causes of common cause failures, these should always be explicitly modeled, as illustrated in Example 10.12. In most cases we will not be able to find high quality input data for the explicitly modeled common causes. Even with low-quality input data, or guesstimates, the result will usually be more accurate than by including the explicit common causes into one of the general (implicit) dependent failure models that were introduced in Chapter 6.

Example 10.12
Consider a parallel system of two pressure sensors that are installed in a pressure vessel. Based on a search for potential causes for common cause failures, we have identified two possible causes: (i) the common tap to the sensors is plugged with solids, and (ii) the sensors are miscalibrated. Other specific causes have not been identified. The two causes for common cause failures may be modeled explicitly as illustrated by the fault tree in Fig. 10.11. In the fault tree the remaining failures of the sensors are said to be independent. If we believe that there are some implicit causes of dependency, in addition to the two explicit causes, this dependency may be modeled by one of the models discussed in Chapter 6, for example, the β-factor model. □

The most commonly used (implicit) model for common cause failures of safety systems is the β-factor model. In the β-factor model we assume that a certain percentage of all failures are common cause failures that will cause all the items to fail at the same time (or, within a very short time interval). The failure rate λ_{DU} with respect to DU failures may therefore be written as

$$\lambda_{DU} = \lambda_{DU}^{(i)} + \lambda_{DU}^{(c)}$$

where $\lambda_{DU}^{(i)}$ is the rate of independent DU failures that only affects one component, and $\lambda_{DU}^{(c)}$ is the rate of common cause DU failures that will cause failure of all the system components at the same time. The common cause factor

$$\beta_{DU} = \frac{\lambda_{DU}^{(c)}}{\lambda_{DU}}$$

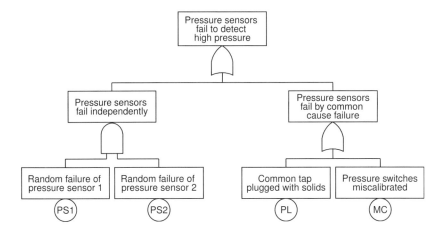

Fig. 10.11 Explicit modeling of common cause failure of a system with two pressure sensors. (Adapted from Summers and Raney 1999.)

is the percentage of common cause DU failures among all DU failures of a component.

Similarly, the spurious trip rate λ_{ST} may be written as

$$\lambda_{ST} = \lambda_{ST}^{(i)} + \lambda_{ST}^{(c)}$$

where $\lambda_{ST}^{(i)}$ is the rate of independent ST failures that only affects one component, and $\lambda_{DU}^{(c)}$ is the rate of common cause ST failures that will cause failure of all the system components at the same time. The common cause factor

$$\beta_{ST} = \frac{\lambda_{ST}^{(c)}}{\lambda_{ST}}$$

is the percentage of common cause ST failures among all ST failures of a component. Since there may be different failure mechanisms leading to DU and ST failures, β_{DU} and β_{ST} need not be equal.

Diagnostic Self-Testing and Common Cause Failures   Common cause failures may be classified in two main types:

1. Multiple failures that occur at the same time due to a common cause
2. Multiple failures that occur due to a common cause, but not necessarily at the same time

As an example of type 2, consider a redundant structure of electronic components that are exposed to a common cause: increased temperature. The components will fail due to the common cause, but usually not at the same time. If we have an SIS with an adequate diagnostic coverage with respect to this type of failure, we may be able to detect the first common cause failure and take action before the system fails.

A system failure due to the common cause may therefore be avoided.

Remark: If the common cause, increased temperature, is due to a cooling fan failure, this should be explicitly modeled as illustrated in Example 10.12. Monitoring the condition of the cooling fan would in this case give an earlier warning than diagnostic testing of the electronic components, and a higher probability of successful shutdown before a system common cause failure occurs. A similar example is discussed in IEC 61508-6 without mentioning any explicit modeling of the cooling fan.□

When we have identified the causes of potential common cause failures (e.g., by applying a checklist), we should carefully split the potential common cause failures in the two types (1 and 2) above. For each cause leading to failures of type 2 we should evaluate the ability of the diagnostic self-testing to reveal the failure (or the failure cause), the time required to take action, and the probability that this action will prevent a system failure.

It seems obvious that the common cause factor β for an SIS good diagnostic coverage should be lower than for a system with no, or a poor, diagnostic coverage. We should therefore be careful and not use estimates for β from old-fashioned systems when analyzing a modern SIS with good diagnostic coverage.

Example 10.13 Parallel System
Reconsider the parallel system of two fire detectors in Example 10.4, and assume that DU failures occur with a common cause factor β_{DU}. The PFD of the parallel system is from (10.10) and (10.13) approximately

$$\text{PFD}(\beta_{DU}) \approx \frac{[(1-\beta_{DU})\lambda_{DU}\tau]^2}{3} + \frac{\beta_{DU}\lambda_{DU}\tau}{2} \quad (10.27)$$

With respect to spurious trips, the system is a series system, and the trip rate is therefore

$$\lambda_{ST}^{1oo2}(\beta_{ST}) = (2-\beta_{ST})\lambda_{ST} \quad (10.28)$$

The rate of spurious trips will therefore decrease when β_{ST} increases.

By using the same data as in Example 10.4, $\lambda_{DU} = 0.21 \cdot 10^{-6}$ hours^{-1} and $\tau = 2190$ hours, and $\beta_{DU} = \beta_{ST} = 0.10$, we get form (10.27)

$$\text{PFD}(\beta_{DU}) \approx 5.71 \cdot 10^{-8} + 2.30 \cdot 10^{-5} \approx 2.31 \cdot 10^{-5}$$

We observe that with realistic estimates of λ_{DU} and τ, PFD$_{DU}$ is dominated by the common cause term in (10.27). We may therefore use the approximation

$$\text{PFD}(\beta_{DU}) \approx \frac{\beta_{DU}\lambda_{DU}\tau}{2}$$

when $\lambda_{DU}\tau$ is small. □

Example 10.14 2-out-of-3 System

The probability of failure on demand for a 2-out-of-3 system is from (10.12) and (10.13)

$$\text{PFD}(\beta_{\text{DU}}) \approx [(1 - \beta_{\text{DU}})\lambda_{\text{DU}}\tau]^2 + \frac{\beta_{\text{DU}}\lambda_{\text{DU}}\tau}{2} \qquad (10.29)$$

With a local alarm on the logic solver we may avoid almost all independent spurious trips. All common cause failures will, on the other hand, result in a system spurious trip, and we therefore have

$$\lambda_{\text{ST}}^{2oo3}(\beta_{\text{ST}}) = \beta_{\text{ST}}\lambda_{\text{ST}} \qquad (10.30)$$

With the same data as in Example 10.13 we get from (10.29)

$$\text{PFD}(\beta_{\text{DU}}) \approx 1.71 \cdot 10^{-7} + 2.30 \cdot 10^{-5} \approx 2.32 \cdot 10^{-5}$$

As in Example 10.13 we observe that with realistic estimates of λ_{DU} and τ, PFD_{DU} is dominated by the common cause term in (10.29). We may therefore use the approximation

$$\text{PFD}(\beta_{\text{DU}}) \approx \frac{\beta_{\text{DU}}\lambda_{\text{DU}}\tau}{2}$$

when $\lambda_{\text{DU}}\tau$ is small. □

In Example 10.13 and Example 10.14 we saw that the $\text{PFD}_{\text{DU}}(\beta_{\text{DU}})$ was dominated by the common cause term of the expressions (10.27) and (10.29), respectively when $\lambda_{\text{DU}}\tau$ is small. It is straightforward to show that the same applies to all $k$oo$n$ systems, where $n \geq 2$, and $k \leq n$. We will therefore have that

$$\text{PFD}^{koon}(\beta_{\text{DU}}) \approx \frac{\beta_{\text{DU}}\lambda_{\text{DU}}\tau}{2} \qquad (10.31)$$

when $\lambda_{\text{DU}}\tau$ is small. When $\beta_{\text{DU}} > 0$, we will therefore get approximately the same result for all types of $k$oo$n$ configurations, and the result is nearly independent of the number n of components, as long as $n \geq 2$. This may not be a realistic feature of the β-factor model. A more realistic alternative to the β-factor model has been proposed as part of the PDS approach that is described in Section 10.7.

IEC 61508 recommends using the β-factor model with a single "plant specific" β that is determined by using a checklist for all voting configurations (see IEC 61508-6, appendix D). This makes a comparison between different voting logics rather meaningless. Corneliussen and Hokstad (2003) have criticized the β-factor model and introduced a multiple β-factor (MBF) model, that is a generalization of the β-factor model.

Remarks

- Some reliability data sources (see Chapter 14) present the total failure rates, while other data sources only present the independent failure rates. The data in

OREDA (2002) are collected from maintenance reports and contain all failures, both independent and common cause failures. The data in MIL HDBK 217F mainly come from laboratory testing of single components and therefore only present the failure rate of independent failures. When using data from reliability data sources in common cause failure models, we should be aware of this difference.

- Some causes of common cause failures, like miscalibration of sensors, will be equally likely for a single component as it is for a system of several components. If we include miscalibration as a cause of common cause failures of n redundant sensors, it should also be included for a single sensor. This problem is further discussed by Summers and Raney (1999).

10.6 IEC 61508

The international standard, IEC 61508 *Functional safety of electrical/electronic/programmable electronic (E/E/PE) safety-related systems* is the main standard for safety instrumented systems. IEC 61508 is a generic, performance-based standard that covers most safety aspects of an SIS. As such, many topics covered in IEC 61508 are outside the scope of this book. In this section we will give a brief presentation of some main aspects of IEC 61508 that are relevant for the theory and methods presented in this book.

IEC 61508 has seven parts:

Part 1:   General requirements
Part 2:   Requirements for E/E/PE safety-related systems
Part 3:   Software requirements
Part 4:   Definitions and abbreviations
Part 5:   Examples of methods for the determination of safety integrity levels
Part 6:   Guidelines on the application of IEC 61508-2 and IEC 61508-3
Part 7:   Overview of techniques and measures

IEC 61508 gives safety requirements to SISs and provides guidance to validation and verification of such systems. The first three parts are normative parts and deal with the assessment of industrial process risk and the SIS hardware and software reliability. The remaining four parts deal with definitions and provide informative annexes to the standard.

Part 1 defines the overall performance based criteria for an industrial process. It mandates the use of an overall safety life cycle model (see Fig. 10.12). Part 2 is directed toward manufacturers and integrators of SISs and presents methods and techniques that can be used to design, evaluate, and certify the hardware reliability of an SIS, and thus its contribution to process risk reduction.

IEC 61508 is a generic standard that is common to several industries. Application-specific standards and guidelines are therefore developed, giving more specific requirements. Among these standards and guidelines are:

- IEC 61511 *Functional safety – Safety instrumented systems for the process industry*. The Instrument Society of America (ISA) has independently developed ANSI/ISA S84.01 *Application of safety instrumented systems for the process industries* that is similar to IEC 61511.

- IEC 62061 *Safety of machinery – Functional safety of electrical, electronic and programmable electronic systems*. This standard was initially developed as a European standard to support the EU Machinery Directive.

- IEC 61513 *Nuclear power plants – Instrumentation and control for systems important to safety – General requirements for systems*.

- EN 50126 *Railway applications – The specification and demonstration of reliability, availability, maintainability, and safety (RAMS)*.

- EN 50128 *Railway applications – Software for railway control and protection systems*.[2]

- EN 50129 *Railway applications – Safety related electronic systems for signalling*.

- OLF guidelines for the application of IEC 61508 and IEC 61511 in the petroleum activities on the Norwegian continental shelf.

- AIChE guidelines for safe automation of chemical processes.

10.6.1 Safety Life Cycle

The requirements in IEC 61508 are related to an overall *safety life cycle* as shown in Fig. 10.12. The figure is reproduced from IEC 61508-1, page 18. The standard covers all safety life cycle activities from initial concept, through hazard analysis and risk assessment, development of the safety requirements, specification, design and implementation, operation and maintenance, and modification, to final decommissioning and/or disposal. Each of the 17 steps in the safety life cycle in Fig. 10.12 are described in detail in IEC 61508-1, section 7.

ANSI/ISA S84.01 follows a similar life cycle model as IEC 61508 and IEC 61511 to identify the need for an SIS.

---

[2] EN 50126 and EN 50128 were based on earlier drafts of IEC 61508.

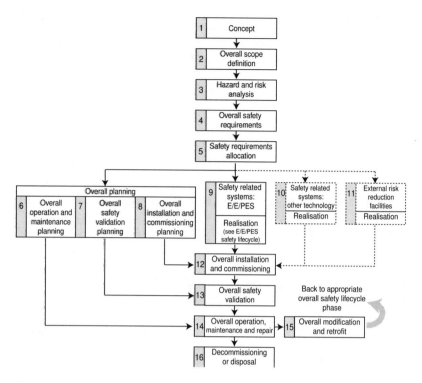

Fig. 10.12 Overall safety life cycle. (From IEC 61508-1, p.18. Reproduced with permission from IEC. Visit www.iec.ch for further information.)

10.6.2 Safety Integrity Level

Safety integrity is a fundamental concept in IEC 61508 and is defined as the *probability of a safety-related system satisfactorily performing the required safety functions under all the stated conditions within a specified period of time* (see IEC 61508-4, sect. 3.5). The safety integrity is classified into four discrete levels called *safety integrity levels* (SIL).

The SIL is in turn defined by the probability of failure on demand. The relation between the SIL and the PFD is shown in Table 10.3.

ANSI/ISA S84.01 uses the same safety integrity levels as presented in Table 10.3 but clearly states that SIL 4 is not relevant in the process industry.

An SIL has to be assigned to each safety instrumented function (SIF). Notice that the safety integrity level is assigned to the safety instrumented *functions*, and not to the safety instrumented *system*, that may comprise several safety instrumented functions.

Assume that process demands for an SIF with low demand mode occur according to an HPP with rate β demands per hour. For each demand, the SIF will fail to perform the required function with a probability PFD. A *critical situation* occurs if a process demand occurs and the SIF fails. Let $N_c(t)$ denote the number of critical situations in the time interval $(0, t)$. The process $\{N_c(t), t > 0\}$ is therefore an HPP with rate

Table 10.3 Safety Integrity Levels for Safety Functions.

Safety Integrity Level (SIL)	Low Demand Mode of Operation[a] (Aver. probability of failure to perform its design function on demand)	High Demand Mode or Continuous Mode of Operation[b] (Probability of a dangerous failure per hour)
4	$\geq 10^{-5}$ to $< 10^{-4}$	$\geq 10^{-9}$ to $< 10^{-8}$
3	$\geq 10^{-4}$ to $< 10^{-3}$	$\geq 10^{-8}$ to $< 10^{-7}$
2	$\geq 10^{-3}$ to $< 10^{-2}$	$\geq 10^{-7}$ to $< 10^{-6}$
1	$\geq 10^{-2}$ to $< 10^{-1}$	$\geq 10^{-6}$ to $< 10^{-5}$

Reproduced from IEC 61508-1, Tables 2 and 3, with the permission from IEC. Visit www.iec.ch for further information.

a) *Low demand mode* means that the frequency of demands for operation of the SIS is not greater than once per year, and not greater than twice the proof-test frequency.

b) *High demand or continuous mode* means that the frequency of demands for operation of the SIS is greater than once per year or greater than twice the proof-test frequency.

$\beta_c = \beta \cdot \text{PFD}$. The probability that n critical situations will occur in the interval $(0, t)$ is

$$\Pr(N_c(t) = n) = \frac{(\beta \cdot \text{PFD} \cdot t)^n}{n!} e^{-\beta \cdot \text{PFD} \cdot t} \quad \text{for } n = 0, 1, 2, \ldots \quad (10.32)$$

The mean time between critical situations is

$$\text{MTBF} = \frac{1}{\beta \cdot \text{PFD}} \quad (10.33)$$

When the mean time between demands is 10^4 hours (≈ 1.15 years), we notice that the mean time between critical situations will be the same for an SIF with low demand mode as for an SIF with high demand mode, with the same SIL. The demand rate β is usually defined as the net demand rate for the SIF, excluding the demands that are effectively taken care of by non-SIS safeguards and other risk reduction facilities (refer to steps 10 and 11 in Fig. 10.12).

The *risk* related to a specified critical event for an SIF with low demand mode is a function of (1) the potential consequences of a critical event and (2) the frequency of the critical event. To select an appropriate SIL, we therefore need to assess

1. The frequency β of demands for the SIS
2. The potential consequences following an occurrence of the critical event

10.6.3 Compliance with IEC 61508

The overall objective of IEC 61508 is to identify the required safety instrumented functions (SIFs), to establish the required SIL for each SIF, and to implement the safety functions in an SIS in order to achieve the desired safety level for the process. IEC 61508 is *risk based* and decisions taken shall be based on criteria related to risk reduction and tolerability of risk.

The objective and the requirements related to each of the life cycle phases in Fig. 10.12 are described in detail in Section 7 of IEC 61508-1. The required actions are discussed, for example, by Stavrianidis and Bhimavarapu (1998). The actions that have to be carried out, and the extent of these actions, will vary with the type and complexity of the system (process). We have proposed a sequence of actions in the following. The described actions should be regarded as a supplement to the detailed requirements in the standard. Our proposed actions do not replace the requirements in the standard, but may hopefully give additional insight. When developing this sequence of actions, we have had the process section of an offshore oil/gas platform in mind. For other processes/applications, some of the actions might be reduced or omitted.

1. *System definition*: We start with a *conceptual design* of the system. The conceptual design is assumed to be a basic design where no safety instrumented functions are implemented. The conceptual design is a (close to) final design that is described by process and instrument diagrams (P&IDs), other flow diagrams, and calculation results.

2. *Definition of EUCs*: The system (process) must be broken down into suitable subsystems. The subsystems are called *equipment under control* (EUC). Guidance on how to define EUCs is given in OLF (2001, Appendix B). Examples of suitable EUCs are pressure vessels, pumping stations, and compressors.

3. *Risk acceptance criteria*: We have to define *risk acceptance criteria*, or *tolerable risk* criteria, for each EUC. In some industries, like the Norwegian offshore industry, risk acceptance criteria have to be defined on the plan (platform) level in the initial phases of a development project. The risk acceptance criteria are qualitative or quantitative criteria related to the risk to humans, the environment, and sometimes also related to material assets and production regularity. Risk acceptance criteria may, for example, be formulated as "the fatal accident rate (FAR)[3] shall be less than nine," and "no release of toxic gas to the atmosphere with a probability of occurrence greater than 10^{-4} in one year."

   The plant risk acceptance criteria have to be broken down and allocated to the various EUCs. The allocation of requirements must be based on criteria related to feasibility, fairness, and cost and is generally not a straightforward task.

4. *Hazard analysis*: A hazard analysis has to be carried out to identify all potential hazards and process demands[4] of each EUC. The hazard analysis may be carried out using methods like:

---

[3] FAR = expected number of fatalities per 10^8 hours of exposure.
[4] A process demand is significant deviation from normal operation that can lead to adverse consequences for humans, the environment, material assets, or production regularity.

- Preliminary hazard analysis
- Hazard and operability analysis (HAZOP) (e.g., see IEC 61882)
- FMECA
- Safety analysis table (SAT) analysis (as described in ISO 10418)
- Checklists

The hazard analysis will provide:

(a) A list of all potential process demands that may occur in the EUC
(b) The direct causes of each process demand
(c) Rough estimates of the frequency of each project demand
(d) A rough assessment of the potential consequences of each process demand
(e) Identification of non-SIS protection layers for each process demand

The hazard analysis shall consider all reasonable, foreseeable circumstances including possible fault conditions, misuse, and extreme environmental conditions. The hazard and risk analysis shall also consider possible human errors and abnormal or infrequent modes of operation of the EUC.

5. *Quantitative risk assessment*: A quantified risk assessment is carried out to quantify the risk caused by the various process demands for the EUC and for the system (process). The risk assessment is carried out by methods like:

   - Fault tree analysis
   - Event tree analysis
   - Consequence analysis (e.g., fire and explosion loads)
   - Simulation (e.g., accident escalation)

   The quantitative risk assessment will provide:

   (a) Estimates of the frequency of the process demands identified in step 4
   (b) Identification of potential consequences of each process demand and assessment of these consequences
   (c) Risk estimates related to each process demand, and for the EUC
   (d) Requirements for risk reduction to meet the tolerable risk criteria for the EUC

Note 1: The traditional quantitative risk analysis (QRA) that is carried out for Norwegian offshore installations (NORSOK Z-013) will generally not meet all the requirements for risk assessment in IEC 61508.

Note 2: The quantitative risk analysis may partly be replaced with a layer of protection analysis (LOPA) (see, Dowell III 1998, and AIChE 2001).

452   RELIABILITY OF SAFETY SYSTEMS

6. *Non-SIS layers of protection*: The required risk reduction identified in step 5 may in some cases be obtained by non-SIS layers of protection. In this step, various non-SIS layers of protection (e.g., mechanical devices, fire walls) are identified and evaluated with respect to EUC risk reduction. Based on this step, we can decide whether or not a safety instrumented function (SIF) is required to meet the risk acceptance criteria.

7. *Determination of SIL*: The required SIL for each safety instrumented function is determined such that the risk reduction defined in step 5 for the EUC may be obtained. Qualitative and quantitative approaches to the determination of SIL are provided in IEC 61508-5.

Note 3: The Norwegian offshore industry has proposed an alternative approach, where the risk assessments and the SIL determinations are carried out for a generic system. Based on these analyses, a minimum SIL is specified for each category of EUCs (see OLF 2001).

8. *Specifications and reliability requirements*: The specifications and reliability requirements of the safety instrumented functions have to be defined.

9. *SIS design*: The SIS has to be designed according to the specifications in step 8. IEC 61511 and ANSI/ISA 84.01 give guidance on building an SIS with specific safety instrumented functions that meet a desired SIL.

10. *PFD calculation*: Reliability models are established and the PFD calculated for the proposed SIS design.

11. *Spurious trip assessment*: The frequency of spurious trip (ST) failures of the proposed SIS design has to be estimated. Other potential, negative effects of the proposed SIS design should be evaluated. (This step is not required in IEC 61508).

12. *Iteration*: We must now check that the proposed SIS design fulfills the criteria in step 7, and that the frequency of spurious trip failures is acceptable. If not, the design has to be modified. Several iterations may be necessary.

13. *System risk evaluation*: The system (process) risk reduction due to the proposed SIS is now assessed.

14. *Verification*: The required modifications and analysis are made to ascertain that the proposed SIS meets the risk reduction (SIL) requirements.

10.7   THE PDS APPROACH

The safety unavailability of an SIS with low demand mode may be assessed by the methods described in Sections 10.2 and 10.3. A more comprehensive approach has, however, been developed by SINTEF[5] as part of the PDS[6] project. The PDS method is used to quantify both the reliability (the safety unavailability and the spurious trip rate) and the life cycle cost of an SIS. A brief introduction to PDS is given by Hansen and Aarø (1997). A more recent and detailed description is given by Corneliussen and Hokstad (2003). The PDS method is compatible with the requirements in IEC 61508 and can be used to verify whether or not a specific SIL requirement is met.

10.8 MARKOV APPROACH

Consider a safety system that is tested periodically with test interval τ. When a failure is detected during a test, the system is repaired. The time required for testing and repair is considered to be negligible.

Let $X(t)$ denote the state of the safety system at time t, and let $\mathcal{X} = \{0, 1, \ldots, r\}$ be the (finite) set of all possible states. Assume that we can split the state space $\mathcal{X}$ in two parts, a set B of functioning states, and a set F of failed states, such that $F = \mathcal{X} - B$. The average probability of failure on demand, PFD(n), of the system in test interval n is

$$\text{PFD}(n) = \frac{1}{\tau} \int_{(n-1)\tau}^{n\tau} \Pr(X(t) \in F) \, dt \qquad (10.34)$$

for $n = 1, 2, \ldots$. If a demand for the safety system occurs in interval n, the (average) probability that the safety system is able to shut down the EUC is PFD(n). The following approach is mainly based on Lindqvist and Amundrustad (1998).

We assume that $\{X(t)\}$ behaves like a homogeneous Markov process (see Chapter 8) with transition rate matrix $\mathbb{A}$ as long as time runs inside a test interval, that is, inside intervals $(n-1)\tau \leq t < n\tau$, for $n = 1, 2, \ldots$. Let $P_{jk}(t) = \Pr(X(t) = k \mid X(0) = j)$ denote the transition probabilities for $j, k \in \mathcal{X}$, and let $\mathbb{P}(t)$ denote the corresponding matrix. Failures detected by diagnostic self-testing and ST failures may occur and be repaired within the test interval.

Let $Y_n = X(n\tau-)$ denote the state of the system immediately before time $n\tau$, that is, immediately before test n. If a malfunctioning state is detected during a test, a repair action is initiated, and changes the state from Y_n to a state Z_n, where Z_n denotes the state of the system just after the test (and possible repair) n. When Y_n is given, we assume that Z_n is independent of all transitions of the system before time $n\tau$. Let

$$\Pr(Z_n = j \mid Y_n = i) = R_{ij} \quad \text{for all } i, j \in \mathcal{X} \qquad (10.35)$$

denote the transition probabilities, and let $\mathbb{R}$ denote the corresponding transition matrix. If the state of the system is $Y_n = i$ just before test n, the matrix $\mathbb{R}$ tells

---

[5]SINTEF is the Norwegian abbreviation for the Foundation of Science and Technology at the Norwegian Institute of Technology.
[6]PDS is a Norwegian abbreviation for "reliability of computer-based safety systems."

us the probability that the system is in state $Z_n = j$ just after test/repair n. The matrix $\mathbb{R}$ depends on the repair strategy, and also on the quality of the repair actions. Probabilities of maintenance-induced failures and imperfect repair may be included in $\mathbb{R}$. The matrix $\mathbb{R}$ is called the *repair matrix* of the system.

Example 10.15
Consider a safety valve that is located in the production tubing in an oil/gas production well. The valve is closed and tested for leakage at regular intervals. When the valve is closed, it may fail to reopen. That is, the failure mode fail to open (FTO) may occur. Experience has shown that a specific type of wells will fail to reopen approximately once every 200 tests. The probability of FTO failure can easily be taken into account in the repair matrix $\mathbb{R}$. □

Let the distribution of the state of the safety system at time $t = 0$, $Z_0 \equiv X(0)$ be denoted by $\rho = [\rho_0, \rho_1, \ldots, \rho_r]$, where $\rho_i = \Pr(Z_0 = i)$, and $\sum_{i=0}^{r} \rho_i = 1$. The distribution of the state of the system just before the first test, at time τ, is

$$\begin{aligned}
\Pr(Y_1 = k) &= \Pr(X(\tau-) = k) \\
&= \sum_{j=0}^{r} \Pr(X(\tau-) = k \mid X(0) = j) \cdot \Pr(X(0) = j) \\
&= \sum_{j=0}^{r} \rho_j \cdot P_{jk}(\tau) = [\rho \cdot \mathbb{P}(\tau)]_k
\end{aligned} \qquad (10.36)$$

for any $k \in \mathcal{X}$, where $[\boldsymbol{B}]_k$ denotes the kth entry of the vector $\boldsymbol{B}$.

Let us now consider a test interval n (≥ 1). Just after test interval n the state of the system is Z_n. We assume that the Markov process in $n\tau \leq t < (n+1)\tau$, given its initial state Z_n, is independent of all transitions that have taken place before time $n\tau$.

$$\begin{aligned}
\Pr(Y_{n+1} = k \mid Y_n = j) &\\
&= \sum_{i=0}^{r} \Pr(Y_{n+1} = k \mid Z_n = i, Y_n = j) \cdot \Pr(Z_n = i \mid Y_n = j) \\
&= \sum_{i=0}^{r} P_{ik}(\tau) R_{ji} = [\mathbb{R} \cdot \mathbb{P}(\tau)]_{jk}
\end{aligned} \qquad (10.37)$$

where $[\mathbb{B}]_{jk}$ denotes the (jk)th entry of the matrix $\mathbb{B}$. It follows that $\{Y_n, n = 0, 1, \ldots\}$ is a discrete-time Markov chain with transition matrix

$$\mathbb{Q} = \mathbb{R} \cdot \mathbb{P}(\tau) \qquad (10.38)$$

In the same way,

$$\Pr(Z_{n+1} = k \mid Z_n = j)$$
$$= \sum_{i=0}^{r} \Pr(Z_{n+1} = k \mid Y_{n+1} = i, Z_n = j) \cdot \Pr(Y_{n+1} = i \mid Z_n = j)$$
$$= \sum_{i=0}^{r} P_{ji}(\tau) \cdot R_{ik} = [\mathbb{P}(\tau) \cdot \mathbb{R}]_{jk} \quad (10.39)$$

and $\{Z_n, n = 0, 1, \ldots\}$ is a discrete-time Markov chain with transition matrix

$$\mathbb{T} = \mathbb{P}(\tau) \cdot \mathbb{R} \quad (10.40)$$

Let $\boldsymbol{\pi} = [\pi_0, \pi_1, \ldots, \pi_r]$ denote the stationary distribution of the Markov chain $\{Y_n, n = 0, 1, \ldots\}$. Then $\boldsymbol{\pi}$ is the unique probability vector satisfying the equation

$$\boldsymbol{\pi} \cdot \mathbb{Q} \equiv \boldsymbol{\pi} \cdot \mathbb{R} \cdot \mathbb{P}(\tau) = \boldsymbol{\pi} \quad (10.41)$$

where π_i is the long-term proportion of times the system is in state i just before a test.

In the same way, let $\boldsymbol{\gamma} = [\gamma_0, \gamma_1, \ldots, \gamma_r]$ denote the stationary distribution of the Markov chain $\{Z_n, n = 0, 1, \ldots\}$. Then $\boldsymbol{\gamma}$ is the unique probability vector satisfying the equation

$$\boldsymbol{\gamma} \cdot \mathbb{T} \equiv \boldsymbol{\gamma} \cdot \mathbb{P}(\tau) \cdot \mathbb{R} = \boldsymbol{\gamma} \quad (10.42)$$

where γ_i is the long-term proportion of times the system is in state i just after a test/repair.

Let F denote the set of all states representing a DU failure in $\mathcal{X}$, and define $\pi_F = \sum_{i \in F} \pi_i$. Then, π_F denotes the long-run proportion of times the system is in a dangerously failed state immediately before a test. If, for example, $\pi_F = 5 \cdot 10^{-3}$, the system will have a critical failure, on the average, in one out of 200 tests. Moreover, $1/\pi_F$ is the mean time, in the long run, between visits to F (measured with time unit τ). The mean time between DU failures is hence

$$\text{MTBF}_{\text{DU}} = \frac{\tau}{\pi_F} \quad (10.43)$$

and the average rate of DU failures is

$$\lambda_{\text{DU}} = \frac{1}{\text{MTBF}_{\text{DU}}} = \frac{\pi_F}{\tau} \quad (10.44)$$

The average probability of failure on demand in interval n, $\text{PFD}(n)$ may now be expressed as

$$\text{PFD}(n) = \frac{1}{\tau} \int_{(n-1)\tau}^{n\tau} \Pr(X(t) \in F) \, dt$$
$$= \frac{1}{\tau} \int_0^{\tau} \sum_{j=0}^{r} \sum_{k \in F} P_{jk}(t) \cdot \Pr(Z_n = j) \, dt \quad (10.45)$$

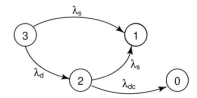

Fig. 10.13 State transition diagram for the failure process described by Hokstad and Frøvig (1996).

Since $\Pr(Z_n = j) \to \gamma_j$ when $n \to \infty$, we get the long-term average probability of failure on demand, PFD, as

$$\text{PFD} = \lim_{n \to \infty} \text{PFD}(n) = \frac{1}{\tau} \int_0^\tau \sum_{j=0}^r \sum_{k \in F} P_{jk}(t) \cdot \gamma_j \, dt = \sum_{j=0}^r \gamma_j Q_j \quad (10.46)$$

where

$$Q_j = \frac{1}{\tau} \int_0^\tau \sum_{k \in F} P_{jk}(t) \, dt$$

is the PFD given that the system is in state j at the beginning of the test interval.

Example 10.16
Hokstad and Frøvig (1996) have studied a single component that is subject to various types of failure mechanisms. In one of their examples, they study a component with the following states:

State	Description
3	Component as good as new
2	Degraded (noncritical) failure
1	Critical failure caused by sudden shock
0	Critical failure caused by degradation

The component is able to perform its intended function when it is in state 3 or state 2 and has a critical failure if it is in state 1 or state 0. State 1 is produced by a random shock, while state 0 is produced by degradation. In state 2 the component is able to perform its intended function but has a specified level of degradation.

It is assumed that the Markov process is defined by the state transition diagram in Fig. 10.13 and the transition rate matrix

$$\mathbb{A} = \begin{pmatrix} 0 & 0 & 0 & 0 \\ 0 & 0 & 0 & 0 \\ \lambda_{dc} & \lambda_s & -(\lambda_{dc} + \lambda_s) & 0 \\ 0 & \lambda_s & \lambda_d & -(\lambda_s + \lambda_d) \end{pmatrix}$$

where λ_s is the rate of failures caused by a random shock, λ_d is the rate of degradation failures, and λ_{dc} is the rate of degraded failures that become critical.

Since no repair is performed within the test interval, the failed states 0 and 1 are absorbing states. Let us assume that we know that the system is in state 3 at time 0, such that $\rho = [1, 0, 0, 0]$. We may now use the methods outlined in Section 8.9 to solve the forward Kolmogorov equations $\boldsymbol{P}(t) \cdot \mathbb{A} = \dot{\boldsymbol{P}}(t)$ and find the distribution $\mathbb{P}(t)$. It is clear that $\mathbb{P}(t)$ can be written as

$$\mathbb{P}(t) = \begin{pmatrix} 1 & 0 & 0 & 0 \\ 0 & 1 & 0 & 0 \\ P_{20}(t) & P_{21}(t) & P_{22}(t) & 0 \\ P_{30}(t) & P_{31}(t) & P_{32}(t) & P_{33}(t) \end{pmatrix}$$

The first two rows of $\mathbb{P}(t)$ are obvious since state 0 and state 1 are absorbing. The entry $P_{23}(t) = 0$ since it is impossible to have a transition from state 2 to state 3. From the state transition diagram the diagonal entries are seen to be

$$P_{22}(t) = e^{-(\lambda_s + \lambda_{dc})t}$$
$$P_{33}(t) = e^{-(\lambda_s + \lambda_d)t}$$

The remaining entries were shown by Lindqvist and Amundrustad (1998) to be

$$P_{20}(t) = \frac{\lambda_{dc}}{\lambda_s + \lambda_{dc}} \left(1 - e^{-(\lambda_s + \lambda_{dc})t}\right)$$

$$P_{21}(t) = \frac{\lambda_s}{\lambda_s + \lambda_{dc}} \left(1 - e^{-(\lambda_s + \lambda_{dc})t}\right)$$

$$P_{30}(t) = \frac{\lambda_d \lambda_{dc}}{(\lambda_d + \lambda_s)(\lambda_s + \lambda_{dc})} + \frac{\lambda_d \lambda_{dc}}{(\lambda_d - \lambda_{dc})(\lambda_d + \lambda_s)} e^{-(\lambda_s + \lambda_d)t}$$
$$+ \frac{\lambda_d \lambda_{dc}}{(\lambda_{dc} - \lambda_d)(\lambda_s + \lambda_{dc})} e^{-(\lambda_s + \lambda_{dc})t}$$

$$P_{31}(t) = \frac{\lambda_s(\lambda_d + \lambda_s + \lambda_{dc})}{(\lambda_d + \lambda_s)(\lambda_s + \lambda_{dc})} + \frac{\lambda_s \lambda_{dc}}{(\lambda_d - \lambda_{dc})(\lambda_d + \lambda_s)} e^{-(\lambda_s + \lambda_d)t}$$
$$+ \frac{\lambda_s \lambda_d}{(\lambda_{dc} - \lambda_d)(\lambda_s + \lambda_{dc})} e^{-(\lambda_s + \lambda_{dc})t}$$

$$P_{32}(t) = \frac{\lambda_d}{\lambda_d - \lambda_{dc}} \left(e^{-(\lambda_s + \lambda_{dc})t} - e^{-(\lambda_s + \lambda_d)t}\right)$$

Several repair policies may be adopted:

1. All failures are repaired after each test, such that system always starts in state 3 after each test.

2. All critical failures are repaired after each test. In this case, the system may have a degraded failure when it starts up after the test.

3. The repair action may be imperfect, meaning that there is a probability that the failure will not be repaired.

All Failures Are Repaired after Each Test In this case all failures are repaired, and we assume that the repair is perfect, such that the system will be in state 3 after each test. The corresponding repair matrix $\mathbb{R}_1$ is therefore

$$\mathbb{R}_1 = \begin{pmatrix} 0 & 0 & 0 & 1 \\ 0 & 0 & 0 & 1 \\ 0 & 0 & 0 & 1 \\ 0 & 0 & 0 & 1 \end{pmatrix}$$

With this policy, all test intervals have the same stochastic properties. The average PFD is therefore given by

$$\text{PFD} = \frac{1}{\tau} \int_0^\tau (P_{31}(t) + P_{30}(t)) \, dt$$

All Critical Failures Are Repaired after Each Test In this case the $\mathbb{R}$ matrix is

$$\mathbb{R}_2 = \begin{pmatrix} 0 & 0 & 0 & 1 \\ 0 & 0 & 0 & 1 \\ 0 & 0 & 1 & 0 \\ 0 & 0 & 0 & 1 \end{pmatrix}$$

Imperfect Repair after Each Test In this case the $\mathbb{R}$ matrix is

$$\mathbb{R}_3 = \begin{pmatrix} r_0 & 0 & 0 & 1 - r_0 \\ 0 & r_1 & 0 & 1 - r_1 \\ 0 & 0 & r_2 & 1 - r_2 \\ 0 & 0 & 0 & 1 \end{pmatrix}$$

The PFD may be found from (10.46). The calculation is straightforward, but the expressions become rather complex and are not included here. Some further results are given by Lindqvist and Amundrustad (1998).

PROBLEMS

10.1 Fig. 10.14 illustrates a part of a smoke detection system. The system comprises two optical smoke detectors (with separate batteries) and a start relay. All components

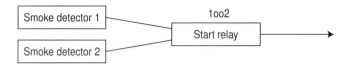

Fig. 10.14 Smoke detector system (simplified).

are assumed to be independent with constant failure rates:

Smoke detector 1 and 2 $\lambda_{SD} = 2 \cdot 10^{-4}$ failures per hour
Start relay $\lambda_{SR} = 5 \cdot 10^{-5}$ failures per hour

The system is tested and, if necessary, repaired after time intervals of equal length $\tau = 1$ month. After each test (repair) the system is considered to be "as good as new." The repair time is assumed to be negligible. Dangerous undetected failures are only detected during tests.

(a) Determine the PFD for the system.

(b) Determine the mean number of test intervals the system passes from $t = 0$ until the first DU failure.

(c) Assume that in a specific test you find that the system has a DU failure. Determine the mean time the system has been in a failed state.

(d) Assume that fires occur as a homogenous Poisson process with intensity $\beta = 5$ fires per year. Find the probability that a fire occurs while a DU failure of the smoke detection system is present, during a period of 2 years.

10.2 Reconsider the 1oo2 system of independent fire detectors in Example 10.4 but assume that the two fire detectors are different and have failure rates $\lambda_{DU,1}$ and $\lambda_{DU,2}$ respectively, with respect to DU failures. The fire detectors are tested at the same time with test interval τ.

(a) Find the PFD for the fire detector system.

(b) Find an approximation to the PFD when $\lambda_{DU,i} \cdot \tau$ is "small" for $i = 1, 2$.

10.3 Reconsider the 2oo3 system of independent fire detectors in Example 10.5 but assume that the three fire detectors are different and have failure rates $\lambda_{DU,1}$, $\lambda_{DU,2}$, and $\lambda_{DU,3}$ respectively, with respect to DU failures. The fire detectors are tested at the same time with test interval τ.

(a) Find the PFD for the fire detector system.

(b) Find an approximation to the PFD when $\lambda_{DU,i} \cdot \tau$ is "small" for $i = 1, 2, 3$.

10.4 You are planning to install a pressure sensor system on a pressure vessel. From past experience you know that the pressure sensors you are planning to use have the following constant failure rates with respect to the actual failure modes:

No signal when the pressure
increases beyond the pressure setting $\lambda_{FTF} = 3.10 \cdot 10^{-6}$ failures/hour

False high pressure signal $\lambda_{FA} = 3.60 \cdot 10^{-6}$ failures/hour

The pressure sensors will be connected to a logic unit (LU). The LU transforms the incoming signals and transmits them to the emergency shutdown (ESD) system. The failure rates of the LU are estimated to be:

Does not transmit
correct signal $\lambda_A = 0.10 \cdot 10^{-6}$ failures/hour per input

False high pressure signal out $\lambda_B = 0.05 \cdot 10^{-6}$ failures/hour

Four different system configurations are considered:

- One single pressure sensor (with LU)
- Two pressure sensors in parallel
- Three pressure sensors as a 2-out-of-3 system
- Four pressure sensors as a 2-out-of-4 system

The pressure sensors and the logic unit will be tested and, if necessary, repaired at the same time once a month. Dangerous undetected failures will only be detected during tests. After a test (repair) all items are assumed to be "as good as new." The time required for testing and repair is assumed to be negligible.

(a) Determine the PFD with respect to DU failures for each of the four system configurations when you assume that all items are independent and the failure rates of cables, and so on, are negligible.

(b) Determine the probability of getting at least one false alarm (FA) from each of the four system configurations during a period of one year.

(c) Which of the four system configurations would you install?

10.5 Consider a parallel structure of n identical components with constant failure rates λ. The system is put into operation at time $t = 0$. The system is tested and if necessary repaired after regular time intervals of length τ. After a test (repair) the system is considered to be "as good as new." The system is exposed to common cause

PROBLEMS   461

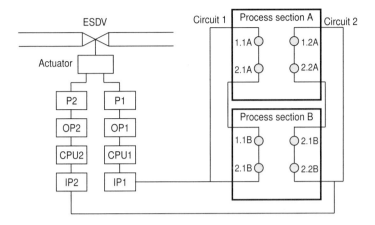

Fig. 10.15 Sketch of an emergency shutdown system.

failures that may be modeled by a β-factor model. Let PFD_n denote the probability of failure on demand of a parallel structure of order n.

(a) Determine PFD_n as a function of λ, τ, and β.

(b) Let $\lambda = 5 \cdot 10^{-5}$ failures per hour, and $\tau = 3$ months, and make a sketch of MFDT_n as a function of β for $n = 2$ and $n = 3$.

(c) With the same data as in question (b), determine the difference between MFDT_2 and MFDT_3 when $\beta = 0$ and $\beta = 0.20$, respectively.

10.6 Fig. 10.15 shows a part of a shutdown system of a process plant. There are two process sections, A and B. If a fire occurs in one of the process sections, the ESD system is installed to close the emergency shutdown valve, ESDV. The ESD valve has a fail-safe hydraulic actuator. The valve is held open by hydraulic pressure. When the hydraulic pressure is bled off, the valve will close.

Each process section has two redundant detector circuits (circuit 1 and circuit 2). Each detector circuit is connected to the ESDV actuator by a pilot valve, which by signal from the detectors opens and bleeds off the hydraulic pressure in the ESDV actuator, and thereby closes the ESD valve. Further, each circuit comprises an input card, a central processing unit (CPU), an output card, and two fire detectors in each process section. When a fire detector is activated, the current in that circuit is broken. When the current to the input card is broken, a "message" is sent to the CPU via the output card to open the pilot valve. It is assumed that minor fires in one of the process sections cannot be detected by the fire detectors in the other process section.

It is assumed that all the components are independent with constant failure rates. Each component has two different failure modes: (i) Fail to function (FTF) (i.e., no reaction when a signal is received) and (ii) False alarm. The system components, their symbols, and FTF failure rates are listed in Table 10.4.

Table 10.4 Failure Rates for the "Fail to Function" Mode.

Component	Symbol	FTF Failure Rate λ (failures per hour)
ESD valve	ESDV	$3.0 \cdot 10^{-6}$
Actuator	Actuator	$5.0 \cdot 10^{-6}$
Pilot valve	P1, P2	$2.0 \cdot 10^{-6}$
Output card	OP1, OP2	$0.1 \cdot 10^{-7}$
Input card	IP1, IP2	$0.1 \cdot 10^{-7}$
CPU	CPU1, CPU2	$0.1 \cdot 10^{-7}$
Fire detector	1.1A, 1.2A, 2.1A, 2.2A 1.1B, 1.2B, 2.1B, 2.2B	$4.0 \cdot 10^{-6}$

(a) Construct a fault tree with respect to the TOP event: "The ESD valve does not close when a fire occurs in process section A."

Write down the extra assumptions you have to make during the fault tree construction. As seen from Table 10.4 the failure rates of the input card, the CPU, and the output card are negligible compared to the failure rates of the other components. To simplify the fault tree construction, you may therefore disregard the input/output cards and the CPU.

Show that the fault tree has the following minimal cut sets:

$$\{\text{Actuator}\}$$
$$\{\text{ESDV}\}$$
$$\{\text{P1, P2}\}$$
$$\{\text{P1, 1.1A, 2.1A}\}$$
$$\{\text{P2, 1.2A, 2.2A}\}$$
$$\{\text{1.1A, 1.2A, 2.1A, 2.2A}\}$$

All the components are tested once a month. FTF failures are normally only detected during tests. The time required for testing and, if necessary, repair is assumed to be negligible compared to the length of the testing interval. In question (b) we shall assume that the testing of the various components are carried out at different, and for us unknown, times.

(b) i. Determine the PFD for each of the relevant components.
   ii. Determine the TOP event probability by the "upper bound approximation," when the basic events of the fault tree are assumed to be independent.
   iii. Discuss the accuracy of the "upper bound approximation" in this case.

iv. Describe other, and more exact methods, to compute the TOP event probability. Discuss pros and cons for each of these methods.

(c) Minor fires are assumed to occur in process section A on the average two times a year, according to a homogeneous Poisson process. A *critical situation* occurs when a fire occurs at the same time as the ESD system has FTF failure (i.e., when the TOP event is present). Find the probability of at least one such *critical situation* during a period of 10 years.

(d) Next consider the subsystem comprising the two fire detectors 1.1A and 2.1A. Determine the PFD of this subsystem when the detectors are tested:
  (i) Once every third month at different and, for us, unknown time points.
  (ii) At the same time once every third month.
  (iii) By staggered testing, where detector 1.1A is tested once every third month and detector 2.1A is also tested once every third month, but always one month later than detector 1.1A.

Which of these testing regimes would you prefer (give pros and cons). Explain why the PFD in case (i) is different from the PFD in case (ii).

(e) Do you consider the suggested system structure to be optimal with respect to avoid "False alarm" failures? Suggest an improved structure and discuss possible positive and negative properties of this structure.

10.7 A downhole safety valve (DHSV) is placed in the oil/gas production tubing on offshore production platforms, approximately 50 to 100 meters below the sea floor. The valve is held open by hydraulic pressure through a 1/16-inch hydraulic pipeline from the platform. When the hydraulic pressure is bled off, the valve will close by spring force. The valve is thus *fail safe close*. The valve is the last barrier against blowouts in case of an emergency situation on the platform. It is very important that the valve is functioning as a safety barrier, and the valve is therefore tested at regular intervals.

There are two main types of DHSVs; wireline retrievable (WR) valves and tubing retrievable (TR) valves. WR valves are locked in a landing nipple in the tubing and may be installed and retrieved by a wireline operation from the platform. A TR valve is an integrated part of the tubing. To retrieve a TR valve, the tubing has to be pulled. Here we shall consider a WR valve. When the WR valve fails, it will be retrieved by a wireline operation and a new valve of the same type will be installed in the same nipple.

The DHSV is tested once a month. During the testing, which requires approximately 1.5 hours, the production has to be closed down. The mean time to repair a failure is estimated to be 9 hours.

The DHSV has four main failure modes:

 FTC: Fail to close on command
 LCP: Leakage in closed position
 FTO: Fail to open on command
 PC: Premature closure

The failure modes FTC and LCP are critical with respect to safety. The failure modes FTO and PC are noncritical with respect to safety, but will stop the production. The three failure modes FTC, LCP, and FTO may only be detected during testing, while PC failures are detected at once since the production from the well closes down.

The following failure mode distribution has been discovered:

$$\begin{array}{ll} \text{FTC:} & 15\% \\ \text{LCP:} & 20\% \\ \text{FTO:} & 15\% \\ \text{PC:} & 50\% \end{array}$$

The failure rates are assumed to be constant with respect to all failure modes. The mean time between valve failures (with respect to all failure modes) has been estimated to be 44 months.

If a critical failure is detected during a test, the well will be unsafe during approximately one third of the repair time. If a noncritical failure is detected, the well will be safe during these operations.

(a) Determine the mean time between FTC failures of a valve.

(b) Determine the probability that a valve survives a test interval without any failure.

(c) Find the PFD. The time required for testing and repair shall be taken into account. Discuss the complications encountered in this calculation due to PC failures.

(d) Find the mean proportion of time the production is shut down due to DHSV testing and failures.

(e) Assume now that an emergency situation occurs on the platform on the average once every 50 platform years, which requires that the DHSV must be closed. A *critical situation* occurs when such an emergency situation occurs when the DHSV is not functioning as a safety barrier. Compute the mean time between this type of *critical situations*.

(f) Consider a platform with 20 production wells, with a DHSV in each well. In an emergency situation all the wells have to be closed down. With the same assumptions as above, determine the mean time between *critical situations* on the platform.

11
Life Data Analysis

11.1 INTRODUCTION

In order to obtain information about a particular life distribution $F(t)$ for an item (component/system), it is often necessary to carry out a life test where n identical, numbered items are activated in order to record their lifetimes. If the test is allowed to run until all the n items have failed and the lifetimes are recorded, the data set thus obtained is said to be *complete*.

Often we have to be satisfied with incomplete data sets. This may be because it is impractical or too expensive to wait until all the items have failed, or because individual items are "lost" for one reason or another, or because in recording lifetime we must make do with stating relatively large time intervals to which the lifetimes belong. In such situations the data set is said to be *censored*. The examples show that such censoring can be planned, but also that circumstances may arise that are beyond control.

Instead of observing lifetimes from a controlled life test we may observe lifetimes of items in actual operation. This type of data is often called *field data*. Assume that we observe n lifetimes of items of the same type that are operated under identical operational and environmental conditions. The *potential lifetime T_i* of item i ($i = 1, 2, \ldots, n$) is the lifetime the item would have if it were allowed to operate until failure. When the data set is censored, we are not able to observe all the potential lifetimes. Throughout this chapter we assume that the potential lifetimes $T_1, T_2, \ldots, T_n$ are independent and identically distributed (i.i.d.) with a continuous life distribution $F(t)$.

In the following we first describe some common types of censoring. Thereafter we discuss analysis of complete and censored data sets in a *nonparametric* setup, that is, without assuming any particular parametric model for the life distribution. We start with the empirical distribution function and the empirical survivor function that provide estimates for the life distribution function $F(t)$ and the survivor function $R(t) = 1 - F(t)$, respectively, when the data set is complete. Thereafter we discuss the *Kaplan-Meier estimator* of the survivor function $R(t)$ and also a graphical method, *hazard plotting*, that can be used to decide whether a set of life data originates from a specified life distribution or not. Both these methods can be used for complete and censored data sets. The last nonparametric method discussed in this chapter is a plotting method based on the *total time on test (TTT) transform*, that can be used both for complete and censored data sets.

In Section 11.4 we discuss estimation in parametric models. We start with providing estimates and confidence intervals in binomial and homogeneous Poisson process (HPP) models. Thereafter we discuss maximum likelihood estimation (MLE) when the life distribution is assumed to be exponential, Weibull, and inverse Gaussian, for complete data sets and for some specific types of censoring. The presentation of the estimation in parametric models is very brief and limited. The reader should therefore consult a textbook on life data analysis for more details. Excellent presentations of life data analysis may be found in, for example, Kalbfleisch and Prentice (1980), Lawless (1982), Crowder et al. (1991), Ansell and Phillips (1994), and Meeker and Escobar (1998).

When considering data from repairable items, we must first verify that we have a renewal process where the lifetimes (interarrival times) are independent and identically distributed. A number of graphical methods and formal tests have been developed to check this assertion. It is always wise to start an analysis of data from repairable items by drawing a Nelson-Aalen plot as described in Chapter 7. If the Nelson-Aalen plot is nonlinear, the methods described in this chapter should *not* be used.

In practice, the assumption of identically distributed lifetimes corresponds to the assumption that the items are nominally identical, that is, of the same type and exposed to approximately the same environmental and operational stresses. The assumption of independence means that the items are not affected by the operation or failure of any other item in the study.

Explanatory variables (covariates, concomitant variables) are not considered in this chapter. Analysis of life data with explanatory variables is briefly discussed in Chapter 12. This topic is thoroughly discussed, for example, by Kalbfleisch and Prentice (1980), Lawless (1982), Cox and Oakes (1984), Crowder et al. (1991), and Ansell and Phillips (1994).

11.2 COMPLETE AND CENSORED DATA SETS

In this section we briefly describe four types of censoring. The presentation is restricted to *right censoring*, meaning that observation of a lifetime may be terminated before the item fails. In this case we know when an item was put into operation, but not necessarily when the item fails. The data set is said to be *left censored* when we do not know when all the items were put into operation. In this case we know that an item is functioning when the observation period starts, but not necessarily how long the item has been functioning. In some data sets we may have both right and left censoring, in which case we say that the data set is *doubly censored*. Left censoring is not further discussed in this book.

Throughout this chapter the censoring mechanism is assumed to satisfy the requirement of *independent censoring*. Briefly, this means that censoring occurs independent of any information gained from previously failed items in the same study.

The assumption of a continuous life distribution implies that the probability of observing two lifetimes that are (exactly) equal is zero. In practice, we do not record the exact lifetimes, but record the lifetimes as number of time units (minutes, hours, days or months). In this case several lifetimes may be recorded as equal, and we say that the data set contains *ties*. A tie in the data set is said to have *order r*, when r lifetimes are recorded with the same value.

Throughout this chapter we let T_i denote the lifetime of item i when the lifetime is considered as a random variable. The observed value of T_i is denoted by t_i.

11.2.1 Complete Data Set

The data set is *complete* when we are able to observe the real times to failure for *all* the n items we are studying. The data set is hence $T_1, T_2, \ldots, T_n$, where T_i denotes the time to failure for item i. We may rearrange the data set in an increasing sequence:

$$T_{(1)} \leq T_{(2)} \leq \cdots \leq T_{(n)}$$

where $T_{(i)}$ is called the ith *order statistic* in the sample.

11.2.2 Type I Censoring

Sometimes, for economical or other reasons, a life test has to be terminated at a specified time t_0. All items are activated at time $t = 0$ and followed until failure or until time t_0 when the experiment is terminated. This is often the case in medical research. After the experiment, only the lifetimes of those items that have failed before t_0 will be known exactly.

This type of censoring is called *censoring of type I*, and the information in the data set obtained then consists of s ($\leq n$) observed, ordered lifetimes:

$$T_{(1)} \leq T_{(2)} \leq \cdots \leq T_{(s)}$$

In addition we know that $(n - s)$ items have survived the time t_0, and this information should also be utilized.

Since the number (S) of items that fail before time t_0 obviously is stochastic, there is a chance that none or relatively few of the items will fail before t_0. This may be a weakness of the design.

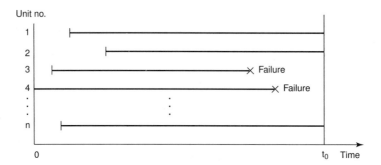

Fig. 11.1 Censored data with staggered entry.

11.2.3 Type II Censoring

If we want to ensure that the resulting data set contains a fixed number r of observed lifetimes and furthermore want to terminate the test as fast as possible, the design must allow for the test to terminate at the rth failure; $0 < r < n$. As before, we assume that all the items are activated at $t = 0$. The information obtained through the experiment consists of the data set

$$T_{(1)} \leq T_{(2)} \leq \cdots \leq T_{(r)}$$

in addition to the fact that $(n - r)$ items have survived the time $T_{(r)}$.

In this case the number (r) of recorded failures is nonstochastic. The price for obtaining this is that the time $T_{(r)}$ to complete the experiment is stochastic. A weakness of this design is therefore that we cannot know beforehand how long the experiment will last.

11.2.4 Type III Censoring

Type III censoring is a combination of the first two types. The test terminates at the time that occurs first, t_0 or the rth failure (t_0 and r must both be fixed beforehand).

11.2.5 Type IV Censoring

In this case n numbered identical items are activated at different given point(s) in time. If the time for censoring of item i, S_i; $i = 1, 2, \ldots, n$, is stochastic, the censoring is said to be of type IV. This model can also be used in the following situation:

Example 11.1
Let us assume that there are compelling reasons why a test must be terminated by time t_0. Further, the activating times for the individual items are stochastic, as, for example, they are in a medical experiment where patients may enter the study more or less randomly; see Fig. 11.1.

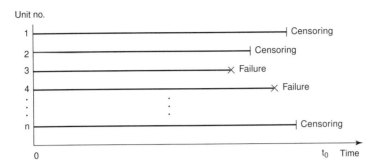

Fig. 11.2 Censored data with staggered entry shifted toward $t = 0$.

If we in this situation set the activating time points for the individual items to $t = 0$, then the censoring time points may be regarded as stochastic; see Fig. 11.2. □

Example 11.2
Suppose that we are studying a series system composed of two different components, A and B. If we are primarily concerned with studying to what degree failure in component A leads to system failure, we can interpret system failure caused by component B, as censoring. □

Example 11.3
Suppose that we have observed a set of field data for a specific type of valves and want to analyze the data to obtain information about the occurrence of a specific failure mode, say, fail to close (FTC). The lifetimes of other failure modes will in this case be interpreted as censored times. □

11.3 NONPARAMETRIC METHODS

11.3.1 Introduction

In this section we do not make any assumptions about the life distribution $F(t)$, other than assuming that $F(t)$ is continuous, and, in some cases, that $F(t)$ is strictly increasing as a function of t. This situation is called *nonparametric*, and we will use a data set (complete or censored) to obtain *nonparametric estimates* of the survivor function $R(t)$ and associated measures of the reliability of the item that is studied.

11.3.2 Sample Measures

Assume that we have observed a *complete* data set $t_1, t_2, \ldots, t_n$, and let the ordered data set be denoted by $t_{(1)}, t_{(2)}, \ldots, t_{(n)}$. Several sample measures may be derived from the data set:

The sample mean

$$\bar{t} = \frac{1}{n}\sum_{i=1}^{n} t_i \qquad (11.1)$$

The sample median

$$t_m = \begin{cases} t_{(k+1)} & \text{if } n = 2k+1 \\ (t_{(k)} + t_{(k+1)})/2 & \text{if } n = 2k \end{cases} \qquad (11.2)$$

The sample variance

$$s^2 = \frac{1}{n-1}\sum_{i=1}^{n}(x_i - \bar{x})^2 = \frac{1}{n-1}\left(\sum_{i=1}^{n} x_i^2 - \frac{1}{n}\left(\sum_{i=1}^{n} x_i\right)^2\right) \qquad (11.3)$$

The sample standard deviation

$$s = \sqrt{\frac{1}{n-1}\sum_{i=1}^{n}(x_i - \bar{x})^2} \qquad (11.4)$$

The sample coefficient of variation

$$\text{CV} = \bar{t}/s \quad \text{i.e., a unit-free measure} \qquad (11.5)$$

Remarks

1. For the exponential distribution, we found in Section 2.9 that the mean was equal to the standard deviation. The sample coefficient of variation, CV, may therefore be used to decide whether or not the exponential model might be a realistic model for the life distribution. If CV is found to be far from 1.0, then the exponential distribution is probably not a realistic distribution.

2. The sample measures presented above are available in standard spreadsheet programs (like Excel). The spreadsheet programs also include several other sample measures, like percentiles, kurtosis, and skewness.

11.3.3 The Empirical Distribution and Survivor Functions

Let $F(t)$ denote the life distribution for a certain type of items. We want to estimate the distribution function $F(t)$ and the survivor function $R(t) = 1 - F(t)$ from a *complete* data set of n independent lifetimes. Let $t_{(1)} \leq t_{(2)} \leq \cdots \leq t_{(n)}$ be the data set arranged in ascending order.

The *empirical distribution function* is defined as

$$F_n(t) = \frac{\text{Number of lifetimes } \leq t}{n} \qquad (11.6)$$

If we assume that there are no ties in the data set, the empirical distribution function may be written

$$F_n(t) = \begin{cases} 0 & \text{for } t < t_{(1)} \\ i/n & \text{for } t_{(i)} \leq t < t_{(i+1)}; \quad i = 1, 2, \ldots, (n-1) \\ 1 & \text{for } t_{(n)} \leq t \end{cases} \quad (11.7)$$

The corresponding *empirical survivor function* is

$$R_n(t) = 1 - F_n(t) = \frac{\text{Number of lifetimes} > t}{n} \quad (11.8)$$

If there are no ties in the data set, the empirical survivor function may also be written

$$R_n(t) = \begin{cases} 1 & \text{for } t < t_{(1)} \\ 1 - \dfrac{i}{n} & \text{for } t_{(i)} \leq t < t_{(i+1)}; \quad i = 1, 2, \ldots, (n-1) \\ 1 & \text{for } t_{(n)} \leq t \end{cases} \quad (11.9)$$

Note that $R_n(t)$, like $F_n(t)$, is continuous from the right. Consider a specified point of time t^*. The number $Z(t^*)$ of the n lifetimes that are $> t^*$ is from Section 2.8 binomially distributed $(n, R(t^*))$. The mean value of $Z(t^*)$ is $n \cdot R(t^*)$, and $R_n(t^*)$ in (11.8) is therefore an *unbiased* estimator for $R(t^*)$. In the same way, $F_n(t^*)$ is an unbiased estimator for $F(t^*)$. A point-wise confidence interval for $F(t^*)$ and $R(t^*)$ may be found from (11.62).

If all observations are distinct, $R_n(t)$ is a step function that decreases by $1/n$ just before each observed failure time. A simple adjustment accommodates any ties present in the data. $R_n(t)$ as a function of t is illustrated in Fig. 11.4. From (11.9) we have

$$R_n(t_{(i)}) = 1 - \frac{i}{n}$$

Let $t_{(i)-}$ denote a time just before time $t_{(i)}$. The empirical survivor function at time $t_{(i)-}$ is

$$R_n(t_{(i)-}) = 1 - \frac{i-1}{n}$$

The average value of R_n evaluated near $t_{(i)}$ is thus

$$\bar{R}_n(t_{(i)}) = \frac{1}{2}\left(R_n(t_{(i)}) + R_n(t_{(i)-})\right) = 1 - \frac{i - \frac{1}{2}}{n} \quad (11.10)$$

Some authors (e.g., Crowder et al. 1991 p. 40) find it natural to plot $\left(t_{(i)}, \bar{R}_n(t_{(i)})\right)$ instead of the empirical survivor function (11.9) since it will produce a more "smooth curve."

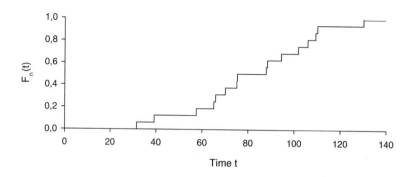

Fig. 11.3 The empirical distribution function $F_n(t)$ for the data in Example 11.4.

Example 11.4

Suppose that $n = 16$ independent lifetimes (given in months) have been observed[1]:

31.7	39.2	57.5	65.0	65.8	70.0	75.0	75.2
87.7	88.3	94.2	101.7	105.8	109.2	110.0	130.0

The sample mean $\bar{t}$ of the lifetimes is

$$\bar{t} = \frac{1}{16} \sum_{i=1}^{16} t_i \approx 81,64$$

the sample median is $M = (t_{(8)} + t_{(9)})/2 = (75.2 + 87.7)/2 = 81.45$, and the sample standard deviation is

$$S = \sqrt{\frac{1}{16-1} \sum_{i=1}^{16} (t_i - \bar{t})^2} \approx 26,78$$

For the exponential distribution we found in Section 2.9 that $E(T) = \text{SD}(T)$. Since $\bar{t} > S$ for this data set, the underlying distribution is probably not exponential.

The empirical distribution function, $F_n(t)$, is illustrated in Fig. 11.3. The empirical survivor function, $\hat{R}(t) = 1 - F_n(t)$, is illustrated in Fig. 11.4. The estimate in (11.10) is plotted in Fig. 11.5. This plot provides the same information as the empirical survivor function in Fig. 11.4. □

---

[1] The data is adapted from an example in Nelson (1972).

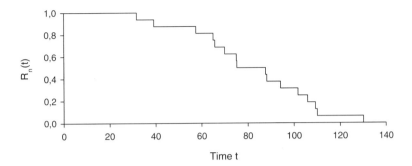

Fig. 11.4 The empirical survivor function $R_n(t) = 1 - F_n(t)$ for the data in Example 11.4.

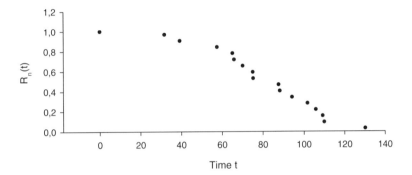

Fig. 11.5 The empirical survivor function for the data in Example 11.4 made by the estimate in (11.10).

11.3.4 Probability Plotting Paper

It is not easy to see whether or not a plot like the one in Fig. 11.5 corresponds to a specified underlying distribution $F(t)$. By an adequate change of the scale of the axes we may, however, make this task rather straightforward.

We will illustrate the procedure by the Weibull distribution. The survivor function of the Weibull distribution with scale parameter λ and shape parameter α is ($R(t) = e^{-(\lambda t)^{\alpha}}$. By taking the logarithm, we get

$$\ln R(t) = -(\lambda t)^{\alpha}$$
$$\ln[-\ln R(t)] = \alpha \ln \lambda + \alpha \ln t \qquad (11.11)$$

If we estimate $R(t)$ by (11.10) and plot $\ln[-\ln \bar{R}(t_{(j)})]$ against $t_{(j)}$ for $j = 1, 2, \ldots, n$, the plotted points should approximately follow a straight line, if the underlying distri-

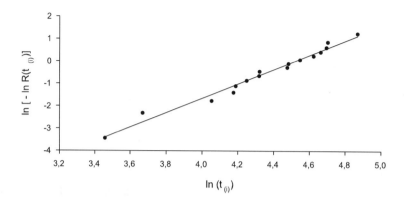

Fig. 11.6 The empirical survivor function for the data in Example 11.4 plotted by using formula (11.10) on a paper scaled according to formula (11.11).

bution is a Weibull distribution. In that case, the scale parameter α may from (11.11) be estimated as the slope of the line, and the scale parameter λ may afterwards be found from the intersection with the y axis. In Fig. 11.6 the plot in Fig. 11.5 is scaled according to (11.11) and a straight line, is fitted by the least square method. The plotted points are seen to be rather close to the line and we may therefore conclude that the data in Example 11.4 probably are from a Weibull distribution.

Special plotting papers for the Weibull distribution have been developed with the scaling in equation (11.11). The paper is called a *Weibull paper* and will usually include graphical aids to estimate the parameters α and λ. The scaling will be different for the various distributions, and we therefore need to have a plotting paper for each specific distribution. Plotting papers for a variety of distributions may be downloaded from www.weibull.com.

11.3.5 The Kaplan-Meier Estimator

We will now show how to estimate the survivor function $R(t)$ from an *incomplete* data set with censoring of type IV (see page 468). The n numbered items are activated at time $t = 0$ and the censoring time for item i, C_i, is stochastic. Associated with item i for $i = 1, 2, \ldots, n$ are two nonnegative random variables, namely the lifetime T_i, that would be observed if item i is not exposed to censoring, called the *potential lifetime*, and the time C_i, when the item is possibly censored. We assume that (T_i, C_i) for $i = 1, 2, \ldots, n$, are i.i.d. with a continuous distribution. Further, we assume that T_i and C_i for $i = 1, 2, \ldots, n$, are independent with continuous marginal distributions.

In this situation, it is only possible to record the smaller of T_i and C_i for item i for $i = 1, \ldots, n$, but at the same time we know whether we are observing a failure or a censoring. Let us introduce

$$Y_i = \min\{T_i, C_i\}$$

and the indicators

$$\delta_i = \begin{cases} 1 & \text{if } T_i \leq C_i \quad \text{(Failure)} \\ 0 & \text{if } T_i > C_i \quad \text{(Censoring)} \end{cases} \quad i = 1, \ldots, n$$

After the life test is terminated, we are left with the data set

$$(Y_1, \delta_1), (Y_2, \delta_2), \ldots, (Y_n, \delta_n)$$

Kaplan and Meier (1958) suggested the following estimation procedure: Fix $t > 0$. Let $t_{(1)} < t_{(2)} < \cdots < t_{(n)}$ denote the recorded functioning times, either until failure or to censoring, ordered according to size. Let J_t denote the set of all indices j where $t_{(j)} \leq t$ and $t_{(j)}$ represents a failure time. Let n_j denote the number of items functioning and in observation immediately before time $t_{(j)}$, $j = 1, 2, \ldots, n$. Then the Kaplan-Meier estimator of $R(t)$ is defined as follows:

$$\hat{R}(t) = \prod_{j \in J_t} \frac{n_j - 1}{n_j} \tag{11.12}$$

When the data set is complete, the Kaplan-Meier estimator (11.12) is seen to be equal to the empirical survivor function $R_n(t)$. Before we give the motivation for the Kaplan-Meier estimator, let us illustrate its use by an example.

Example 11.5
We change the situation given in Example 11.4 so that only the recorded lifetimes that are not starred (∗) in Table 11.1 represent the times to failure. The remaining times (with a ∗) represent censored times. In Table 11.1 we have calculated $\hat{R}(t)$ from (11.12). In Table 11.2 the Kaplan-Meier estimate is determined as a function of time. In the time interval $(0, 31.7)$ until the first failure it is reasonable to set $\hat{R}(t) = 1$. The estimate is displayed graphically in Fig. 11.7. The diagram in Fig. 11.7 is called a *Kaplan-Meier plot*. □

We see from (11.12) and also from Fig. 11.7 that $\hat{R}(t)$ is a step function, continuous from the right, that equals 1 at $t = 0$. $\hat{R}(t)$ drops by a factor of $(n_j - 1)/n_j$ at each failure time $t_{(j)}$. The estimator $\hat{R}(t)$ does not change at the censoring times. The effect of the censoring is, however, influencing the values of n_j and hence the size of the steps in $\hat{R}(t)$.

A slightly problematic point is that $\hat{R}(t)$ never reduces to zero when the longest time $t_{(n)}$ recorded is a censored time. For this reason $\hat{R}(t)$ is usually taken to be undefined for $t > t_{(n)}$. This problem is further discussed by Kalbfleisch and Prentice (1980, p. 12).

Table 11.1 Computation of the Kaplan-Meier Estimate.

Rank j	Inverse rank $(n - j + 1)$	Ordered Failure and Censoring Times $t_{(j)}$	$\hat{p}_j$	$\hat{R}(t_{(j)})$
0	–	–	1	1.000
1	16	31.7	$\frac{15}{16}$	0.938
2	15	39.2	$\frac{14}{15}$	0.875
3	14	57.5	$\frac{13}{14}$	0.813
4	13	65.0*	1	0.813
5	12	65.8	$\frac{11}{12}$	0.745
6	11	70.0	$\frac{10}{11}$	0.677
7	10	75.0*	1	0.677
8	9	75.2*	1	0.677
9	8	87.5*	1	0.677
10	7	88.3*	1	0.677
11	6	94.2*	1	0.677
12	5	101.7*	1	0.677
13	4	105.8	$\frac{3}{4}$	0.508
14	3	109.2*	1	0.508
15	2	110.0	$\frac{1}{2}$	0.254
16	1	130.0*	1	0.254

Note: Censoring times are starred (*)

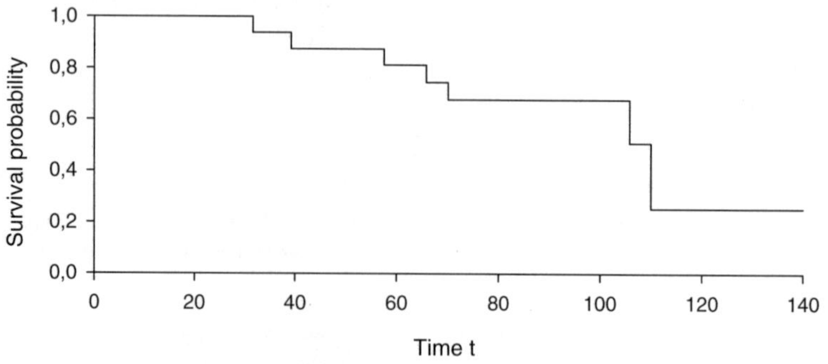

Fig. 11.7 Kaplan-Meier plot for the data in Example 11.5.

NONPARAMETRIC METHODS   477

Table 11.2 The Kaplan-Meier Estimate as a Function of Time.

t			$\hat{R}(t)$	
0	$\leq t <$	31.7		= 1.000
31.7	$\leq t <$	39.2	$\frac{15}{16}$	= 0.938
39.2	$\leq t <$	57.5	$\frac{15}{16} \cdot \frac{14}{15}$	= 0.875
57.5	$\leq t <$	65.8	$\frac{15}{16} \cdot \frac{14}{15} \cdot \frac{13}{14}$	= 0.813
65.8	$\leq t <$	70.0	$\frac{15}{16} \cdot \frac{14}{15} \cdot \frac{13}{14} \cdot \frac{11}{12}$	= 0.745
70.0	$\leq t <$	105.8	$\frac{15}{16} \cdot \frac{14}{15} \cdot \frac{13}{14} \cdot \frac{11}{12} \cdot \frac{10}{11}$	= 0.677
105.8	$\leq t <$	110.0	$\frac{15}{16} \cdot \frac{14}{15} \cdot \frac{13}{14} \cdot \frac{11}{12} \cdot \frac{10}{11} \cdot \frac{3}{4}$	= 0.508
110.0	$\leq t$		$\frac{15}{16} \cdot \frac{14}{15} \cdot \frac{13}{14} \cdot \frac{11}{12} \cdot \frac{10}{11} \cdot \frac{3}{4} \cdot \frac{1}{2}$	= 0.254

Justification for the Kaplan-Meier Estimator The motivation for the Kaplan-Meier estimator is as follows[2]: Let the time period $[0, \infty)$ be divided into small time intervals $(u_j, u_{j+1}]$ for $j = 0, 1, \ldots$ where $u_0 = 0$ and the intervals are so short that, based on the continuity assumptions about T and C, we can disregard the following possibilities:

1. Two or more items fail in the *same* interval.

2. One item fails and another is censored in the *same* interval.

Now let $t \in (u_m, u_{m+1}]$. Then obviously

$$\begin{aligned} R(t) &= \Pr(T > t) \\ &= \Pr(T > u_0) \cdot \Pr(T > u_1 \mid T > u_0) \cdots \Pr(T > t \mid T > u_m) \end{aligned}$$
(11.13)

Since $F(t)$ is assumed to be a continuous life distribution for all $t \geq 0$, then $\Pr(T > u_0) = \Pr(T > 0) = 1$. Hence

$$\begin{aligned} R(t) &= \Pr(T > u_1 \mid T > u_0) \cdot \Pr(T > u_2 \mid T > u_1) \cdots \Pr(T > t \mid T > u_m) \\ &= \prod_{j=0}^{m} p_j \end{aligned}$$
(11.14)

---

[2] The following derivation is slightly different from the derivation presented by Kaplan and Meier (1958). See also Kalbfleisch and Prentice (1980), Lawless (1982), Cox and Oakes (1984), and Cocozza-Thivent (1997) for further alternatives.

where

$$p_j = \Pr(T > u_{j+1} \mid T > u_j) \text{ for } j = 0, 1, \ldots, (m-1)$$
$$p_m = \Pr(T > t \mid T > u_m)$$

Kaplan-Meier's idea is to estimate each single factor on the right-hand side of (11.14) and thereafter use the product of these estimators as an estimator of $R(t)$. What will now be a reasonable estimator of $p_j = \Pr(T > u_{j+1} \mid T > u_j)$?

1. If neither failure nor censoring occurs in $(u_j, u_{j+1}]$, then the same number of items will be active at the start and at the end of this interval. In this case, it seems reasonable to use the estimator

$$\hat{p}_j = \Pr(T > u_{j+1} \mid T > u_j) = 1$$

2. Next suppose that censoring occurs in $(u_j, u_{j+1}]$. Then due to the assumption about short intervals, we may ignore the possibility that failure occurs in the same interval. Accordingly, in this case we have recorded no failures in the interval, and it seems reasonable to use the same estimator as in condition 1.

3. But suppose that failures occur in $(u_j, u_{j+1}]$. Due to the assumption about short intervals, we may ignore the possibility of more than one failure occurring in this interval. Let n_j denote the number of items at risk (i.e., that are functioning and in observation) at the beginning of the interval $(u_j, u_{j+1}]$. The number of items at risk at the end of the same interval is then $(n_j - 1)$. Since $(n_j - 1)$ items out of n_j survive the interval $(u_j, u_{j+1}]$, a natural estimator of $p_j = \Pr(T > u_{j+1} \mid T > u_j)$ is

$$\hat{p}_j = \frac{n_j - 1}{n_j} \tag{11.15}$$

A reasonable estimator of $\Pr(T > t \mid T > u_m)$ is found in the same way. Thus the only intervals where the estimator $\hat{p}_j$ is different from 1 are the intervals where a failure occurs. By increasing the number of intervals such that the length of each interval, except the last, approaches zero, we see that the estimator $\hat{p}_j$ is different from 1 only "at" the failure times.

As stated above, we may partly disregard the intervals where no failures occur. To simplify the notation, we therefore redefine the n_j's:

n_j = number of items at risk (functioning and in observation) immediately before time $t_{(j)}$ for $j = 1, 2, \ldots, n$

The probabilities p_j may now be estimated for infinitesimal intervals around the $t_{(j)}$'s by

$$\hat{p}_j = \begin{cases} 1 & \text{if a censoring occurred at } t_{(j)} \\ \dfrac{n_j - 1}{n_j} & \text{if a failure occurred at } t_{(j)} \end{cases} \quad j = 1, 2, \ldots, n \tag{11.16}$$

and $\hat{p}_0 = 1$.

The Kaplan-Meier estimator of the survivor function $R(t)$ is then given by [see (11.14)]

$$\hat{R}(t) = \prod_{j=0}^{n} \hat{p}_j \qquad (11.17)$$

Let J_t denote the set of all integers j such that $t_{(j)}$ is a failure time and $t_{(j)} \leq t$. The Kaplan-Meier estimator[3] in (11.17) may then be written

$$\hat{R}(t) = \prod_{j \in J_t} \frac{n_j - 1}{n_j} \qquad (11.18)$$

Let $t_{(1)} \leq t_{(2)} \leq \cdots \leq t_{(n)}$ denote the recorded functioning times, until either failure or censoring, arranged by size. If two or more of these coincide, they are arranged in random order. If a failure time and a censoring time are recorded as equal, the convention is often adopted (see Cox and Oakes 1984, p. 49) that censoring times are considered to be infinitesimally larger than the failure times. This makes sense since an item that is censored at time t almost certainly survives past t.

Some Properties of the Kaplan-Meier Estimator A thorough discussion of the properties of the Kaplan-Meier estimator $\hat{R}(t)$ may be found in Kalbfleisch and Prentice (1980), Lawless (1982), and Cox and Oakes (1984). We will here only summarize a few properties without proofs:

1. The Kaplan-Meier estimator $\hat{R}(t)$ can be derived as a nonparametric maximum likelihood estimator (MLE). This derivation was originally given by Kaplan and Meier (1958).

2. The Kaplan-Meier estimator may be slightly modified to data sets with *ties*. If we assume that d_i items fail at time $t_{(i)}$ for $i = 1, 2, \ldots, n$, the Kaplan-Meier estimator becomes (e.g., see Kalbfleisch and Prentice 1980, p. 12):

$$\hat{R}(t) = \prod_{j \in J_t} \frac{n_j - d_j}{n_j} \qquad (11.19)$$

3. $\hat{R}(t)$ is a consistent estimator of $R(t)$ under quite general conditions with estimated asymptotic variance (Kalbfleisch and Prentice 1980, p. 14):

$$\widehat{\text{var}}(\hat{R}(t)) = (\hat{R}(t))^2 \sum_{j \in J_t} \frac{d_j}{n_j(n_j - d_j)} \qquad (11.20)$$

Expression (11.20) is known as *Greenwood's formula*.

---

[3] The estimator is also called the *product limit* (PL) estimator.

4. The Kaplan-Meier estimator has an asymptotic normal distribution since it is a maximum likelihood estimator. Hence confidence limits for $R(t)$ can be determined using normal approximation. For details see Cox and Oakes (1984, pp. 51–52)

11.3.6 Nelson's Estimator of the Cumulative Failure Rate

Let $R(t)$ denote the survivor function for a certain type of items, and assume that the distribution is continuous with probability density $f(t) = R'(t)$, where $f(t) > 0$ for $t > 0$. No further assumptions are made about the distribution (nonparametric model).

The failure rate function [force of mortality (FOM)] was defined in Section 2.5 as

$$z(t) = \frac{f(t)}{R(t))} = -\frac{d}{dt} \ln R(t) \tag{11.21}$$

The cumulative failure rate function is

$$Z(t) = \int_0^t z(u)\,du = -\ln R(t) \tag{11.22}$$

The survivor function may therefore be written

$$R(t) = e^{-Z(t)}$$

Assume that we have an incomplete data set with stochastic censoring (type IV), with the same properties as described on page 468.

A natural estimator of the cumulative failure rate $Z(t)$ is then deducted from the Kaplan-Meier estimator, $\hat{R}(t)$, as

$$\hat{Z}(t) = -\ln \hat{R}(t) = -\ln \prod_{j=0}^{n} \hat{p}_j \tag{11.23}$$

where $\hat{p}_j$ for $j = 1, 2, \ldots, n$ is defined in (11.17). If we, as before, let J_t be the set of integers such that $t_{(j)}$ is a failure time and $t_{(j)} \le t$, then (11.23) may be written

$$\begin{aligned}
\hat{Z}(t) &= -\ln \prod_{j=0}^{n} \hat{p}_j = -\ln \prod_{j \in J_t} \left(\frac{n_j - 1}{n_j}\right) = -\sum_{j \in J_t} \ln\left(1 - \frac{1}{n_j}\right) \\
&= \sum_{j \in J_t} \left(\frac{1}{n_j} + \frac{1}{2n_j^2} + \cdots\right)
\end{aligned} \tag{11.24}$$

An alternative estimator of $Z(t)$ was proposed by Nelson (1969). To derive this estimator we need some lemmas.

Lemma 11.1 Let T be continuously distributed with strictly increasing distribution function $F(t)$. Then

1. $U = F(T)$ has a uniform distribution over the interval $(0, 1)$ and
2. $Z(T) = -\ln(1 - F(T))$ is exponentially distributed with parameter 1.

Proof

1. Let $u \in (0, 1]$. Then
$$\Pr(U \leq u) = \Pr(F(T) \leq u) = \Pr(T \leq F^{-1}(u)) = F(F^{-1}(u)) = u$$

2. Let $z > 0$. Then
$$\begin{aligned} \Pr(Z \leq z) &= \Pr(-\ln(1 - F(T)) \leq z) \\ &= \Pr(F(T) \leq 1 - e^{-z}) = \Pr(U \leq 1 - e^{-z}) = 1 - e^{-z} \end{aligned}$$

□

From Lemma 11.1 we easily obtain

Corolary 11.1 Let $T_1, T_2, \ldots, T_n$ be independent and identically continuously distributed with strictly increasing distribution function $F(t)$. Denote the corresponding order statistic
$$T_{(1)} < T_{(2)} < \cdots < T_{(n)}$$

Let
$$Z(T_{(j)}) = -\ln(1 - F(T_{(j)})) \quad \text{for } j = 1, 2, \ldots, n \tag{11.25}$$

and replace $Z(T_{(j)})$ by $Z_{(j)}$. Then
$$Z_{(1)} < Z_{(2)} < \cdots < Z_{(n)}$$

can be interpreted as the order statistic of n independent, identical, exponentially distributed variables with parameter 1.

Lemma 11.2 Let the assumptions be the same as in Corollary 11.1, and $Z_{(j)}$ be defined as in (11.25). Then
$$E(Z_{(j)}) = E(Z(T_{(j)})) = \frac{1}{n} + \frac{1}{n+1} + \cdots + \frac{1}{n-j+1} \tag{11.26}$$

A proof of Lemma 11.2 may be found, for example, in Barlow and Proschan (1975, p. 60). From (11.26), Nelson (1969) proposed to estimate the cumulative failure $Z(t)$ based on a *complete* data set by

$$\hat{Z}(t) = \begin{cases} 0 & \text{for } t < T_{(1)} \\ \sum_{j=1}^{r} \frac{1}{(n-j+1)} & \text{for } T_{(r)} \leq t < T_{(r+1)} \end{cases} \tag{11.27}$$

where $r = 1, 2, \ldots, n - 1$. The estimator $\hat{Z}(t)$ is called the *Nelson estimator* for a complete data set.

Having estimated the cumulative failure rate by $\hat{Z}(t)$, it seems natural to estimate the survivor function by

$$R^*(t) = e^{-\hat{Z}(t)} \qquad (11.28)$$

Before giving a justification for the estimator $\hat{Z}(t)$, we will illustrate its use by an example.

Example 11.6 Complete Data Set
Reconsider the (complete) data set in Example 11.4. The Nelson estimate $\hat{Z}(t)$ may be calculated from (11.27) for $t = t_{(j)}$ for $j = 1, \ldots, 16$. The corresponding Nelson estimate $R^*(t_{(j)})$ for the survivor function may then be calculated from (11.28). The result is shown in Table 11.3. On the right-hand side of the table the corresponding Kaplan-Meier estimate $\hat{R}(t_{(j)})$ is shown. □

Censored Data Sets Assume now that we have a data set that is subject to censoring of type IV. In this situation, Nelson proposed the following procedure for estimating the cumulative failure rate and the survivor function: As before, let

$$t_{(1)} \leq t_{(2)} \leq \cdots \leq t_{(n)}$$

denote the recorded times until either failure or censoring, and they are ordered according to size.[4] Let the index ν run through the integers j, where $t_{(j)}$ $j = 1, 2, \ldots$ denotes the times to failure such that $t_{(j)} < t$.

The *Nelson estimator* of the cumulative failure rate is then

$$\hat{Z}(t) = \sum_{\nu} \frac{1}{n - \nu + 1} = \sum_{j \in J_t} \frac{1}{n_j} \qquad (11.29)$$

Note that the Nelson estimator $\hat{Z}(t)$ is a first order approximation to the estimator (11.24) derived from the Kaplan-Meier estimator. The Nelson estimator of the survivor function at time t is

$$R^*(t) = e^{-\hat{Z}(t)} \qquad (11.30)$$

Before we give a justification for these estimators, we will use them in an example.

Example 11.7
Reconsider the censored (type IV) data set in Example 11.5. The Nelson estimate $\hat{Z}(t)$ may be calculated from (11.29) for the failure times $t_{(1)}, t_{(2)}, t_{(3)}, t_{(5)}, t_{(6)}, t_{(13)}$, and $t_{(15)}$ (hence ν runs only through the values 1, 2, 3, 5, 6, 13, and 15). Then $R^*(t)$

---
[4] If two or more of the observations coincide, they are arranged in random order.

Table 11.3 Nelson Estimator for the Complete Data Set in Example 11.6, Compared with the Kaplan-Meier Estimator.

Rank j	Lifetime $t_{(j)}$	Inverse of Number at Risk $(n-j+1)^{-1}$	Nelson Estimate $\hat{Z}(t_{(j)})$	Survivor Function Estimate	
				Nelson $R^*(t_{(j)})$	Kaplan-Meier $\hat{R}(t_{(j)})$
1	31.7	$\frac{1}{16}$	0.0625	0.939	0.938
2	39.2	$\frac{1}{15}$	0.1292	0.879	0.875
3	57.5	$\frac{1}{14}$	0.2006	0.818	0.813
4	65.0	$\frac{1}{13}$	0.2775	0.758	0.750
5	65.8	$\frac{1}{12}$	0.3609	0.697	0.688
6	70.0	$\frac{1}{11}$	0.4518	0.637	0.625
7	75.0	$\frac{1}{10}$	0.5518	0.576	0.563
8	75.2	$\frac{1}{9}$	0.6629	0.515	0.500
9	87.5	$\frac{1}{8}$	0.7879	0.455	0.448
10	88.3	$\frac{1}{7}$	0.9307	0.394	0.375
11	94.2	$\frac{1}{6}$	1.0974	0.334	0.313
12	101.7	$\frac{1}{5}$	1.2974	0.273	0.250
13	105.8	$\frac{1}{4}$	1.5474	0.213	0.188
14	109.2	$\frac{1}{3}$	1.9807	0.138	0.125
15	110.0	$\frac{1}{2}$	2.4807	0.084	0.063
16	130.0	$\frac{1}{1}$	3.4807	0.031	0.000

is determined from (11.30). The results are shown in Table 11.4. On the right-hand side of Table 11.4, the corresponding Kaplan-Meier estimate $\hat{R}(t)$ is shown.

As we can see, there is good "agreement" between the Kaplan-Meier estimates and the Nelson estimates for the survivor function in this data set. □

Justification for the Nelson Estimator We now make the same considerations as we did earlier when we justified the Kaplan-Meier estimator. The time axis is divided into small time intervals $(u_j, u_{j+1}]$ for $j = 0, 1, \ldots$, where the intervals are so short that one can disregard these possibilities:

1. Two or more items fail in the *same* interval.

2. One item fails and another is censored in the *same* interval.

Table 11.4 Nelson Estimator for the Censored Data Set in Example 11.7, Compared with the Kaplan-Meier Estimator.

j	v	Time to Failure	Nelson Estimate $\hat{Z}(t_j)$		Nelson $R^*(t_{(j)})$	Kaplan-Meier $\hat{R}(t_{(j)})$
				= 0.0000	1.000	1.000
1	1	31.7	$\frac{1}{16}$	= 0.0625	0.939	0.938
2	2	39.2	$\frac{1}{16}+\frac{1}{15}$	= 0.1292	0.879	0.875
3	3	57.5	$\frac{1}{16}+\frac{1}{15}+\frac{1}{14}$	= 0.2006	0.818	0.813
4						
5	5	65.8	$\frac{1}{16}+\frac{1}{15}+\frac{1}{14}+\frac{1}{12}$	= 0.2839	0.753	0.745
6	6	70.0	$\frac{1}{16}+\frac{1}{15}+\cdots+\frac{1}{11}$	= 0.3748	0.687	0.677
7						
8						
9						
10						
11						
12						
13	13	105.8	$\frac{1}{16}+\cdots+\frac{1}{11}+\frac{1}{4}$	= 0.6248	0.535	0.508
14						
15	15	110.0	$\frac{1}{16}+\cdots+\frac{1}{4}+\frac{1}{2}$	= 1.1248	0.320	0.254
16						

In addition, we assume that the intervals are so short that the failure rate in the interval $(u_j, u_{j+1}]$ may be considered constant and equal to λ_j for $j = 1, 2, \ldots$.

Now suppose that $t \in (u_m, u_{m+1}]$. As on page 477

$$R(t) = \Pr(T > u_1 \mid T > u_0) \cdots \Pr(T > t \mid T > u_m) \qquad (11.31)$$

As before, the idea is to estimate each single factor on the right-hand side of (11.31) and use the product of these estimators as an estimator of $R(t)$. What will now be a reasonable estimator of $p_j = \Pr(T > u_{j+1} \mid T > u_j)$? With the same approach we used for justifying the Kaplan-Meier estimator, the only intervals for which it will be natural to estimate p_j with something other than 1, will be the intervals $(u_j, u_{j+1}]$ where a *failure* occurs. If we denote the number of items that have either failed or have been censored in the course of $(0, u_j]$ with $(r - 1)$, there will be $(n - r + 1)$ active items at the beginning of the interval. The total functioning time in such an interval will be approximately equal to $(n - r + 1)(u_{j+1} - u_j)$, and hence a natural estimator of λ_j will be

$$\hat{\lambda}_j = \frac{1}{(n - r + 1)(u_{j+1} - u_j)} \qquad (11.32)$$

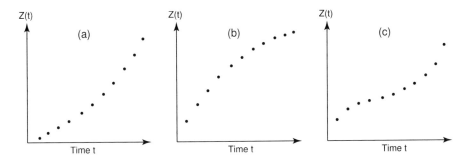

Fig. 11.8 Estimated cumulative failure rate $\hat{Z}(t)$ indicating (a) increasing failure rate (IFR), (b) decreasing failure rate (DFR), and (c) bathtub-shaped failure rate.

A natural estimator of p_j, when a failure occurs in $(u_j, u_{j+1}]$ is therefore

$$\hat{p}_j = \exp\left(-\hat{\lambda}_j(u_{j+1} - u_j)\right) = \exp\left(-\frac{1}{n - r + 1}\right) \qquad (11.33)$$

If we insert these estimators in (11.31), this leads to the estimator $R^*(t)$ given in (11.30).

Nelson Plot From (11.21) and (11.22) it follows that

$$z(t) \text{ increasing in } t \iff Z(t) \text{ convex} \qquad (11.34)$$

Correspondingly

$$z(t) \text{ decreasing in } t \iff Z(t) \text{ concave} \qquad (11.35)$$

If we plot the points $(t_{(i)}, \hat{Z}(t_{(i)}))$ on a rectangular coordinate system and if the pattern of the plot is as shown in Fig. 11.8(a), this indicates that $Z(t)$ is convex, that again means that we are dealing with an increasing failure rate (IFR) life distribution function. Similarly a plot such as the one depicted in Fig. 11.8(b) indicates a decreasing failure rate (DFR) life distribution, while the plot in Fig. 11.8(c) indicates a life distribution with bathtub-shaped failure rate function. In Fig. 11.9, the points $(t_{(j)}, \hat{Z}(t_{(j)}))$ in Table 11.4 (Example 11.7) are plotted. The diagram is called a *Nelson plot* since it was suggested by Nelson (1969)[5]. The plot is also called a *hazard plot*, because the failure rate function (FOM) is sometimes called a hazard function.

Often we are interested in checking whether it is reasonable or not to assume that a specified life distribution (normal distribution, exponential distribution or Weibull distribution) is the basis for the observed life data. Special graph papers for the actual distributions have been developed for this purpose. The graph paper for distribution $F(t)$ is so designed that if we plot $(t_{(j)}, \hat{Z}(t_{(j)}))$ on this paper, the pattern of the

---

[5]In his original paper Nelson (1969) suggested plotting $(\hat{Z}(t_{(j)}), t_{(j)})$ instead of $(t_{(j)}, \hat{Z}(t_{(j)}))$.

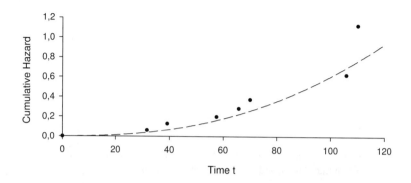

Fig. 11.9 Nelson plot of the data in Example 11.7, together with an overlay curve (dashed line) for the Weibull distribution with $\alpha = 2.38$ and $\lambda = 8.12 \cdot 10^{-3}$.

plot will be approximately a straight line if the life data really originates from the distribution $F(t)$. If so, the parameters of the distribution may be read directly from the plot.

Another way to check whether the life distribution $F(t)$ is appropriate is the following two steps: First, we estimate the parameters of the distribution $F(t)$, for example, by using the maximum likelihood principle. Next, we draw the estimated cumulative failure rate of the distribution in the same coordinate system as the plot of $(t_{(j)}, \hat{Z}(t_{(j)}))$. If the plotted values are close to the estimated cumulative failure rate function, this indicates that the distribution $F(t)$ is the basis for the observed data.

For the data in Example 11.7 the MLE (see Section 11.4) of the parameters α and λ of the Weibull distribution are $\hat{\alpha} = 2.38$ and $\hat{\lambda} = 8.12 \cdot 10^{-3}$. The estimated cumulative failure rate (hazard) function $\hat{Z}(t) = (\hat{\lambda}t)^{\hat{\alpha}}$ is drawn as an overlay curve to the plot in Fig. 11.9. We may check the goodness of fit to the Weibull distribution by visual inspection or by using a formal test.

11.3.7 Total Time on Test Plot for Complete Data Sets

Assume that we have a complete data set of n independent lifetimes with continuous distribution function $F(t)$ that is strictly increasing for $F^{-1}(0) = 0 < t < F^{-1}(1)$. Further, it is assumed that the distribution has finite mean μ.

Definition 11.1 The total time on test at time t, $\mathcal{T}(t)$, is defined as

$$\mathcal{T}(t) = \sum_{j=1}^{i} T_{(j)} + (n-i)t \qquad (11.36)$$

where i is such that

$$T_{(i)} \leq t < T_{(i+1)} \quad \text{for } i = 0, 1, \ldots, n$$

and $T_{(0)}$ is defined to be equal to 0 and $T_{(n+1)} = +\infty$. □

The total time on test $\mathcal{T}(t)$ denotes the total observed lifetime of the n items. We assume that all the n items are put into operation at time $t = 0$ and that the observation is terminated at time t. In the time interval $(0, t]$, a number, i, of the items have failed. The total functioning time of these i items is $\sum_{j=0}^{i} T_{(j)}$. The remaining $n - i$ items survive the time interval $(0, t]$. The total functioning time of these $n - i$ items is thus $(n - i)t$.

The total time on test at the ith failure is

$$\mathcal{T}(T_{(i)}) = \sum_{j=1}^{i} T_{(j)} + (n-i)T_{(i)} \quad \text{for } i = 1, 2, \ldots, n \tag{11.37}$$

In particular,

$$\mathcal{T}(T_{(n)}) = \sum_{j=1}^{n} T_{(j)} = \sum_{j=1}^{n} T_j$$

The total time on test at the ith failure, $\mathcal{T}(T_{(i)})$, may be scaled by dividing by $\mathcal{T}(T_{(n)})$. The *scaled total time on test* at time t is defined as $\mathcal{T}(t)/\mathcal{T}(T_{(n)})$.

If we plot the points

$$\left(\frac{i}{n}, \frac{\mathcal{T}(T_{(i)})}{\mathcal{T}(T_{(n)})} \right) \quad \text{for } i = 1, 2, \ldots, n$$

we obtain the *TTT plot* of the data set.

Example 11.8
Suppose that we have activated 10 identical items and observed their lifetimes (in hours):

6.3	11.0	21.5	48.4	90.1
120.2	163.0	182.5	198.0	219.0

Let us construct the TTT plot for this data set. First, we calculate the quantities we are going to need and put them in a table as done in Table 11.5. The TTT plot for this (complete) data set is shown in Fig. 11.10. □

To be able to interpret the shape of the TTT plot, we need the following theorem that is stated here without proof (see Barlow and Campo, 1975).

Theorem 11.1 Let $U_1, U_2, \ldots, U_{n-1}$ be independent random variables with a uniform distribution over $(0, 1]$. If the underlying life distribution is *exponential*, the random variables

$$\frac{\mathcal{T}(T_{(1)})}{\mathcal{T}(T_{(n)})}, \frac{\mathcal{T}(T_{(2)})}{\mathcal{T}(T_{(n)})}, \ldots, \frac{\mathcal{T}(T_{(n-1)})}{\mathcal{T}(T_{(n)})} \tag{11.38}$$

Table 11.5 TTT Estimates for the Data in Example 11.8.

i	$T_{(i)}$	$\sum_{j=1}^{i} T_{(j)}$	$\sum_{j=1}^{i} T_{(j)} + (n-i)T_{(i)} = \mathcal{T}(T_{(i)})$		$\dfrac{i}{n}$	$\dfrac{\mathcal{T}(T_{(i)})}{\mathcal{T}(T_{(n)})}$
1	6.3	6.3	6.3 + 9·6.3	= 63.0	0.1	0.06
2	11.0	17.3	17.3 + 8·11.0	= 105.3	0.2	0.10
3	21.5	38.8	38.8 + 7·21.5	= 189.3	0.3	0.18
4	48.4	87.2	87.2 + 6·48.4	= 377.6	0.4	0.36
5	90.1	177.3	177.3 + 5·90.1	= 627.8	0.5	0.59
6	120.2	297.5	297.5 + 4·120.2	= 778.3	0.6	0.73
7	163.0	460.5	460.5 + 3·163.0	= 949.5	0.7	0.90
8	182.5	643.0	643.0 + 2·182.5	= 1008.0	0.8	0.95
9	198.0	841.0	841.0 + 1·198.0	= 1039.0	0.9	0.98
10	219.0	1060.0	1060.0 + 0	= 1060.0	1.0	1.00

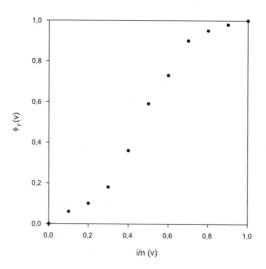

Fig. 11.10 TTT plot of the data in Example 11.8.

have the same joint distribution as the $(n-1)$ ordered variables $U_{(1)}, U_{(2)}, \ldots, U_{(n-1)}$.

From this theorem follows:

Corolary 11.2 If the underlying life distribution $F(t)$ is exponential, then

1. $\text{var}(\mathcal{T}(T_i)/\mathcal{T}(T_n))$ is finite.
2. $\text{E}(\mathcal{T}(T_i)/\mathcal{T}(T_n)) = 1/n \quad$ for $i = 1, 2, \ldots, n$.

If the underlying life distribution is exponential, we should, from Corolary 11.2(2), expect that for large n

$$\frac{\mathcal{T}(T_{(i)})}{\mathcal{T}(T_{(n)})} \approx \frac{i}{n} \quad \text{for } i = 1, 2, \ldots, (n-1)$$

As this is not the case for the TTT plot in Fig. 11.10, we can conclude that the underlying life distribution for the data in Example 11.8 is probably not exponential.

To decide from a TTT plot whether or not the corresponding life distribution is IFR or DFR, we need a little more theory. We will be content with a heuristic argument[6].

We claim that

$$\mathcal{T}(T_{(i)}) = n \int_0^{T_{(i)}} (1 - F_n(u)) \, du \tag{11.39}$$

where $F_n(t)$ denotes the empirical distribution function (11.7). Assertion (11.39) can be proved in the following way (remember that per definition $T_{(0)} = 0$):

$$n \int_0^{T_{(i)}} (1 - F_n(u)) \, du$$

$$= n \left(\sum_{j=1}^{i} \int_{T_{(j-1)}}^{T_{(j)}} (1 - \frac{j-1}{n}) \, du \right)$$

$$= \sum_{j=1}^{i} (n - j + 1)(T_{(j)} - T_{(j-1)})$$

$$= nT_{(1)} + (n-1)(T_{(2)} - T_{(1)}) + \cdots + (n-i+1)(T_{(i)} - T_{(i-1)})$$

$$= \sum_{j=1}^{i} T_{(j)} + (n-i)T_{(i)} = \mathcal{T}(T_{(i)})$$

We now come to the heuristic part of the argument. First let n equal $2m + 1$, where m is an integer. Then $T_{(m+1)}$ is the median of the data set. What happens to the integral

$$\int_0^{T_{(m+1)}} (1 - F_n(u)) \, du \quad \text{when } m \to \infty$$

---

[6] A more rigorous treatment is found, for example, in Barlow and Campo (1975).

When $m \to \infty$, we can expect that
$$F_n(u) \to F(u)$$
and that
$$T_{(m+1)} \to \{\text{median of } F\} = F^{-1}(1/2)$$
and therefore that
$$\frac{1}{n} \mathcal{T}(T_{(m+1)}) \to \int_0^{F^{-1}(1/2)} (1 - F(u))\, du \tag{11.40}$$

Next, let n equal $4m + 3$. In this case $T_{(2m+2)}$ is the median of the data, and $T_{(m+1)}$ and $T_{(3m+3)}$ are the lower and upper quartiles, respectively.

When $m \to \infty$, by arguing as we did above, we can expect the following:
$$\begin{aligned} \frac{1}{n} \mathcal{T}(T_{(m+1)}) &\to \int_0^{F^{-1}(1/4)} (1 - F(u))\, du \\ \frac{1}{n} \mathcal{T}(T_{(2m+2)}) &\to \int_0^{F^{-1}(1/2)} (1 - F(u))\, du \\ \frac{1}{n} \mathcal{T}(T_{(3m+3)}) &\to \int_0^{F^{-1}(3/4)} (1 - F(u))\, du \end{aligned} \tag{11.41}$$

In addition, according to (2.12)
$$E(T) = \mu = \int_0^\infty (1 - F(u))\, du = \int_0^{F^{-1}(1)} (1 - F(u))\, du \tag{11.42}$$

When $n \to \infty$, we can therefore expect that
$$\frac{1}{n} \sum_{i=1}^n T_i = \frac{1}{n} \mathcal{T}(T_{(n)}) \to \int_0^{F^{-1}(1)} (1 - F(u))\, du \tag{11.43}$$

The integrals that we obtain as limits by this approach, seem to be of interest and we will look at them more closely. They are all of the type
$$\int_0^{F^{-1}(v)} (1 - F(u))\, du \quad \text{for } 0 \leq v \leq 1$$

We call this integral, $\int_0^{F^{-1}(v)} (1 - F(u))\, du$, the *TTT transform* of the distribution $F(t)$ and denote it by $H_F^{-1}(v)$.

Definition 11.2 The TTT transform of the distribution F is
$$H_F^{-1}(v) = \int_0^{F^{-1}(v)} (1 - F(u))\, du \quad \text{for } 0 \leq v \leq 1 \tag{11.44}$$

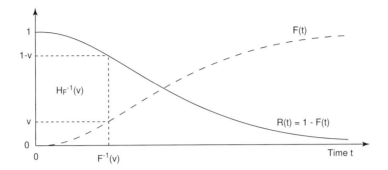

Fig. 11.11 The TTT transform of the distribution F.

The TTT transform of the distribution $F(t)$ is illustrated in Fig. 11.11. Notice that $H_F^{-1}(v)$ is the area under $R(t) = 1 - F(t)$ between $t = 0$ and $t = F^{-1}(v)$.

It can be shown under assumptions of a general nature that there is a one-to-one correspondence between a distribution $F(t)$ and its TTT transform $H_F^{-1}(v)$ (see Barlow and Campo, 1975).

We see from (11.44) and (11.42) that

$$H_F^{-1}(1) = \int_0^{F^{-1}(1)} (1 - F(u))\,du = \mu \tag{11.45}$$

The *scaled TTT transform* of $F(t)$ is defined as

$$\varphi_F(v) = \frac{H_F^{-1}(v)}{H_F^{-1}(1)} = \frac{1}{\mu} H_F^{-1}(v) \quad \text{for } 0 \le v \le 1 \tag{11.46}$$

Example 11.9 Exponential Distribution
The distribution function of the exponential distribution is

$$F(t) = 1 - e^{-\lambda t} \quad \text{for } t \ge 0,\ \lambda > 0$$

and hence

$$F^{-1}(v) = -\frac{1}{\lambda} \ln(1 - v) \quad \text{for } 0 \le v \le 1$$

Thus the TTT transform of the exponential distribution is

$$\begin{aligned}
H_F^{-1}(v) &= \int_0^{(-\ln(1-v))/\lambda} e^{-\lambda u}\,du = -\frac{1}{\lambda} e^{-\lambda u} \Big|_0^{-\frac{1}{\lambda}\ln(1-v)} \\
&= \frac{1}{\lambda} - \frac{1}{\lambda} e^{\lambda \ln(1-v)/\lambda} \\
&= \frac{1}{\lambda} - \frac{1}{\lambda}(1 - v) = \frac{v}{\lambda} \quad \text{for } 0 \le v \le 1
\end{aligned}$$

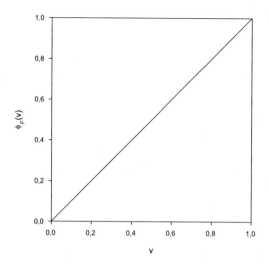

Fig. 11.12 Scaled TTT transform of the exponential distribution.

Further

$$H_F^{-1}(1) = \frac{1}{\lambda}$$

The corresponding scaled TTT transform is therefore

$$\frac{v/\lambda}{1/\lambda} = v \quad \text{for } 0 \leq v \leq 1 \tag{11.47}$$

The scaled TTT transform of the exponential distribution is thus a straight line from $(0, 0)$ to $(1, 1)$, as illustrated in Fig. 11.12. □

Example 11.10 Weibull Distribution
It is usually not straightforward to determine the TTT transform of a life distribution. We will illustrate this by trying to determine the TTT transform of the Weibull distribution

$$F(t) = 1 - e^{-(\lambda t)^\alpha} \quad \text{for } t \geq 0, \ \lambda > 0, \ \alpha > 0$$

The inverse function of F is

$$F^{-1}(v) = \frac{1}{\lambda}(-\ln(1-v))^{1/\alpha} \quad \text{for } 0 \leq v \leq 1$$

The TTT transform of the Weibull distribution is

$$H_F^{-1}(v) = \int_0^{F^{-1}(v)} (1 - F(u))\, du = \int_0^{(-\ln(1-v))^{1/\alpha}/\lambda} e^{-(\lambda u)^\alpha}\, du$$

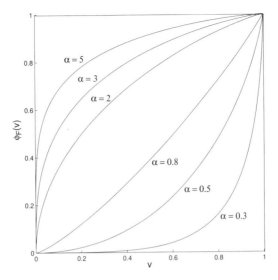

Fig. 11.13 Scaled TTT transforms of the Weibull distribution for some selected values of α.

By substituting $x = (\lambda u)^\alpha$ we obtain

$$H_F^{-1}(v) = \frac{1}{\alpha\lambda} \int_0^{-\ln(1-v)} x^{1/\alpha - 1} e^{-x}\, dx \qquad (11.48)$$

which shows that the TTT transform of the Weibull distribution may be expressed by the incomplete gamma function. However, several approximation formulas are available.

The mean time to failure is obtained by inserting $v = 1$ in $H_F^{-1}(v)$.

$$H_F^{-1}(1) = \frac{1}{\alpha\lambda} \int_0^\infty x^{1/\alpha + 1} e^{-x}\, dx = \frac{1}{\alpha\lambda}\Gamma\left(\frac{1}{\alpha}\right) = \frac{1}{\lambda}\Gamma\left(\frac{1}{\alpha} + 1\right)$$

which coincides with the result we obtained in (2.37). Note that the scaled TTT transform of the Weibull distribution depends only on the shape parameter α and is independent of the scale parameter λ. □

Computer programs able to compute the TTT transforms of the most common life distributions and to plot scaled TTT transforms are available. Scaled TTT transforms of the Weibull distribution for some selected values of the shape parameter α are illustrated in Fig. 11.13.

We will now prove the following theorem:

Theorem 11.2 If $F(t)$ is a continuous life distribution that is strictly increasing for $F^{-1}(0) = 0 < t < F^{-1}(1)$, then

$$\frac{d}{dv} H_F^{-1}(v)\big|_{v=F(t)} = \frac{1}{z(t)} \qquad (11.49)$$

where $z(t)$ is the failure rate of the distribution $F(t)$.

Proof
Since

$$\begin{aligned}
\frac{d}{dv}H_F^{-1}(v) &= \frac{d}{dv}\int_0^{F^{-1}(v)}(1-F(u))\,du \\
&= [1-F(F^{-1}(v))]\frac{d}{dv}F^{-1}(v) \\
&= (1-v)\frac{1}{f(F^{-1}(v))}
\end{aligned}$$

then

$$\frac{d}{dv}H_F^{-1}(v)|_{v=F(t)} = (1-F(t))\cdot\frac{1}{f(t)} = \frac{1}{z(t)}$$

From Theorem 11.2 we can now prove:

Theorem 11.3 If $F(t)$ is a continuous life distribution, strictly increasing for $F^{-1}(0) = 0 < t < F^{-1}(1)$, then

1. $F \sim \text{IFR} \iff H_F^{-1}(v)$ concave; $0 \leq v \leq 1$.
2. $F \sim \text{DFR} \iff H_F^{-1}(v)$ convex; $0 \leq v \leq 1$.

The arguments, used to prove properties 1 and 2 are completely analogous. We therefore prove only property 1.

Proof

$$\begin{aligned}
F \sim \text{IFR} &\iff z(t) \text{ is nondecreasing in } t \\
&\iff \frac{1}{z(t)} \text{ is nonincreasing in } t \\
&\iff \frac{d}{dv}H_F^{-1}(v)|_{v=F(t)} \text{ is nonincreasing in } t \\
&\iff \frac{d}{dv}H_F^{-1}(v) \text{ is nonincreasing in } v \\
&\quad \text{since } F(t) \text{ is strictly increasing} \\
&\iff H_F^{-1}(v) \text{ is concave, } 0 \leq v \leq 1
\end{aligned}$$

If we are going to estimate the scaled TTT transform of $F(t)$ for different v values on the basis of the observed lifetimes, it is natural to use the estimator

$$\frac{\int_0^{F_n^{-1}(v)}(1-F_n(u))\,du}{\int_0^{F_n^{-1}(1)}(1-F_n(u))\,du} \quad \text{for } v = \frac{i}{n}, \ i = 1, 2, \ldots, n \qquad (11.50)$$

Introducing the notation

$$H_n^{-1}(v) = \int_0^{F_n^{-1}(v)} (1 - F_n(u))\, du \quad \text{for } v = \frac{i}{n}, \ i = 1, 2, \ldots, n \quad (11.51)$$

this estimator can be written

$$\frac{H_n^{-1}(v)}{H_n^{-1}(1)} \quad \text{for } v = \frac{i}{n}, \ i = 1, 2, \ldots, n \quad (11.52)$$

By comparing (11.52) with (11.39), it seems natural to call $H_n^{-1}(v)/H_n^{-1}(1)$ the empirical, scaled TTT transform of the distribution $F(t)$.

The following theorem is useful when we wish to exploit the TTT plot to provide information about the life distribution $F(t)$:

Theorem 11.4 If $F(t)$ is a continuous life distribution function, strictly increasing for $F^{-1}(0) = 0 < t < F^{-1}(1)$, then

$$\frac{H_n^{-1}(\frac{i}{n})}{H_n^{-1}(1)} = \frac{\mathcal{T}(T_{(i)})}{\mathcal{T}(T_{(n)})} \quad \text{for } i = 1, 2, \ldots, n \quad (11.53)$$

where $\mathcal{T}(T_{(i)})$, as before, denotes the total time on test at time $T_{(i)}$.

Proof
According to (11.51) and (11.7), for $i = 1, 2, \ldots, n$,

$$H_n^{-1}(\frac{i}{n}) = \int_0^{F_n^{-1}(\frac{i}{n})} (1 - F_n(u))\, du$$

$$= \int_0^{T_{(i)}} (1 - F_n(u))\, du = \frac{1}{n} \mathcal{T}(T_{(i)})$$

while

$$H_n^{-1}(1) = \int_0^{F_n^{-1}(1)} (1 - F_n(u))\, du$$

$$= \int_0^{\infty} (1 - F_n(u))\, du = \frac{1}{n} \mathcal{T}(T_{(n)}) = \frac{1}{n} \sum_{i=1}^n T_i$$

By introducing these results in (11.52) we get (11.53). □

Hence the scaled total time on test at time $T_{(i)}$ seems to be a natural estimator of the scaled TTT transform of $F(t)$ for $v = i/n$; $i = 1, 2, \ldots, n$. One way of obtaining an estimate for the scaled TTT transform for $(i-1)/n < v < i/n$ is by applying linear interpolation between the estimate for $v = (i-1)/n$ and $v = i/n$. In the following we will use this procedure.

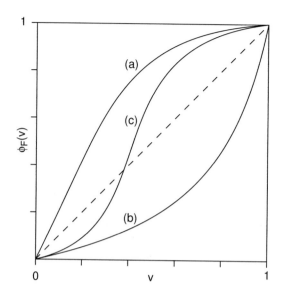

Fig. 11.14 TTT plots indicating (a) increasing failure rate (IFR), (b) decreasing failure rate (DFR), and (c) bathtub-shaped failure rate.

Now suppose that we have carried out a life test as described in the introduction to this chapter. We first determine $\mathcal{T}(T_{(i)})/\mathcal{T}(T_{(n)})$ for $i = 1, 2, \ldots, n$ as we did in Example 11.8, plot the points $[i/n, \mathcal{T}(T_{(i)})/\mathcal{T}(T_{(n)})]$, and join pairs of neighboring points with straight lines. The curve obtained is an estimate for $H_F^{-1}(v)/H_F^{-1}(1) = \frac{1}{\theta} H_F^{-1}(v)$; $0 \leq v \leq 1$.

We may now assess the shape of the curve [the estimate for $H_F^{-1}(v)$] in the light of Theorem 11.3, and in this way obtain information about the underlying distribution $F(t)$.

A plot like the one shown in Fig. 11.14(a), indicates that $H_F^{-1}(v)$ is concave. The plot therefore indicates that the corresponding life distribution $F(t)$ is IFR.

Using the same type of argument, the plot in Fig. 11.14(b) indicates that $H_F^{-1}(v)$ is convex, so that the corresponding life distribution $F(t)$ is DFR. Similarly, the plot in Fig. 11.14(c) indicates that $H_F^{-1}(v)$ "is first convex" and "thereafter concave." In other words, the failure rate of the corresponding lifetime distribution has a bathtub shape.

The TTT plot obtained in Example 11.8, therefore indicates that these data originate from a life distribution with bathtub-shaped failure rate.

Example 11.11
The following data from Lieblein and Zelen (1956) are the numbers of millions of revolutions to failure for each of 23 ball bearings. The original data have been put in numerical order for convenience.

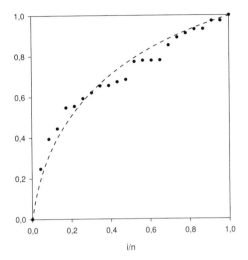

Fig. 11.15 TTT plot of the ball bearing data in Example 11.11 together with an overlay curve of the TTT transform of the Weibull distribution with shape parameter $\alpha = 2.10$.

17.88	28.92	33.00	41.52	42.12	45.60	48.40
51.84	51.96	54.12	55.56	67.80	68.64	68.64
68.88	84.12	93.12	98.64	105.12	105.84	127.92
128.04	173.40					

The TTT plot of the ball bearing data is presented in Fig. 11.15. The TTT plot indicates an increasing failure rate. We may try to fit a Weibull distribution to the data. The Weibull parameters α and λ are estimated to be $\hat{\alpha} = 2.10$ and $\hat{\lambda} = 1.22 \cdot 10^{-2}$. The TTT transform of the Weibull distribution with these parameters is plotted as an overlay curve to the TTT plot in Fig. 11.15.

Example 11.12 Age Replacement

A well-known application of the TTT transform and the TTT plot is the age replacement problem that is discussed in Section 9.6. Here an item is replaced at a cost $c+k$ at *failure* or at a cost c at a *planned replacement* when the item has reached a certain age t_0.

The average replacement cost per time unit of this policy was found to be

$$C(t_0) = \frac{c + k \cdot F(t_0)}{\int_0^{t_0} (1 - F(t)) \, dt} \qquad (11.54)$$

The objective is now to determine the value of t_0 that minimizes $C(t_0)$. If the distribution function $F(t)$ and all its parameters are known, it is a straightforward task to determine the optimal value of t_0. One way to solve this problem is to apply the TTT transform.

By introducing the TTT transform (11.54) as

$$C(t_0) = \frac{c + k \cdot F(t_0)}{H_F^{-1}(F(t_0))} = \frac{1}{H_F^{-1}(1)} \frac{c + k \cdot F(t_0)}{\varphi_F(F(t_0))}$$

where $H_F^{-1}(1)$ is the mean time to failure (MTTF) of the item, and $\varphi_F(v) = H_F^{-1}(v)/H_F^{-1}(1)$ is the scaled TTT transform of the distribution function $F(t)$.

The optimal value of t_0 may be determined by first finding the value $v_0 = F(t_0)$ that minimizes

$$C_1(v_0) = \frac{c + k \cdot v_0}{\varphi_F(v_0)}$$

and thereafter determine t_0 such that $v_0 = F(t_0)$. The minimizing value of v_0 may be found by setting the derivative of $C_1(v_0)$ with respect to v_0 equal to zero, and solve the equation for v_0:

$$\frac{d}{v_0} C_1(v_0) = \frac{\varphi_F(v_0) \cdot k - \varphi_F'(v_0)(c + k \cdot v_0)}{\varphi_F(v_0)^2} = 0$$

This implies that

$$\varphi_F'(v_0) = \frac{\varphi_F(v_0)}{c/k + v_0} \tag{11.55}$$

The optimal value of v_0, and hence t_0, may now be determined by the following simple graphical method.

1. Draw the scaled TTT transform in a 1×1 coordinate system.

2. Identify the point $(-c/k, 0)$ on the abcissa axis.

3. Draw a tangent from $(-c/k, 0)$ to the TTT transform.

The optimal value of v_0 can now be read as the abcissa of the point where the tangent touches the TTT transform. If $v_0 = 1$, then $t_0 = \infty$, and no preventive replacements should be performed. The procedure is illustrated in Fig. 11.16.

When a set of times to failure of the actual type of item has been recorded, we may use this data set to obtain the empirical, scaled TTT transform of the underlying distribution function $F(t)$, and draw a TTT plot. The optimal replacement age t_0 may now be determined by the same procedure as described above. This is illustrated in Fig. 11.17. The procedure is further discussed, for example, by Bergman and Klefsjö (1982, 1984). □

11.3.8 Total Time on Test Plot for Censored Data Sets

When the data set is incomplete and the censoring is of type IV (stochastic), we may argue as follows to obtain a TTT plot: The TTT transform, as defined in (11.44), is

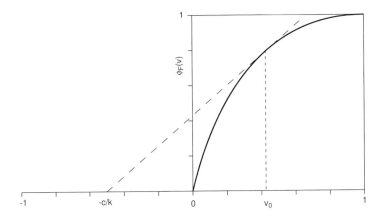

Fig. 11.16 Determination of the optimal replacement age from the scaled TTT transform.

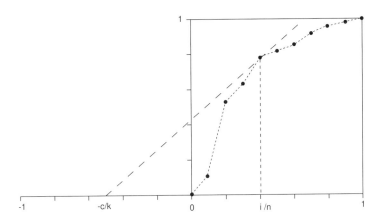

Fig. 11.17 Determination of the optimal replacement age from a TTT plot.

valid for a wide range of distribution functions $F(t)$, also for step functions. Instead of estimating the TTT transform $H_F^{-1}(t)$ by introducing the empirical distribution function $F_n(t)$ as we did in (11.51), we could estimate $F(t)$ by $(1 - \hat{R}(t))$, where $\hat{R}(t)$ is the Kaplan-Meier estimator of $R(t)$.

Technically, the plot is obtained as follows: Let $T_{(1)}, T_{(2)}, \ldots, T_{(k)}$ denote the k ordered failure times among $T_1, T_2, \ldots, T_n$ and let

$$v_{(i)} = 1 - \hat{R}(T_{(i)}) \text{ for } i = 1, 2, \ldots, k \tag{11.56}$$

Define

$$\hat{H}^{-1}(v_{(i)}) = \int_0^{T_{(i)}} \hat{R}(u)\, du = \sum_{j=1}^{i-1}(T_{(j+1)} - T_{(j)})\hat{R}(T_{(j)}) \tag{11.57}$$

where $T(0) = 0$.

The TTT plot is now obtained by plotting the points

$$\left(\frac{v_{(i)}}{v_{(k)}}, \frac{\hat{H}^{-1}(v_{(i)})}{\hat{H}^{-1}(v_{(k)})} \right) \quad \text{for } 1 = 1, 2, \ldots, k \tag{11.58}$$

Note that when $k = n$, that is, when the data set is complete, then

$$v_{(i)} = \frac{i}{n}$$
$$\hat{H}^{-1}(v_{(i)}) = \mathcal{T}(T_{(i)})$$

and we get the same TTT plot as we got for complete data sets.

11.3.9 A Brief Comparison

In this section we have presented three nonparametric estimation and plotting techniques that may be applied to both complete and censored data. (The empirical survivor function is equal to the Kaplan-Meier estimate when the data set is complete and is therefore considered as a special case of the Kaplan-Meier approach). The estimates obtained by using the Kaplan-Meier, and the Nelson (hazard plotting) approach are rather similar, so it is not important which of these is chosen. The nature of the estimate based on TTT transform is different from the other two estimates and may provide supplementary information.

The plots may also be used as a basis for selection of an adequate parametric distribution $F(t)$. In this respect the three plots provide somewhat different information. The Kaplan-Meier plot is very sensitive to variations in the early and middle phases of a item's lifetime but is not very sensitive in the right tail of the distribution. The Nelson plot is not at all sensitive in the early part of the life distribution, since the plot is "forced" to start in (0, 0). The TTT plot is very sensitive in the middle phase of the life distribution but less sensitive in the early phase and in the right tail, since the plot is "forced" to start in (0, 0) and end up in (1, 1). To get adequate information about the whole distribution all the three plots should be studied.

11.4 PARAMETRIC METHODS

In this section we study maximum likelihood (ML) estimation in some parametric models. First we briefly discuss two discrete models, the binomial and the Poisson models. Thereafter we study the exponential, the Weibull, and the inverse Gaussian models for complete data sets, and some selected types of censoring. An introduction to ML estimation is given in Appendix E. The presentation in this section is very brief. The reader is advised to consult textbooks on life data analysis for further details. Adequate references are listed in Section 11.1.

11.4.1 Binomial Data

A variety of situations, like startup of a fire pump and switching of redundant equipment, may be realistically modeled by the binomial model. The binomial model was discussed in Section 2.8. A random variable X has a binomial distribution (n, p) if

$$\Pr(X = x) = \binom{n}{x} p^x (1-p)^{n-x} \quad \text{for } x = 1, 2, \ldots, n \tag{11.59}$$

The mean is $E(X) = np$, and the variance is $\text{var}(X) = np(1-p)$. A natural *unbiased* estimator of p is

$$\hat{p} = \frac{X}{n} \tag{11.60}$$

The variance of $\hat{p}$ is

$$\text{var}(\hat{p}) = \frac{p(1-p)}{n} \tag{11.61}$$

A $1 - \varepsilon$ confidence interval for p is given by

$$\left(\frac{x}{x + (n-x+1) f_{1-\varepsilon/2, 2(n-x+1), 2x}}, \frac{(x+1) f_{1-\varepsilon/2, 2(x+1), 2(n-x)}}{n - x + (x+1) f_{1-\varepsilon/2, 2(x+1), 2(n-x)}} \right) \tag{11.62}$$

where $f_{\varepsilon, \nu_1, \nu_2}$ denotes the $100\varepsilon\%$ percentile of the Fisher distribution with ν_1 and ν_2 degrees of freedom (see Sverdrup 1967, p. 289).

Normal Approximation When np and $n(1-p)$ are both large, we may alternatively use normal approximation to find the confidence interval:

$$\frac{X - np}{\sqrt{np(1-p)}} = \frac{n(\hat{p} - p)}{\sqrt{np(1-p)}} \approx \mathcal{N}(0, 1)$$

This approximation is usually good when np and $n(1-p)$ are both greater than 5. It implies that

$$\Pr\left(-u_{\varepsilon/2} \leq \frac{n(\hat{p} - p)}{\sqrt{np(1-p)}} \leq u_{\varepsilon/2} \right) \approx 1 - \varepsilon$$

where $u_{\varepsilon/2}$ denotes the upper $100\varepsilon/2\%$ percentile of the standard normal distribution $\mathcal{N}(0, 1)$. The expression within the parentheses is equivalent to

$$n(\hat{p} - p)^2 \leq u_{\varepsilon/2}^2 \cdot p(1-p)$$

and

$$(n + u_{\varepsilon/2}^2) p^2 - (2n\hat{p} + u_{\varepsilon/2}^2) p + n\hat{p}^2 \leq 0$$

Since $n + u_{\varepsilon/2}^2 > 0$, the left-hand side of the above inequality is negative when $p_1 < p < p_2$, where p_1 and p_2 are the roots of

$$(n + u_{\varepsilon/2}^2)p^2 - (2n\hat{p} + u_{\varepsilon/2}^2)p + n\hat{p}^2 = 0$$

These roots are easily determined to be

$$\frac{1}{2a}\left(-b \pm \sqrt{b^2 - 4ac}\right)$$

where $a = n + u_{\varepsilon/2}^2$, $b = 2n\hat{p} + u_{\varepsilon/2}^2$, and $c = n\hat{p}^2$.

Hence if p_1 and p_2 are the smaller and the larger of these two roots, respectively, then

$$p_1 < p < p_2$$

constitutes an approximate $1 - \varepsilon$ confidence interval for p.

11.4.2 Data from a Homogeneous Poisson Process

Assume that we observe a population of independent and identical items with constant failure rate λ, during a total time in service t. Under the assumptions made in Section 7.2, failures will occur according to a homogeneous Poisson process. The number of failures, X, observed during this period, will have a Poisson distribution with parameter λt:

$$\Pr(X = x) = \frac{(\lambda t)^x}{x!} e^{-\lambda t} \quad \text{for } x = 0, 1, \ldots \tag{11.63}$$

The mean of X is $E(X) = \lambda t$, and the variance is $\text{var}(X) = \lambda t$. An *unbiased* estimator for λ is

$$\tilde{\lambda} = \frac{X}{t} \tag{11.64}$$

The variance of $\hat{\lambda}$ is

$$\text{var}\left(\hat{\lambda}\right) = \frac{\lambda}{t} \tag{11.65}$$

A $1 - \varepsilon$ confidence interval for λ is given by (e.g., Cocozza-Thivent 1997, p. 63, or Cox and Oakes 1984, p. 41)[7]

$$\left(\frac{1}{2t} z_{1-\varepsilon/2, 2X}, \frac{1}{2t} z_{\varepsilon/2, 2(X+1)}\right) \tag{11.66}$$

---

[7]The same confidence intervals were presented in Section 7.2 where we discussed the homogeneous Poisson process.

where $z_{\varepsilon,\nu}$ denotes the upper $100\varepsilon\%$ percentile of the χ^2 distribution with ν degrees of freedom. A table of $z_{\varepsilon,\nu}$ for some values of ε and ν is given in Appendix F.

In some situations it is of interest to give an upper $(1-\varepsilon)$ confidence limit for λ. Such a limit is obtained through the one-sided confidence interval given by

$$\left(0, \; \frac{1}{2t} z_{\varepsilon, 2(X+1)}\right) \tag{11.67}$$

Note that this interval is also applicable when no failure $(X=0)$ is observed during the total time in service t.

Normal Approximation When λt is large (say $\lambda t > 15$), the Poisson distribution may be approximated by the normal distribution $\mathcal{N}(\lambda t, \lambda t)$

$$\frac{X - \lambda t}{\sqrt{\lambda t}} \sim \mathcal{N}(0, 1)$$

Hence if $u_{\varepsilon/2}$ denotes the upper $100\varepsilon/2\%$ percentile of the standard normal distribution $\mathcal{N}(0,1)$,

$$\Pr\left(-u_{\varepsilon/2} \leq \frac{X - \lambda t}{\sqrt{\lambda t}} \leq u_{\varepsilon/2}\right) \approx 1 - \varepsilon$$

that is

$$\Pr\left(\left(\frac{X - \lambda t}{\sqrt{\lambda t}}\right)^2 \leq u_{\varepsilon/2}^2\right) \approx 1 - \varepsilon$$

but

$$\left(\frac{X - \lambda t}{\sqrt{\lambda t}}\right)^2 \leq u_{\varepsilon/2}^2 \Leftrightarrow (\lambda t)^2 - \lambda t(2X + u_{\varepsilon/2}^2) + X^2 \leq 0$$

which implies that

$$P[\lambda_1(X) \leq \lambda \leq \lambda_2(X)] \approx 1 - \varepsilon$$

where $\lambda_1(X)$ and $\lambda_2(X)$ respectively denote the lower and the upper roots of the equation

$$(\lambda t)^2 - (2X + u_{\varepsilon/2}^2)\lambda t + X^2 = 0$$

that is

$$\lambda_1(X) = \frac{1}{t}\left(X + \frac{1}{2}u_{\varepsilon/2}^2 - u_{\frac{\varepsilon}{2}}\sqrt{X + \frac{1}{4}u_{\varepsilon/2}^2}\right)$$

$$\lambda_2(X) = \frac{1}{t}\left(X + \frac{1}{2}u_{\varepsilon/2}^2 + u_{\varepsilon/2}\sqrt{X + \frac{1}{4}u_{\varepsilon/2}^2}\right)$$

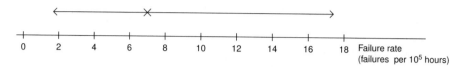

Fig. 11.18 Estimate and 90% confidence interval for the data in Example 11.13.

Example 11.13
Consider a population of independent and identical items with constant failure rate λ. We have observed $x = 3$ during an accumulated time in service of 5 years $= 5 \cdot 8760$ hours $= 43,800$ hours. The estimate (11.64) is

$$\tilde{\lambda} = \frac{3}{43,800} \text{ (hours)}^{-1} \approx 6.85 \cdot 10^{-5} \text{ (hours)}^{-1}$$

and a 90% confidence interval (11.66) is [unit = (hours)$^{-1}$]

$$\left(\frac{1}{2 \cdot 43,800} z_{0.95,6}, \frac{1}{2 \cdot 43,800} z_{0.05,8} \right) = \left(1.87 \cdot 10^{-5}, 17.71 \cdot 10^{-5} \right)$$

where the percentiles $z_{0.95,6}$ and $z_{0.05,8}$ are found in Appendix F. The estimate and the confidence interval are illustrated in Fig. 11.18. □

The length of the confidence interval is seen to shorter the more failures (x) that are observed and the longer the accumulated time (t) in service. From Fig. 11.18 we notice that the distance from the estimate to the upper bound of the interval is longer than the distance to the lower bound. This is a general feature of the confidence interval (11.66).

11.4.3 Exponentially Distributed Lifetimes: Complete Sample

Assume that we have recorded a complete data set of n actual lifetimes $T_1, T_2, \ldots, T_n$ that are independent and identically exponentially distributed with unknown failure rate λ. The likelihood function is then

$$L(\lambda; t_1, t_2, \ldots t_n) = \lambda^n e^{-\lambda \sum_{j=1}^n t_j} \quad \text{for } \lambda > 0, \ t_j > 0, \ j = 1, \ldots, n \quad (11.68)$$

The MLE of λ is

$$\lambda^* = \frac{n}{\sum_{j=1}^n T_{(j)}} = \frac{n}{\mathcal{T}(T_{(n)})} \quad (11.69)$$

where $\mathcal{T}(T_{(n)})$, as before, denotes the total time on test at the last failure. Let us study the properties of this estimator and first find out whether it is unbiased or not.

Since T_j is exponentially distributed with parameter λ, $2\lambda T_j$ will be χ^2 distributed with two degrees of freedom for $j = 1, \ldots, n$ (Dudewicz and Mishra, 1988, p. 276). Since the T_j's are independent, then $2\lambda \sum_{j=1}^n T_j$ will be χ^2 distributed with $2n$ degrees of freedom.

Therefore

$$\lambda^* = \frac{n}{\sum_{j=1}^n T_{(j)}} = \frac{2n\lambda}{2\lambda \sum_{j=1}^n T_{(j)}}$$

has the same distribution as $2n\lambda/Z$, where Z is χ^2 distributed with $2n$ degrees of freedom. Accordingly

$$E(\lambda^*) = 2n\lambda\, E\left(\frac{1}{Z}\right)$$

But

$$\begin{aligned}
E\left(\frac{1}{Z}\right) &= \int_0^\infty \frac{1}{z} \cdot \frac{1}{2^n\, \Gamma(n)} \cdot z^{n-1} e^{-z/2}\, dz \\
&= \frac{1}{2(n-1)} \int_0^\infty \frac{1}{2^{n-1}\Gamma(n-1)} z^{n-2} e^{-z/2}\, dz \\
&= \frac{1}{2(n-1)}
\end{aligned}$$

Therefore

$$E(\lambda^*) = 2n\lambda \cdot \frac{1}{2(n-1)} = \frac{n}{n-1} \cdot \lambda$$

The estimator λ^* is accordingly not unbiased. The estimator $\hat{\lambda}$, given by

$$\hat{\lambda} = \frac{n-1}{n} \cdot \lambda^* = \frac{n-1}{\sum_{j=1}^n T_j} = \frac{n-1}{\mathcal{T}(T_{(n)})}$$

is easily seen to be unbiased. Let us determine $\text{var}(\hat{\lambda})$:

$$\text{var}\left(\hat{\lambda}\right) = \left(\frac{n-1}{n}\right)^2 \cdot \text{var}(\lambda^*) = 4(n-1)^2 \lambda^2 \text{var}\left(\frac{1}{Z}\right)$$

where Z has the same meaning as above. Now

$$\text{var}\left(\frac{1}{Z}\right) = E\left(\frac{1}{Z^2}\right) - \left[E\left(\frac{1}{Z}\right)\right]^2$$

and

$$E\left(\frac{1}{Z^2}\right) = \int_0^\infty \frac{1}{z^2} \frac{1}{2^n\, \Gamma(n)} z^{n-1} e^{-z/2} dz = \frac{1}{4(n-1)(n-2)}$$

Hence

$$\begin{aligned}
\text{var}\left(\hat{\lambda}\right) &= 4(n-1)^2 \lambda^2 \cdot \left(\frac{1}{4(n-1)(n-2)} - \frac{1}{4(n-1)^2}\right) \\
&= (n-1)\lambda^2 \cdot \left(\frac{1}{n-2} - \frac{1}{n-1}\right) = \frac{\lambda^2}{n-2}
\end{aligned}$$

The estimator

$$\hat{\lambda} = \frac{n-1}{\mathcal{T}(T_{(n)})} \qquad (11.70)$$

is therefore unbiased and has variance

$$\text{var}(\hat{\lambda}) = \frac{\lambda^2}{n-2} \qquad (11.71)$$

To establish a $1 - \varepsilon$ confidence interval for λ, we use the fact that $2\lambda \sum_{j=1}^{n} T_j$ is χ^2 distributed with $2n$ degrees of freedom. Hence

$$\Pr\left(z_{1-\varepsilon/2, 2n} \leq 2\lambda \sum_{j=1}^{n} T_j \leq z_{\varepsilon/2, 2n}\right) = 1 - \varepsilon$$

and

$$\Pr\left(\frac{z_{1-\varepsilon/2, 2n}}{2\sum_{j=1}^{n} T_j} \leq \lambda \leq \frac{z_{\varepsilon/2, 2n}}{2\sum_{j=1}^{n} T_j}\right) = 1 - \varepsilon$$

Thus a $1 - \varepsilon$ confidence interval for λ is

$$\left(\frac{z_{1-\varepsilon/2, 2n}}{2\sum_{j=1}^{n} T_j}, \frac{z_{\varepsilon/2, 2n}}{2\sum_{j=1}^{n} T_j}\right) \qquad (11.72)$$

To find out whether or not the failure rate λ is less than λ_0, we may formulate this problem as a problem in hypothesis testing. We test

$$H_0 : \lambda \geq \lambda_0 \quad \text{against} \quad H_1 : \lambda < \lambda_0$$

As a first step let us derive a test for

$$H_0' : \lambda = \lambda_0 \quad \text{against} \quad H_1 : \lambda < \lambda_0$$

Then it seems reasonable to reject H_0' when $\hat{\lambda} \leq k$, where k is determined to give the test the significance level ε:

$$\Pr(\hat{\lambda} \leq k \mid H_0') \leq \varepsilon$$

Now

$$\hat{\lambda} \leq k \iff \frac{1}{\hat{\lambda}} \geq \frac{1}{k}$$

$$\iff \frac{1}{(n-1)} \sum_{j=1}^{n} T_j \geq \frac{1}{k}$$

$$\iff 2\lambda_0 \sum_{j=1}^{n} T_j \geq \frac{2(n-1)}{k} \cdot \lambda_0$$

By introducing $2\lambda_0(n-1)/k = c$, the test can be written as

$$\text{Reject } H_0' \text{ when } 2\lambda_0 \sum_{j=1}^{n} T_j \geq c$$

Under H_0', $2\lambda_0 \sum_{j=1}^{n} T_j$ is χ^2 distributed with $2n$ degrees of freedom, and accordingly c is chosen equal to $z_{\varepsilon,2n}$.

Intuitively, the same test can also be used when H_0 is to be tested against H_1. The test then has the power function

$$\Pr\left(2\lambda_0 \sum_{j=1}^{n} T_j \geq z_{\varepsilon,2n} \mid \lambda\right) = \Pr\left(2\lambda \sum_{j=1}^{n} T_j \geq \frac{\lambda}{\lambda_0} z_{\varepsilon,2n}\right)$$

$$= 1 - \Gamma_{2n}\left(z_{\varepsilon,2n} \cdot \frac{\lambda}{\lambda_0}\right) \quad (11.73)$$

Here $\Gamma_{2n}(z)$ denotes the distribution function of the χ^2 distribution with $2n$ degrees of freedom, and is thus nondecreasing in z. Accordingly $1 - \Gamma_{2n}(z_{\varepsilon,2n} \cdot \lambda/\lambda_0)$ is nonincreasing as a function of λ and

$$\Pr\left(2\lambda_0 \sum_{j=1}^{n} T_j \geq z_{\varepsilon,2n} \mid \lambda\right) \leq \varepsilon \text{ for } \lambda \geq \lambda_0$$

with the equality sign valid for $\lambda = \lambda_0$.

For testing

$$H_0 : \lambda \geq \lambda_0 \text{ against } H_1 : \lambda < \lambda_0 \quad (11.74)$$

we therefore use the test criterion:

$$\text{Reject } H_0 \text{ when } 2\lambda_0 \sum_{j=1}^{n} T_j \geq z_{\varepsilon,2n} \quad (11.75)$$

The power function of the test is given by (11.73).

11.4.4 Exponentially Distributed Lifetimes: Censored Data

Assume that n independent and identical items with constant failure rate λ have been observed until either failure or censoring. Let U be the index set of uncensored functioning times, meaning that if $j \in U$ then the time T_j is a time to failure, for $j = 1, 2, \ldots, n$. Similarly let C denote the index set of censoring times. The likelihood function is in this situation

$$L(\lambda; t_1, t_2, \ldots, t_n) = \prod_{j \in U} f(t_j; \lambda) \prod_{i \in C} R(t_i; \lambda)$$

$$= \prod_{j \in U} \lambda e^{-\lambda t_j} \prod_{i \in C} e^{-\lambda t_i} \quad (11.76)$$

Censoring of Type II With censoring of type II the life test is terminated as soon as r failures have been observed. The data set will therefore contain r times to failure, and $n - r$ censored times. From Theorem D.2 (see Appendix D) and (11.76) the likelihood function for this situation is

$$L(\lambda; t_1, \ldots, t_r) \propto \frac{n!}{(n-r)!} \lambda^r e^{-\lambda[\sum_{j=1}^r t_j + (n-r)t_r]}$$

$$= \frac{n!}{(n-r)!} \lambda^r e^{-\lambda \mathcal{T}(t_r)} \quad \text{for } 0 < t_1 < \cdots < t_r$$

and we find the maximum likelihood estimator λ_{II}^* of λ in the usual way:

$$\lambda_{\text{II}}^* = \frac{r}{\mathcal{T}(T_{(r)})} \tag{11.77}$$

Which properties does this estimator have? Let us first see what we can say about the probability distribution of λ_{II}^*. If D_j denotes the time interval from the $(j-1)$th to the jth failure, then

$$\begin{aligned} T_1 &= D_1 \\ T_2 &= D_1 + D_2 \\ &\vdots \\ T_r &= D_1 + D_2 + \cdots + D_r \end{aligned}$$

and

$$\sum_{j=1}^n T_{(j)} = rD_1 + (r-1)D_2 + \cdots + D_r$$

Furthermore

$$(n-r)T_{(r)} = (n-r)(D_1 + D_2 + \cdots + D_r)$$

Therefore, the total time on test at time $T_{(r)}$ is

$$\begin{aligned} \mathcal{T}(T_{(r)}) &= nD_1 + (n-1)D_2 + \cdots + (n-(r-1))D_r \\ &= \sum_{j=1}^r (n-(j-1))D_j \end{aligned}$$

Introducing

$$D_j^* = (n - (j-1))D_j \quad \text{for } j = 1, 2, \ldots, r$$

we know from Theorem D.4 in Appendix D that $2\lambda D_1^*, 2\lambda D_2^*, \ldots, 2\lambda D_r^*$ are independent and χ^2 distributed, each with 2 degrees of freedom. Hence $2\lambda \mathcal{T}(T_{(r)})$ is χ^2 distributed with $2r$ degrees of freedom, and we can utilize this to find $E(\lambda_{\text{II}}^*)$:

$$E(\lambda_{\text{II}}^*) = E\left(\frac{r}{T(T_{(r)})}\right) = 2\lambda r \cdot E\left(\frac{1}{2\lambda T(T_{(r)})}\right) = 2\lambda r \cdot E\left(\frac{1}{Z}\right)$$

where Z is χ^2 distributed with $2r$ degrees of freedom. This implies that (see page 505)

$$E\left(\frac{1}{Z}\right) = \frac{1}{2(r-1)}$$

Hence

$$E\left(\lambda_{\mathrm{II}}^*\right) = 2\lambda r \cdot \frac{1}{2r-1} = \lambda \cdot \frac{r}{(r-1)}$$

The estimator λ_{II}^* is accordingly not unbiased. However,

$$\hat{\lambda}_{\mathrm{II}} = \frac{r-1}{\mathcal{T}(T_{(r)})} \qquad (11.78)$$

is easily seen to be unbiased. By the method used on page 505, we find that

$$\mathrm{var}\left(\hat{\lambda}_{\mathrm{II}}\right) = \frac{\lambda^2}{r-2}$$

Confidence intervals, as well as tests for standard hypotheses about λ, may now be derived from the fact that $2\lambda \mathcal{T}(T_{(r)})$ is χ^2 distributed with $2r$ degrees of freedom. The procedure is the same as the one used on page 506.

Censoring of Type I The fact that the number (S) of items failing before time t_0 is stochastic makes this situation more difficult to deal with from a probabilistic point of view. We will therefore confine ourselves to suggesting an intuitive estimator of λ.

First notice that the estimators for λ, derived in the case of complete data sets and of type II censored data, both could be written as a fraction with numerator equal to "number of recorded failures -1" and denominator equal to "total time on test at the termination of the test." It seems intuitively reasonable to use the same fraction when we have type I censoring.

In this case the number of failures is S, while the total time on test is

$$\mathcal{T}(t_0) = \sum_{j=1}^{S} T_{(j)} + (n-S)t_0 \qquad (11.79)$$

Hence

$$\hat{\lambda}_{\mathrm{I}} = \frac{S-1}{\mathcal{T}(t_0)}$$

seems to be a reasonable estimator of λ.

It can be shown that this estimator is biased for small samples. However, asymptotically it will have the same properties as $\hat{\lambda}_{\mathrm{II}}$ has (see Mann et al. 1974, p. 173).

11.4.5 Weibull Distributed Lifetimes

Complete Sample Let $T_1, T_2, \ldots, T_n$ be a complete sample of lifetimes that are independent and identically Weibull distributed with probability density

$$f_T(t) = \alpha \lambda^\alpha t^{\alpha-1} e^{-(\lambda t)^\alpha} \quad \text{for } t > 0, \alpha > 0, \lambda > 0$$

The likelihood function is

$$L(\alpha, \lambda; t_1, t_2, \ldots, t_n) = \alpha^n \lambda^{\alpha n} \prod_{j=1}^{n} t_j^{\alpha-1} e^{-(\lambda t_j)^\alpha} \quad (11.80)$$

and the log likelihood is

$$\ln L(\alpha, \lambda; t_1, t_2, \ldots, t_n) = \sum_{j=1}^{n} \left(\alpha \ln \lambda + \ln \alpha + (\alpha - 1) \ln t_j - (\lambda t_j)^\alpha \right)$$

$$= n\alpha \ln \lambda + n \ln \alpha + \sum_{j=1}^{n} (\alpha - 1) \ln t_j - \lambda^\alpha \sum_{j=1}^{n} t_j^\alpha$$

The likelihood equations now become

$$\frac{\partial \ln L}{\partial \lambda} = \sum_{j=1}^{n} \left(\frac{\alpha}{\lambda} - \alpha \lambda^{\alpha-1} \cdot t_j^\alpha \right) = 0 \quad (11.81)$$

$$\frac{\partial \ln L}{\partial \alpha} = n \ln \lambda + \frac{n}{\lambda} + \sum_{j=1}^{n} \ln t_j - \sum_{j=1}^{n} (\lambda t_j)^\alpha \ln(\lambda t_j) = 0 \quad (11.82)$$

Solution of these equations give the MLE, α^* and λ^*. From (11.81) we obtain

$$\lambda^* = \left(\frac{n}{\sum_{j=1}^{n} t_j^{\alpha^*}} \right)^{1/\alpha^*} \quad (11.83)$$

By inserting (11.83) into (11.82) we obtain the equation

$$\frac{n}{\alpha^*} + \sum_{j=1}^{n} \ln t_j - \frac{n \sum_{j=1}^{n} t_j^{\alpha^*} \ln t_j}{\sum_{j=1}^{n} t_j^{\alpha^*}} = 0 \quad (11.84)$$

The estimate α^* may be determined from (11.84) by numerical methods. This estimate is next inserted into (11.83) to determine the estimate λ^*.

Censoring of Type II With censoring of type II the life test data set contains r times to failure, and $n - r$ censored times, and the censoring takes place at time $t_{(r)}$.

Analogous with (11.76) the likelihood function is proportional with

$$L(\alpha, \lambda; t) \propto \prod_{j=1}^{r} \alpha \lambda^{\alpha} t_j^{\alpha-1} e^{-(\lambda t_j)^{\alpha}} \left[e^{-(\lambda t_{(r)})^{\alpha}} \right]^{n-r}$$

$$= \alpha^r \lambda^{\alpha r} \prod_{j=1}^{r} t_j^{\alpha-1} e^{-(n-r)(\lambda t_{(r)})^{\alpha}}$$

where t denotes the observed data set, that is the r times to failure, and the $n-r$ censoring times that are all equal to $t_{(r)}$. The log likelihood is

$$\ln L(\alpha, \lambda; t)$$
$$= r \ln \alpha + r\alpha \ln \lambda + (\alpha - 1) \sum_{j=1}^{r} \ln t_j - \sum_{j=1}^{r} (\lambda t_j)^{\alpha} - (n - r)(\lambda t_{(r)})^{\alpha}$$

Analogous with complete data situation we can determine the MLE estimates α^* and λ^* from

$$\lambda^* = \left(\frac{r}{\sum_{j=1}^{r} t_j^{\alpha^*} + (n-r) t_{(r)}^{\alpha^*}} \right)^{1/\alpha^*} \quad (11.85)$$

and

$$\frac{r}{\alpha^*} + \sum_{j=1}^{r} \ln t_j - \frac{r \sum_{j=1}^{r} t_j^{\alpha^*} \ln t_j + (n-r) t_{(r)}^{\alpha^*} \ln t_{(r)}}{\sum_{j=1}^{r} t_j^{\alpha^*} + (n-r) t_{(r)}^{\alpha^*}} = 0 \quad (11.86)$$

11.4.6 Inverse Gaussian Distributed Lifetimes

Assume that we have recorded a *complete* data set of n actual lifetimes $T_1, T_2, \ldots, T_n$ that are independent inverse Gaussian distributed with unknown parameters μ and λ that are both positive. Let T_j for $j = 1, \ldots, k$, be IG$(\mu, n_j \lambda)$ where $n_1, \ldots, n_k$ are known, positive integers, $\sum_{j=1}^{k} n_j = N$, and $\mu > 0$, $\lambda > 0$, but otherwise unknown. Then the likelihood function becomes

$$L(\mu, \lambda; t_1, \ldots, t_k) = \prod_{j=1}^{k} \left(\sqrt{\frac{n_j}{2\pi t_j^3}} \cdot \lambda^{1/2} \cdot \exp\left(-\frac{\lambda n_j (t_j - \mu)^2}{2\mu^2 t_j} \right) \right) \quad (11.87)$$

and the maximum likelihood estimators of μ and λ, μ^* and λ^* are given by

$$\mu^* = \frac{1}{N} \sum_j n_j T_j, \quad (11.88)$$

and

$$\frac{1}{\lambda^*} = \frac{1}{k} \sum_{j=1}^{k} \frac{n_j (T_j - \mu^*)^2}{\mu^{*2} T_j} = \frac{1}{k} \sum_{j=1}^{k} n_j \left(\frac{1}{T_j} - \frac{1}{\mu^*} \right). \quad (11.89)$$

Furthermore

$$E(\mu^*) = \mu, \quad \mathrm{var}(\mu^*) = \frac{\mu^3}{\lambda N}.$$

$$E\left(\frac{1}{\lambda^*}\right) = \frac{k-1}{k} \cdot \frac{1}{\lambda}, \quad \mathrm{var}\left(\frac{1}{\lambda^*}\right) = \frac{2(k-1)}{k} \cdot \frac{1}{\lambda^2} \qquad (11.90)$$

Now, let $T_1, \ldots, T_k$ be i.i.d. IG(μ, λ), where $\mu > 0$ and $\lambda > 0$, but otherwise unknown. Then the MLE for μ and λ, μ^* and λ^* are given by

$$\mu^* = \frac{1}{k}\sum_{j=1}^{k} T_j = \overline{T}, \quad \frac{1}{\lambda^*} = \frac{1}{k}\sum_{j=1}^{k}\left(\frac{1}{T_j} - \frac{1}{\overline{T}}\right) \qquad (11.91)$$

Schrödinger proposed the estimators in (11.91) as early as 1915 and denoted them "Die Wahrscheinlichste" (Schrödinger 1915).

In the following we restrain ourselves to the case where $T_1, T_2, \ldots, T_k$ are i.i.d. IG(μ, λ), where μ and λ are both positive, but otherwise unknown.

By utilizing the well-known result that the MLE of a function $g(\mu, \lambda)$ is $g(\mu^*, \lambda^*)$, where μ^* and λ^* are the MLEs of μ and λ, respectively, one gets the MLE for var(T):

$$\widehat{\mathrm{var}(T)} = \frac{\mu^{*3}}{\lambda^*} \qquad (11.92)$$

Similarly the MLE for the survivor function $R(t; \mu, \lambda)$ is

$$\hat{R}(t; \mu^*, \lambda^*) = \Phi\left(-\frac{\sqrt{\lambda^*}}{\mu^*}\sqrt{t} - \sqrt{\lambda^*} \cdot \frac{1}{\sqrt{t}}\right)$$
$$- e^{2\lambda^*/\mu^*}\Phi\left(-\frac{\sqrt{\lambda^*}}{\mu^*}\sqrt{t} - \sqrt{\lambda^*} \cdot \frac{1}{\sqrt{t}}\right) \qquad (11.93)$$

and the MLE of the failure rate $z(t; \mu, \lambda)$ is

$$\hat{z}(t; \mu^*, \lambda^*)$$
$$= \frac{\sqrt{\lambda^*/(2\pi t^3)}\exp[-\lambda^*(t-\mu^*)^2/(2\mu^{*2}t)]}{\Phi(-\sqrt{\lambda^*t}/\mu^* + \sqrt{\lambda^*}/\sqrt{t}) + \exp[2\lambda^*/\mu^*]\Phi(-\sqrt{\lambda^*t}/\mu^* - \sqrt{\lambda^*}/\sqrt{t})} \qquad (11.94)$$

Exponentiality, Completeness, and Sufficiency. Let $T_1, T_2, \ldots, T_n$ be i.i.d. IG(μ, λ), where μ and λ are positive but unknown. Then the joint density of $T_1, T_2, \ldots, T_n$

can be written

$$f_{T_1,\ldots,T_n}(t_1,\ldots,t_n;\mu,\lambda)$$
$$= \left(\frac{\lambda}{2\pi}\right)^{n/2} \cdot \prod_{j=1}^{n} t_j^{-3/2} \cdot \exp\left(-\frac{\lambda}{2\mu^2}\sum_{j=1}^{n}\frac{(t_j-\mu)^2}{t_j}\right)$$
$$= \exp\left(\frac{n}{2}\ln\frac{\lambda}{2\pi} - \sum_{j=1}^{n}\ln t_j^{3/2} - \frac{\lambda}{2\mu^2}\sum_{j=1}^{n}t_j + \frac{\lambda n}{\mu} - \frac{\lambda}{2}\sum_{j=1}^{n}\frac{1}{t_j}\right)$$
$$= \exp\left(\sum_{i=1}^{2}c_i(\mu,\lambda)\cdot Y_i(t) + d(\mu,\lambda) + s(t)I_{A(T)}\right)$$

where

$$c_1(\mu,\lambda) = -\frac{1}{2\mu^2}, \quad c_2(\mu,\lambda) = -\frac{\lambda}{2}, \quad d(\mu,\lambda) = \frac{n}{2}\ln\frac{\lambda}{2\pi} + \frac{\lambda n}{\mu}$$

$$S(t) = -\sum \ln t_j^{\frac{3}{2}}, \quad A(t) = \{t;\ t_j > 0,\ j=1,2,\ldots,n\}$$

$$Y_1(t) = \sum_j t_j \quad \text{and} \quad Y_2(t) = \sum_j \frac{1}{t_j}$$

$I_{A(t)}$ is the indicator for $A(t) \subset R^n$

Hence (e.g., see Bickel and Doksum 1977, p. 72), the $IG(\mu,\lambda)$ family constitutes a two-parameter exponential family, and $(\sum_{j=1}^{n}T_j, \sum_{j=1}^{n}1/T_j)$ is a natural sufficient statistic for this family.

Since the range of $c(\mu,\lambda) = (c_1(\mu,\lambda), c_2(\mu,\lambda))$ contains an open rectangle, the statistic $(\sum_{j=1}^{n}T_j, \sum_{j=1}^{n}1/T_i)$ is complete as well as sufficient (e.g., see Bickel and Doksum, 1977, p. 123).

Theorem 11.5 Let $T_1, T_2, \ldots, T_k$ be i.i.d. $IG(\mu,\lambda)$, $\mu > 0, \lambda > 0$, but otherwise unknown. Then the estimators

$$\mu^* = \overline{T} \quad \text{and} \quad \frac{1}{\hat{\lambda}} = \frac{1}{(k-1)}\sum_{j=1}^{k}\left(\frac{1}{T_j} - \frac{1}{\overline{T}}\right) \quad (11.95)$$

are UMVU[8] for μ and $1/\lambda$ respectively.

---

[8]UMVU = uniformly minimum variance unbiased (e.g., Bickel and Doksum 1977, p. 119).

Proof:
From Theorem 9.4 follows that μ^* and $1/\hat{\lambda}$ are unbiased esimators of μ and $1/\lambda$ respectively. Furthermore they are functions of the observations through the complete sufficient statistic $(\sum_{j=1}^{k} T_j, \sum_{j=1}^{k} 1/T_j)$. Hence they are UMVU for μ and $1/\lambda$. (e.g., Bickel and Doksum, 1977, p. 122). □

Chhikara and Folks (1974) have derived UMVU estimators of $F_T(t_0; \mu, \lambda)$ in the situation where

- μ is known, λ is unknown
- μ is unknown, λ is known
- μ and λ are both unknown

These estimators can be used directly to find UMVU estimators of the survivor function $R(t; \mu, \lambda) = 1 - F_T(t; \mu, \lambda)$.

Example 11.14
The following data, consisting of the times to failure, given in 1000 hours, of a new class H insulation on trial in a motorette test at 260°C, are taken from Nelson (1971).

260°C : 0.600, 0.744, 0.744, 0.744, 0.912, 1.128, 1.320, 1.464, 1.608, 1.896

In his analysis, Wayne Nelson assumed the lifetimes to be independent and *lognormally* distributed. For the purpose of illustration, we will instead assume them to be i.i.d. IG(μ, λ) where μ and λ both positive, but otherwise unknown.

We obtain the following MLE of μ and λ:

$$\mu^* = 1.116 \frac{1}{\lambda^*} = \frac{1}{10}\sum_{j=1}^{10}\left(\frac{1}{t_j} - \frac{1}{\bar{t}}\right) = 0.13113$$

$$\lambda^* = 7.62602$$

Furthermore the MLE of var(T) is

$$\widehat{\text{var}(T)} = \frac{\mu^{*3}}{\lambda^*} = 0.18226$$

As a check, let us also estimate var(T) by the common estimator

$$S_{k-1}^2 = \frac{1}{k-1}\sum_{i=1}^{k}(T_i - \bar{T})^2$$

This leads to the estimate 0.19293 of var(T) which is in good correspondence with the MLE of var(T) above. Furthermore $\lim_{t\to\infty} z(t; 260°\text{C})$ may be estimated by $\lambda^*/2\mu^{*2} = 3.06$. In Wayne Nelson's model, this limit is *zero*.

One may now ask the question: Which one of these two limiting values appears to be the most realistic one in this situation? The choice between the lognormal and the inverse Gaussian model should then be made accordingly. □

11.5 MODEL SELECTION

To use parametric methods we have to select an adequate model (life distribution) for the items that are tested. In some situations we may select the class of life distributions based on engineering judgment of the deterioration mechanisms the items are exposed to. For a special type of fatigue, we may, for example, decide that an inverse Gaussian distribution provides an adequate model.

In other cases we may plot the data by the nonparametric methods described in Section 11.3 and use the plots to decide which class of life distribution to use for the parametric analysis. As discussed on page 500, the three plotting techniques presented in Section 11.3 are sensitive to variation in the data in different phases of the life of the item. All the three plotting techniques should therefore be used.

Assume that we, from the plots, find that the corresponding failure rate function $z(t)$ is increasing and seems to be close to the failure rate function of a Weibull distribution. An adequate approach would then be to:

1. Assume a Weibull distribution and find the MLE of the parameters α and λ from the data set.

2. Plot the corresponding MLE of $R(t)$, $Z(t)$, and the TTT transform as overlay curves in the Kaplan-Meier plot, the Nelson plot, and the TTT plot, respectively, and study the goodness of fit by visual inspection.

3. If the goodness of fit is deemed to be adequate, accept the Weibull model and use this model for the further analysis.

In some situations it may be of interest to establish a formal statistical test to decide whether a suggested distribution is adequate or not. Such tests are called goodness-of-fit tests and are thoroughly discussed in the textbooks on life data analysis listed in Section 11.1. See also the thorough discussion in Blischke and Murthy (2000, Chapter 11). Goodness-of-fit tests are not further discussed in this book, except for Barlow-Proschan's test that is used to test whether or not the underlying life distribution is exponential.

11.5.1 Barlow-Proschan's Test

Barlow and Proschan (1969) proposed a test based on the test statistic W defined below. The test statistic W is so designed that it has a tendency to become large (small) when the underlying distribution has an increasing (decreasing) failure rate. For a *complete* data set $(T_1, T_2, \ldots, T_n)$ the Barlow-Proschan statistic W is simply the sum of the scaled total time on test values at the failure times:

$$W = U_1 + U_2 + \cdots + U_n \tag{11.96}$$

where

$$U_i = \frac{\mathcal{T}(T_{(i)})}{\mathcal{T}(T_{(n)})} \text{ for } i = 1, 2, \ldots, n \tag{11.97}$$

denote the scaled total time on test at the ith failure.

For a *censored* data set (of type IV) the test statistic W is modified as follows: Let $T_{(1)}, T_{(2)}, \ldots, T_{(k)}$ denote the ordered *failure* times in the sample of n functioning times. Note that "withdrawals" may occur between $T_{(i)}$ and $T_{(i+1)}$ and that k in general is a random variable. Let further S_i for $i = 1, 2, \ldots, k$ be the total time on test between the $(i-1)$th and the ith failure, that is, between time $X_{(i-1)}$ and time $X_{(i)}$. Thus S_i is a sum with one term for each item that was functioning in the relevant time interval, the contribution from each item being its functioning time in that interval.

Barlow-Proschan's test statistic W for censored data is defined as

$$W = \frac{\sum_{i=1}^{k-1}(k-i)S_i}{\sum_{i=1}^{k} S_i} \tag{11.98}$$

When the failure rate is constant, it is shown in Barlow and Proschan (1969) that W may be written as:

$$W = U_1 + U_2 + \cdots + U_{k-1}$$

where U_i ($i = 1, 2, \ldots, k-1$) are independent uniform random variables on $[0, 1]$. Hence, when the failure rate is constant (λ_0):

$$E_{\lambda_0}(W) = \frac{k-1}{2} \quad \text{and} \quad \text{var}_{\lambda_0}(W) = \frac{k-1}{12}$$

When the failure rate is constant (λ_0), and k is large, we may use the normal approximation

$$W \approx \mathcal{N}\left(\frac{k-1}{2}, \frac{k-1}{12}\right)$$

that is

$$\frac{W - (k-1)/2}{\sqrt{(k-1)/12}} \approx \mathcal{N}(0, 1)$$

Example 11.15
Let us illustrate the use of Barlow-Proschan's test statistic W for the data set (measured in 10^4 hours):

$$\begin{array}{cccccc} 0.35 & 0.50^* & 0.75^* & 1.00 & 1.30 & 1.80 \\ 3.00^* & 3.15^* & 4.85^* & 5.50 & 5.50^* & 6.25^* \end{array}$$

Censored times are starred ($*$). From these data we compute

$S_1 = 12 \cdot 0.35 = 4.20$
$S_2 = (0.50 - 0.35) + (0.75 - 0.35) + 9(1.00 - 0.35) = 6.40$
$S_3 = 8(1.30 - 1.00) = 2.40$
$S_4 = 7(1.80 - 1.30) = 3.50$
$S_5 = (3.00 - 1.80) + (3.15 - 1.80) + (4.85 - 1.80) + 3(5.50 - 1.80) = 16.70$

Table 11.6 Critical Values for Barlow and Proschan's Test Statistic W.

	α				
$k-1$	0.100	0.050	0.025	0.010	0.005
2	1.553	1.684	1.776	1.859	1.900
3	2.157	2.331	2.469	2.609	2.689
4	2.753	2.953	3.120	3.300	3.411
5	3.339	3.565	3.754	3.963	4.097
6	3.917	4.166	4.376	4.610	4.762
7	4.489	4.759	4.988	5.244	5.413
8	5.056	5.346	5.592	5.869	6.053
9	5.619	5.927	6.189	6.487	6.683
10	6.178	6.504	6.781	7.097	7.307
11	6.735	7.077	7.369	7.702	7.924
12	7.289	7.647	7.953	8.302	8.535

The table is adapted from Barlow and Proschan (1969).

Thus

$$W = \frac{4S_1 + 3S_2 + 2S_3 + S_4}{S_1 + S_2 + S_3 + S_4 + S_5} = \frac{44.3}{33.2} = 1.33$$

As an illustration, let us use Barlow-Proschan's test to test the null hypothesis (H_0): "The failure rate is constant" against the alternative hypothesis (H_1): "The failure rate is increasing on the average (IFRA)."

Barlow-Proschan's test criterion is

> Reject H_0 when $W \geq w_\alpha$, where the critical value w_α is determined so as to get significance level α.

The critical values w_α are given in Table 11.6 for selected values of α and number of failures k. The table is adapted from Barlow and Proschan (1969). When $k \geq 13$, normal approximation may be used. Then

$$w_\alpha = u_\alpha \sqrt{\frac{(k-1)}{12}} + \frac{(k-1)}{2}$$

where u_α denotes the upper $100\alpha\%$ percentile of the standard normal distribution (e.g., $u_{0.05} = 1.645$).

Let us now use Barlow and Proschan's test to check whether the failure rate is constant or increasing, with a significance level of $\alpha = 0.10$. In this case $(k-1) = 4$, and the critical value is from Table 11.6 equal to $w_{0.10} = 2.753$. The test statistic, W, was computed to be $W = 1.33$ which is less than $w_{0.10} = 2.753$. There is thus *no reason to reject the null hypothesis* of constant failure rate and accept the alternative hypothesis of increasing failure rate average (IFRA).

We may also use the test statistic W to test the null hypothesis (H_0): "The failure rate is constant" against the alternative hypothesis (H_1): "The failure rate is on the average decreasing (DFRA)." The test criterion of Barlow-Proschan's test now becomes

Reject H_0 when $(k - 1 - W) > w_\alpha$, where the critical value w_α is determined so as to give significance level α.

For the data in this example, $(k - 1 - W) = 5 - 1 - 1.33 = 2.67$ and $w_{0.10} = 3.339$. Hence, these data give *no reason to reject the null hypothesis* of constant failure rate and accept the alternative hypothesis of DFRA. □

A test of H_0: "The failure rate is constant", against Weibull alternatives is discussed by Cox and Oakes (1984, p. 43).

PROBLEMS

11.1 Assume that you have determined the lifetimes for a total of 12 identical items and obtained the following results (given in hours):

10.2, 89.6, 54.0, 96.0, 23.3, 30.4, 41.2, 0.8, 73.2, 3.6, 28.0, 31.6

(a) Find the sample mean and the sample standard deviation for the data set. Can you draw any conclusions about the underlying distribution $F(t)$ by comparing the sample mean and the sample standard deviation?

(b) Construct the empirical survivor function for the data set.

(c) Plot the data on a Weibull paper.[9] What conclusions can you draw from the plot?

(d) Construct the TTT plot for the data set. What conclusion can you draw from the TTT plot about the corresponding life distribution?

11.2 Failure time data from a compressor were discussed in Example 7.2. All compressor failures at a certain process plant in the time period from 1968 until 1989 have been recorded. In this period a total of 90 critical failures occurred. In this context, a critical failure is defined to be a failure causing compressor downtime. The compressor is very important for the operation of the process plant, and every effort is taken to restart a failed compressor as soon as possible. The 90 repair times (in hours) are presented chronologically in Table 11.7. The repair time associated to the first failure was 1.25 hours, the second repair time was 135.00 hours, and so on.

(a) Plot the repair times in chronological order to check whether or not there is a trend in the repair times. Is there any reason to claim that the repair times increase with the age of the compressor?

---

[9] Weibull paper may be downloaded from www.weibull.com.

Table 11.7 Repair Times (Hours) in Chronological Order.

1.25	135.00	0.08	5.33	154.00	0.50	1.25	2.50	15.00
6.00	4.50	32.50	9.50	0.25	81.00	12.00	0.25	1.66
5.00	7.00	39.00	106.00	6.00	5.00	17.00	5.00	2.00
2.00	0.33	0.17	0.50	18.00	2.50	0.33	0.50	2.00
0.33	4.00	20.00	6.00	6.30	15.00	23.00	4.00	5.00
28.00	16.00	11.50	0.42	38.33	10.50	9.50	8.50	17.00
34.00	0.17	0.83	0.75	1.00	0.25	0.25	2.25	13.50
0.50	0.25	0.17	1.75	0.50	1.00	2.00	2.00	38.00
0.33	2.00	40.50	4.28	1.62	1.33	3.00	5.00	120.00
0.50	3.00	3.00	11.58	8.50	13.50	29.50	29.50	112.00

(b) Assume now that the repair times are independent and identically distributed. Construct the empirical distribution function for the repair times

(c) Plot the repair times on a lognormal plotting paper.[10] Is there reason to believe that the repair times are lognormally distributed?

11.3 Consider the set of material strength data presented by Crowder et al. (1991, p. 46). An experiment has been carried out to gain information on the strength of a certain type of braided cord: 48 pieces of cord were investigated; 7 cords were damaged during the experiment, implying right-censored strength values.

26.8*	29.6*	33.4*	35.0*	36.3	40.0*	41.7	41.9*	42.5*
43.9	49.9	50.1	50.8	51.9	52.1	52.3	52.3	52.4
52.6	52.7	53.1	53.6	53.6	53.9	53.9	54.1	54.6
54.8	54.8	55.1	55.4	55.9	56.0	56.1	56.5	56.9
57.1	57.1	57.3	57.7	57.8	58.1	58.9	59.0	59.1
59.6	60.4	60.7						

(a) Establish a Kaplan-Meier plot of the material strength data.

(b) Establish a TTT plot of the material strength data.

(c) Discuss the effect of this type of censoring.

(d) Describe the form of the failure rate function.

11.4 Establish a graph paper such that the Nelson plot of Weibull distributed life data is close to a straight line. Describe how the Weibull parameters α and λ may be estimated from the plot.

---

[10]Lognormal plotting paper may be downloaded from www.weibull.com.

11.5 Let X be binomially distributed $(20, p)$. X is observed and found to be equal to 3.

(a) Determine the corresponding exact 90% confidence interval for p.

(b) Determine an approximate 90% confidence interval for p based on formulas (11.59) to (11.61).

11.6 Let X be binomially distributed $(25, p)$. A test for the hypothesis

$$H_0 : p = 0.10 \quad \text{against the alternative} \quad H_1 : p \neq 0.10$$

is wanted. Use an exact test to find out whether or not you would reject H_0 when X is observed and found equal to 1 (choose significance level $\alpha = 0.05$).

11.7 Consider a homogeneous Poisson process (HPP) with intensity λ. Let $N(t)$ denote the number of failures (events) in a time interval of length t. $N(t)$ is hence Poisson distributed with parameter λt. Assume that the process is observed in a time interval of length $t = 2$ years. In this time period a total of 7 failures have been observed.

(a) Find an estimate of λ.

(b) Determine a 90% confidence interval for λ.

11.8 Let X have a Poisson distribution with parameter λ. A test for the hypothesis

$$H_0 : \lambda = 3 \quad \text{against the alternative} \quad H_1 : \lambda \neq 3$$

is wanted. Use an exact test to find out whether or not you would reject H_0 when X is observed and found equal to 6 (choose significance level $\alpha = 0.05$).

11.9 Let X have a Poisson distribution with parameter λ.

(a) Determine an exact 90% confidence interval for λ when X is observed and found equal to 6. For comparison, also determine an approximate 90% confidence interval for λ, using the approximation of the Poisson distribution to $\mathcal{N}(\lambda, \lambda)$.

(b) Solve the same problem as stated in (a) when X is observed and found equal to 14.

11.10 Denote the distribution function of the Poisson distribution with parameter λ by $\mathcal{P}_o(x; \lambda)$, and the distribution function of the χ^2 distribution with ν degrees of freedom by $\Gamma_\nu(z)$.

(a) Show that $\mathcal{P}_o(x; \lambda) = 1 - \Gamma_{2(x+1)}(2\lambda)$. (Hint: First show that $1 - \Gamma_{2(x+1)}(2\lambda) = \int_{2\lambda}^{\infty} \frac{u^x}{x!} e^{-u}\, du$, and next apply repeated partial integrations to the integral.)

(b) Let $\lambda_1(X)$ and $\lambda_2(X)$ be defined by

$$\mathcal{P}_o(x; \lambda_1(x)) = \frac{\alpha}{2}$$

$$\mathcal{P}_o(x-1; \lambda_2(x)) = 1 - \frac{\alpha}{2}$$

Use the result of (a) to show that

$$\lambda_1(x) = \frac{1}{2} z_{\alpha/2, 2x}$$

$$\lambda_2(x) = \frac{1}{2} z_{1-\alpha/2, 2(x+1)}$$

where $z_{\varepsilon, \nu}$ denotes the upper $100\varepsilon\%$ percentile of the χ^2 distribution with ν degrees of freedom.

11.11 Suppose that we have experienced that the lifetime T for a certain type of item is exponentially distributed with unknown failure rate $\lambda > 0$. Furthermore, suppose that we have recorded lifetimes for a total of 12 such items and interpret the result as n independent observations of T. The observed lifetimes (in hours) are:

10.2, 89.6, 54.0, 96.0, 23.3, 30.4, 41.2, 0.8, 73.2, 3.6, 28.0, 31.6

(a) Estimate λ.

(b) Test the hypothesis $\lambda \geq 0.025$ against $\lambda < 0.025$ (choose significance level $= 0.05$).

(c) Determine a 95% confidence interval for λ.

11.12 Reconsider the situation in Example 11.11, but now assume that the times to failure are those that are not starred:

| 31.7 | 39.2* | 57.5 | 65.5 | 65.8* | 70.0 | 75.0* | 75.2* |
| 87.5* | 88.3* | 94.2 | 101.7* | 105.8* | 109.2 | 110.0 | 130.0* |

(a) Calculate the Kaplan-Meier estimate $\hat{R}(t)$ and display it graphically.

(b) Calculate Nelson's estimate $R*(t)$ for the survivor function and display it graphically.

11.13 Suppose that the data set in Problem 11.11 was obtained by simultaneously activating 20 identical items, but that the test was terminated at the 12th failure.

(a) What type of censoring is this?

(b) Estimate λ in this situation.

(c) Calculate a 95% confidence interval for λ.

(d) Compare the results with those derived in Problem 11.11.

11.14 Establish a graph paper such that the Nelson plot of normally distributed ($\mathcal{N}(\mu, \sigma^2)$) life data is close to a straight line. Describe how the parameters μ and σ may be estimated from the plot.

11.15 Let $T_1, T_2, \ldots, T_n$ be independent and identically distributed IG(μ, ν). Assume λ to be known.

(a) Show that the MLE of μ, μ^* is given by

$$\mu^* = \frac{1}{n}\sum_{j=1}^{n} T_j = \overline{T}$$

(b) Determine a $1 - \alpha$ confidence interval for μ. (Hint: $\frac{\lambda n}{\mu^2 \mu^*}(\mu^* - \mu)^2$ is χ^2 distributed with one degree of freedom).

11.16 Let $T_1, T_2, \ldots, T_n$ be independent and identically distributed IG(μ, ν). Assume μ to be known.

(a) Show that the MLE of λ, λ^*, is given by

$$\frac{1}{\lambda^*} = \frac{1}{n}\sum_{j=1}^{n} \frac{(T_j - \mu)}{\mu^2 T_j}$$

and that

$$E\left(\frac{1}{\lambda^*}\right) = \frac{1}{\lambda}$$

$$\mathrm{var}\left(\frac{1}{\lambda^*}\right) = \frac{2}{n\lambda^2}$$

(b) Determine a $1 - \varepsilon$ confidence interval for λ. (Hint: Use the hint given in the previous problem).

11.17 Lieblein and Zelen (1956) analyzed data on the endurance of deep groove ball bearings. They registered the number X_i of million revolutions before failure for 23 such ball bearings. Their distinct ordered failure times are

17.88	28.92	33.00	41.52	42.12	45.60	48.48	51.84
51.96	54.12	55.56	67.80	68.64	68.64	68.88	84.12
93.12	98.64	105.12	105.84	127.92	128.04	173.40	

Previously, these data have been analyzed assuming that they follow a Weibull distribution or a lognormal distribution, respectively. Now let us analyze this data set under the assumption that they are independent IG(x_j, μ, λ).

(a) Determine the MLE μ^* of μ and the MLE λ^* of λ.

(b) What would be your estimate of the variance of this distribution? Compare this estimate with the general estimate $S^2 = \sum_{j=1}^{n}(X_j - \overline{X})^2/(n-1)$.

(c) To test whether or not the inverse Gaussian distribution gives a good model for this data set, determine a percentage-percentage (P-P) plot of the data.

To get this P-P plot first estimate $F(x_{(j)})$ by the empirical distribution function

$$F(x_{(j)}) = \frac{j - (1/2)}{n} \quad \text{for } j = 1, 2, \ldots, n$$

Next determine $F(x_{(j)})$ by $\text{IG}(x_{(j)}, \mu^*, \lambda^*)$. Plot $F(x_{(j)})$ against $F(x_{(j)})$ by $\text{IG}(x_{(j)}, \mu^*, \lambda^*)$ for $j = 1, 2, \ldots, n$.

If the model fits, the curve $\left(F(x_{(j)}), \text{IG}(x_{(j)}, \mu^*, \lambda^*)\right)$ will approximately fall along a straight line. Does it? What is your conclusion?

12
Accelerated Life Testing

12.1 INTRODUCTION

Many of the devices produced today for complex technical systems have very high reliability under normal use conditions. Such devices may have a mean time to failure of 100,000 hours ($\approx$ 11.5 years) or more. The time involved in a life test such as those described in Section 11.1 would therefore be exorbitant. Furthermore the device is likely to be out of date and therefore of no interest by the time the test is completed. The questions then arise of how to make the optimal choice between several types or designs of a device and how to collect information about the corresponding life distributions under normal use conditions.

A common way of tackling this problem is to expose the device to sufficient overstress to bring the mean time to failure down to an acceptable level. Thereafter one tries to "extrapolate" from the information obtained under overstress to normal use conditions. This approach is called *accelerated life testing* (ALT) or overstress testing. Books describing statistical methods, test plans and data analysis for accelerated life testing include Mann et al. (1974), Kalbfleisch and Prentice (1980), Lawless (1982), Jensen and Petersen (1982), Cox and Oakes (1984), Viertl (1988), and Nelson (1990).

Depending on the kind of device in question, the accelerated testing conditions may involve a higher level of temperature, pressure, voltage, load, vibration, and so on, than the corresponding levels occurring in normal use conditions. These variables are called *stressors* (stress variables or covariates). In a specific situation there may be one or several (m) stressors $s_1, s_2, \ldots, s_m$ acting simultaneously. The vector $s = (s_1, s_2, \ldots, s_m)$ is called *the stress vector*.

In simple situations there is only one stressor, s, occurring on two levels, $s^{(1)}$ and $s^{(2)}$, where $s^{(1)} < s^{(2)}$. Let $s^{(0)}$ ($\leq s^{(1)}$) denote normal stress. The situation becomes somewhat more complicated when m stressors $s_1, s_2, \ldots, s_m$ are involved and stressor s_j occurs on n_j levels:

$$s_j^{(1)} < s_j^{(2)} < \cdots < s_j^{(n_j)} \quad \text{for } j = 1, 2, \ldots, m$$

Let $s_j^{(0)}$ ($\leq s_j^{(1)}$) denote normal stress for stressor j, for $j = 1, 2, \ldots, m$. The situation becomes even more complicated when the stressors are continuously increasing with time, as illustrated in Fig. 12.3. The first two cases lead to *step-stress accelerated tests* (SALT), the last one leads to *progressive-stress accelerated tests* (PALT).

12.2 EXPERIMENTAL DESIGNS FOR ALT

Let us for the sake of simplicity suppose that there is only *one* stressor s. The testing experiment can be conducted according to different designs. Three such designs are:

Design I The experiment involves use of k stress levels $s^{(1)} < s^{(2)} < \cdots < s^{(k)}$ as illustrated un Fig. 12.1. Let $s^{(0)}$ ($\leq s^{(1)}$) denote normal stress. A (large) number of test items are assumed to be available for the experiment, and n_j of these are to be exposed to the stress $s^{(j)}$. Censoring of type II (see Section 11.2) is applied. The experiment is then carried out as follows:

1. One stress level $s^{(i)}$ is *chosen at random* among $s^{(1)}, s^{(2)}, \ldots, s^{(k)}$, and n_i test items are chosen at random among the test items at hand. These n_i items are then exposed to stress level $s^{(i)}$. The test is terminated when $r_i (\leq n_i)$ failures have occurred. Let $T_{i1}, T_{i2}, \ldots, T_{in_i}$ denote the times to failure or censoring.

2. Another stress level $s^{(j)}$ is chosen at random among the remaining levels; n_j test items are chosen at random among the remaining items and exposed to stress level $s^{(j)}$. The test is terminated when $r_j (\leq n_j)$ failures have occurred. Let $T_{j1}, T_{j2}, \ldots, T_{jn_j}$ denote the times to failure or censoring. This procedure is continued until k stress levels have been selected.

If the number of test items at hand is large compared to $n = \sum_{j=1}^{k} n_j$ it seems reasonable to assume that $T_{01}, T_{02}, \ldots, T_{k r_k}$ are independent, which simplifies the analysis.

Design II Fix k points of time $0 < t_1 < t_2 < \cdots < t_k < t$ (Fig. 12.2). Put n randomly chosen test items on test at time 0. In the time interval $(0, t_1]$ the items are subject to stress $s^{(1)}$. In the interval $(t_1, t_2]$ the items that have not failed by time t_1 are kept in operation under stress $s^{(2)}$. In the next interval $(t_2, t_3]$ the items that still have not failed by time t_2 are kept in operation under stress $s^{(3)}$, and so on. In the time interval $(t_k, \infty]$ the items that have not failed by time t_k are kept in operation under stress $s^{(k+1)}$ until they have all failed (hence no censoring). The lifetimes of the n test items are denoted $T_1, T_2, \ldots, T_n$.

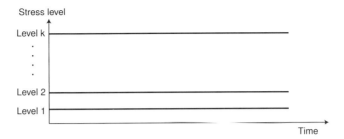

Fig. 12.1 Design I for accelerated tests.

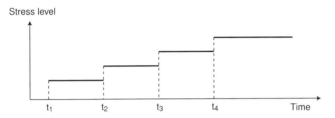

Fig. 12.2 Design II for accelerated tests.

Fig. 12.3 Design III for accelerated tests.

Design III A number n of test items are chosen at random among the test items at hand and exposed to a stress $s(t)$, which is increasing with time until the items have all failed. The stress $s(t)$ as a function of time t, that is illustrated in Fig. 12.3, is assumed to be known. The lifetimes of the n test items are observed and denoted $T_1, T_2, \ldots, T_n$. If n is small compared to the number of items at hand and if the n items are operating independently, it seems reasonable to assume that $T_1, T_2, \ldots, T_n$ are independent in design II and design III.

12.3 PARAMETRIC MODELS USED IN SALT

The first step in any statistical analysis usually consists of formulating a stochastic model of the situation based on a priori knowledge and the experimental design used.

In this case we are concerned with life distributions. Hence the data obtained through the SALT is supposed to give information about:

1. the life distribution function

$$F_T(t; s) = \Pr(T \le t; s)$$

2. the survivor function

$$R_T(t; s) = 1 - F_T(t; s)$$

3. or, the failure rate function

$$z(t; s)$$

which are all more or less unknown.

The first question to be asked is: What do we know a priori about the life distribution under normal use conditions? For example, do we have reason to believe that it will belong to the exponential, the Weibull, the lognormal, or the inverse Gaussian family of distributions? Are we able to derive this life distribution from the physical conditions at hand? One way of searching for a suitable model is to display the data on different kinds of probability paper and pick the family of distributions accordingly. Another way to find a suitable model is to try different models and select the one that best fits the data. The question of model discrimination has been discussed in Webster and Van Parr (1965), Hunter and Reiner (1965), Box and Hill (1967), Mann et al. (1974, Chapter 7), and Box et al. (1978, Chapter 16), among others.

For the sake of simplicity, let us suppose that we succeed in establishing an appropriate *parametric* family of lifetime distributions under normal use conditions. The next question is: How does overstress affect this family of distributions? For example, will the life distribution under overstress belong to the same parametric family as the one obtained under normal stress? If so, the only effect of stress on the life distribution is that different stress levels lead to different parameter vectors in this family.

In an early paper on ALT, Levenbach (1957) was able to conclude that the life distributions occurring in the case he was studying belonged to the lognormal family under normal stress as well as under overstress at certain levels. Let us suppose we are as lucky as Levenbach and find that our lifetime distributions belong to a specified parametric family. Then the next question which needs to be answered is: In which way does the parameter vector of this family depend on the stress vector s? Here we only intend to give an introduction to the subject of ALT and hence will content ourselves with considering two simple examples.

Example 12.1 Design I

Suppose that the experiment is carried out as described in design I where only one stressor s has been used, and that the family of life distributions is the exponential with mean $\theta(s)$, and hence failure rate $\lambda(s) = \theta(s)^{-1}$. What is required, is the function $\lambda(s)$ describing the relation between the stress and failure rate. In principle any function $\lambda(s)$ may do. However, three of the most commonly used relations are

$$\begin{aligned} \theta(s) &= cs^{-a} && \text{(power rule model)} \\ \lambda(s) &= ce^{-b/s} && \text{(Arrhenius model)} \\ \lambda(s) &= cse^{-b/s} && \text{(simple Eyring model)} \end{aligned} \qquad (12.1)$$

where a, b, and c denote constants which are more or less unknown. These models may be derived from the physics of failure of the device in question. The power rule model has mostly been applied to dielectric breakdown of capacitors and fatigue testing of materials. The Arrhenius model has been applied to thermal aging and is also applicable to semiconductor materials. The simple Eyring model has been applied to devices exposed to constant thermal stress. A more thorough discussion of these three models and of other more general models may be found in Mann et al. (1974). □

The constants $a, b, \ldots$ appearing in these models have to be estimated on the basis of the recorded life lengths under overstress. This may be done by inserting the appropriate expression for $\lambda(s)$ in the lifetime distribution. Given the stress level $s^{(j)}$ for $j = 1, 2, \ldots, k$, $F_T(t_j; s^{(j)})$ is now known except for the values of the constants $a, b, \ldots$. The occurring constants may then be estimated from the data, for example, by applying the maximum likelihood principle or the least squares principle. Let us denote the estimates by $a^*, b^*, \ldots$.

The final step then consists of inserting these estimates together with the normal stress $s^{(0)}$ in the lifetime distribution function. The result is an "estimate" of the *normal stress* life distribution, $\hat{F}_T(t; s^{(0)})$ based on the *overstress* life data.

Before we specify the relation between the stressor s and the failure rate $\lambda(s)$ in our simple example, let us remind ourselves of a few results about exponentially distributed random variables. The total information which is obtained through the SALT is expressed by $(s^{(j)}, n_j, r_j, T_{j1}, T_{j2}, \ldots, T_{jn_j})$, for $j = 1, 2, \ldots, k$. From what we learned in Section 11.4, it seems natural to summarize this and say that the information obtained by the SALT is expressed by

$$(s^{(j)}, n_j, r_j, T_j) \quad \text{for } j = 1, 2, \ldots, k \qquad (12.2)$$

where

$$T_j = \mathcal{T}(T_{n_j}) \quad \text{for } j = 1, 2, \ldots, k \qquad (12.3)$$

is the total time on test at stress level $s^{(j)}$.[1]

---

[1] $T = (T_1, T_2, \ldots, T_k)$ is a complete, sufficient statistic in our model. $E(T_j/\lambda_j) = \lambda(s_j)$, for $j = 1, 2, \ldots, k$. Using the Lehmann-Scheffé theorem, it follows that T_j/λ_j is a uniformly minimum variance unbiased estimate for $\lambda(s_j)$. See, for example, Bickel and Doksum (1977, p. 122). Hence no information is lost by summarizing this way.

Let $\theta(s_j)$ denote the mean time to failure at stress level $s^{(j)}$, and let $\lambda(s^{(j)})$ denote the corresponding failure rate. Then we know from Section 11.4 that $Z_j = 2\lambda(s^{(j)})T_j$ is χ^2 distributed with $2r_j$ degrees of freedom, $j = 1, 2, \ldots, k$:

$$f_{Z_j}(z_j) = \frac{1}{2^{r_j}\Gamma(r_j)} z_j^{r_j-1} e^{-z_j/2} \quad \text{for } z_j > 0, \ j = 1, 2, \ldots, k \tag{12.4}$$

Accordingly

$$\begin{aligned} f_{T_j}(t_j) &= \frac{1}{2^{r_j}\Gamma(r_j)} (2\lambda(s^{(j)})t_j)^{r_j-1} e^{-\lambda(s^{(j)})t_j} \cdot 2\lambda(s^{(j)}) \\ &= \frac{1}{\Gamma(r_j)} \cdot \lambda(s^{(j)})^{r_j} t_j^{r_j-1} e^{-\lambda(s^{(j)})t_j} \quad \text{for } t_j > 0, \ j = 1, 2, \ldots, k \end{aligned}$$

Hence

$$f_{T_1,\ldots,T_k}(t_1, \ldots, t_k) = \prod_{j=1}^{k} \left(\frac{1}{\Gamma(r_j)} \cdot \lambda(s^{(j)})^{r_j} \cdot t_j^{r_j-1} e^{-\lambda(s^{(j)})t_j} \right)$$

$$\text{for } t_j > 0, \ j = 1, 2, \ldots, k \tag{12.5}$$

As an example, let us consider the case where the relation between the stressor s and the mean $\theta(s)$ is described by the power rule model:

$$\theta(s^{(j)}) = c(s^{(j)})^{-a} \quad \text{for } j = 1, 2, \ldots, k \tag{12.6}$$

Then

$$\lambda(s^{(j)}) = \frac{1}{c} \left(s^{(j)} \right)^a \tag{12.7}$$

If we amend the power rule slightly, without changing its basic character, to

$$\lambda(s^{(j)}) = \left(\frac{s^{(j)}}{\dot{s}} \right)^a \tag{12.8}$$

where $\dot{s}$ is the weighted geometric mean of the s_j's

$$\dot{s} = \prod_{j=1}^{k} \left(s^{(j)} \right)^{r_j / \sum_{i=1}^{k} r_i} \tag{12.9}$$

it will later on turn out that the maximum likelihood estimator (MLE) of a and c, a^* and c^* becomes asymptotically independent. Hence this change is worthwhile.

Inserting (12.8) into (12.5) leads to

$$f_{T_1,\ldots,T_k}(t_1, \ldots, t_k, a, c)$$

$$= \prod_{j=1}^{k} \frac{1}{\Gamma(r_j)} \left[\frac{1}{c} \left(\frac{s^{(j)}}{\dot{s}} \right)^a \right]^{r_j} t_j^{r_j-1} e^{-\left(s^{(j)}/\dot{s}\right)^a \cdot t_j / c} \tag{12.10}$$

The corresponding likelihood function is

$$L(a, c; t_1, \ldots, t_k) = \prod_{j=1}^{k} \frac{1}{\Gamma(r_j)} \left[\frac{1}{c} \left(\frac{s^{(j)}}{\dot{s}} \right)^a \right]^{r_j} t_j^{r_j - 1} e^{-(s^{(j)}/\dot{s})^a t_j / c} \quad (12.11)$$

and the log likelihood function is

$$\ln L(a, c; t_1, \ldots, t_k)$$
$$= \sum_{j=1}^{k} \left[-\ln \Gamma(r_j) - r_j \ln c + a r_j \ln \left(\frac{s^{(j)}}{\dot{s}} \right) \right.$$
$$\left. + (r_j - 1) \ln t_j - \frac{1}{c} \left(\frac{s^{(j)}}{\dot{s}} \right)^a \cdot t_j \right] \quad (12.12)$$

The MLE of a and c, a^* and c^* are obtained by solving the two equations:

$$\frac{\partial \ln L}{\partial a} = \sum_{j=1}^{k} -r_j (\ln s^{(j)} - \ln \dot{s}) - \sum_{j=1}^{k} \frac{1}{c} \left(\frac{s^{(j)}}{\dot{s}} \right)^a \ln \left(\frac{s^{(j)}}{\dot{s}} \right) t_j = 0 \quad (12.13)$$

and

$$\frac{\partial \ln L}{\partial c} = \sum_{j=1}^{k} -\frac{r_j}{c} + \sum_{j=1}^{k} \left(\frac{s^{(j)}}{\dot{s}} \right)^a \cdot t_j \cdot \frac{1}{c^2} = 0 \quad (12.14)$$

with respect to a and c.

From (12.7) we realize that

$$\ln \dot{s} = \sum_{j=1}^{k} \frac{r_j}{\sum_{i=1}^{k} r_i} \ln s^{(j)} \quad (12.15)$$

That is

$$\sum_{j=1}^{k} r_j \ln \dot{s} = \sum_{j=1}^{k} r_j \ln s^{(j)} \quad (12.16)$$

or

$$\sum_{j=1}^{k} r_j (\ln s^{(j)} - \ln \dot{s}) = 0 \quad (12.17)$$

Hence (12.13) is reduced to

$$\sum_{j=1}^{k} \left(\frac{s_j}{\dot{s}} \right)^a \ln \left(\frac{s^{(j)}}{\dot{s}} \right) \cdot t_j = 0 \quad (12.18)$$

which determines the MLE of a, a^*. The parameter c is then determined by (12.14):

$$c^* = \frac{1}{\sum_{i=1}^{k} r_i} \sum_{j=1}^{k} \left(\frac{s^{(j)}}{\dot{s}}\right)^{a^*} t_j \qquad (12.19)$$

Equations (12.18) and (12.19) do not allow us to determine a^* and c^* analytically. Iterative procedures must be used.

It can be shown (see Mann et al. 1974, p. 426) that the asymptotic variances of a^* and c^* are

$$\text{as var}(a^*) = \left[\sum_{j} r_j \left(\ln \frac{s^{(j)}}{\dot{s}}\right)^2\right]^{-1} \qquad (12.20)$$

$$\text{as var}(c^*) = c^2 \left[\sum_{j=1}^{k} r_j\right]^{-1} \qquad (12.21)$$

and that the asymptotic covariance is

$$\text{as cov}(a^*, c^*) = 0 \qquad (12.22)$$

Furthermore it can be shown that (a^*, c^*) are asymptotically distributed as bivariate normal variables. See Mann et al. (1974, p. 83). Hence a^* and c^* are asymptotically independent.

A reasonable estimate of the failure rate under normal stress $s^{(0)}$ is then

$$\lambda_0^* = \frac{1}{c^*} \left(\frac{s^{(0)}}{\dot{s}}\right)^{a^*} \qquad (12.23)$$

Therefore the density of the lifetime under normal stress may be estimated by

$$f_T(t) = \lambda_0^* e^{-\lambda_0^* t} \quad \text{for } t > 0$$

Table 12.1 gives some references where other SALT-parametric models have been studied.

Example 12.2 Design III

Let us consider experiments where n identical items, operating independently, are put on test at time 0. In the time interval $(0, t]$ the items are subject to stress $s^{(0)}$, while in the interval (t, ∞) the items that have not failed by time t are kept in operation under stress $s^{(1)}$ $(> s^{(0)})$ until they all have failed. Typically $s^{(0)}$ corresponds to normal stress, $s^{(1)}$ to accelerated stress.

Suppose furthermore that the accumulated fatigue in the material subject to wear is modeled as a Wiener process $\{W_0(y), y \geq 0\}$ with drift $\eta > 0$ and diffusion constant

Table 12.1 Some References to Papers Where SALT-Parametric Models Have Been Studied.

Life Distribution		References
$F_T(t, s) = 1 - e^{-\lambda(s) \cdot t}$	$\lambda(s) = cs^a$	Mann et al. (1974);
	$\lambda(s) = ce^{-b/s}$	Singpurwalla (1973)
	$\lambda(s) = cse^{-b/s}$	
$F_T(t; s) = 1 - e^{-[\lambda(s) \cdot t]^\alpha}$	$\lambda(s) = cs^a$	Mann et al. (1974);
	α constant	Nelson (1975);
		Singpurwalla and
		Al-Khayyal (1977)
$\ln T \sim \mathcal{N}(\nu(s), \tau^2)$	$\nu(s) = \alpha + \beta s$	Nelson and
	τ^2 constant	Kielpinski (1976)

$\delta^2 > 0$. Failure occurs when the fatigue process $W_0(y)$ crosses a critical boundary ω. [The Wiener process $W_0(y)$ is defined to be an independent increment Gaussian process with $W_0(0) = 0$ and mean $E(W_0(y)) = \eta y$. Moreover, each increment $(W_0(y_2) - W_0(y_1))$ for $0 < y_1 < y_2$, has variance $\delta^2(y_2 - y_1)$.]

The basic result, whose history and proof can be found in Chhikara and Folks (1989), is that if we define the fatigue failure time Y as the first time that the fatigue process $W_0(y)$ crosses the critical boundary ω, and if we set $\mu = \omega/\mu$ and $\lambda = \omega^2/\sigma^2$, then Y has the inverse Gaussian distribution $IG(y, \mu, \lambda)$ with probability density function

$$f_Y(y; \mu, \lambda) = \frac{\lambda}{\sqrt{2\pi y^3}} e^{-(\lambda/\mu^2)[(y-\mu)^2/y]} \text{ for } y > 0, \mu > 0, \lambda > 0$$

We now make the assumption that the fatigue process changes from one Wiener process to another at the stress change point t. More precisely, in the interval $(0, t]$, we suppose that the failure occurs if the process $W_0(y)$ crosses the critical boundary $\omega > 0$, where $W_0(y)$ is a Wiener process with drift $\eta > 0$ and diffusion constant $\delta^2 > 0$. At the stress change point t, if $W_0(y)$ has not yet crossed ω in $(0, t]$, the stress is changed from $s^{(0)}$ to $s^{(1)}$, and a new Wiener process starts out at the point $(t, W_0(t))$. This is illustrated in Fig. 12.4. We assume that

$$W_1(y) = W_0(t + \alpha(y - t)) \text{ for } y > t, \alpha > 1$$

□

Hence our SALT fatigue process is

$$W(y) = \begin{cases} W_0(y) & \text{for } y \leq t \\ W_0(t + \alpha(y - t)) & \text{for } y > t \end{cases}$$

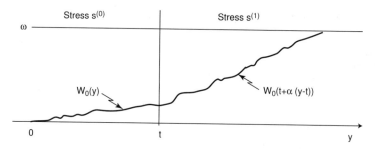

Fig. 12.4 A fatigue process $W(y)$ with stress level increased from $s^{(0)}$ to $s^{(1)}$ at time t.

In Doksum and Høyland (1992) it is shown that the distribution $F_T(t)$ of the stress failure time T in this situation is

$$F_T(y) = \begin{cases} F_0(y) & \text{for } 0 \leq y \leq t \\ F_0(t + \alpha(y - t)) & \text{for } y > t \end{cases}$$

where $F_0(y) = IG(y; \mu, \lambda)$ is the inverse Gaussian distribution whose probability density is given by (2.62).

Let $y_1, y_2, \ldots, y_n$ be the observed failure times, and introduce

$$y_j(\alpha) = y_j \quad \text{for } y_j \leq t$$

and

$$y_j(\alpha) = t + \alpha(y_j - t) \quad \text{for } y_j > t \qquad (12.24)$$

In this α known case, the likelihood function, which we label $L_\alpha(\mu, \lambda)$ can be written

$$L_\alpha(\mu, \lambda) = \alpha^m \prod_{j=1}^n f_0(y_j(\alpha)), \quad m = \sum_{s=1}^n I(y_j > t)$$

where m is the number of y's greater than t.

We note that the likelihood is proportional to the usual one-sample inverse Gaussian situation with the only change that y_j is replaced by $y_j(\alpha)$ for $j = 1, 2, \ldots, n$. Using (11.39), we see that the MLE for μ and $1/\lambda$ are given by

$$\mu_\alpha^* = \frac{1}{n} \sum_{j=1}^n y_j(\alpha)$$

$$\frac{1}{\lambda_\alpha^*} = \frac{1}{n} \sum_{j=1}^n \left(\frac{1}{y_j(\alpha)} - \frac{1}{\mu^*(\alpha)} \right) \qquad (12.25)$$

Hence an estimate of the life distribution of the failure fatigue time T under stress $s^{(0)}$ (normal stress) is $IG(t; \mu_\alpha^*, \lambda_\alpha^*)$.

It also follows that μ_α^* has the IG($y; \mu, n\lambda$) distribution, that $n\lambda/\lambda_\alpha^*$ has a χ^2 distribution with $(n-1)$ degrees of freedom, and that μ_α^* and λ_α^* generally have all the desirable properties of one-sample inverse Gaussian estimates. In particular, the estimator μ_α^* of the mean time to failure under stress $s^{(0)}$ has variance $\mu^3/n\lambda$.

A more realistic situation would be the one where α is unknown. The case is discussed in Doksum and Høyland (1992). They also discuss step-stress models with $m(> 2)$ stress levels, and models with continuously increasing stress. The case of stochastic censoring is also discussed.

12.4 NONPARAMETRIC MODELS USED IN ALT

Two main assumptions need to be true if a successful parametric model is to be established for ALT:

1. The life distributions have to belong to the *same* known parametric family under normal stress as well as under overstress.

2. The relation between the parameter vector of this family and the stressors has to be known.

If these conditions are met, the solution of the problem is considerably simplified. Hence there is a danger that these assumptions may be made too often for mathematical convenience only and with no basis in reality. To prevent this from happening, considerable efforts have been made in recent years to develop models and methods that do not require assumptions about the functional form of the lifetime distribution in question. We will conclude this chapter with some comments on such models and methods and with references to literature that contains more detailed information.

It may be argued in certain situations—even in situations where we do not know the family of life distributions in question—that the ratio between the failure rates $z(t, s^{(i)})$ and $z(t, s^{(j)})$, corresponding to any two stress levels $s^{(i)}$ and $s^{(j)}$, is constant over time:

$$\frac{z(t; s^{(i)})}{z(t; s^{(j)})} = g(s^{(i)}, s^{(j)}) \qquad (12.26)$$

where g is a function of $(s^{(i)}, s^{(j)})$ only. Such models are denoted *proportional hazards* (PH) models. If we replace $z(t; s^{(0)})$ by $z_0(t)$ and $g(s^{(1)}, s^{(0)})$ by $g_1(s)$, a PH model is characterized by the relation

$$z(t; s) = z_0(t) \cdot g_1(s) \qquad (12.27)$$

where g_1 is a function of s only.

Then the survivor function may be written [see (2.9)] as follows:
For stress $s^{(0)}$,

$$R(t; s^{(0)}) = e^{-\int_0^t z_0(u)\,du} \qquad (12.28)$$

For stress s,

$$R(t; s) = e^{-\int_0^t z_0(u) g_1(s)\, du} \tag{12.29}$$

Accordingly in a PH model, the survivor functions $R(t; s)$ and $R(t; s^{(0)})$ have to satisfy the relation

$$R(t; s) = R(t; s^{(0)})^{g_1(s)} \tag{12.30}$$

[A parametric model where (12.27) is satisfied is the two-parameter Weibull family with constant shape parameter α (independent of s) and a scale parameter λ, which may depend on the stressor.]

If the life distribution function is only known to be continuous but is otherwise unspecified, the failure rate $\lambda_0(t)$ of the PH model is likewise unspecified. The corresponding nonparametric PH model is quite flexible and may be applicable to many situations occurring in practice. For a more thorough discussion of these models and their applications, the reader is referred to Viertl (1988), Nelson (1990), and Meeker and Escobar (1998).

As already indicated in (12.27) the PH models are characterized by the relation

$$z(t; s) = z_0(t) \cdot g_1(s)$$

Note that $z_0(t)$ and $g(s)$ may contain unknown parameters. Cox (1972) discussed a special subclass of nonparametric PH models, now usually referred to as Cox models, which also are of particular interest in biomedical applications.

Cox considers a situation with m different stressors $s_1, s_2, \ldots, s_m$. In order to come into line with the notation used in connection with Cox models, the notation used in the introduction to this chapter must be altered slightly.

Let stressor s_1 occur at level x_1
s_2 occur at level x_2
$\vdots$

and so on. Then Cox makes the assumption that in the PH model

$$g(x) = \exp\left(\sum_{i=1}^m \beta_i x_i\right) = \exp(\boldsymbol{\beta}' x) \tag{12.31}$$

where $x = (x_1, x_2, \ldots, x_n)$ represents an $(m \times 1)$ vector of stressor levels, and $\boldsymbol{\beta}' = (\beta_1, \beta_2, \ldots, \beta_m)$ a vector of regression coefficients. Since Cox introduces parameters in one *part* only of the nonparametric model, his model is sometimes denoted *semiparametric*.

Note that the x_i's may be transformed values of the real stress levels. Hence Cox models are very general and may be applicable to a large number of practical situations.

With this assumption (12.29) may be rewritten as

$$R(t; x, \boldsymbol{\beta}) = R\left(t, s^{(0)}\right)^{\exp(\boldsymbol{\beta}' x)} \tag{12.32}$$

Suppose n randomly chosen items have been exposed to the m stressors and the corresponding life lengths of the items are denoted $T_1, T_2, \ldots, T_n$. Then associated with each lifetime T_i is a regression vector $x_i = (x_{1i}, x_{2i}, \ldots, x_{mi})$ of data values of the m stressors, $i = 1, 2, \ldots, n$. The question is then how to estimate $F_0(t)$ and the regression coefficients.

An application of the maximum likelihood principle turns out to be complicated. Cox suggests the following approach:

First, he defines on a heuristic basis what he calls a "partial likelihood" which is not a proper likelihood function, and which in a PH model does not involve the failure rate. The partial likelihood is then used for estimating the regression coefficients. He then reads the estimated values $\hat{\beta}_1, \hat{\beta}_2, \ldots, \hat{\beta}_m$ as true values of the regression coefficients and concludes by estimating $F_0(t)$ from the recorded life data, using the Kaplan-Meier approach (see Section 11.3).

Application of this procedure is rather complicated. Readers are referred to Kalbfleisch and Prentice (1980, pp. 76–78) or Lawless (1982, pp. 345–347).

Estimates for the β's in the Cox model may be calculated by using various computer codes. These codes usually allow for two important generalizations of the Cox model (12.32):

1. The assumption of a simple underlying failure rate for all the life data is relaxed. The code allows for several subgroups within the recorded data.

2. The stressors x are allowed to vary with time.

PROBLEMS

12.1 (The sole purpose of considering the following rather unrealistic situation is to illustrate how accelerated life data may be analyzed in a very simple situation.)

The device in question is supposed to have an exponential life distribution with failure rate $\lambda(s) = c \cdot s$ when exposed to the stress s, for all stresses. c is an unknown positive constant, and the purpose of the study is to estimate $\lambda(s)^{(0)} = c \cdot s^{(0)}$, where $s^{(0)}$ denotes the normal use stress. The expected life length is expected to be very long under stress s_0. Therefore an accelerated life test is carried out according to design I, where the selected stresses $s^{(1)} < s^{(2)} < \cdots < s^{(k)}$ are very much larger than $s^{(0)}$, to obtain short life lengths under these stresses.

The experiment leads to the result in Table 12.2. T_{ij}, for $i = 1, 2, \ldots, k$, $j = 1, 2, \ldots, n_i$ are all assumed to be independent.

(a) Show that $2cs^{(i)} \sum_{j=1}^{n_i} T_{ij}$, $i = 1, 2, \ldots, k$ are independent and χ^2 distributed with $2n_i$ degrees of freedom.

(b) Use the result in (a) to derive an estimator $\hat{c}$ of c.

(c) Finally, estimate the expected life length of the device in question under normal stress $s^{(0)}$.

Table 12.2 Observed Lifetimes at the Various Stress Levels.

Stress Level	Observed Lifetimes
$s^{(1)}$	$T_{11}, T_{12}, \ldots, T_{1n_1}$
$s^{(2)}$	$T_{21}, T_{22}, \ldots, T_{2n_2}$
$\vdots$	$\vdots$
$s^{(k)}$	$T_{k1}, T_{k2}, \ldots, T_{kn_k}$

13
Bayesian Reliability Analysis

13.1 INTRODUCTION

The first step in almost any statistical analysis is to establish a stochastic model of the situation at hand. The observations to be collected are then considered to be realizations of random variables $X_1, X_2, \ldots, X_n$. So far in this book, we have assumed it to be possible to derive the joint distribution function, $F(x_1, x_2, \ldots, x_n; \theta_1, \theta_2, \ldots, \theta_r)$ of the X_i's through basic scientific knowledge of the phenomenon to be analyzed, information obtained from exploratory data, and possibly some simplifying assumptions. Here $\boldsymbol{\theta} = (\theta_1, \theta_2, \ldots, \theta_r)$ denotes a vector of constants, belonging to some subspace Ω of the r-dimensional Euclidean space. In this model no vector $\boldsymbol{\theta}$ in Ω is more likely to occur than any other.

A natural question to ask is whether or not this approach always is the most appropriate one when we want to express a priori knowledge of the phenomenon. When following this line of action, essential parts of a priori knowledge may not be taken into account. Suppose, for example, that p denotes the reliability of a certain component at time t. Then p will be assumed to belong to [0, 1], but no values of p in this interval is given preference, even if one is quite certain that p is close to 1, say. This a priori knowledge easily get lost in the model.

In Bayesian inference one can introduce this kind of knowledge into the model by interpreting p as a random variable with some density $f(p)$, expressing what one thinks (believes) about the occurring value of p. In Section 13.7 we discuss possible interpretations of such distributions. For the time being we will only study the immediate consequences of such models.

The intention of this chapter is mainly to illustrate the Bayesian philosophy and some of its consequences. Those who want a comprehensive presentation of Bayesian inference are referred to Martz and Waller (1982), Berger (1985), Box and Tiao (1992), and Gelman et al. (1995). Here, we will restrict ourselves to considering Bayesian estimation in some simple situations.

A central tool required for the application of Bayesian methods is Bayes' theorem.

Theorem 13.1 (Bayes' Theorem) Let $B_1, B_2, \ldots$ be mutually exclusive and exhaustive events contained in a sample space S, that is

$$\Pr\left(\bigcup_{i=1}^{\infty} B_i\right) = 1$$
$$B_i \cap B_j = \emptyset \qquad \text{for } i \neq j$$
$$\Pr(B_i) > 0 \qquad \text{for each } i$$

and let A be an event in S such that $\Pr(A) > 0$. Then for each k,

$$\Pr(B_k \mid A) = \frac{\Pr(A \mid B_k) \Pr(B_k)}{\sum_{i=1}^{\infty} \Pr(A \mid B_i) \Pr(B_i)} \tag{13.1}$$

□

Bayes' theorem is named after the Reverend Thomas Bayes (1702–1761), who used the theorem in the fundamental paper, *An Essay toward Solving a Problem in the Doctrine of Chances*, which was published in 1763.

Proof
Using the definition of conditional probability we can write

$$\Pr(B_k \mid A) = \frac{\Pr(A \cap B_k)}{\Pr(A)} = \frac{\Pr(A \mid B_k) \Pr(B_k)}{\Pr(A)}$$

The denominator $\Pr(A)$ can be expanded as

$$\Pr(A) = \sum_{i=1}^{\infty} \Pr(A \cap B_i) = \sum_{i=1}^{\infty} \Pr(A \mid B_i) \Pr(B_i)$$

□

The expression

$$\Pr(A) = \sum_{i=1}^{\infty} \Pr(A \mid B_i) \Pr(B_i) \tag{13.2}$$

is called the *law of total probability*.

13.2 BASIC CONCEPTS

In our non-Bayesian setup, let X be a random variable with probability density function $f(x, \theta)$, $\theta \in \Omega$. According to the Bayesian point of view, θ is interpreted as a realization of a random variable Θ in Ω with some density $f_\Theta(\theta)$. The density $f_\Theta(\theta)$ expresses what one thinks (believes) about the occurring value of Θ, *before* any observation has been taken, that is, *a priori*. $f_\Theta(\theta)$ is called the *prior density* of Θ. $f(x, \theta)$ is then read as the conditional density of X, given $\Theta = \theta$, and rewritten as $f_{X|\Theta}(x \mid \theta)$.

With this interpretation, the joint density of X and Θ, $f_{X,\Theta}(x, \theta)$, is given by

$$f_{X,\Theta}(x, \theta) = f_{X|\Theta}(x \mid \theta) \cdot f_\Theta(\theta) \tag{13.3}$$

Proceeding on this basis, the marginal density of X, $f_X(x)$ is

$$\begin{aligned} f_X(x) &= \int_\Omega f_{X,\Theta}(x, \theta) \, d\theta \\ &= \int_\Omega f_{X|\Theta}(x \mid \theta) \cdot f_\Theta(\theta) \, d\theta \end{aligned} \tag{13.4}$$

The conditional density of Θ, given $X = x$, becomes

$$f_{\Theta|X}(\theta \mid x) = \frac{f_{X,\Theta}(x, \theta)}{f_X(x)} \tag{13.5}$$

or

$$f_{\Theta|X}(\theta \mid x) = \frac{f_{X|\Theta}(x \mid \theta) \cdot f_\Theta(\theta)}{f_X(x)} \tag{13.6}$$

which is seen to be a simple form of Bayes's theorem (13.1).

By $f_{\Theta|X}(\theta \mid x)$ we express our *belief* concerning the distribution of Θ *after* having observed $X = x$, that is, *a posteriori*, and $f_{\Theta|X}(\theta \mid x)$ is therefore called the *posterior density* of Θ. Note that when X is observed, $f_X(x)$ occurs in (13.6) as a constant. Hence $f_{\Theta|X}(\theta \mid x)$ is always *proportional* to $f_{X|\Theta}(x \mid \theta) f_\Theta(\theta)$, which we write as:

$$f_{\Theta|X}(\theta \mid x) \propto f_{X|\Theta}(x \mid \theta) \cdot f_\Theta(\theta) \tag{13.7}$$

The Bayesian approach may be characterized as an updating process of our information about the parameter θ. First, a probability density for Θ is assigned before any observations of X is taken. Then, as soon as the first X is observed and becomes available, the prior distribution of Θ is updated to the posteriori distribution of Θ, given $X = x$. The observed value of X has therefore changed our belief regarding the value of Θ. This process may be repeated. In the next step our posterior distribution of Θ, given $X = x$, is chosen as the new prior distribution, another X is being observed, and one is lead to a second posterior distribution, and so on. This updating process is illustrated in Fig. 13.1.

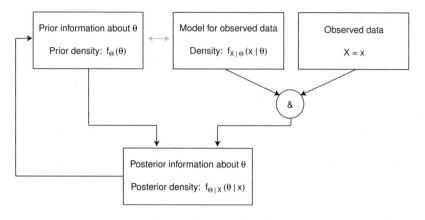

Fig. 13.1 Bayesian "updating" process.

Example 13.1
A nonrepairable valve is assumed to have constant failure rate λ. Experience (or belief) leads us to think that the failure rate is a random variable Λ which is *gamma distributed* with parameters α and β. The *prior* density of Λ is therefore

$$f_\Lambda(\lambda) = \frac{\beta^\alpha}{\Gamma(\alpha)} \lambda^{\alpha-1} e^{-\beta\lambda} \quad \text{for } \lambda > 0 \tag{13.8}$$

The mean value of Λ is

$$E(\Lambda) = \frac{\alpha}{\beta} \tag{13.9}$$

The gamma distribution is further discussed in Section 2.11. The probability density function of the time to failure T of the valve, when the failure rate λ is known, is

$$f_{T|\Lambda}(t \mid \lambda) = \lambda e^{-\lambda t} \quad \text{for } t > 0, \; \lambda > 0 \tag{13.10}$$

We will now assume that we can test n valves of the same type one by one. Before the first test, we assume the prior distribution of the failure rate Λ to be gamma distributed with parameters $\alpha_1 = 2$ and $\beta_1 = 1$,

$$f_\Lambda(\lambda) = \lambda e^{-\lambda} \quad \text{for } \lambda > 0 \tag{13.11}$$

The (first) prior density of Λ is illustrated in Fig. 13.2. Let T_1 denote the time to failure of the first valve tested. The joint density of T_1 and Λ becomes

$$\begin{aligned} f_{T_1,\Lambda}(t_1,\lambda) &= \lambda e^{-\lambda t_1} \cdot \lambda e^{-\lambda} \\ &= \lambda^2 e^{-\lambda(t_1+1)} \quad \text{for } t_1 > 0, \; \lambda > 0 \end{aligned} \tag{13.12}$$

The marginal density of T_1 is

$$\begin{aligned} f_{T_1}(t_1) &= \int_0^\infty \lambda^2 e^{-\lambda(t_1+1)} d\lambda \\ &= \frac{\Gamma(3)}{(t_1+1)^3} = \frac{2}{(t_1+1)^3} \quad \text{for } t > 0 \end{aligned} \tag{13.13}$$

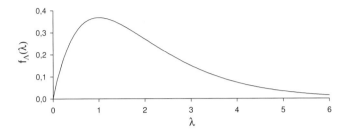

Fig. 13.2 The gamma density with parameters (2,1).

The conditional density of Λ, given $T_1 = t_1$, the posterior density, is

$$f_{\Lambda|T_1}(\lambda \mid t_1) = \frac{\lambda^2 e^{-\lambda(t_1+1)}}{2}(t_1+1)^3$$
$$= \frac{(t_1+1)^3}{\Gamma(3)}\lambda^{3-1}e^{-\lambda(t_1+1)} \quad \text{for } \lambda > 0 \qquad (13.14)$$

which also represents a gamma density, now with parameters α_2 and β_2 where

$$\alpha_2 = 3 = \alpha_1 + 1 \quad \text{since} \quad \alpha_1 = 2$$
$$\beta_2 = (t_1 + 1) = \beta_1 + t_1 \quad \text{since} \quad \beta_1 = 1$$

The procedure may now be repeated with (13.14) as our new prior distribution. Then we observe the lifetime $T_2 = t_2$ of a similar valve and are lead to a new posterior distribution which is a gamma distribution with parameters:

$$\alpha_3 = \alpha_2 + 1 = \alpha_1 + 2$$
$$\beta_3 = \beta_2 + t_2 = \beta_1 + (t_1 + t_2)$$

and so on.

The posterior density (13.14) could also have been derived directly using (13.7)

$$f_{\Lambda|T_1}(\lambda \mid t_1) \propto f_{T_1|\Lambda}(t_1 \mid \lambda) \cdot f_\Lambda(\lambda)$$
$$\propto \lambda e^{-\lambda t_1} \cdot \lambda e^{-\lambda}$$
$$\propto \lambda^2 e^{-\lambda(t_1+1)} \qquad (13.15)$$

Hence

$$f_{\Lambda|T_1}(\lambda \mid t_1) = k(t_1) \cdot \lambda^2 e^{-\lambda(t_1+1)} \quad \text{for } t_1 > 0 \qquad (13.16)$$

and since (13.16) is a density, $k(t_1)$ is easily determined to be $(1+t_1)^3/2$. This leads to the same posterior density as we derived in (13.14).

From (13.9) it follows that

$$
\left.\begin{array}{rcl}
E(\Lambda) & = & \dfrac{2}{1} \\[2mm]
E(\Lambda \mid T_1 = t_1) & = & \dfrac{2+1}{1+t_1} \\[2mm]
E(\Lambda \mid T_1 = t_1, T_2 = t_2) & = & \dfrac{2+1+1}{1+t_1+t_2} \\[2mm]
& \vdots &
\end{array}\right\} \tag{13.17}
$$

Equation (13.17) illustrates how we update our belief about the mean of Λ, as observations of T become available. □

13.3 BAYESIAN POINT ESTIMATION

Let us return to the general setup in Section 13.2. Recall that θ is a realization of a random variable $\Theta \in \Omega$ with some prior density $f_\Theta(\theta)$, and that X is a random variable with continuous density given $\Theta = \theta$, $f_{X|\Theta}(x \mid \theta)$. Our task is now to estimate the value θ of Θ that belongs to an observed value x of X. We will denote this estimator by $\hat{\theta}(X)$.

As is usual, we will prefer an estimator that minimizes the mean quadratic loss:

$$E\left[\left(\hat{\theta}(X) - \Theta\right)^2\right]$$

Such an estimator will be denoted a *Bayesian estimator* (of θ) (with minimum expected quadratic loss). Note that in the Bayesian framework, X and Θ are both random variables. How should $\hat{\theta}(X)$ be chosen?

$$E\left[(\hat{\theta}(X) - \Theta)^2\right] = \int_{-\infty}^{+\infty} \int_\Omega (\hat{\theta}(X) - \theta)^2 \, f_{X,\Theta}(x, \theta) \, dx \, d\theta$$

By using (13.3) we get

$$E\left[(\hat{\theta}(X) - \Theta)^2\right] = \int_{-\infty}^{+\infty} f_X(x) \left[\int_\Omega (\theta - \hat{\theta}(X))^2 f_{\Theta|X}(\theta \mid x) \, d\theta \right] dx$$

Obviously $E[(\hat{\theta}(X) - \Theta)^2]$ becomes minimized if, for each x, $\hat{\theta}(x)$ is chosen to minimize

$$\int_\Omega (\theta - \hat{\theta}(x))^2 f_{\Theta|X}(\theta \mid x) \, d\theta$$

In probability theory the following lemma is well known:

Lemma 13.1 Let Y be a random variable with density $f_Y(y)$ and finite variance τ^2. Then

$$\psi(\eta) = \int_{-\infty}^{+\infty} (y - \eta)^2 f_Y(y) \, dy$$

is minimized when η is chosen as $E(Y)$. □

This lemma, applied to our problem, tells us that $E[\hat{\theta}(X) - \Theta]^2$ is minimized for

$$\hat{\theta}(X) = E(\Theta \mid X) \tag{13.18}$$

Hence we can conclude that *the Bayesian estimator of θ is the mean of the posterior distribution of Θ*.

Example 13.1 (Cont.)
If we apply Lemma 13.1 to the iteration in Example 13.1, the Bayesian estimate of Λ, after having observed $T_1 = t_1$, is

$$\hat{\lambda}(t_1) = \frac{3}{1 + t_1}$$

□

Let us return to our Bayesian model where θ represents a realization of a random variable $\Theta \in \Omega$ with some prior density $f_\Theta(\theta)$. We are now considering a situation where our data $D = (x_1, x_2, \ldots, x_n)$ consist of observations of n random variables $X_1, X_2, \ldots, X_n$, assumed to be independent and identically distributed, *conditional on θ*, with density $f_{X \mid \Theta}(x \mid \theta)$. Then

$$f_{X_1, X_2, \ldots, X_n \mid \Theta}(x_1, x_2, \ldots, x_n \mid \theta) = \prod_{j=1}^{n} f_{X \mid \Theta}(x_j \mid \theta) \tag{13.19}$$

The posterior distribution of Θ, given $X_1, X_2, \ldots, X_n$, may now be obtained by the same procedure as we used for one single X, and we get

$$f_{\Theta \mid X_1, X_2, \ldots, X_n}(\theta \mid x_1, x_2, \ldots, x_n) \propto \left[\prod_{j=1}^{n} f_{X \mid \Theta}(x_j \mid \theta) \right] \cdot f_\Theta(\theta) \tag{13.20}$$

Considering the right-hand side of (13.20) as a function of θ, given $x_1, x_2, \ldots, x_n$, this can also be written

$$f_{\Theta \mid X_1, X_2, \ldots, X_n}(\theta \mid x_1, x_2, \ldots, x_n) \propto L(\theta \mid x_1, x_2, \ldots, x_n) f_\Theta(\theta) \tag{13.21}$$

where $L(\theta \mid x_1, x_2, \ldots, x_n)$ denotes the likelihood function in the usual meaning. For brevity in the following discussion we will write $L(\theta \mid D)$ instead of $L(\theta \mid x_1, x_2, \ldots, x_n)$.

Example 13.1 (Cont.)
Reconsider the valve in Example 13.1, with constant (unknown) failure rate λ, where λ represents a realization of a random variable Λ with (prior) density (13.8), and let $T_1, T_2, \ldots, T_n$ denote the lifetimes of n such valves. Assume $T_1, T_2, \ldots, T_n$ to be independent and identically distributed, conditional on λ, with density

$$f_{T_i|\Lambda}(t_i \mid \lambda) = \lambda e^{-\lambda t_i} \quad \text{for } \lambda > 0, \ i = 1, \ldots, n$$

We want to determine the Bayesian estimate of λ, based on these lifetimes.

In this case

$$f_{T_1,T_2,\ldots,T_n|\Lambda}(t_1, t_2, \ldots, t_n \mid \lambda) = \prod_{j=1}^{n} \lambda e^{-\lambda t_i} = \lambda^n e^{-\lambda \sum_{i=1}^{n} t_i}$$

and

$$\begin{aligned} f_{\Lambda|T_1,T_2,\ldots,T_n}(\lambda \mid t_1, t_2, \ldots, t_n) &\propto \lambda^n e^{-\lambda \sum_{i=1}^{n} t_i} \cdot \lambda e^{-\lambda} \\ &\propto \lambda^{n+1} e^{-\lambda(1+\sum_{i=1}^{n} t_i)} \end{aligned} \quad (13.22)$$

To be a proper density, the right-hand side of (13.22) must be multiplied by the constant

$$\frac{\left(1 + \sum_{i=1}^{n} t_i\right)^{n+2}}{\Gamma(n+2)}$$

Hence

$$\begin{aligned} &f_{\Lambda|T_1,T_2,\ldots,T_n}(\lambda \mid t_1, t_2, \ldots, t_n) \\ &= \frac{\left(1 + \sum_{i=1}^{n} t_i\right)^{n+2}}{\Gamma(n+2)} \lambda^{n+1} \cdot e^{-\lambda(1+\sum_{i=1}^{n} t_i)} \quad \text{for } \lambda > 0 \end{aligned}$$

which we recognize as a gamma distribution with parameters $\alpha = (n+2)$ and $\beta = (1 + \sum_{i=1}^{n} t_i)$. Since the mean of this gamma distribution is α/β, the Bayesian estimator of λ is

$$\hat{\lambda}(T_1, T_2, \ldots, T_n) = \frac{2+n}{1 + \sum_{i=1}^{n} T_i} \quad (13.23)$$

□

13.4 CREDIBILITY INTERVALS

A credibility interval is the Bayesian analogue to a confidence interval. A credibility interval for Θ, at level $(1 - \varepsilon)$, is an interval $(a(D), b(D))$ such that the conditional probability, given the data D, satisfies

$$\Pr(a(D) < \Theta < b(D) \mid D) = \int_{a(D)}^{b(D)} f_{\Theta|D}(\theta \mid D) \, d\theta = 1 - \varepsilon \quad (13.24)$$

Then the interval $(a(D), b(D))$ is an interval estimate of θ in the sense that the conditional probability of Θ belonging to the interval, given the data, is equal to $(1-\varepsilon)$.

13.5 CHOICE OF PRIOR DISTRIBUTION

13.5.1 Conjugate Families of Distributions

First we state a useful definition:

Definition 13.1 A parametric family $\mathcal{P}$ of distributions $f_\Theta(\theta)$ is said to be closed in sampling with respect to a family $\mathcal{F}$ of distributions $f_{X|\Theta}(x \mid \theta)$ if

$$f_\Theta(\theta) \in \mathcal{P} \Rightarrow f_{\Theta|X}(\theta \mid x) \in \mathcal{P} \qquad (13.25)$$

In that case $\mathcal{P}$ is also said to be a conjugate family to $\mathcal{F}$ or, for short, conjugate to $\mathcal{F}$. □

Example 13.2
Let $\mathcal{F}$ be the family of exponential distributions defined by the probability density function

$$f_{T|\Lambda}(t \mid \lambda) = \lambda e^{-\lambda t} \quad \text{for } t > 0 \qquad (13.26)$$

Let us show that the class $\mathcal{P}$ of gamma distributions, defined by (13.8)

$$f_\Lambda(\lambda) = \frac{\beta}{\Gamma(\alpha)} (\beta\lambda)^{\alpha-1} e^{-\beta\lambda} \quad \text{for } \lambda > 0$$

is conjugate to $\mathcal{F}$. That is, we have to show that the corresponding $f_{\Lambda|T}(\lambda \mid t)$ is a gamma distribution. In this case

$$\begin{aligned} f_{T,\Lambda}(t, \lambda) &= \lambda e^{-\lambda t} \cdot \frac{\beta^\alpha}{\Gamma(\alpha)} \lambda^{\alpha-1} e^{-\lambda\beta} \\ &= \frac{\beta^\alpha}{\Gamma(\alpha)} \lambda^\alpha e^{-\lambda(\beta+t)} \quad \text{for } t > 0, \ \lambda > 0 \end{aligned} \qquad (13.27)$$

Furthermore

$$\begin{aligned} f_T(t) &= \int_0^\infty \frac{\beta^\alpha}{\Gamma(\alpha)} \lambda^\alpha e^{-\lambda(\beta+t)} d\lambda \\ &= \frac{\alpha \beta^\alpha}{(\beta+t)^{\alpha+1}} \quad \text{for } t > 0 \end{aligned} \qquad (13.28)$$

Hence

$$\begin{aligned} f_{\Lambda|T}(\lambda \mid t) &= \frac{\beta^\alpha \left(\lambda^\alpha e^{-\lambda(\beta+t)}\right) / \Gamma(\alpha)}{\alpha \beta^\alpha / (\beta+t)^{\alpha+1}} \\ &= \frac{(\beta+t)^{\alpha+1}}{\Gamma(\alpha+1)} \lambda^\alpha e^{-\lambda(\beta+t)} \quad \text{for } t > 0 \end{aligned} \qquad (13.29)$$

which is a gamma density with parameters $\alpha + 1$ and $\beta + t$. □

The result in Example 13.2 may be stated more formally as:

Theorem 13.2 The family of gamma distributions (α, β) is conjugate to the family of exponential distributions. □

Clearly the assumption of a gamma density as a prior distribution in connection with an exponential distribution is mathematically convenient. Our intention with this prior distribution, however, is to express a priori knowledge of λ, and we raise the question whether or not this purpose is taken care of by using the gamma density as prior. The answer to this question is that the gamma distribution (α, β) is a very flexible distribution. It may take on a wide variety of shapes through varying the parameters α and β. Almost any conceivable shape of the density of Λ can essentially be obtained by proper choice of α and β.

Example 13.3
Consider a plant which has a specified number of identical and independent valves with constant failure rate λ, where λ represents a realization of a random variable Λ with the gamma prior density

$$f_\Lambda(\lambda) = \frac{\beta}{\Gamma(\alpha)} (\beta\lambda)^{\alpha-1} e^{-\beta\lambda} \quad \text{for } \lambda > 0 \tag{13.30}$$

The parameters α and β of the prior distribution is usually "estimated" based on prior experience with the same type of valves, combined with information gained from various reliability data sources (see Chapter 14).

When a valve fails, it will be replaced with a valve of the same type. The associated downtime is considered to be negligible. Valve failures are assumed to occur according to a homogeneous Poisson process with intensity λ. The number of valve failures $N(t)$ during an accumulated time t in service thus has the Poisson distribution

$$\Pr(N(t) = n \mid \Lambda = \lambda) = \frac{(\lambda t)^n}{n!} e^{-\lambda t} \quad \text{for } n = 0, 1, \ldots \tag{13.31}$$

The marginal distribution of $N(t)$ then becomes

$$\begin{aligned}
\Pr(N(t) = n) &= \int_0^\infty \Pr(N(t) = n \mid \Lambda = \lambda) \cdot f_\Lambda(\lambda) \, d\lambda \\
&= \int_0^\infty \frac{(\lambda t)^n}{n!} e^{-\lambda t} \cdot \frac{\beta^\alpha}{\Gamma(\alpha)} \lambda^{\alpha-1} e^{-\beta\lambda} \, d\lambda \\
&= \frac{\beta^\alpha t^n}{\Gamma(\alpha) n!} \int_0^\infty \lambda^{\alpha+n-1} e^{-(\beta+n)\lambda} \, d\lambda \\
&= \frac{\beta^\alpha t^n}{\Gamma(\alpha) n!} \frac{\Gamma(n+\alpha)}{(\beta+t)^{n+\alpha}}
\end{aligned} \tag{13.32}$$

By combining (13.30), (13.31), and (13.32) we get the posterior density of Λ, given $N(t) = n$,

$$f_{\Lambda|N(t)}(\lambda \mid n) = \frac{(\beta + t)^{\alpha+n}}{\Gamma(\alpha + n)} \lambda^{\alpha+n-1} e^{-(\beta+t)\lambda} \tag{13.33}$$

which is recognized as the gamma distribution with parameters $(\alpha + n)$ and $(\beta + t)$.
□

Hence we have shown the following theorem:

Theorem 13.3 The family of gamma distributions (α, β) is conjugate to the family of Poisson distributions. □

Example 13.3 (Cont.)
The Bayesian estimate of λ when $N(t) = n$ is

$$\hat{\lambda} = E(\Lambda \mid N(t) = n) = \frac{\alpha + n}{\beta + t} \tag{13.34}$$

Furthermore the conditional distribution [given $N(t) = n$] of the variable $Z = 2(\beta + t)\Lambda$ is

$$f_{Z|N(t)}(z \mid n) = \frac{1}{2^{\alpha+n}\Gamma(\alpha + n)} z^{\alpha+n-1} e^{-z/2} \quad \text{for } z > 0$$

which is recognized as the χ^2 distribution with $2(\alpha + n)$ degrees of freedom.
A $1 - \varepsilon$ credibility interval for the failure rate is obtained as

$$\Pr\left(\frac{z_{1-\varepsilon/2, 2(\alpha+n)}}{2(\beta + t)} < \Lambda < \frac{z_{\varepsilon/2, 2(\alpha+n)}}{2(\beta + t)} \mid N(t) = n\right) = 1 - \varepsilon \tag{13.35}$$

where $z_{\varepsilon, \nu}$ denotes the upper $100\varepsilon\%$ percentile of the χ^2 distribution with ν degrees of freedom; that is, $\Pr(Z > z_{\varepsilon, \nu}) = \varepsilon$ when $Z \sim \chi_\nu^2$. The upper $100\varepsilon\%$ percentile of the χ_ν^2 is listed in Appendix F for some values of ν and ε.
A one-sided upper credibility interval for the failure rate is obtained in the same way:

$$\Pr\left(\Lambda < \frac{z_{\varepsilon, 2(\alpha+n)}}{2(\beta + t)} \mid N(t) = n\right) = 1 - \varepsilon \tag{13.36}$$

Assume that we have estimated the parameters α and β of the prior gamma distribution to be (all time units in hours)

$$\alpha = 3$$
$$\beta = 1 \cdot 10^4$$

The prior mean and standard deviation is thus

$$E(\Lambda) = 3 \cdot 10^{-4} \quad \text{(prior mean)}$$
$$SD(\Lambda) \approx 1.73 \cdot 10^{-4} \quad \text{(prior standard deviation)}$$

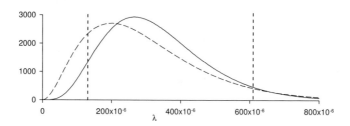

Fig. 13.3 Prior (dotted line) and posterior density of the failure rate in Example 13.3, and a 90% credibility interval for the failure rate.

Further, assume that $n = 2$ failures have been observed during a total time in service $t = 5 \cdot 10^3$ hours.

The Bayesian estimate of λ is thus from (13.34)

$$\hat{\lambda} = \frac{\alpha + n}{\beta + t} = \frac{3 + 2}{10^4 + 5 \cdot 10^3} \approx 3.33 \cdot 10^{-4}$$

A 90% credibility interval for Λ with these data is

$$\Pr\left(\frac{z_{0.95, 10}}{2(\beta + t)} < \Lambda < \frac{z_{0.05, 10)}}{2(\beta + t)} \mid N(t) = 2\right) = 0.90$$

$$\Pr\left(\frac{3.94}{2 \cdot 1.5 \cdot 10^4} < \Lambda < \frac{18.31}{2 \cdot 1.5 \cdot 10^4} \mid N(t) = 2\right) = 0.90$$

$$\Pr(1.31 \cdot 10^{-4} < \Lambda < 6.10 \cdot 10^{-4} \mid N(t) = 2) = 0.90$$

The prior and posterior distribution of Λ for these data are presented in Fig. 13.3, with a 90% credibility interval for Λ. □

Example 13.4
Let $\mathcal{F}$ be the binomial distribution defined by

$$f_{X|\Theta}(x \mid \theta) = \binom{n}{x} \theta^x (1 - \theta)^{n-x}, \quad x = 0, 1, \ldots, n, \; 0 \leq \theta \leq 1 \quad (13.37)$$

Let us show that the class $\mathcal{P}$ of *beta distributions*, defined by the density

$$f_\Theta(\theta) = \frac{\Gamma(r + s)}{\Gamma(r) \cdot \Gamma(s)} \theta^{r-1} (1 - \theta)^{s-1}, \quad 0 \leq \theta \leq 1 \; r > 0, \; s > 0 \quad (13.38)$$

is a conjugate family to $\mathcal{F}$.

To prove this statement, we have to show that the corresponding density $f_{\Theta|X}(\theta \mid x)$ represents a beta distribution. To do so, we make use of (13.7):

$$f_{\Theta|X}(\theta \mid x) \propto \binom{n}{x} \theta^x (1-\theta)^{n-x}$$
$$\cdot \frac{\Gamma(r+s)}{\Gamma(r)+\Gamma(s)} \theta^{r-1}(1-\theta)^{s-1}, \quad 0 < \theta < 1 \quad (13.39)$$

Put differently,

$$f_{\Theta|X}(\theta \mid x) \propto \theta^{x+r-1}(1-\theta)^{n-x+s-1}$$

Hence

$$f_{\Theta|X}(\theta \mid x) = k(x) \cdot \theta^{(x+r)-1}(1-\theta)^{(n-x+s)-1}, \quad 0 < \theta < 1 \quad (13.40)$$

Knowing (13.40) to be a probability density, the "constant" $k(x)$ has to be

$$k(x) = \frac{\Gamma(n+r+s)}{\Gamma(x+r)\Gamma(n-x+s)} \quad (13.41)$$

Equation (13.41) introduced into (13.40), gives the beta density with parameters $(x+r)$, and $(n-x+s)$. □

We have thus proved the following theorem:

Theorem 13.4 The family of beta distributions (r, s) is conjugate to the family of binomial distributions. □

The assumption of a beta prior in connection with a binomial distribution is mathematically convenient. Furthermore the beta distribution is a very flexible distribution. Its density can take on a wide variety of shapes by proper choice of r and s. Note that by choosing $r = 1, s = 1$, the beta density represents a uniform density over $[0, 1]$ which corresponds to no a priori preference for any θ in $[0, 1]$. (In this case we have a *noninformative* prior distribution for θ.)

The mean of the beta distribution (r, s) is easily found to be $r/(r+s)$ (e.g., see Dudewicz and Mishra (1988, p. 224). The prior mean is from (13.38)

$$E(\Theta) = \frac{r}{r+s}$$

and the Bayesian estimate of the probability θ is from (13.40),

$$\hat{\theta}(x) = \frac{r+x}{r+s+n} \quad (13.42)$$

13.5.2 Noninformative Prior Distribution

Example 13.5
A certain type of valves is assumed to have constant (unknown) failure rate λ, where λ represents a realization of a random variable Λ. Let $T_1, T_2, \ldots, T_n$ denote the observed lifetimes of n such valves. Assume $T_1, T_2, \ldots, T_n$ to be independent and identically distributed, conditional on λ, with density

$$f_{T_i|\Lambda}(t_i \mid \lambda) = \lambda e^{-\lambda t_i} \quad \text{for } t_i > 0, \; \lambda > 0$$

When there is no prior information available about the true value of Λ, a noninformative prior distribution seems appropriate.

A noninformative prior distribution is characterized by giving no preference to any of the possible parameter values. Hence, if the possible parameter values constitute a finite interval, the parameter is assumed to be uniformly distributed over that interval. Hence a noninformative prior distribution for the failure rate Λ in our case may, for example, be given by the uniform density

$$f_\Lambda(\lambda) = \begin{cases} \dfrac{1}{M} & \text{for } 0 \leq \lambda \leq M \\ 0 & \text{otherwise} \end{cases} \tag{13.43}$$

where M is taken to be a very large number, say 10^{10}. Then the posterior density of Λ, given the data D, is by (13.22) approximately

$$f_{\Lambda|T_1,T_2,\ldots,T_n}(\lambda \mid t_1, t_2, \ldots, t_n) \propto L(\lambda \mid D) \cdot f_\Lambda(\lambda)$$

$$f_{\Lambda|T_1,T_2,\ldots,T_n}(\lambda \mid t_1, t_2, \ldots, t_n) \propto \lambda^n e^{-\lambda \sum_{i=1}^n t_i} \cdot \frac{1}{M} \tag{13.44}$$

$$f_{\Lambda|T_1,T_2,\ldots,T_n}(\lambda \mid t_1, t_2, \ldots, t_n) \propto \lambda^n e^{-\lambda \sum_{i=1}^n t_i}$$

To become a proper density, the right-hand side of (13.45) must be multiplied by $(\sum_{i=1}^n t_i)^{n+1}/\Gamma(n+1)$. Hence

$$f_{\Lambda|T_1,T_2,\ldots,T_n}(\lambda \mid t_1, t_2, \ldots, t_n) = \frac{\left(\sum_{i=1}^n t_i\right)^{n+1}}{\Gamma(n+1)} \cdot \lambda^n e^{-\lambda \sum_{i=1}^n t_i} \quad \text{for } \lambda > 0 \tag{13.45}$$

which we recognize as the density of a gamma distribution with parameters $(n+1)$ and $\sum_{i=1}^n t_i$.

Hence the Bayesian estimator of the failure rate (with minimum expected quadratic loss) and a noninformative prior becomes

$$\hat{\lambda}(T_1, T_2, \ldots, T_n) = \frac{n+1}{\sum_{i=1}^n T_i} \tag{13.46}$$

The MLE of the failure rate λ in this situation was in Chapter 11 determined to be

$$\lambda^*(T_1, T_2, \ldots, T_n) = \frac{n}{\sum_{i=1}^n T_i}$$

Note that the MLE of λ, $\lambda^*(T_1, T_2, \ldots, T_n)$ coincides with the mode of the posterior distribution (13.45) □

13.6 BAYESIAN LIFE TEST SAMPLING PLANS

The density

$$f_\Theta(\theta) = \frac{b^a}{\Gamma(a)} \cdot \left(\frac{1}{\theta}\right)^{a+1} e^{-b/\theta} \quad \text{for } \theta > 0, \ a > 0, \ b > 0 \qquad (13.47)$$

is called the *inverted gamma density* with parameters a and b, since the variable Θ^{-1} then has an ordinary gamma density with parameters a and $1/b$. Since $\text{var}(\Theta)$ does not exist when $a \le 2$, we will in the following assume that $a > 2$. Then

$$E(\Theta) = \frac{b}{a-1} = \theta_0 \quad \text{(prior mean)}$$

$$\text{var}(\Theta) = \frac{b^2}{(a-1)^2(a-2)} = \sigma_0^2 \quad \text{(prior variance)} \qquad (13.48)$$

The inverted gamma distribution (13.47) can be shown to be conjugate to the family of exponential distributions, defined by the density

$$f_{T|\Theta}(t \mid \theta) = \frac{1}{\theta} e^{-t/\theta} \quad \text{for } t > 0, \ \theta > 0 \qquad (13.49)$$

Equation (13.49) is often used as a prior distribution when estimating θ, which represents the mean time to failure (MTTF) of (13.49). The inverted gamma density is rather flexible and can take on a wide variety of shapes by proper choice of a and b.

13.6.1 Complete Data Sets

Suppose n units are life tested until failure, and let $T_1, T_2, \ldots, T_n$ denote the observed lifetimes. Hence the observed *data* is $D = (t_1, t_2, \ldots, t_n)$. Given $\Theta = \theta$, $T_1, T_2, \ldots, T_n$ are assumed to be independent and identically distributed with density (13.49). We select (13.47) as prior distribution of Θ. Then according to (13.20)

$$f_{\Theta|T_1,T_2,\ldots,T_n}(\theta \mid (t_1, t_2, \ldots, t_n))$$
$$\propto \left(\frac{1}{\theta}\right)^n e^{-(\sum_{i=1}^n t_i)/\theta} \frac{b^a}{\Gamma(a)} \left(\frac{1}{\theta}\right)^{a+1} e^{-b/\theta} \quad \text{for } \theta > 0$$

which implies that

$$f_{\Theta|T_1,T_2,\ldots,T_n}(\theta \mid t_1, t_2, \ldots, t_n)$$
$$\propto \left(\frac{1}{\theta}\right)^{n+a+1} e^{-(b+\sum_{i=1}^n t_i)/\theta} \quad \text{for } \theta > 0 \qquad (13.50)$$

In this case $\sum_{i=1}^{n} t_i$ expresses the *total time on test* at the last failure:

$$\sum_{i=1}^{n} t_i = \mathcal{T}(t_n)$$

Hence (13.50) may be written

$$f_{\Theta|T_1,T_2,\ldots,T_n}(\theta \mid t_1, t_2, \ldots, t_n)$$
$$\propto \left(\frac{1}{\theta}\right)^{n+a+1} \cdot e^{-(b+\mathcal{T}(t_n))/\theta} \quad \text{for } \theta > 0 \qquad (13.51)$$

To become a proper density, the right-hand side of (13.51) has to be multiplied by $[b + \mathcal{T}(t_n)]^{n+a}/\Gamma(n+a)$.

Therefore

$$f_{\Theta|T_1,T_2,\ldots,T_n}(\theta \mid t_1, t_2, \ldots, t_n)$$
$$= \frac{(b+\mathcal{T}(t_n))^{n+a}}{\Gamma(n+a)} \left(\frac{1}{\theta}\right)^{n+a+1} \cdot e^{-(b+\mathcal{T}(t_n))/\theta} \quad \text{for } \theta > 0 \qquad (13.52)$$

which is recognized as the density of an inverted gamma distribution with parameters $(n+a)$ and $(b+\mathcal{T}(t_n))$. The Bayesian estimator of MTTF $= \theta$ (minimizing the mean quadratic loss) coincides with the mean value of (13.52):

$$\hat{\theta}(T_1, T_2, \ldots, T_n) = \frac{b + \mathcal{T}(t_n)}{n+a-1} = \frac{b + n\bar{T}}{n+a-1} \qquad (13.53)$$

The maximum likelihood estimator of the MTTF, θ, is earlier determined to be

$$\theta^*(T_1, T_2, \ldots, T_n) = \frac{1}{n}\sum_{j=1}^{n} T_j = \bar{T}$$

We note that

$$\hat{\theta}(T_1, T_2, \ldots, T_n) = (1-w) \cdot \frac{b}{a-1} + w\bar{T}$$

or

$$\hat{\theta}(T_1, T_2, \ldots, T_n) = (1-w) \cdot \theta_0 + w\theta^* \qquad (13.54)$$

where θ_0 denotes the prior mean (13.48) and

$$w = \frac{n}{n+a-1}$$

Hence the Bayesian estimator of θ is a weighted average of the prior mean θ_0 and the MLE θ^*. We also note that

$$\hat{\theta}(T_1, T_2, \ldots, T_n) \to \theta^* \text{ as } n \to \infty$$

In words, the influence of the prior mean tends to zero as $n \to \infty$.

13.6.2 Type II Censored Data Sets

Let the situation be as above, except for the fact that the life test is terminated at the rth failure, which corresponds to censoring of type II (see Section 11.2). Let $T_{(1)} \leq T_{(2)} \leq \cdots \leq T_{(r)}$ denote the recorded lifetimes.

The joint probability density function of $T_{(1)} \leq T_{(2)} \leq \cdots \leq T_{(r)}$, given $\Theta = \theta$, is

$$f_{T_{(1)},T_{(2)},\ldots,T_{(r)}|\Theta}(t_1, t_2, \ldots, t_r \mid \theta)$$
$$= \frac{n!}{(n-1)!} \cdot \frac{1}{\theta^r} e^{-(\sum_{i=1}^{r} t_i + (n-r)t_r)/\theta} \quad \text{for } t_1 \leq \cdots \leq t_r \quad (13.55)$$

Hence

$$f_{T_{(1)},T_2,\ldots,T_{(r)}|\Theta}(t_1, t_2, \ldots, t_r \mid \theta)$$
$$= \frac{n!}{(n-1)!} \cdot \frac{1}{\theta^r} e^{-(\sum_{i=1}^{r} t_i + (n-r)t_r)/\theta} \frac{b^a}{\Gamma(a)} \cdot \left(\frac{1}{\theta}\right)^{a+1} e^{-b/\theta}$$
$$\text{for } t_1 \leq \ldots \leq t_r, \; \theta > 0, \; a > 2, \; b > 0 \quad (13.56)$$

which implies that

$$f_{\Theta|T_{(1)},T_{(2)},\ldots,T_{(r)}}(\theta \mid t_1, t_2, \ldots, t_r)$$
$$\propto \left(\frac{1}{\theta}\right)^{r+a+1} e^{-(b+\sum_{i=1}^{r} t_i + (n-r)t_r)/\theta} \quad (13.57)$$

If we introduce the total time on test concept $\mathcal{T}(t_r)$ (see Section 11.3), then (13.57) can be written

$$f_{\Theta|T_{(1)},T_{(2)},\ldots,T_{(r)}}(\theta \mid t_1, t_2, \ldots, t_r)$$
$$\propto \left(\frac{1}{\theta}\right)^{r+a+1} e^{-(b+\mathcal{T}(t_r))/\theta} \quad \text{for } \theta > 0 \quad (13.58)$$

We immediately notice the similarity between (13.58) and (13.51), derived for complete data sets and are able to conclude that the Bayesian estimator of the MTTF $= \theta$, in the case of type II censoring is

$$\hat{\theta}(T_1, T_2, \ldots, T_r) = \frac{b + \mathcal{T}(T_r)}{r + a - 1} \quad (13.59)$$

when the inverted gamma density with parameters a and b is used as prior distribution.

13.7 INTERPRETATION OF THE PRIOR DISTRIBUTION

We will confine ourselves to discussing two essentially different interpretations of the prior distribution and illustrate them by discussing an example.

13.7.1 Prior Distribution Based on Empirical Data

Example 13.6

Suppose a commodity is delivered to customers in lots of size a. The relative number of defect units in a lot obviously will vary from one lot to another, and we may interpret the relative number of defect units as a random variable Θ. Every lot is controlled by taking a random sample of size n of units from the lot, and recording the number X of defect units in the sample.

Given $\Theta = \theta$, X will, if a is not too small, be approximately binomially distributed:

$$p_{X|\Theta}(x \mid \theta) = \binom{n}{x} \theta^x (1-\theta)^{n-x}, \quad x = 0, 1, \ldots, n, \quad 0 \le \theta \le 1 \quad (13.60)$$

As already indicated, Θ is likely to vary from lot to lot. However, if the production process is "in control," the θ values will show a statistical regularity that may be expressed by some probability density $p_\Theta(\theta)$. Looking at the situation in this way, *two* random variables X and Θ are attached to every lot.

In principle we can imagine that over a certain period of time we have effectively carried out 100% control of the lot, and thereby observed a large number of realizations of Θ's. The histogram of these values gives us a picture of the possible density of Θ. Even in the case where we have only observed a few such X's, this data set supplies us with some information about the distribution of Θ.

The main point in this situation is that the prior distribution of Θ is empirically motivated on the basis of observed data. The procedure is then the following: First, a prior distribution of Θ is selected from the conjugate family of the binomial distribution (13.60), namely a beta distribution with unspecified parameters r and s. Then these parameters have to be estimated, one way or another, based on the observed Θ-values (exploratory data). The beta distribution with these estimated parameters is then chosen to be the prior distribution of Θ.

Presumably few statisticians will have essential objections to Bayesian inference on this basis, in particular since the beta distribution is very flexible. □

13.7.2 Subjective Bayesian Inference

Example 13.6 (Cont.)

The relative number of defect units in a certain "big" lot is denoted by θ, and the number of defect units in a sample of size n from this lot, is denoted by X. Then approximately

$$\Pr(X = x) = \binom{n}{x} \theta^x (1-\theta)^{n-x} \quad \text{for } x = 0, 1, \ldots, n, \quad 0 \le \theta \le 1$$

In subjective Bayesian inference, the appearing value of θ is an unknown constant in [0, 1]. The statistician (the Bayesian analyst), however, considers certain θ-values in [0, 1] to be more likely to occur than others, and he expresses this subjective belief on which θ value he expects through a (prior) distribution of Θ, $p_\Theta(\theta)$. The successive statistical reasoning hence is based on a *subjective* concept of probability, based on his degree of belief and cannot be given a frequency interpretation. In principle there is nothing illogical in a situation where two (Bayesian) statisticians reach different conclusions as a consequence of different choices of priors. For mathematical convenience, even the subjectivist chooses his prior from the conjugate family of distributions to the binomial. Note, however, that the subjectivist even has to choose the parameters r and s in the beta distribution on a subjective basis. This subjective approach is far more controversial among statisticians than the one where the prior is based on exploratory data. □

Situations we meet in analysis of reliability and risk are sometimes characterized by lack of data. In spite of this lack of data, decisions have to be made. In such situations, it may be necessary to exploit all a priori insight in the matter, even personal judgments, expert opinions, and the like. Then the subjective approach may be the best (and only) solution to the decision problem.

13.8 THE PREDICTIVE DENSITY

Let T and Θ be two random variables where the conditional density of T, given $\Theta = \theta$ is $f_{T|\Theta}(t \mid \theta)$. Let $T_1, T_2, \ldots, T_n$ be n observations of T. Given $\Theta = \theta$, these are assumed to be independent.

If a prior density of Θ, $f_\Theta(\theta)$ is assumed, then the joint density of $T_1, T_2, \ldots, T_n$ and Θ is

$$f_{T_1,T_2,\ldots,T_n,\Theta}(t_1, t_2, \ldots, t_n, \theta) = \left[\prod_{i=1}^{n} f(t_i \mid \theta)\right] \cdot f_\Theta(\theta) \qquad (13.61)$$

Let us for short denote the data set $T_1, T_2, \ldots, T_n$ by D. After having observed D, how should we predict the next value of T?

We may argue as follows: In the Bayesian setup we determine the posterior density of Θ, given D.

$$f_{\Theta|D}(\theta \mid D)$$

Then we define *the predictive density* of T, given D as

$$f_{T|D}(T \mid D) = \int_0^\infty f(t \mid \theta) \cdot f_{\Theta|D}(\theta \mid D)\, d\theta \qquad (13.62)$$

Example 13.1 (Cont.)
Reconsider the valves in Example 13.1. Given $\Lambda = \lambda$, the lifetimes $T_1, T_2, \ldots, T_n$ of the n valves are independent and exponentially distributed with probability density function

$$f_{T|\Lambda}(t \mid \lambda) = \lambda e^{-\lambda t} \quad \text{for } t > 0, \lambda > 0$$

where λ represents a realization of a random variable Λ with prior density

$$f_\Lambda(\lambda) = \lambda e^{-\lambda} \quad \text{for } \lambda > 0 \qquad (13.63)$$

Suppose that we have observed the lifetimes $(T_1, T_2, \ldots, T_n)$ of n such valves. According to (13.22) the posterior density of Λ, given D, is

$$f_{\Lambda|D}(\lambda \mid D) = \frac{(1 + \sum_{i=1}^{n} t_i)^{n+2}}{\Gamma(n+2)} \lambda^{n+1} e^{-\lambda(1+\sum_{i=1}^{n} t_i)} \quad \text{for } \lambda > 0 \qquad (13.64)$$

Hence our guess, based on D, is that the next observation T has density

$$\begin{aligned} f_{T|D}(t \mid D) &= \int_0^\infty \lambda e^{-\lambda t} \cdot \frac{(1 + \sum_{i=1}^{n} t_i)^{n+2}}{\Gamma(n+2)} \lambda^{n+1} e^{-\lambda(1+\sum_{i=1}^{n} t_i)} d\lambda \\ &= \frac{(1 + \sum_{i=1}^{n} t_i)^{n+2}}{\Gamma(n+2)} \int_0^\infty \lambda^{n+2} e^{-\lambda((1+\sum_{i=1}^{n} t_i)+t)} d\lambda \\ &= \frac{(n+2)(1 + \sum_{i=1}^{n} t_i)^{n+2}}{(1 + \sum_{i=1}^{n} t_i + t)^{n+3}} \quad \text{for } t > 0 \end{aligned} \qquad (13.65)$$

Hence our guess is that the survival function for a given new valve of the same type is

$$\begin{aligned} \Pr(T > t \mid D) &= \int_t^\infty \frac{(n+2)(1 + \sum_{i=1}^{n} t_i)^{n+2}}{(1 + \sum_{i=1}^{n} t_i + t)^{n+3}} du \\ &= \left(\frac{1 + \sum_{i=1}^{n} t_i}{1 + \sum_{i=1}^{n} t_i + t} \right)^{n+2} \\ &= \left(1 + \frac{t}{1 + \sum_{i=1}^{n} t_i} \right)^{-(n+2)} \quad \text{for } t > 0 \end{aligned} \qquad (13.66)$$

□

PROBLEMS

13.1 Show that the Bayesian estimator of θ which minimizes the mean absolute error loss $E(|\hat{\theta}(X) - \Theta|)$ is equal to the *median* of the posterior distribution of Θ (given $X = x$).

13.2 Determine the mean and the variance of the inverted gamma distribution (13.47).

13.3 Assume that X has a binomial distribution (n, p), where p represents a realization of a random variable P. The prior distribution of P is $f_P(p) = 1$ for $0 \leq p \leq 1$. Determine the posterior density of P when $X = x$ is observed, and determine the Bayesian estimate for p.

13.4 (Kapur and Lamberson, 1977, p. 402). Seven automobiles are each run over a 36,000 kilometer test schedule. The testing produced a total of 19 failures. Assuming an exponential failure distribution and a gamma prior with parameters $\alpha = 3$ and $\beta = 30,000$, answer the following:

(a) What is the Bayesian point estimate for the MTTF?

(b) What is the 90% lower confidence (credibility) limit on the 10,000-kilometer reliability?

13.5 Let $X_1, X_2, \ldots, X_n$ be independent and identically distributed $\mathcal{N}(\theta, \sigma_0^2)$, where σ_0^2 is known, and θ represents a realization of a random variable Θ with normal distribution $\mathcal{N}(\mu_0, \tau_0^2)$ where μ_0 and τ_0^2 are known.

Show that the Bayesian estimate of Θ (minimizing the mean quadratic loss) is a weighted average of the prior mean and the MLE of θ:

$$\hat{\theta}(X_1, X_2, \ldots, X_n) = \frac{n/\sigma_0^2}{n/\sigma_0^2 + 1/\tau_0^2} \cdot \overline{X} + \frac{1/\tau_0^2}{n/\sigma_0^2 + 1/\tau_0^2} \cdot \mu_0$$

Note that the Bayesian estimate of Θ is a weighted average of the hypothetical estimates of Θ based on the following:

- Data alone (i.e., the standard estimator $\overline{X}$)

- Prior information of Θ but no data, μ_0 (i.e., the Bayesian estimator of μ before any observations are taken)

Again note that the influence of the prior mean μ_0 tends to zero as $n \to \infty$.

13.6 Let $X_1, X_2, \ldots, X_n$ be independent and identically distributed $\mathcal{N}(0, \sigma^2)$.

(a) Show that the joint density of $X_1, X_2, \ldots, X_n$ can be written

$$C \tau^r e^{-\tau \sum_{i=1}^n x_i^2} \text{ where } r = n/2, \ \tau = 1/(2\sigma^2)$$

(b) Choose the gamma distribution (k, λ) with density

$$\frac{\lambda}{\Gamma(k)} (\lambda \tau)^{k-1} e^{-\lambda \tau} \text{ for } \tau > 0$$

as prior density of τ.

Show that the posterior density of τ, given $X_1, X_2, \ldots, X_n$ then becomes a gamma distribution $(k + r, \lambda + \sum_{i=1}^n x_i^2)$ with density

$$C(x_1, x_2, \ldots, x_n) \cdot \tau^{r+k-1} e^{-\tau(\lambda + \sum_{i=1}^n x_i^2)} \text{ for } \tau > 0$$

(c) Use the result in (b) to show that the Bayesian estimator of σ^2 (with minimum expected quadratic loss) becomes

$$\frac{\lambda + \sum_{i=1}^n X_i^2}{n + 2k - 2}$$

(Hint: Since $2\sigma^2 = 1/\tau$, the Bayesian estimator of $2\sigma^2$ is the posterior expectation of $1/\tau$). This problem is based on an example in Lehmann (1983, p. 246).

13.7 Let X have a binomial distribution (n, θ) where θ represents a realization of a random variable Θ with a beta distribution (r, s). Denote the prior mean of Θ by θ_0.

Show that the Bayesian estimate of Θ (minimizing the mean quadratic loss) is a weighted average of the prior mean and the MLE of θ:

$$\hat{\theta}(X) = \frac{n}{r+s+n} \cdot \frac{X}{n} + \frac{r+s}{r+s+n} \cdot \theta_0$$

Note that the Bayesian estimate of Θ is a weighted average of the hypothetical estimates of Θ based on

- Data D alone (i.e., the standard estimator of θ, X/n)
- Prior information of Θ, but no data, θ_0 (i.e., the Bayesian estimator of θ before any observations are taken)

Note that the influence of the prior mean θ_0 tends to zero as $n \to \infty$.

13.8 (Sequential Binomial Sampling) Consider a sequence of binomial trials with success probability θ where the number of trials may depend on the observations. The stopping rule is assumed to be *closed* (see Lehmann, 1983, pp. 93, 243).

A priori θ is assumed to represent a realization of a random variable Θ with a beta distribution (r, s). Let the number of successes, the number of failures, and the total number of trials at the moment when the sampling stops be denoted by X, Y, and N, respectively. Show that the posterior distribution of Θ, given X and Y, is a beta distribution $(r + s, s + n - x)$, and consequently that the Bayesian estimator of θ, given X and Y, is the same, regardless of the closed stopping rule.

14
Reliability Data Sources

14.1 INTRODUCTION

Several types of data are required to model and analyze the reliability of a system. *Technical data* are needed to understand the functions and the functional requirements and to establish a system model. Technical data are usually supplied by the equipment manufacturers. *Operational and environmental data* are necessary to establish component and system models. *Maintenance data*, in the form of procedures, resources, quality, and durations, are necessary to establish the system model and to be able to determine the system availability. Operational, environmental, and maintenance data are plant/system specific and can usually not be found in any data sources. An exception is OREDA (2002) where we can find repair times (man-hours) and downtimes related to the various equipment failure modes. Last but not least, we need various types of *reliability data*. By reliability data we mean information about failure/error modes and time to failure distributions for hardware, software, and humans. When humans are active system operators, we may also need information about their ability to correct failures and restore functions.

This chapter gives a brief survey of some *hardware* reliability data sources, with focus on data sources that are commercially available. A more thorough survey is given by Blischke and Murthy (2000, Section 20.4). A detailed listing of reliability data sources is provided on this book's web page. Please consult this web page for updated Internet addresses and references to the various data sources. In the following we will use the term *database* to denote any type of data source, from a brief data handbook to a comprehensive computerized database.

We start by describing three main categories of hardware reliability databases. We then focus on two commercially available databases, the military handbook MIL-HDBK 217F that provides data for electronic components, and especially the OREDA handbook (OREDA 2002) that provides data for equipment used in offshore oil and gas production and processing. These two databases are selected as examples to highlight the main features of commercially available reliability databases. We end the chapter by discussing some problems related to reliability databases.

14.2 TYPES OF RELIABILITY DATABASES

Hardware reliability databases can generally be classified in three types:

1. Component failure event databases
2. Accident and incident databases
3. Component reliability databases

Each type is briefly described in the following.

14.2.1 Component Failure Event Databases

Many companies are maintaining a component failure event database as part of their computerized maintenance recording system. Failures and maintenance actions are recorded related to the various components. The data are used in maintenance planning and as a basis for system modifications. In some sectors the various companies are exchanging information recorded in their component failure report databases. An example is the Government Industry Data Exchange Program (GIDEP) in the United States.

Some industries have implemented a failure reporting analysis and corrective action system (FRACAS), as described in MIL-STD 2155 and FRACAS (1999). By using FRACAS or similar approaches, failures are formally analyzed and classified before the reports are stored in the failure report database. Several computer programs supporting FRACAS are available.

14.2.2 Accident and Incident Databases

Accidents and incident databases contain information about accidents and near accidents (incidents) within specified categories. The databases are operated by various organizations, consulting companies, and official bodies. Some of the databases are very detailed, while others only contain a brief description of the accident/incident. Component failure information may sometimes be deduced from the accident/incident descriptions. Examples of detailed accident and incident databases include:

- The *Major Accident Reporting System* (MARS) that is operated by the Joint Research Centre in Ispra, Italy, on behalf of the European Union (EU). The MARS database was established to support the EU Seveso (II) directive [96/82/EC] "on the control of major-accident hazards involving dangerous substances." Seveso II plants in Europe have to report all accidents and incidents to the MARS database using a rather detailed format.

- The *Process Safety Incident Database* (PSID) operated by the Center for Chemical Process Safety of the American Institute of Chemical Engineers (AIChE).

In most cases the operator of the database has to actively search for information to store in the database. This is, for example, the case for the World Offshore Accident Database (WOAD) operated by Det Norske Veritas, and the offshore blowout database operated by SINTEF.

Some databases are commercially available; other may require that you are a member of a specific organization or group.

14.2.3 Component Reliability Databases

A wide range of component reliability databases are commercially available. The component reliability databases provide estimates of failure rates for single components. Some databases may also give failure mode distributions and repair times. Databases containing information about manufacturer and make of the various components are usually confidential to people outside a specific company, or a group of companies. An example is the Offshore Reliability Database (OREDA) where the detailed, computerized database is only available to the companies participating in OREDA.

A *generic* component reliability database is a database where the components are classified in broad groups without information about manufacturer, make, and component specifications. An example is the OREDA handbook (OREDA 2002), which contains an extract from the OREDA database. In OREDA (2002) the components are classified as "centrifugal pump; oil processing," "gas turbine; aeroderivative (3000–10000 kW)," and the like.

The failure rate estimates in generic databases may be based on:

1. Recorded failure events
2. Expert judgment
3. Laboratory testing

or, a combination of these.

An important source of reliability data is the Reliability Analysis Center (RAC), which is a U.S. Department of Defense Information Analysis Center managed by Rome Laboratory, New York. The RAC collects, analyzes, and disseminates reliability data for a wide range of items, with a focus on electronic equipment and components. The RAC has also been responsible for the *Military Handbook: Reliability Prediction of Electronic Equipment* (MIL-HDBK 217F). The main data sources from RAC are:

564 RELIABILITY DATA SOURCES

- Electronic Parts Reliability Data (EPRD)
- Nonelectronic Parts Reliability Data (NPRD)
- Nonoperating Reliability Data (NONOP)
- Failure Mode/Mechanism Distributions (FMD)

Another important source of reliability data is the Center for Chemical Process Safety of the AIChE, who operates the Process Equipment Reliability Database (PERD).

14.2.4 Common Cause Failure Data

Analysis of common cause failures requires data that are usually not available in component reliability databases. Very few data sources for common cause failures are available. An example of such a database is the international common cause data exchange (ICDE) program, operated by the Nuclear Energy Agency (NEA) on behalf of nuclear industry authorities in several countries.

14.3 GENERIC RELIABILITY DATABASES

A wide range of generic reliability databases are listed on the book's web page. Several databases have similar formats and give the same type of information. In this section we will describe two different data sources, MIL-HDBK 217F and OREDA, to illustrate the differences in approach. Most of the commercially available reliability data sources are based on the assumption of constant failure rates. Some sources distinguish between different failure modes and present failure rate estimates for each failure mode, while other sources only present a total failure rate covering all failure modes. Some few sources also give estimates of the repair times associated to the various failure modes.

14.3.1 MIL-HDBK 217F

Military handbook, MIL-HDBK 217F, *Reliability Prediction of Electronic Equipment*[1] contains failure rate estimates for the various part types used in electronic systems, such as integrated circuits, transistors, diodes, resistors, capacitors, relays, switches, and connectors. The estimates are mainly based on laboratory testing with controlled environmental stresses. The failure rates in MIL-HDBK 217F are thus only related to component specific (primary) failures. Failures due to external stresses and common cause failures are not included. The handbook gives formulas and data to adjust the failure rate of a component to a specified environment. The data are not related to specific failure modes.

---

[1] The current version of the handbook is MIL-HDBK 217F (Release 2).

MIL-HDBK 217F employs a rather complex method to estimate the failure rate of a component. The method is called the *part stress analysis prediction technique* and is based on detailed stress analysis information as well as environment, quality applications, maximum ratings, complexity, temperature, construction, and a number of other application-related factors. The failure rate estimate has the form

$$\lambda_P = \lambda_B \cdot \pi_Q \cdot \pi_E \cdot \pi_A \cdots$$

where λ_B is the basic failure rate that is estimated from reliability tests performed on components under standard environmental conditions; λ_B is thus given for standardized stresses (e.g., voltage and humidity) and temperature conditions; $\pi_Q, \pi_E, \pi_A, \ldots$ are often called *influence* factors and take into account impact of part quality, equipment environment, application stress, and so on. The values of the basic failure rates and the various factors in the handbook are kept up to date by analysis of failure data on components and systems.

Until the U.S. Department of Defense (DoD) acquisition reform in the mid 1990s, the use of MIL-HDBK 217F was contractually required on many U.S. Government contracts developing electronic systems. Producers of nonmilitary electronic equipment, such as instruments and avionic gear, also often elect to adhere to the handbook, because it offers a convenient and standard way of estimating reliability. Detailed studies have shown that the MIL-HDBK 217F data often are too pessimistic for commercial devices; see, for example, Bodsberg (1987) and O'Connor (2002).

MIL-HDBK 217F also describes a special method for predicting the reliability of a system. The method is called *parts count reliability prediction* and assumes that system success can be achieved only if all the system components are operating, that is, if the system is a series structure. The system failure rate λ_S is obtained by adding the failure rates of the n system components:

$$\lambda_S = \sum_{i=1}^{n} \lambda_i$$

When the system is not a series system, λ_S will give an upper bound of the failure rate. The parts count method has been heavily criticized, for example, by Luthra (1990).

A number of other methods with varying degrees of similarity to MIL-HDBK 217F models have been developed. An example is the Telcordia (previously Bellcore) document *Reliability Prediction Procedure for Electronic Equipment* (SR-332). A wide range of computer programs have been developed to support MIL-HDBK 217F, Telcordia, and similar databases.

Although MIL-HDBK 217F remains an active DoD handbook, it is no longer being actively maintained or updated. The Reliability Analysis Center (RAC), which maintained the handbook on behalf of DoD, is instead promoting a computerized system called Prism.

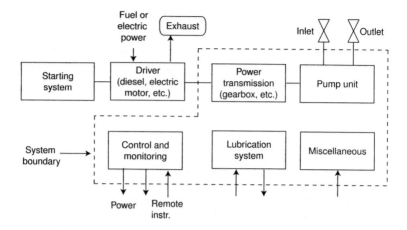

Fig. 14.1 Pumps, boundary definition in OREDA (2002).

14.3.2 OREDA

The Offshore Reliability Data (OREDA) handbooks contain data from a wide range of components and systems used on offshore installations, collected from installations in several geographic areas (see www.oreda.com). Four handbooks have been published in 1984, 1992, 1997, and 2002. OREDA (2002) is based on actual field data collected in the time period 1993 to 2000. The data are classified under the following main headings:

- *Machinery* (compressors, gas turbines, pumps, combustion engines, and turboexpanders)

- *Electric equipment* (electric generators, and motors)

- *Mechanical equipment* (heat exchangers, vessels, heaters, and boilers)

- *control and safety equipment* (Fire and gas detectors, process sensors, and valves)

- *subsea equipment* (Control systems, manifolds, flowlines, risers, wellheads, and X-mas trees)

An important feature in OREDA is the specification (including a drawing) of the physical boundaries of the system. An example of the boundary definition of a pump is shown in Fig. 14.1. The lowest level items in the system hierarchy at which preventive maintenance is carried out are called *maintainable items*. The maintainable items of the pump in Fig. 14.1 are listed in Table 14.1.

An example of how the data is presented in OREDA (2002) is shown in Fig. 14.2, which presents data from 449 ("population") pumps that are installed on 61 ("installations") different offshore platforms. The accumulated (calendar) time in service is $19.0224 \cdot 10^6$ hours, which is approximately 2171.5 pump-years. The accumulated

Table 14.1 Pumps, Subdivision in Maintainable Items in OREDA.

		Pump		
Power Transmission	Pump	Control/Monitoring	Lubrication	Miscellaneous
Gearbox/var. drive	Support	Instruments	Instruments	Purge air
Bearing	Casing	Cabling and	Reservoir	Cooling/heating
Seals	Impeller	boxes	w/heating syst.	system
Lubrication	Shaft	Control unit	Pump w/motor	Filter, cyclone
Coupling to driver	Radial bearing	Actuating device	Filter	Pulsation damper
Coupling to driven	Thrust bearing	Monitoring	Cooler	
unit	Seals	Internal power	Valves/piping	
Instruments	Cylinder liner	supply	Oil	
	Piston	Valves	Seals	
	Diaphragm			
	Instruments			

Reproduced with kind permission of the OREDA Participants.

operational time is $8.6743 \cdot 10^6$ hours, meaning that the pumps, on the average, have been running 45.6% of the time. The pumps have been started up 11,200 times ("no. of demands").

The failure modes are classified in three groups:

1. *Critical*: A failure that causes immediate and complete loss of a system's capability of providing its output.

2. *Degraded*: A failure that is not critical, but that prevents the system from providing its output within specifications. Such a failure would usually, but not necessarily, be gradual or partial, and may develop into a critical failure in time.

3. *Incipient*: A failure that does not immediately cause loss of a system's capability of providing its output, but which, if not attended to, could result in a critical or degraded failure in the near future.

The failure modes are listed on the left-hand side of Fig. 14.2. Only two degraded failure modes and no incipient failure modes are shown in this figure. The remaining failure modes are found on the following pages in the handbook.

The failure rate is estimated for each failure mode together with 90% confidence intervals. The estimate is denoted "mean," while the confidence interval is given by the "lower" and "upper" bounds. The estimates and confidence intervals are presented for both calendar and operational time. The repair time is presented for each failure mode as minimum, mean, and maximum number of man-hours. The average downtime for each failure mode of the pump is presented as "active repair hours."

The failure data are mainly collected from maintenance records. This means that both component specific failures (primary failures) and common cause failures are included. It also implies that spurious failures such as false alarms may not be included in full detail, since such failures do not always require a work order to be corrected. Repair times are recorded whenever possible. For some of the component types, only man-hours were available.

Taxonomy no 1.3		Item Machinery Pumps								
Population	Installations	Aggregated time in service (10⁶ hours)					No of demands			
449	61	Calendar time * 19.0224		Operational time † 8.6743			11200			
Failure mode		No of failures	Failure rate (per 10⁶ hours)				Active rep.hrs	Repair (manhours)		
			Lower	Mean	Upper	SD	n/τ		Min	Mean	Max
Critical		524*	0.00	20.52	108.44	49.34	27.55	37.3	1.0	53.1	1025.0
		524†	1.14	65.40	204.64	72.93	60.41				
Breakdown		45*	0.00	1.27	6.56	5.17	2.37	16.1	3.0	52.5	766.0
		45†	0.01	3.85	15.72	5.95	5.19				
Erratic output		2*	0.00	0.14	0.72	0.58	0.11	19.8	11.0	39.5	68.0
		2†	0.00	0.38	2.00	0.91	0.23				
External leakage - Process medium		86*	0.00	2.38	12.29	9.53	4.52	28.4	2.0	38.3	444.0
		86†	0.00	7.07	33.87	13.94	9.91				
External leakage - Utility medium		46*	0.00	1.20	5.04	5.60	2.42	16.0	2.0	29.8	90.0
		46†	0.00	3.59	16.82	6.84	5.30				
Fail to start on demand		50*	0.01	2.52	9.77	3.62	2.63	52.0	1.0	56.6	551.0
		50†	0.08	13.75	48.28	17.83	5.76				
Fail to stop on demand		2*	0.00	0.10	0.21	0.54	0.11	3.5	3.0	3.5	4.0
		2†	0.00	0.26	1.30	0.56	0.23				
High output		3*	0.00	0.67	3.51	2.44	0.16	-	1.0	3.3	6.0
		3†	0.00	2.31	12.00	5.32	0.35				
Internal leakage		8*	0.00	0.34	1.39	0.52	0.42	95.5	3.0	48.3	188.0
		8†	0.16	0.98	2.37	0.72	0.92				
Low output		46*	0.00	2.50	3.96	15.25	2.42	35.4	3.0	41.2	508.0
		46†	0.00	4.57	13.58	22.90	5.30				
Noise		6*	0.15	0.33	0.56	0.13	0.32	23.3	16.0	60.5	122.0
		6†	0.01	1.03	3.73	1.38	0.69				
Other		8*	0.00	0.57	2.99	2.43	0.42	275.5	2.0	424.5	734.0
		8†	0.00	1.53	7.57	3.21	0.92				
Overheating		5*	0.00	0.27	0.95	0.35	0.26	183.2	3.0	265.0	1025.0
		5†	0.00	6.41	32.56	14.04	0.58				
Parameter deviation		18*	0.00	0.66	3.49	2.31	0.95	11.0	1.0	20.8	88.0
		18†	0.14	1.96	5.66	1.87	2.08				
Spurious stop		133*	0.00	5.69	27.65	11.50	6.99	37.5	1.0	42.1	714.0
		133†	1.57	19.07	53.52	17.47	15.33				
Structural deficiency		33*	0.00	0.41	0.51	4.91	1.73	20.6	5.0	40.5	211.0
		33†	0.00	1.24	3.74	6.18	3.80				
Unknown		1*	0.00	0.05	0.15	0.05	0.05	-	-	-	-
		1†	0.00	0.11	0.33	0.12	0.12				
Vibration		32*	0.00	1.67	7.70	3.10	1.68	81.2	5.0	118.3	896.0
		32†	0.47	5.11	14.03	4.53	3.69				
Degraded		754*	0.00	44.20	210.34	86.32	39.64	20.2	0.3	26.4	798.0
		754†	11.39	238.41	714.72	239.40	86.92				
Abnormal instrument reading		9*	0.00	0.80	4.56	2.45	0.47	9.0	2.0	16.0	65.0
		9†	0.00	2.53	11.22	4.42	1.04				
Erratic output		23*	0.00	2.27	12.50	6.03	1.21	14.8	2.0	16.8	65.0
		23†	0.00	7.88	35.25	13.95	2.65				
Comments											(cont.)

Fig. 14.2 Example of data from OREDA (2002). (Reprinted with the kind permission of the OREDA Participants.)

The OREDA project is still running and is a forum for coordination of reliability data for the oil and gas industry. The detailed data collected during the project is stored in a computerized database that is available to the OREDA Participants. The data in the computerized database is much more detailed that the data presented in OREDA (2002).

14.4 DATA ANALYSIS AND DATA QUALITY

A significant effort has been devoted to the collection and processing of reliability data during the last 20 years. Despite this great effort, the quality of the data available is still not good enough.

The quality of the data presented in the databases obviously depends on the way the data are collected and analyzed. Several guidelines and standards have been issued to obtain high quality in data collection and analysis. Among these are:

1. ISO 14224 *Petroleum and Natural Gas Industries – Collection and Exchange of Reliability and Maintenance Data for Equipment.* This standard may be considered as a spin-off of the OREDA project.

2. *Guidelines for Improving Plant Reliability Through Data Collection and Analysis* (AIChE 1998)

3. *Reliability Data Quality Handbook* (ESReDA)

In the following we briefly discuss problems related to data analysis and reliability databases.

14.4.1 Constant Failure Rates

Almost all commercially available reliability databases provide only constant failure rates, even for mechanical equipment that degrade due to mechanisms like erosion, corrosion, and fatigue. Based on knowledge about the deteriorating mechanisms, the failure rate of such equipment should be increasing. The data available for the analysis is usually the number n of failures during a total time t in service. The failure rate estimated by n/t will thus be an "average failure rate." System failure data are usually collected from a rather limited time period, that may be called the *observation window*.

Assume that the failed components are replaced or restored to an "as good as new" condition, such that we have a renewal process. A number of components are observed during a specified observation window. The observation window may, for example, be from 1 January 2000 till 1 January 2003. In this period we only record the number (n) of failures and the accumulated time (t) in service. A constant failure rate λ is estimated by $\hat{\lambda} = n/t$. If the (real) life distribution is Weibull distribution with an increasing failure rate function, $z(t)$, and we use a constant failure rate estimate, we overestimate the failure rate in the early phase of the component's

570 RELIABILITY DATA SOURCES

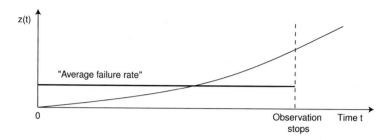

Fig. 14.3 The real failure rate and the erroneously estimated constant failure rate.

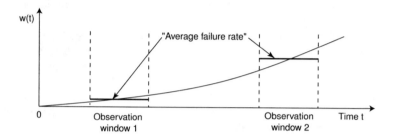

Fig. 14.4 Average failure rates estimated in two different observation windows.

life and underestimate the failure rate in the last part of its life. This is illustrated in Fig. 14.3. The result will especially be wrong if we extrapolate the estimated constant failure rate beyond the time interval where we have collected data.

People who analyze life data are not always aware of the difference between the concepts failure rate function (FOM) and rate of occurrence of failures (ROCOF) as discussed in Chapters 2 and 7. Assume that we have a system with an increasing ROCOF $w(t)$. If we collect failure data in an observation window in an early phase of the system's life, the resulting "average failure rate" is often very different from what we would get in a later observation window. This is illustrated in Fig. 14.4. This effect has been seen in several offshore data collection projects, for example, for downhole safety valves. When a valve has failed, it has been replaced with a new valve of the same type, and we have (erroneously) believed that we had a renewal process. The environmental conditions in the well had, however, changed with time and produced a more hostile environment.

14.4.2 Multiple Samples

The pump failure data in Fig. 14.2 were collected from 449 pumps on 61 different installations. In generic databases, failure rate estimates for generic items are presented. The individual components that are classified within the same generic item do not need to be identical and do not need to be exposed to exactly the same environmental and operational conditions. This is also the case for the pumps in Fig. 14.2. The type

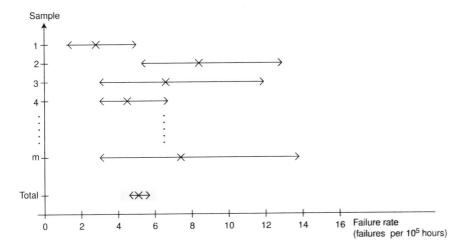

Fig. 14.5 Estimates and confidence intervals for inhomogeneous samples.

of pumps and the operational and environmental conditions may vary within each installation, and especially between the various installations. The data collected will therefore not be a homogeneous sample.

Assume that we have m homogeneous samples of failure data. In sample i we have recorded n_i failures during a total time in operation t_i. The components in the sample are assumed to have constant failure rate λ_i, for $i = 1, 2, \ldots, m$. The failure rate λ_i can be estimated by (11.64):

$$\hat{\lambda}_i = \frac{n_i}{t_i}$$

and a 90% confidence is given by (11.66)

$$\left(\frac{1}{2t_i} z_{0.95, 2n_i} \, , \, \frac{1}{2t_i} z_{0.05, 2(n_i+1)} \right)$$

The estimates and the confidence intervals for the m samples are illustrated in Fig. 14.5. If we (erroneously) assume that all samples have the same failure rate λ, the estimate will be

$$\hat{\lambda} = \frac{\sum_{i=1}^{m} n_i}{\sum_{i=1}^{m} t_i} \qquad (14.1)$$

Since the total number of failures is relatively large, and the total time in operation is relatively long, the confidence interval will be rather short, as illustrated by "total" in Fig. 14.5. It is seen from Fig. 14.5 that the "total" confidence interval does not reflect the uncertainty of the failure rates.

We should therefore carefully check that the samples are homogeneous before we merge the samples. In many databases, the samples are merged without any checking.

In OREDA (2002) an alternative approach is used. The failure rate λ is assumed to be a random variable, that can take different values for the different samples. An estimate of the standard deviation (SD) of the distribution of λ is presented together with the failure rate estimates for each failure mode. A high value of SD indicates that the samples are inhomogeneous. The (average) failure rate is estimated as a weighted average of the failure rate estimates for each sample, following a semi-Bayesian approach. The approach is described in detail in OREDA (2002), and in Lydersen and Rausand (1989). The column n/τ in Fig. 14.2 is estimated by (14.1). A big difference between n/τ and the "mean" failure rate in Fig. 14.2 indicates that the samples are inhomogeneous.

Another approach to handle inhomogeneous samples is presented in Molnes et al. (1986), where failure data from safety valves in oil wells are analyzed. The valves are installed in wells with different characteristics, called *stressors*. The stressors are factors like flowrate, gas/oil ratio, CO_2 content, H_2S content, and sand content. Some main valve characteristics, like diameter and equalizing principle, are also defined as stressors. The failure rate is modeled as a function of the stressors, as proportional hazards models, and analyzed by Cox regression, as mentioned in Chapter 12. In this case we obtain estimates based on a physical modeling of the difference between the samples.

As indicated by the examples presented above, collection and analysis of field data are often difficult tasks, where it is easy to make mistakes. A thorough discussion of reliability databases and problems connected to the collection and analysis of reliability data may be found in Amendola and Keller (1985), Flamm and Luisi (1992), Aupied (1994), and Cooke (1996).

Appendix A
The Gamma and Beta Functions

A.1 THE GAMMA FUNCTION

The gamma function $\Gamma(\alpha)$ is defined for all real $\alpha > 0$ by the integral

$$\Gamma(\alpha) = \int_0^\infty t^{\alpha-1} e^{-t} \, dt \qquad (A.1)$$

By partial integration it is easy to show that

$$\Gamma(\alpha + 1) = \alpha \, \Gamma(\alpha) \quad \text{for all } \alpha > 0 \qquad (A.2)$$

In probabilistic and statistical applications, there is a particular interest for the value of the gamma function when $\alpha = m/2$, where m is a positive integer.

Let k denote a positive integer. Then if $m = 2k$, repeated use of (A.2) leads to

$$\Gamma(k+1) = k \cdot (k-1) \cdots 2 \cdot 1 \cdot \Gamma(1) \qquad (A.3)$$

However

$$\Gamma(1) = \int_0^\infty e^{-t} \, dt = 1$$

Hence
$$\Gamma(k+1) = k! \tag{A.4}$$

Next consider the case $m = (2k+1)/2$. By repeated use of (A.2), it leads to

$$\Gamma\left(\frac{2k+1}{2}\right) = \frac{2k-1}{2} \cdot \frac{2k-3}{2} \cdots \frac{1}{2} \cdot \Gamma\left(\frac{1}{2}\right) \tag{A.5}$$

However,
$$\Gamma\left(\frac{1}{2}\right) = \int_0^\infty t^{-1/2} e^{-t}\, dt$$

By introducing $u = \sqrt{2t}$ as a new variable of integration, we get

$$\int_0^\infty t^{-1/2} e^{-t}\, dt = \sqrt{2} \int_0^\infty e^{-u^2/2}\, du = \frac{1}{\sqrt{2}} \int_{-\infty}^\infty e^{-u^2/2}\, du = \sqrt{\pi}$$

Hence
$$\Gamma\left(\frac{2k+1}{2}\right) = \frac{2k-1}{2} \cdot \frac{2k-3}{2} \cdots \frac{1}{2} \cdot \sqrt{\pi} \tag{A.6}$$

In Table A.1 the Gamma function $\Gamma(\alpha)$ is given for values of α between 1.00 and 2.00.

A.2 THE BETA FUNCTION

The beta function $B(r, s)$ is defined for all real $r > 0, s > 0$ by the integral

$$B(r, s) = \int_0^1 u^{r-1} (1-u)^{s-1}\, du \tag{A.7}$$

It may be shown (e.g., see Cramér 1946, p. 127) that

$$B(r, s) = \frac{\Gamma(r) \cdot \Gamma(s)}{\Gamma(r+s)}$$

Hence
$$f_U(u; r, s) = \frac{\Gamma(r) \cdot \Gamma(s)}{\Gamma(r+s)} u^{r-1} (1-u)^{s-1}\, du$$
$$0 \leq u \leq 1,\ r > 0,\ s > 0 \tag{A.8}$$

may obviously be interpreted as a probability density since $\int_0^1 f_U(u; r, s)\, du = 1$.
A random variable with density (A.8) is said to be beta distributed (r, s). Then

$$E(U) = \frac{r}{r+s} \tag{A.9}$$

$$\text{var}(U) = \frac{rs}{(r+s)^2(r+s+1)} \tag{A.10}$$

Table A.1 Gamma Function $\Gamma(\alpha)$ for α between 1.00 and 2.00.

α	$\Gamma(\alpha)$	α	$\Gamma(\alpha)$	α	$\Gamma(\alpha)$	α	$\Gamma(\alpha)$
1.00	1.00000	1.25	0.90640	1.50	0.88623	1.75	0.91906
1.01	0.99433	1.26	0.90440	1.51	0.88659	1.76	0.92137
1.02	0.98884	1.27	0.90250	1.52	0.88704	1.77	0.92376
1.03	0.98355	1.28	0.90072	1.53	0.88757	1.78	0.92623
1.04	0.97844	1.29	0.89904	1.54	0.88818	1.79	0.92877
1.05	0.97350	1.30	0.89747	1.55	0.88887	1.80	0.93138
1.06	0.96874	1.31	0.89600	1.56	0.88964	1.81	0.93408
1.07	0.96415	1.32	0.89464	1.57	0.89049	1.82	0.93685
1.08	0.95973	1.33	0.89338	1.58	0.89142	1.83	0.93969
1.09	0.95546	1.34	0.89222	1.59	0.89243	1.84	0.94261
1.10	0.95135	1.35	0.89115	1.60	0.89352	1.85	0.94561
1.11	0.94740	1.36	0.89018	1.61	0.89468	1.86	0.94869
1.12	0.94359	1.37	0.88931	1.62	0.89592	1.87	0.95184
1.13	0.93993	1.38	0.88854	1.63	0.89724	1.88	0.95507
1.14	0.93642	1.39	0.88785	1.64	0.89864	1.89	0.95838
1.15	0.93304	1.40	0.88725	1.65	0.90012	1.90	0.96177
1.16	0.92980	1.41	0.88676	1.66	0.90167	1.91	0.96523
1.17	0.92670	1.42	0.88636	1.67	0.90330	1.92	0.96877
1.18	0.92373	1.43	0.88604	1.68	0.90500	1.93	0.97240
1.19	0.92089	1.44	0.88581	1.69	0.90678	1.94	0.97610
1.20	0.91817	1.45	0.88566	1.70	0.90864	1.95	0.97988
1.21	0.91558	1.46	0.88560	1.71	0.91057	1.96	0.98374
1.22	0.91311	1.47	0.88563	1.72	0.91258	1.97	0.98768
1.23	0.91075	1.48	0.88575	1.73	0.91467	1.98	0.99171
1.24	0.90852	1.49	0.88595	1.74	0.91683	1.99	0.99581
						2.00	1.00000

Note: $\Gamma(\alpha)$ for other positive values of α may be calculated from formula (A.2).

The beta density may exhibit a large number of shapes, including the uniform distribution (choose $r = 1$ and $s = 1$), U-shaped densities, unimodal right-skewed, and unimodal left-skewed densities.

Appendix B
Laplace Transforms

Let $f(t)$ be a function that is defined on the interval $(0, \infty)$. The Laplace[1] transform $f^*(s)$ of the function $f(t)$ is defined by

$$f^*(s) = \int_0^\infty e^{-st} f(t)\, dt$$

where s is a real number. In more advanced treatments of the Laplace transform, s is permitted to be a complex number. All functions do not have a Laplace transform. For instance, if $f(t) = e^{t^2}$, the integral diverges for all values of s.

The Laplace transform of $f(t)$ is also often denoted by $\mathcal{L}[f(t)]$:

$$\mathcal{L}[f(t)] = f^*(s) = \int_0^\infty e^{-st} f(t)\, dt \tag{B.1}$$

to indicate the relation between the functions f and f^*. When $f(t)$ is the probability density of a nonnegative random variable T, the Laplace transform of $f(t)$ is seen to be equal to the expected value of the random variable e^{-sT}.

$$E(e^{-sT}) = \int_0^\infty e^{-st} f(t)\, dt = f^*(s)$$

---

[1] Named after the French mathematician Pierre-Simon Laplace (1749–1827).

LAPLACE TRANSFORMS

The function $f(t)$ is called the inverse Laplace transform of $f^*(s)$, and is written

$$f(t) = \mathcal{L}^{-1}[f^*(s)] \tag{B.2}$$

Theorem B.1 Let $f(t)$ be a function that is piecewise continuous on every finite interval in the range $t \geq 0$ and satisfies

$$|f(t)| \leq M e^{\alpha t} \quad \text{for all } t \geq 0$$

and for some constants α and M. Then the Laplace transform of $f(t)$ exists for all $s > \alpha$. □

Example B.1
Consider the function $f(t) = e^{\alpha t}$, where α is a constant. We have

$$\begin{aligned}
f^*(s) &= \int_0^\infty e^{-st} e^{\alpha t} \, dt = \int_0^\infty e^{-t(s-\alpha)} \, dt \\
&= \lim_{\tau \to \infty} \left[\frac{-1}{s-\alpha} e^{-t(s-\alpha)} \right]_0^\tau \\
&= \frac{1}{s-\alpha} \quad \text{when } s > \alpha
\end{aligned}$$

Thus

$$\mathcal{L}[e^{\alpha t}] = \frac{1}{s-\alpha} \quad \text{when } s > \alpha$$

□

Some important properties of the Laplace transform are listed below. The proofs are left to the reader who may consult standard textbooks on mathematical analysis.

1. $\mathcal{L}[f_1(t) + f_2(t)] = \mathcal{L}[f_1(t)] + \mathcal{L}[f_2(t)]$
2. $\mathcal{L}[\alpha f(t)] = \alpha \mathcal{L}[f(t)]$
3. $\mathcal{L}[f(t-\alpha)] = e^{-\alpha s} \mathcal{L}[f(t)]$
4. $\mathcal{L}[e^{\alpha t} f(t)] = f^*(s-\alpha)$
5. $\mathcal{L}[f'(t)] = s\mathcal{L}[f(t)] - f(0)$
6. $\mathcal{L}[\int_0^t f(u) \, du] = \frac{1}{s} \mathcal{L}[f(t)]$
7. $\mathcal{L}[\int_0^t f_1(t-u) f_2(u) \, du] = \mathcal{L}[f_1(t)] \cdot \mathcal{L}[f_2(t)]$
8. $\lim_{s \to \infty} s f^*(s) = \lim_{t \to 0} f(t)$
9. $\lim_{s \to 0} s f^*(s) = \lim_{t \to \infty} f(t)$

Table B.1 Some Laplace Transforms.

$f(t),\ t \geq 0$	$f^*(s) = \mathcal{L}[f(t)]$
1	$\dfrac{1}{s}$
t	$\dfrac{1}{s^2}$
t^2	$\dfrac{2!}{s^3}$
t^n	$\dfrac{n!}{s^{n+1}}$ for $\alpha > -1$ for $n = 0, 1, 2, \ldots$
t^α	$\dfrac{\Gamma(\alpha + 1)}{s^{\alpha+1}}$ for $\alpha > -1$
$e^{\alpha t}$	$\dfrac{1}{s - \alpha}$
$e^{\alpha t} t^n$	$\dfrac{n!}{(s - \alpha)^{n+1}}$
$\cos \omega t$	$\dfrac{s}{s^2 + \omega^2}$
$\sin \omega t$	$\dfrac{\omega}{s^2 + \omega^2}$
$\cosh \alpha t$	$\dfrac{s}{s^2 - \alpha^2}$
$\sinh \alpha t$	$\dfrac{\alpha}{s^2 - \alpha^2}$

Appendix C
Kronecker Products

Let $\mathbb{A}$ be a matrix with dimension $m_A \times n_A$, and $\mathbb{B}$ be a matrix with dimension $m_B \times n_B$. The *Kronecker product*[1] of $\mathbb{A}$ and $\mathbb{B}$, written $\mathbb{A} \otimes \mathbb{B}$ is defined as

$$\mathbb{A} \otimes \mathbb{B} = \begin{pmatrix} a_{11}\mathbb{B} & a_{12}\mathbb{B} & \cdots & a_{1n_A}\mathbb{B} \\ a_{21}\mathbb{B} & a_{22}\mathbb{B} & \cdots & a_{2n_A}\mathbb{B} \\ \vdots & \vdots & \ddots & \vdots \\ a_{m_A1}\mathbb{B} & a_{m_A2}\mathbb{B} & \cdots & a_{m_An_A}\mathbb{B} \end{pmatrix}$$

The Kronecker product is also known as the Zehfuss product and the tensor product.

A *Kronecker sum* is an ordinary sum of Kronecker products. The Kronecker sum, written $\mathbb{A} \oplus \mathbb{B}$, is defined for square matrices $\mathbb{A}$ and $\mathbb{B}$ as

$$\mathbb{A} \oplus \mathbb{B} = \mathbb{A} \otimes \mathbb{I}_{n_B} + \mathbb{I}_{n_A} \otimes \mathbb{B}$$

where n_A is the size of the square matrix $\mathbb{A}$, n_B is the size of the square matrix $\mathbb{B}$, and $\mathbb{I}$ is the identity matrix.

Some Properties of the Kronecker Product

---

[1] Named after the German/Polish mathematician Leopold Kronecker (1823–1891).

1. Associativity:

$$\mathbb{A} \otimes (\mathbb{B} \otimes \mathbb{C}) = (\mathbb{A} \otimes \mathbb{B}) \otimes \mathbb{C}$$

2. Distributivity over ordinary matrix addition:

$$(\mathbb{A} + \mathbb{B}) \otimes (\mathbb{C} + \mathbb{D}) = \mathbb{A} \otimes \mathbb{C} + \mathbb{B} \otimes \mathbb{C} + \mathbb{A} \otimes \mathbb{D} + \mathbb{B} \otimes \mathbb{D}$$

3. Compatibility with ordinary matrix multiplication:

$$\mathbb{AB} \otimes \mathbb{CD} = (\mathbb{A} \otimes \mathbb{C})(\mathbb{B} \otimes \mathbb{D})$$

4. Compatibility with ordinary matrix inversion:

$$(\mathbb{A} \otimes \mathbb{B})^{-1} = \mathbb{A}^{-1} \otimes \mathbb{B}^{-1}$$

Further details about the Kronecker product may be found in Graham (1981). Kronecker products are available in many mathematical tools, like MATLAB and Maple.

Appendix D
Distribution Theorems

First we refer (without proof) two important theorems from distribution theory:

Theorem D.1 Let X be a continuous random variable with probability density $f_X(x)$ and sample space S_X. Furthermore let $a(x)$ be strictly monotonous in x and differentiable with respect to x for all x. Then $y = a(x)$ is a one-to-one transformation on x with inverse $x = b(y)$ which maps S_X into S_Y, and the density of $Y = a(X)$ is given by

$$f_Y(y) = f_X(b(y)) |b'(y)| \tag{D.1}$$

□

The extension of this theorem to multivariate distributions is given in the next theorem.

Theorem D.2 Let $X_1, X_2, \ldots, X_n$ be continuously distributed with joint probability density $f_{X_1, X_2, \ldots, X_n}(x_1, x_2, \ldots, x_n)$ and sample space $S_{X_1, X_2, \ldots, X_n}$. If

$$y_i = a_i(x_1, x_2, \ldots, x_n) \quad \text{for } i = 1, 2, \ldots, n \tag{D.2}$$

is a one-to-one transformation on the $x's$ with inverse

$$x_i = b_i(y_1, y_2, \ldots, y_n) \quad \text{for } i = 1, 2, \ldots, n \tag{D.3}$$

that maps the sample space $S_{X_1, X_2, \ldots, X_n}$ into $S_{Y_1, Y_2, \ldots, Y_n}$, then the joint density of

$$Y_i = a_i(X_1, X_2, \ldots, X_n) \quad \text{for } i = 1, 2, \ldots, n$$

is given by

$$f_{Y_1,Y_2,\ldots,Y_n}(y_1, y_2, \ldots, y_n)$$
$$= f_{X_1,X_2,\ldots,X_n}(b_1(y_1, y_2, \ldots, y_n), \ldots, b_n(y_1, y_2, \ldots, y_n)) \cdot |J| \quad \text{(D.4)}$$

where J denotes the Jacobian

$$J = \begin{vmatrix} \dfrac{\partial b_1(y_1, y_2, \ldots, y_n)}{\partial y_1} & \cdots & \dfrac{\partial b_1(y_1, y_2, \ldots, y_n)}{\partial y_n} \\ \vdots & & \vdots \\ \dfrac{\partial b_n(y_1, y_2, \ldots, y_n)}{\partial y_1} & \cdots & \dfrac{\partial b_n(y_1, y_2, \ldots, y_n)}{\partial y_n} \end{vmatrix} \quad \text{(D.5)}$$

□

We will now use Theorems D.1 and D.2 to derive some useful theorems.

Theorem D.3 Let $X_1, X_2, \ldots, X_n$ be independently and identically distributed with common distribution function $F_X(x)$ and probability density $f_X(x)$. Denote the observations ordered according to magnitude by $X_{(1)} < X_{(2)} < \cdots < X_{(r)}$, and let r be any integer such that $r \leq n$. Then the joint density of $X_{(1)}, X_{(2)}, \ldots, X_{(n)}$ is given by

$$f_{X_{(1)},X_{(2)},\ldots,X_{(r)}}(x_1, x_2, \ldots, x_r) = \frac{n!}{(n-r)!} [1 - F_X(x_r)]^{n-r} \cdot \prod_{i=1}^{r} f_X(x_i)$$
$$\text{for } 0 < x_1 < x_2 < \cdots < x_r \quad \text{(D.6)}$$

□

The proof may be based on a multinomial argument.

Theorem D.4 Let $X_1, X_2, \ldots, X_n$ be independently and identically distributed with density $f_X(x) = \lambda e^{-\lambda x}$ for $x > 0$, $\lambda > 0$. Let the corresponding order statistic be denoted by $X_{(1)} < X_{(2)} < \cdots < X_{(n)}$. Now introduce the random variables

$$D_j = X_{(j)} - X_{(j-1)} \text{ for } j = 1, 2, \ldots, n. \; X_{(0)} \stackrel{\text{def}}{=} 0 \quad \text{(D.7)}$$

Then

1. $D_1, D_2, \ldots, D_n$ are independent random variables.
2. D_j is exponentially distributed with parameter $(n-j+1)\lambda$ for $j = 1, 2, \ldots, n$.
3. $D_j^* = (n - j + 1)D_j$ for $j = 1, 2, \ldots, n$ are exponentially distributed with parameter λ.

Proof. According to Theorem D.3

$$f_{X_{(1)},X_{(2)},\ldots,X_{(n)}}(x_1, x_2, \ldots, x_n) = n! \lambda^n e^{-\lambda \sum_{i=1}^{n} x_i} \quad \text{(D.8)}$$
$$\text{for } 0 < x_1 < x_2 < \cdots < x_n$$

Furthermore the mapping $D_j = X_{(j)} - X_{(j-1)}$ for $j = 1, 2, \ldots, n$ is one-to-one, since $X_{(j)} = \sum_{i=1}^{j} D_i$ for $j = 1, 2, \ldots, n$.

The Jacobian of this mapping is easily shown to be 1. Furthermore

$$\sum_{j=1}^{n} X_j = \sum_{j=1}^{n} X_{(j)} = nD_1 + (n-1)D_2 + \cdots + (n-j+1)D_j + \cdots + D_n$$

$$= \sum_{j=1}^{n} (n-j+1)D_j$$

Then according to Theorem D.2

$$f_{D_1, D_2, \ldots, D_n}(d_1, d_2, \ldots, d_n) = n! \, \lambda^n \, e^{-\lambda \sum_{j=1}^{n}(n-j+1)d_j}$$

$$= \prod_{j=1}^{n} [(n-j+1)\lambda \, e^{-\lambda(n-j+1)d_j}] = \prod_{j=1}^{n} f_{D_j}(d_j)$$

for $d_j > 0$ and $j = 1, 2, \ldots, n$

where we have introduced

$$(n-j+1)\lambda \, e^{-\lambda(n-j+1)d_j} = f_{D_j}(d_j)$$

Hence properties 1 and 2 of Theorem D.4 are proved. Property 3 follows directly, using the transform

$$D_j^* = (n-j+1)D_j \quad \text{for } j = 1, 2, \ldots, n$$

The D_j^* for $j = 1, 2, \ldots, n$ are sometimes called the *normed time differences* of the process. □

Appendix E
Maximum Likelihood Estimation

An important general method for constructing estimators is based on the *maximum likelihood* principle. Although as early as 1821 the German mathematician C. F. Gauss was first to apply the idea, the method is usually credited to the English statistician R. A. Fisher, who introduced the idea in 1921 in a short paper, and later on, in a series of papers, investigated the properties of the estimators, so obtained. Here we shall content ourselves with giving a short presentation of the method and restrict ourselves to regular parametric models.

To fix the idea, let $X_1, X_2, \ldots, X_n$ denote n independent, identically distributed random variables with density (alternatively frequency function) $f(x; \theta_1, \theta_2, \ldots, \theta_m)$, where f is of known form and $\boldsymbol{\theta} = (\theta_1, \theta_2, \ldots, \theta_m)$ belongs to a subset Θ of m-dimensional space, but otherwise is unknown. $X_1, X_2, \ldots, X_n$ may, for example, represent the lifetimes of n identical units of some kind.

Let us first consider the case of *no censoring*. Then the joint density (frequency function) of $X_1, X_2, \ldots, X_n$ is given by $\prod_{i=1}^{n} f(x_i; \boldsymbol{\theta})$.

Now consider this expression as a function of $\boldsymbol{\theta}$ for fixed $x_1, x_2, \ldots, x_n$ and denote this function

$$L(\boldsymbol{\theta}; x_1, x_2, \ldots, x_n) = L(\boldsymbol{\theta}; \boldsymbol{x}) \tag{E.1}$$

Then $L(\theta; x)$ is called the *likelihood function*. If X has a discrete distribution, $L(\theta; x)$ directly expresses the probability of observing the values $x_1, x_2, \ldots, x_n$ for a given θ and hence indicates "how likely" it is to obtain the observations $x_1, x_2, \ldots, x_n$ for each given θ. A similar argument applies in the case where X has a continuous distribution.

The idea of the maximum likelihood principle is now as follows:
Search for the value $\hat{\theta}(x_1, x_2, \ldots, x_n)$ which is most likely to have produced the observations $x_1, x_2, \ldots, x_n$,

$$L(\hat{\theta}(x); x) \geq L(\theta; x) \quad \text{for } \theta \in \Theta \tag{E.2}$$

$\hat{\theta}(X_1, X_2, \ldots, X_n)$ is then called the *maximum likelihood estimator* of θ (MLE of θ).

In the case of *censored* observations, the likelihood function is somewhat modified. Suppose some of the observations are left censored and some are right censored. Then we can split the observation numbers, $1, 2, \ldots, n$, into three disjoint sets, U, say, corresponding to the uncensored observations, C_R corresponding to right censored and C_L corresponding to left censored observations. In this situation, the likelihood function is defined as

$$L(\theta; x) = \prod_{j \in C_L} F(x_j; \theta) \prod_{j \in U} f(x_j; \theta) \prod_{j \in C_R} R(x_j; \theta) \tag{E.3}$$

Thus the modification of the likelihood function is as follows: For the left censored observations, replace the corresponding densities by the distribution function (F), for the right censored observations, replace the densities by the survivor function $(R = 1 - F)$.

Since $\ln L(\theta; x)$ attains its maximum for the same value of θ as does $L(\theta; x)$, $\hat{\theta}(x)$ may also be found from

$$\ln L(\hat{\theta}(x); x) \geq \ln L(\theta; x) \quad \text{for } \theta \in \Theta \tag{E.4}$$

Usually it is more convenient from a mathematical point of view to work with $\ln L(\theta; x)$ than with $L(\theta; x)$.

In commonly occurring situations, the first step in the search for the MLE of θ, is to solve the likelihood equations

$$U_j = \frac{\partial \ln L(\theta; x)}{\partial \theta_j} = 0 \quad \text{for } j = 1, 2, \ldots, m \tag{E.5}$$

This approach will frequently involve numerical methods such as Newton and quasi-Newton algorithms for solving equations.

Suppose that we succeed in determining the MLE $\hat{\theta}_1, \hat{\theta}_2, \ldots, \hat{\theta}_m$ of $\theta_1, \theta_2, \ldots, \theta_m$, and that we are interested in estimating some function $\psi(\theta) = g(\theta_1, \theta_2, \ldots, \theta_m)$, where g is a specified one-to-one function. Then the MLE $\hat{\psi}$ of ψ will be given by $g(\hat{\theta}_1, \hat{\theta}_2, \ldots, \hat{\theta}_m)$.

Under mild regularity conditions, $\hat{\theta}$ is a consistent estimator of θ. Asymptotic properties of $\hat{\theta}$ are discussed in several textbooks in statistics, for example, by Mann et al. (1974), Kalbfleish and Prentice (1980), Lawless (1982), and Ansell and Phillips (1994). The reader is referred to one of these.

Example E.1

Let $T_1, T_2, \ldots, T_k$ be independent and identically distributed with an inverse Gaussian distribution IG(μ, λ), where $\mu > 0$ and $\lambda > 0$, but otherwise unknown.

The ML estimators μ^* and λ^* for μ and λ may be derived in the following way: The likelihood function becomes in this case

$$L(\mu, \lambda; t_1, t_2, \ldots, t_k) = \lambda^{k/2} \prod_{j=1}^{k} \sqrt{\frac{1}{2\pi t_j^3}} \exp\left(-\frac{\lambda}{2} \frac{(t_j - \mu)^2}{\mu^2 t_j}\right)$$

and

$$\ln L(\mu, \lambda; t_1, t_2, \ldots, t_k) = \frac{k}{2} \ln \lambda - \frac{1}{2} \sum_{j=1}^{k} \ln 2\pi t_j^3 - \frac{\lambda}{2} \sum_{j=1}^{k} \frac{(t_j - \mu)^2}{\mu^2 t_j}$$

The ML estimates of μ and λ, μ^* and λ^* may now be determined as the solutions of the equations:

$$\frac{\partial \ln L}{\partial \mu} = 0 \tag{E.6}$$

and

$$\frac{\partial \ln L}{\partial \lambda} = 0 \tag{E.7}$$

Equation (E.6) becomes

$$-\frac{\lambda}{2} \sum_{j=1}^{k} \frac{-2(t_j - \mu)\mu^2 t_j - 2\mu t_j (t_j - \mu)^2}{\mu^4 t_j^2} = 0$$

$$\sum_{j=1}^{k} \frac{-2\mu^2 t_j^2 + 2\mu^3 t_j - 2\mu t_j^3 + 4\mu^2 t_j^2 - 2\mu^3 t_j}{\mu^4 t_j} = 0$$

that is,

$$\sum_{j=1}^{k} t_j = k\mu$$

Hence

$$\mu^* = \frac{1}{k} \sum_{j=1}^{k} t_j = \bar{t}$$

We next introduce $\mu = \bar{t}$ into (E.7) and differentiate with respect to λ. Then (E.7) becomes

$$\frac{k}{2} \cdot \frac{1}{\lambda} - \frac{1}{2} \sum_{j=1}^{k} \frac{(t_j - \bar{t})^2}{\bar{t}^2 t_j} = 0$$

that is,

$$\frac{1}{\lambda} = \frac{1}{k} \sum_{j=1}^{k} \frac{(t_j - \bar{t})^2}{\bar{t}^2 t_j} = \frac{1}{k} \sum_{j=1}^{k} \left(\frac{1}{t_j} - \frac{1}{\bar{t}} \right)$$

Hence λ^* is given by

$$\frac{1}{\lambda^*} = \frac{1}{k} \sum_{j=1}^{k} \left(\frac{1}{t_j} - \frac{1}{\bar{t}} \right)$$

It may be verified that the likelihood function attains its maximum for $\mu = \mu^*$ and $\lambda = \lambda^*$. Hence μ^* and λ^* are the MLE of μ and λ. □

In some cases the likelihood equations (E.5) are nonlinear and have to be iteratively solved by use of a computer.

Appendix F
Statistical Tables

Table F.1 The Cumulative Standard Normal Distribution
$$\Phi(z) = P(Z \le z) = \int_{-\infty}^{z} \frac{1}{\sqrt{2\pi}} e^{-u^2/2} \, du$$

z	0.00	0.01	0.02	0.03	0.04	0.05	0.06	0.07	0.08	0.09
0.0	.500	.504	.508	.512	.516	.520	.524	.528	.532	.536
0.1	.540	.544	.548	.552	.556	.560	.564	.567	.571	.575
0.2	.579	.583	.587	.591	.595	.599	.603	.606	.610	.614
0.3	.618	.622	.626	.629	.633	.637	.641	.644	.648	.652
0.4	.655	.659	.663	.666	.670	.674	.677	.681	.684	.688
0.5	.691	.695	.698	.702	.705	.709	.712	.716	.719	.722
0.6	.726	.729	.732	.736	.739	.742	.745	.749	.752	.755
0.7	.758	.761	.764	.767	.770	.773	.776	.779	.782	.785
0.8	.788	.791	.794	.797	.800	.802	.805	.808	.811	.813
0.9	.816	.819	.821	.824	.826	.829	.831	.834	.836	.839
1.0	.841	.844	.846	.849	.851	.853	.855	.858	.860	.862
1.1	.864	.867	.869	.871	.873	.875	.877	.879	.881	.883
1.2	.885	.887	.889	.891	.893	.894	.896	.898	.900	.901
1.3	.903	.905	.907	.908	.910	.911	.913	.915	.916	.918
1.4	.919	.921	.922	.924	.925	.926	.928	.929	.931	.932
1.5	.933	.934	.936	.937	.938	.939	.941	.942	.943	.944
1.6	.945	.946	.947	.948	.949	.951	.952	.953	.954	.954
1.7	.955	.956	.957	.958	.959	.960	.961	.962	.962	.963
1.8	.964	.965	.966	.966	.967	.968	.969	.969	.970	.971
1.9	.971	.972	.973	.973	.974	.974	.975	.976	.976	.977
2.0	.977	.978	.978	.979	.979	.980	.980	.981	.981	.982
2.1	.982	.983	.983	.983	.984	.984	.985	.985	.985	.986
2.2	.986	.986	.987	.987	.987	.988	.988	.988	.989	.989
2.3	.989	.990	.990	.990	.990	.991	.991	.991	.991	.992
2.4	.992	.992	.992	.992	.993	.993	.993	.993	.993	.994
2.5	.994	.994	.994	.994	.994	.995	.995	.995	.995	.995
2.6	.995	.995	.996	.996	.996	.996	.996	.996	.996	.996
2.7	.997	.997	.997	.997	.997	.997	.997	.997	.997	.997
2.8	.997	.998	.998	.998	.998	.998	.998	.998	.998	.998
2.9	.998	.998	.998	.998	.998	.998	.999	.999	.999	.999
3.0	.999	.999	.999	.999	.999	.999	.999	.999	.999	.999

$\Phi(-z) = 1 - \Phi(z)$

Table F.2 Percentage Points of the Chi-Square (χ^2) Distribution
$P(Z > z_{\alpha,\nu}) = \alpha$

$\nu \backslash \alpha$	0.995	0.990	0.975	0.950	0.10	0.05	0.025	0.010	0.005
1	0.00	0.00	0.00	0.00	2.71	3.84	5.02	6.63	7.88
2	0.01	0.02	0.05	0.10	4.61	5.99	7.38	9.21	10.60
3	0.07	0.11	0.22	0.35	6.25	7.81	9.35	11.34	12.84
4	0.21	0.30	0.48	0.71	7.78	9.49	11.14	13.28	14.86
5	0.41	0.55	0.83	1.15	9.24	11.07	12.38	15.09	16.75
6	0.68	0.87	1.24	1.64	10.64	12.59	14.45	16.81	18.55
7	0.99	1.24	1.69	2.17	12.02	14.07	16.01	18.48	20.28
8	1.34	1.65	2.18	2.73	13.36	15.51	17.53	20.09	21.96
9	1.73	2.09	2.70	3.33	14.68	16.92	19.02	21.67	23.59
10	2.16	2.56	3.25	3.94	15.99	18.31	20.48	23.21	25.19
11	2.60	3.05	3.82	4.57	17.28	19.68	21.92	24.72	26.76
12	3.07	3.57	4.40	5.23	18.55	21.03	23.34	26.22	28.30
13	3.57	4.11	5.01	5.89	19.81	22.36	24.74	27.69	29.82
14	4.07	4.66	5.63	6.57	21.06	23.68	26.12	29.14	31.32
15	4.60	5.23	6.27	7.26	22.31	25.00	27.49	30.58	32.80
16	5.14	5.81	6.91	7.96	23.54	26.30	28.85	32.00	34.27
17	5.70	6.41	7.56	8.67	24.77	27.59	30.19	33.41	35.72
18	6.26	7.01	8.23	9.39	25.99	28.87	31.53	34.81	37.16
19	6.84	7.63	8.91	10.12	27.20	30.14	32.85	36.19	38.58
20	7.43	8.26	9.59	10.85	28.41	31.41	34.17	37.57	40.00
21	8.03	8.90	10.28	11.59	29.62	32.67	35.48	38.93	41.40
22	8.64	9.54	10.98	12.34	30.81	33.92	36.78	40.29	42.80
23	9.26	10.20	11.69	13.09	32.01	35.17	38.08	41.64	44.18
24	9.89	10.86	12.40	13.85	33.20	36.42	39.36	42.98	45.56
25	10.52	11.52	13.12	14.61	34.38	37.65	40.65	44.31	46.93
26	11.16	12.20	13.84	15.38	35.56	38.89	41.92	45.64	48.29
27	11.81	12.88	14.57	16.15	36.74	40.11	43.19	46.96	49.64
28	12.46	13.56	15.31	16.93	37.92	41.34	44.46	48.28	50.99
29	13.12	14.26	16.05	17.71	39.09	42.56	45.72	49.59	52.34
30	13.79	14.95	16.79	18.49	40.26	43.77	46.98	50.89	53.67
40	20.71	22.16	24.43	26.51	51.81	55.76	59.34	63.69	66.77
50	27.99	29.71	32.36	34.76	63.17	67.50	71.42	76.15	79.49
60	35.53	37.48	40.48	43.19	74.40	79.08	83.30	88.38	91.95
70	43.28	45.44	48.76	51.74	85.53	90.53	95.02	100.42	104.22
80	51.17	53.54	57.15	60.39	96.58	101.88	106.63	112.33	116.32
90	59.20	61.75	65.65	69.13	107.6	113.14	118.14	124.12	128.30
100	67.33	70.06	74.22	77.93	118.5	124.34	129.56	135.81	140.17

Acronyms

AIChE	American Institute of Chemical Engineers
ALT	accelerated life testing
ARI	arithmetic reduction of intensity
BBN	Bayesian belief network
CBM	condition based maintenance
CM	corrective maintenance
DD	dangerous detected (failure)
DFR	decreasing failure fate
DFRA	decreasing failure rate average
DHSV	downhole safety valve
DU	dangerous undetected (failure)
EIReDA	European industry reliability data
EN	European norm
ESD	emergency shutdown
ESDV	emergency shutdown valve
EUC	equipment under control
FAST	function analysis system technique

FFA	functional failure analysis
FMEA	failure modes and effects analysis
FMECA	failure modes, effects, and criticality analysis
FOM	force of mortality
FRACAS	failure reporting analysis and corrective action system
FSI	functional significant item
FTA	fault tree analysis
FTF	fail to function
GIDEP	Government Industry Data Exchange Program
HAZOP	hazard and operability study
HPP	homogeneous Poisson process
IDEF	integrated definition language
IEC	International Electrotechnical Commission
IEEE	Institute of Electrical and Electronic Engineers
IFR	increasing failure rate
IFRA	increasing failure rate average
i.i.d.	independent and identically distributed
ISO	International Organization for Standardization
LCC	life cycle cost
LCP	life cycle profit
LOPA	layer of protection analysis
MCSI	maintenance cost significant item
MDT	mean downtime
MFDT	mean fractional deadtime
MLE	maximum likelihood estimator
MRL	mean residual life
MSI	maintenance significant item
MTBF	mean time between failures
MTBM	mean time between maintenance
MTBR	mean time between replacements/renewals
MTTF	mean time to failure
MTTR	mean time to repair
MUT	mean up-time
NBU	new better than used

NBUE	new better than used in expectation
NHPP	nonhomogeneous Poisson process
NTNU	Norwegian University of Science and Technology
NUREG	title of reports from U.S. NRC (Nuclear Regulatory Commission)
NWU	new worse than used
OEE	overall equipment efficiency
OREDA	Offshore Reliability Data
PALT	progressive stress accelerated test
PFD	probability of failure on demand
PM	preventive maintenance
QRA	quantitative risk analysis
RAM	reliability, availability, and maintainability
RAMS	reliability, availability, maintainability, and safety
RBD	reliability block diagram
RCM	reliability centered maintenance
ROCOF	rate of occurrence of failures
SADT	system analysis and design technique
SAE	Engineering Society for Advancing Mobility in Land Sea Air and Space
SALT	step-stress accelerated test
SIF	safety instrumented function
SIL	safety integrity level
SINTEF	Foundation of Science and Technology at the Norwegian Institute of Technology
SIS	safety instrumented system
TPM	total productive maintenance
TQM	total quality management

Glossary

Accelerated test   A test in which the applied stress level is chosen to exceed that stated in the reference conditions in order to shorten the time required to observe the stress response of the item, or magnify the responses in a given time. To be valid, an accelerated test shall not alter the basic modes and mechanisms of failure or their relative prevalence (BS 4778).

Accident   An unintended event or sequence of events that causes death, injury, environmental, or material damage (DEF-STAN 00-56)

Active redundancy   That redundancy wherein all means for performing a required function are intended to operate simultaneously [IEC 50(191)].

Availability   The ability of an item (under combined aspects of its reliability, maintainability, and maintenance support) to perform its required function at a stated instant of time or over a stated period of time (BS 4778).

Basic event   The bottom or "leaf" events of a fault tree. The limit of resolution of the fault tree. Examples of basic events are component failures and human errors (NASA 2002).

Coherent system   A system whose structure function is nondecreasing as a function of all state variables. *Note:* A system is coherent when:
 * if all the components are in a failed state, the system is in a failed state,
 * if all the components are functioning, the system is functioning,
 * when the system is in a failed state, no additional component failures will cause the system to function,
 * when the system is functioning, no component repair will cause the system to fail.

Common cause failure Multiple component faults that occur at the same time or that occur in a relatively small time window and that are due to a common cause (NASA 2002).

Failure, which is the result of one or more events, causing coincident failures of two or more separate channels in a multiple channel system, leading to system failure (IEC 61508, Part 4).

Corrective maintenance The actions performed, as a result of failure, to restore an item to a specified condition [MIL-STD-2173(AS)].

The maintenance carried out after a failure has occurred and intended to restore an item to a state in which it can perform its required function (BS 4778).

Dependability The collective term used to describe the availability performance and its influencing factors: reliability performance, maintainability performance, and maintenance support performance (IEC 60300).

Design review A formal, documented, comprehensive, and systematic examination of a design to evaluate the design requirements and the capability of the design to meet these requirements and to identify problems and propose solutions (ISO 8402).

Distribution function Consider a random variable X. The distribution function of X is

$$F_X(x) = \Pr(X \leq x)$$

Downtime The period of time during which an item is not in a condition to perform its required function (BS 4778).

Equipment under control (EUC) Equipment, machinery, apparatus, or plant used for manufacturing, process, transportation, medical, or other activities (IEC 61508, Part 4).

Fail safe A design property of an item that prevents its failures being critical failures (BS 4778).

A design feature that ensures the system remains safe or, in the event of a failure, causes the system to revert to a state that will not cause a mishap (MIL-STD 882D).

Failure The termination of its ability to perform a required function (BS 4778).

An unacceptable deviation from the design tolerance or in the anticipated delivered service, an incorrect output, the incapacity to perform the desired function (NASA 2002).

A cessation of proper function or performance; inability to meet a standard; non-performance of what is requested or expected (NASA 2000).

Failure cause The physical or chemical processes, design defects, quality defects, part misapplication, or other processes which are the basic reason for failure or which initiate the physical process by which deterioration proceeds to failure (MIL-STD-1629A).

The circumstances during design, manufacture, or use which have led to a failure [IEC 50(191)].

Failure effect The consequence(s) a failure mode has on the operation, function, or status of an item (MIL-STD-1629).

Failure mode The effect by which a failure is observed on the failed item (EuReDatA, 1983).

Failure mode and effect analysis (FMEA) A procedure by which each potential failure mode in a system is analyzed to determine the results or effects thereof on the system and to classify each potential failure mode according to its severity (MIL-STD-1629A).

Failure rate The rate at which failures occur as a function of time. If T denotes the time to failure of an item, the failure rate $z(t)$ is defined as

$$z(t) = \lim_{\Delta t \to \infty} \frac{\Pr(t < T \leq t + \Delta t \mid T > t)}{\Delta t}$$

The failure rate is sometimes called "force of mortality (FOM)."

Failure symptom An identifiable physical condition by which a potential failure can be recognized [MIL-STD-2173(A)].

Fatigue Reduction in resistance to failure of a material over time, as a result of repeated or cyclic applied loads [MIL-STD-2173(AS)].

Fatigue life For an item subject to fatigue, the total to functional failure failure of the item [MIL-STD-2173(AS)].

Fault A defect, imperfection, mistake, or flaw of varying severity that occurs within some hardware or software component or system. "Fault" is a general term and can range from a minor defect to a failure (NASA 2002).

Abnormal condition that may cause a reduction in, or loss of, the capability of a functional unit to perform a required function (IEC 61508, Part 4).

Fault tolerance Ability of a functional unit to continue to perform a required function in the presence of faults or errors (IEC 61508, Part 4).

Force of mortality Same as "Failure rate."

Functional unit Entity of hardware or software, or both, capable of accomplishing a specified purpose (IEC 61508, Part 4).

Functioning state The state when an item is performing a required function (see also "Operating state").

Gradual failure Failure that could be anticipated by prior examination or monitoring (BS 4778).

Hazard rate Same as "Failure rate".

Hidden failure A failure not evident to the crew or operator during the performance of normal duties [MIL-STD-2173(AS)].

Infant mortality The relatively high conditional probability of failure during the period immediately after an item enters service. Such failures are due to defects in manufacturing not detected by quality control [MIL-STD-2173(AS)].

Inspection Activities such as measuring, examining, testing, gauging one or more characteristics of a product or service, and comparing these with specified requirements to determine conformity (ISO 8402).

Intermittent failure Failure of an item for a limited period of time, following which the item restores its required function without being subjected to any external corrective action (BS 4778).

Life cycle cost (LCC) The cost of acquisition and ownership of a product over a defined period of its life cycle. It may include the cost of development, acquisition, operation, support, and disposal of the product (IEC 60300).

Maintainability The ability of an item, under stated conditions of use, to be retained in, or restored to, a state in which it can perform its required functions, when maintenance is performed under stated conditions and using prescribed procedures and resources (BS 4778).

Maintenance The combinations of all technical and corresponding administrative actions, including supervision actions, intended to retain an entity in, or restore it to, a state in which it can perform its required function [IEC 50(191)].

Maintenance support performance The ability of a maintenance organization, under given conditions, to provide upon demand, the resources required to maintain an entity, under a given maintenance policy [IEC 50(191)].

Mean time to failure (MTTF) Let T denote the time to failure of an item, with probability density $f(t)$ and survivor function $R(t)$. The mean time to failure is the mean (expected) value of T which is given by

$$\text{MTTF} = \int_0^\infty t \cdot f(t)\, dt = \int_0^\infty R(t)\, dt$$

Mean time to repair (MTTR) Let D denote the down time (or repair time) after a failure of an item. Let $f_D(d)$ denote the probability density of D, and let $F_D(d)$ denote the distribution function of D. The mean time to repair is the mean (expected) value of D which is given by

$$\text{MTTR} = \int_0^\infty t \cdot f_D(t)\, dt = \int_0^\infty (1 - F_D(t))\, dt$$

MTTR is sometimes called the *mean down time (MDT)* of the item. In some situations MTTR is used to denote the mean *active* repair time instead of the mean down time of the item.

Operating state The state when an entity is performing a required function [IEC 50(191)].

Percentile Let X be a random variable with distribution function $F(x)$. The upper $100\epsilon\%$ percentile x_ϵ of the distribution $F(x)$ is defined such that

$$\Pr(X > x_\epsilon) = \epsilon$$

Preventive maintenance The maintenance carried out at predetermined intervals or corresponding to prescribed criteria and intended to reduce the probability of failure or the performance degradation of an item (BS 4778).

Probability density Consider a random variable X. The probability density function $f_X(x)$ of X is

$$f_X(x) = \frac{dF_X(x)}{dx} = \lim_{\Delta x \to \infty} \frac{\Pr(x < X \leq x + \Delta x)}{\Delta x}$$

where $F_X(x)$ denotes the distribution function of X.

Quality The totality of features and characteristics of a product or service that bear on its ability to satisfy stated or implied needs (ISO 8402).

Redundancy In an entity, the existence of more than one means for performing a required function [IEC 50(191)].

Existence of means, in addition to the means which would be sufficient for a functional unit to perform a required function or for data to represent information (IEC 61508, Part 4).

Reliability The ability of an item to perform a required function, under given environmental and operational conditions and for a stated period of time (ISO 8402).

Reliability centered maintenance (RCM) A disciplined logic or methodology used to identify preventive maintenance tasks to realize the inherent reliability of equipment at least expenditure of resources [MIL-STD-2173(AS)].

Repair The part of corrective maintenance in which manual actions are performed on the entity [IEC 50(191)].

Required function A function, or a combination of functions, of an entity, which is considered necessary to provide a given service [IEC 50(191)].

Safety Freedom from those conditions that can cause death, injury, occupational illness, or damage to or loss of equipment or property (MIL-STD-882D).

The expectation that a system does not, under defined conditions, lead to a state in which human life is endangered (DEF-STAN 00-56).

Safety integrity Probability of a safety-related system satisfactorily performing the required safety functions under all the stated conditions within a specified period of time (IEC 61508, Part 4).

Safety integrity level (SIL) Discrete level (one out of a possible four) for specifying the safety integrity requirement of the safety functions to be allocated to the E/E/PE safety-related systems, where safety integrity level 4 has the highest level of safety integrity and safety integrity level 1 has the lowest (IEC 61508, Part 4).

Security Dependability with respect to the prevention of unauthorized access and/or handling of information (Laprie, 1992).

Severity The consequences of a failure mode. Severity considers the worst potential consequence of a failure, determined by the degree of injury, property damage, or system damage that could ultimately occur (MIL-STD-1629A).

Single failure point The failure of an item which would result in failure of the system and is not compensated for by redundancy or alternative operational procedure (MIL-STD-1629).

State variable A variable $X(t)$ associated with an item such that:

$$X(t) = \begin{cases} 1 & \text{if the item is functioning at time } t \\ 0 & \text{if the item is in a failed state at time } t \end{cases}$$

State vector A vector $X(t) = (X_1(t), X_2(t), \ldots, X_n(t))$ of the state variables of the n components comprising the system.

Step stress test A test consisting of several stress levels applied sequentially for periods of equal duration to one sample. During each period a stated stress level is applied and the stress level is increased from one period to the next (BS 4778).

Structure function A variable $\phi(X(t))$ associated with a system (with state vector $X(t)$) such that:

$$\phi(X(t)) = \begin{cases} 1 & \text{if the system is functioning at time } t \\ 0 & \text{if the system is in a failed state at time } t \end{cases}$$

Sudden failure A failure that could not be anticipated by prior examination or monitoring [IEC 50(191)].

Survivor function Let T denote the time to failure of an item. The survivor function $R(t)$ of the item is

$$R(t) = \Pr(T > t) \quad \text{for } t \geq 0$$

$R(t)$ is sometimes called the *reliability function* or the *survival probability at time t* of the item.

System A bounded physical entity that achieves in its domain a defined objective through interaction of its parts (DEF-STAN 00-56).

Set of elements which interact according to a design, where an element of a system can be another system, called subsystem, which may be a controlling system or a controlled system and may include hardware, software and human interaction (IEC 61508, Part 4).

Systematic failure Failure related in a deterministic way to a certain cause, which can only be eliminated by a modification of the design or of the manufacturing process, operational procedures, documentation, or other factors (IEC 61508, Part 4).

Test frequency The number of tests of the same type per unit time interval; the inverse of the test interval (IEEE Std. 352).

Test interval The elapsed time between the initiation of identical tests on the same sensor, channel, etc. (IEEE Std. 352).

Wear-out failure A failure whose probability of occurrence increases with the passage of time, as a result of processes inherent in the entity [IEC 50(191)].

References

1. Aalen, O. O. 1975. Statistical Inference for a Family of Counting Processes. Ph.D. Dissertation. Dept. of Statistics, University of California, Berkeley.

2. Aalen, O. O. 1978. Non-parametric inference for a family of counting processes. *Annals of Statistics* **6**:701–726.

3. AIChE. 1985. *Guidelines for Hazard Evaluation Procedures*. American Institute of Chemical Engineers, Center for Chemical Process Safety, New York.

4. AIChE. 1989. *Guidelines for Chemical Process Quantitative Risk Analysis*. American Institute of Chemical Engineers, Center for Chemical Process Safety, New York.

5. AIChE. 1993. *Guideline for Safe Automation of Chemical Processes*. American Institute of Chemical Engineers, Center for Chemical Process Safety, New York.

6. AIChE. 1998. *Guidelines for Improving Plant Reliability Through Data Collection and Analysis*. American Institute of Chemical Engineers, Center for Chemical Process Safety, New York.

7. AIChE. 2001. *Layer of Protection Analysis, Simplified Process Risk Assessment*. American Institute of Chemical Engineers, Center for Chemical Process Safety, New York.

8. Akersten, P. A. 1998. Imperfect repair models. In *Safety and Reliability. Proceeedings of the European Conference on Safety and Reliability - ESREL'98* S.

Lydersen, G. K. Hansen, and H. A. Sandtorv, eds. A. A. Balkema, Rotterdam, pp. 369–372.

9. Akersten, P. A. 1991. *Repairable Systems Reliability, Studied by TTT-Plotting Techniques*. Ph.D. thesis. Division of Quality Technology, Department of Mechanical Engineering. Linköping University, Sweden.

10. Amendola, A., and A. Z. Keller. 1985. *Reliability Data Bases*. Reidel, Dordrecht, The Netherlands.

11. Amoia, V., and M. Santomauro. 1977. Computer oriented reliability analysis of large electrical systems. *Summer Symposium on Circuit Theory*, Czechoslovac Academy of Science, Prague, pp. 317–325.

12. Andersen, P., Ø. Borgan, R. Gill, and N. Keiding. 1993. *Statistical Models Based on Counting Processes*, Springer, New York.

13. Andersen, P. K., and Ø. Borgan. 1985. Counting process models for life history data: A review. *Scandinavian Journal of Statistics* **12**:97–158.

14. Ansell, J. I., and M. J. Phillips. 1994. *Practical Methods for Reliability Data Analysis*. Oxford University Press, Oxford, UK.

15. ANSI/ISA-S84.01. 1996. *Application of Safety Instrumented Systems for the Process Industries*. Instrument Society of America S84.01 Standard. Research Triangle Park, NC.

16. Apostolakis, G., and P. Moieni. 1987. The foundation of models of dependence in probabilistic safety assessment. *Reliability Engineering* **18**:177–195.

17. Ascher, H., and H. Feingold. 1984. *Repairable Systems Reliability. Modeling, Inference, Misconceptions and Their Causes*. Marcel-Dekker, New York.

18. Atwood, C. L. 1986. The binomial failure rate common cause model. *Technometrics* **28**:139–148.

19. Atwood, C. L. 1992. Parametric estimation of time-dependent failure rates for probabilistic risk assessment. *Reliability Engineering and System Safety* **37**:181–194.

20. Aupied, J. 1994. *Retour d'expérience appliqué à la sûreté de fonctionnement des matériels en exploitation*. Editions Eyrolles, Paris.

21. Aven, T. 1986. Reliability/availability evaluation of coherent systems based on minimal cut sets. *Reliability Engineering* **13**:93–104.

22. Aven, T. 1992. *Reliability and Risk Analysis*. Elsevier Science, New York.

23. Bain, L. J., M. Engelhardt, and F. T. Wright. (1985). Tests for an increasing trend in the intensity of a poisson process. *Journal of the American Statistical Association* **80**:419–422.

24. Barlow, R. E. 1998. *Engineering Reliability*. SIAM, Philadelphia.

25. Barlow, R. E., and R. Campo. 1975. Total time on test processes and applications to failure data analysis. In *Reliability and Fault Tree Analysis*. R. E. Barlow, J. B. Fussell and N. D. Singpurwalla, eds. SIAM, Philadelphia.

26. Barlow, R. E., and L. Hunter. 1961. Optimum preventive maintenance policies. *Operations Research* **8**:90–100.

27. Barlow, R. E., and H. E. Lambert. 1975. Introduction to fault tree analysis. In *Reliability and Fault Tree Analysis*. R. E. Barlow, J. B. Fussell, and N. D. Singpurwalla eds. SIAM, Philadelphia.

28. Barlow, R. E., and F. Proschan. 1965. *Mathematical Theory of Reliability*. Wiley, New York.

29. Barlow, R. E., and F. Proschan. 1969. A note on tests for monotone failure rate. *Annals of Mathematical Statistics* **40**:595–600.

30. Barlow, R.E., and F. Proschan. 1975. *Statistical Theory of Reliability and Life Testing, Probability Models*. Holt, Rinehart, and Winston, New York.

31. Barlow, R. E., J. B. Fussell, and N. D. Singpurwalla. 1975. *Reliability and Fault Tree Analysis*. SIAM, Philadelphia.

32. Beckman, L. 1998. Determining the required safety integrity level for your process. *ISA Transactions* **37**:105–111.

33. Bedford, T., and R. Cooke. 2001. *Probabilistic Risk Analysis: Foundations and Methods*. Cambridge University Press, Cambridge, UK.

34. Bendell, A., and L. A. Walls. 1985. Exploring reliability data. *Quality and Reliability Engineering International* **1**:37–51.

35. Bérenguer, C., A. Grall, L. Dieulle, and M. Roussignol. 2003. Maintenance policy for a continuously monitored deteriorating system. *Probability in Engineering and Informational Sciences* **17**:235–250.

36. Berger, J. O. 1985. *Statistical Decision Theory and Bayesian Analysis*. Springer, New York.

37. Bergman, B., and B. Klefsjö. 1982. A graphical method applicable to age replacement problems. *IEEE Transactions on Reliability* **31**:478–481.

38. Bergman, B., and B. Klefsjö. 1984. The total time on test concept and its use in reliability theory. *Operations Research* **32**:596–606.

39. Bergman, B., and B. Klefsjö. 1994. *Quality; From Customer Needs to Customer Satisfaction*. McGraw-Hill, London.

40. Bhattacharayya, G. K., and A. Fries. 1982. Fatigue failure models—Birnbaum-Saunders vs. inverse Gaussian. *IEEE Transactions on Reliability* **31**:439–441.

41. Bickel, P. J., and K. A. Doksum. 1977. *Mathematical Statistics: Basic Ideas and Selected Topics*. Holden-Day, San Francisco.

42. Billinton, R., and R. N. Allen. 1983. *Reliability Evaluation of Engineering Systems: Concepts and Techniques*. Longman Scientific & Technical, Essex, England.

43. Birnbaum, Z. W. 1969. On the importance of different components in a multicomponent system. In *Multivariate Analysis*. P. R. Krishnaiah, ed. Academic, San Diego, pp. 581–592.

44. Birnbaum, Z. W., and S. C. Saunders. 1969. A new family of life distributions. *Journal of Applied Probability* **6**:319-327.

45. Blakey, K. B., R. J. Art, and R. J. Bosnak. 1998. ASME risk-based in-service inspection and testing: An outlook to the future. *Risk Analysis* **18**:407-421.

46. Blanche, K. M., and A. B. Shrivastava. 1994. Defining failure of manufacturing machinery and equipment. In *Proceedings from the Annual Reliability and Maintainability Symposium*, 1994:69–75.

47. Blischke, W. R., and D. N. P. Murthy. 2000. *Reliability; Modeling, Prediction, and Optimization*. Wiley, New York.

48. Block, H. W., W. S. Borges, and T. H. Savits. 1985. Age-dependent minimal repair. *Journal of Applied Probability* **22**:370–385.

49. Bodsberg, L. 1987. Failure Rate Prediction of Electronic Components. SINTEF Report STF75 A87006. SINTEF, Trondheim, Norway.

50. Bodsberg, L., and P. Hokstad. 1988. Reliability of Safety Shutdown Systems. Models for Dependent and Undetected Failures. SINTEF Report STF75 A88011, SINTEF, Trondheim, Norway.

51. Bon, J-L. 1995. *Fiabilité des Systemès, Méthodes Mathématiques*. Masson, Paris.

52. Bourne, A. J., G. T. Edwards, D. M. Hunns, D. R. Poulter, and I. A. Watson. 1981. *Defences against Common-Mode Failures in Redundancy Systems*. SRD R 196. UK Atomic Energy Authority, Warrington.

53. Box, G. E. P., and W. J. Hill. 1967. Discrimination among mechanistic models. *Technometrics* **9**:57–71.

54. Box, G. E. P., W. G. Hunter, and J. S. Hunter. 1978. *Statistics for Experimenters*. Wiley, New York.

55. Box, G. E. P., and G. C. Tiao. 1992. *Bayesian Inference in Statistical Analysis*. Wiley, New York.

56. Brown, M., and F. Proschan. 1983. Imperfect repair. *Journal of Applied Probability* **20**:851–859.

57. BS 4778. British Standard: *Glossary of Terms Used in Quality Assurance Including Reliability and Maintainability Terms*. British Standards Institution, London.

58. BS 5760. 1996. British Standard: *Reliability of Constructed or Manufactured Products, Systems, Equipments and Components*. British Standards Institution, London.

59. BS 5760-5. 1991. British Standard: *Reliability of Systems, Equipments, and Components*. Part 5. Guide to failure modes, effects and criticality analysis (FMEA and FMECA). British Standards Institution, London.

60. CARA FaultTree. Computer program for fault tree analysis. Available from Sydvest Programvare AS. Internet address: http://www.sydvest.com.

61. Castanier, B. 2001. Modélisation Stochastique et Optimisation de la Maintenance Conditionelle des Systèmes à Dégradation Graduelle. PhD thesis, Université de Technologie de Troyes, France.

62. CCIP. 1997. *Critical Foundations – Protecting America's Infrastructures*. Report from the President's commission on critical infrastructure protection. U.S. Department of Homeland Security, Washington, DC.

63. Chan, J-K., and L. Shaw. 1993. Modeling repairable systems with failure rates that depend on age and maintenance. *IEEE Transactions on Reliability* **42**(4):566–571.

64. Chatterjee, P. 1975. Modularization of fault trees; A method to reduce the cost of analysis. In *Reliability and Fault Tree Analysis*. R. E. Barlow, J. B. Fussell, and N. D. Singpurwalla, eds. SIAM, Philadelphia.

65. Cheng Leong, A., and R. K. L. Gay. 1993. IDEF0 modelling for project risk assessment. *Computers in Industry* **22**:227–230.

66. Cheok, M. C., G. W. Parry, and R. S. Sherry. 1998. Use of importance measures in risk-informed regulatory applications. *Reliability Engineering and System Safety* **60**:213–226.

67. Chhikara, R. S., and J. L. Folks. 1977. The inverse Gaussian distribution as a lifetime model. *Technometrics* **19**:461–468.

68. Chhikara, R. S., and J. L. Folks. 1989. *The Inverse Gaussian Distribution. Theory, Methodology and Applications*. Marcel Dekker, New York.

69. Cho, D. I., and M. Parlar. 1991. A survey of maintenance models for multi-unit systems. *European Journal of Operational Research* **51**:1–23.

70. Christer, A. H. 1992. Delay-time models of industrial inspection maintenance problems. *Journal of Operational Research* **51**:1–23.

71. Christer, A. H. 1999. Developments in delay time analysis for modelling plant maintenance. *Journal of the Operational Research Society* **50**:1120–1137.

72. Christer, A. H. 2002. A review of delay time analysis for modelling plant maintenance. In *Stochastic Models in Reliability and Maintenance*. S. Osaki, ed. Springer, Berlin, pp. 89–123.

73. Cocozza-Thivent, C. 1997. *Processus Stochastiques et Fiabilité des Systèmes*, Springer, Berlin.

74. Cooke, R. M. 1996. The design of reliability data bases. Part I and II. *Reliability Engineering and System Safety* **51**:137–146 and 209–223.

75. Corneliussen, K., and P. Hokstad. 2003. *Reliability Prediction Method for Safety Instrumented Systems; PDS Method Handbook, 2003 Edition*. SINTEF Report STF38A02420, SINTEF, Trondheim, Norway

76. Cox, D. R. 1962. *Renewal Theory*. Methuen, London.

77. Cox, D. R. 1972. Regression models and life tables (with discussion). *Journal of the Royal Statistical Society* **B 21**:411–421.

78. Cox, D. R., and V. Isham. 1980. *Point Processes*. Chapman and Hall, London.

79. Cox, D. R., and P. A. Lewis. 1966. *The Statistical Analysis of Series of Events*. Methuen, London.

80. Cox, D. R., and H. D. Miller. 1965. *The Theory of Stochastic Processes*. Methuen, London.

81. Cox, D. R., and D. Oakes. 1984. *Analysis of Survival Data*. Chapman and Hall, London.

82. Cramér, H. 1946. *Mathematical Methods of Statistics*. Princeton University Press, Princeton.

83. Crow, L. H. 1974. Reliability analysis of complex repairable systems. In *Reliability and Biometry*. F. Proschan, and R. J. Serfling, eds. SIAM, Philadelphia, pp. 379–410.

84. Crowder, M. J., A. C. Kimber, R. L. Smith, and T. J. Sweeting. 1991. *Statistical Analysis of Reliability Data*. Chapman and Hall, London.

85. DEF-STD 02-45 (NES 45). 2000. *Requirements for the Application of Reliability-Centred Maintenance Techniques to HM Ships, Submarines, Royal Fleet Auxiliaries and other Naval Auxiliary Vessels.* Defense Standard. U.K. Ministry of Defence, Bath, England.

86. DEF-STD 00-56. 1996. *Safety Management Requirements for Defence Systems.* UK Defence Standardization, Glasgow.

87. Dekker, R. 1996. Application of maintenance optimization models: A review and analysis. *Reliability Engineering and System Safety* **51**:229–240.

88. Dohi, T., N. Kaio, and S. Osaki. 2002. Renewal processes and their computational aspects. In *Stochastic Models in Reliability and Maintenance.* S. Osaki, ed. Springer, Berlin.

89. Doksum, K. A., and A. Høyland. 1992. Models for variable-stress accelerated life testing experiments based on Wiener processes and the inverse Gaussian distribution. *Technometrics* **34**:74–82.

90. Dowell III, A. M. 1998. Layer of protection analysis for determining safety integrity level. *ISA Transactions* **37**:155–165.

91. Doyen, L., and O. Gaudoin. 2002a. Modelling and assessment of maintenance efficiency for repairable systems. Proceedings from ESREL Conference, 19–21 March, 2002, Lyon, France.

92. Doyen, L., and O. Gaudoin. 2002b. Models for assessing maintenance efficiency. Proceedings from the Third International Conference on Mathematical Methods in Reliability, 17–20 June, 2002, Trondheim, Norway.

93. DNV-RP-A203. 2001. *Qualification Procedures for New Technology.* Recommended Practice, Det Norske Veritas, Høvik, Norway.

94. Drenick, R. F. 1960. The failure law of complex equipment. *Journal of the Society for Industrial Applied Mathematics* **8**:680–690.

95. Duane, J. T. 1964. Learning curve approach to reliability monitoring. *IEEE Transactions on Aerospace* **2**:563–566.

96. Dudewicz, E. J., and S. A. Mishra. 1988. *Modern Mathematical Statistics.* Wiley, New York.

97. Ebeling. C. E. 1997. *Reliability and Maintainability Engineering.* McGraw-Hill, New York.

98. Edwards, G. T., and I. A. Watson. 1979. A study of common-mode failures. *UKAEA SRD R 146.*

99. EIReDA. 1998. *European Industry Reliability Data.* Crete University Press, Crete.

100. Elvebakk, G. 1999. *Analysis of Repairable Systems Data: Statistical Inference for a Class of Models Involving Renewals, Heterogeneity and Time Trends.* PhD thesis, Department of Mathematical Sciences, Norwegian University of Science and Technology, Trondheim, Norway.

101. EN 50126. 1999. *Railway Applications – The Specification and Demonstration of Reliability, Availability, Maintainability and Safety (RAMS)*, Cenelec, Brussels.

102. EN 50128. 2001. *Railway Applications – Software for Railway Control and Protection Systems.* Cenelec, Brussels.

103. EN 50129. 1999. *Railway Applications – Safety Related Electronic Systems for Signalling.* Cenelec, Brussels.

104. Endrenyi, J. 1978. *Reliability Modeling in Electric Power Systems.* Wiley, New York.

105. EPRI NP-5777 (G .L. Clellin, J. E. Mott, and A. M. Smith). 1988. *Defensive Strategies for Reducing Susceptibility to Common Cause Failures* Vol. 1 *Defensive Strategies*, Vol. 2 *Data Analysis.* Electric Power Research Institute.

106. EPRI (1995). *PSA Application Guide*, EPRI TR-105396, Electric Power Research Institute.

107. Esary, J. D., F. Proschan, and D. W. Walkup. 1967. Association of random variables. *Annals of Mathematical Statistics* **38**:1466–1474.

108. ESReDA. *Reliability Data Quality Handbook*, European Safety, Reliability and Data Association. Available from Det Norske Veritas, Veritasveien 1, NO 1322 Høvik, Norway.

109. Feller, W. 1968. *An Introduction to Probability Theory and Its Applications, Vol. I.* Wiley, New York.

110. Flamm, J., and T. Luisi, eds. 1992. *Reliability Data Collection and Analysis.* Kluwer, Deventer, The Netherlands.

111. Fleming, K. N. 1974. A reliability model for common mode failures in redundant safety systems. General Atomic Report, GA-13284, Pittsburgh, PA.

112. Fleming, K. N., A. Mosleh, and R. K. Deremer. 1986. A systematic procedure for the incorporation of common cause events into risk and reliability models. *Nuclear Engineering and Design* **93**:245–279.

113. Fox, J. 1993. *Quality Through Design. The Key to Successful Product Delivery.* McGraw-Hill, London.

114. FRACAS. 1999. *Failure Reporting, Analysis and Corrective Action System (FRACAS) Application Guidelines.* Reliability Analysis Center (RAC), 201 Mill Street, Rome, NY.

115. Fussell, J. B. 1975. How to handcalculate system safety and reliability characteristics. *IEEE Transactions on Reliability* **24**:169–174.

116. Gaudoin, O. 1992. Optimal properties of the Laplace trend test for software-reliability models. *IEEE Transactions on Reliability* **41**:525–532.

117. Gelman, A., J. B. Carlin, H. S. Stern, and D. B. Rubin. 1995. *Bayesian Data Analysis*. Chapman and Hall, New York.

118. Gertsbakh, I. 2000. *Reliability Theory with Applications to Preventive Maintenance*. Springer, New York.

119. Gertsbakh, I. 1989. *Statistical Reliability Theory*. Marcel Dekker, New York.

120. Graham, A. 1981. *Kronecker Products and Matrix Calculations with Applications*. Wiley, New York.

121. Grall, A., L. Dieulle, C. Bérenguer, and M. Roussignol. 2002. Continuous-time predictive-maintenance scheduling for a deteriorating system. *IEEE Transactions on Reliability* **51**:141–150.

122. Gumbel, E. J. 1958. *Statistics of Extremes*. Columbia University Press, New York.

123. Hammer, W. 1972. *Handbook of System and Product Safety*. Prentice-Hall, Englewood Cliffs, NJ.

124. Hansen, G. K., and R. Aarø. 1997. *Reliability Quantification of Computer-Based Safety Systems. An Introduction to PDS*. SINTEF report STF38 A97434, SINTEF, Trondheim, Norway

125. Harris, B. 1986. Stochastic models for common failures. In *Reliability and Quality Control*. A. P. Basu, ed. Elsevier Science, New York, pp. 185–200.

126. Henley, E. J., and H. Kumamoto, 1981. *Reliability Engineering and Risk Assessment*. Prentice-Hall, Englewood Cliffs, NJ. (Reprinted and distributed by IEEE Press, 1991 as *Probabilistic Risk Assessment—Reliability Engineering, Design, and Analysis*.)

127. Hoch, R. 1990. A practical application of reliability centered maintenance. The American Society of Mechanical Engineers, 90-JPGC/Pwr-51, Joint ASME/IEEE Power Generation Conference, Boston, MA, 21–25 October.

128. Hokstad, P. 1988. A shock model for common-cause failures. *Reliability Engineering and System Safety* **23**:127–145.

129. Hokstad, P. 1993. Common cause and dependent failure models. In *New Trends in System Reliability Evaluation*. K. B. Misra, ed. Elsevier, Amsterdam.

130. Hokstad, P. 1997. The failure intensity process and the formulation of reliability and maintenance models. *Reliability Engineering and System Safety* **58**:69–82.

131. Hokstad, P., and A. T. Frøvig. 1996. The modelling of degraded and critical failures for components with dormant failures. *Reliability Engineering and System Safety* **51**:189–199.

132. Holen, A. T., A. Høyland, and M. Rausand. 1988. *Reliability Analysis* (in Norwegian). Tapir, Trondheim.

133. Huang, G. Q., ed. 1996. *Design for X: Concurrent Engineering Imperatives*. Chapman and Hall, London.

134. Humphreys, R. A. 1987. Assigning a numerical value to the beta factor common cause evaluation. *Reliability '87*. Proceedings paper 2C.

135. Hunter and Box. 1965.

136. Hunter, W. G., and A. M. Reiner, 1965. Designs for discrimination between two rival models. *Technometrics* **7**:307–323.

137. ICDE. 1994. International Common Cause Failure Data Exchange (ICDE) project. Nuclear Energy Agency.
Internet address: http://www.nea.fr/html/jointproj/icde.html.

138. IEC 50(191). 1990. *International Electrotechnical Vocabulary (IEV) – Chapter 191 – Dependability and Quality of Service.* International Electrotechnical Commission, Geneva.

139. IEC 60300. 1992. *Dependability Management*. International Electrotechnical Commission, Geneva.

140. IEC 60300-3-9. 1999. *Dependability Management – Application Guide: Risk Analysis of Technological systems*. International Electrotechnical Commission, Geneva.

141. IEC 60300-3-11. 1999. *Dependability Management – Application Guide: Reliability Centered Maintenance*. International Electrotechnical Commission, Geneva.

142. IEC 60812. 1985. *Analysis Techniques for System Reliability – Procedures for Failure Mode and Effect Analysis (FMEA)*. International Electrotechnical Commission, Geneva.

143. IEC 61025. 1990. *Fault Tree Analysis (FTA)*. International Electrotechnical Commission, Geneva.

144. IEC 61078. 1991. *Analysis Techniques for Dependability – Reliability Block Diagram Method*. International Electrotechnical Commission, Geneva.

145. IEC 61508. 1997. *Functional Safety of Electrical/Electronic/Programmable Electronic Safety-Related Systems.* Part 1–7. International Electrotechnical Commission, Geneva.

146. IEC 61511. 2003. *Functional Safety – Safety Instrumented Systems for the Process Industry*. International Electrotechnical Commission, Geneva.

147. IEC 61513. 2001. *Nuclear power plants – Instrumentation and Control for Systems Important to Safety – General Requirements for Systems*. International Electrotechnical Commission, Geneva.

148. IEC 61882. 2001. *Hazard and Operability Studies (HAZOP-Studies) – Application Guide*. International Electrotechnical Commission, Geneva.

149. IEC 62061. 2002 (draft). *Safety of Machinery – Functional Safety of Electrical, Electronic and Programmable Electronic Systems*. International Electrotechnical Commission, Geneva.

150. IEEE Std. 352. 1982. *IEEE Guide for General Principles of Reliability Analysis of Nuclear Power Generating Station Protection Systems*. IEEE, New York

151. IEEE Std. 500. 1984. *IEEE Guide to the Collection and Presentation of Electrical, Electronic, Sensing Component, and Mechanical Equipment Reliability Data for Nuclear Power Generating Stations*. Wiley, New York.

152. Ishikawa, K. 1986. *Guide to Quality Control*. Productivity Press, Cambridge, MA.

153. ISO 8402. 1986. *Quality Vocabulary*. International Standards Organization, Geneva.

154. ISO 9000. 1994. *Quality Management and Quality Assurance Standards. Guidelines for Selection and Use*. International Standards Organization, Geneva.

155. ISO 10418. 1993. *Petroleum and Natural Gas Industries – Offshore Production Platforms – Analysis, Design, Installation and Testing of Basic Surface Safety Systems*. International Standards Organization, Geneva.

156. Jelinski, Z., and P. B. Moranda. 1972. Software Reliability Research. In *Statistical Computer Performance Evaluation*. W. Freiberger, ed. Academic, San Diego.

157. Jensen, F. V. 2001. *Bayesian Networks and Decision Graphs*. Springer, New York.

158. Jensen, F., and N. E. Petersen, (1982). *Burn-in: An Engineering Approach to the Design and Analysis of Burn-in Procedures*. Wiley, New York.

159. Johnson, N. L., and S. Kotz. 1970. *Distributions in Statistics. Continuous Univariate Distributions*, Vols. 1–2. Houghton Mifflin, Boston.

160. Kalbfleisch, J. D., and R. L. Prentice. 1980. *The Statistical Analysis of Failure Time Data*. Wiley, New York.

161. Kaminskiy, M., and V. A. Krivstov. 1998. *A Monte Carlo Approach to Repairable System Reliability Analysis*. Springer, New York.

162. Kaplan, E. L., and P. Meier. 1958. Nonparametric estimation from incomplete observations. *Journal of the American Statistical Association* **53**:457–481.

163. Kaplan, S. 1990. Bayes is for eagles. *IEEE Transactions on Reliability* **39**:130–131.

164. Kapur, K. C., and L. R. Lamberson. 1977. *Reliability in Engineering Design*. Wiley, New York.

165. Kawauchi, Y., and M. Rausand. 2002. A new approach to production regularity assessment in the oil and chemical industries. *Reliability Engineering and System Safety* **75**:379–388.

166. Kijima, M., and N.A. Sumita. 1986. A useful generalization of renewal theory; counting process governed by non-negative Markovian increments. *Journal of Applied Probability* **23**:71–88.

167. Klefsjö, B., and U. Kumar. 1992. Goodness-of-fit tests for the power-law process based on the TTT-plot. *IEEE Transactions on Reliability* **41**:593–598.

168. Knight, C. R. 1991. Four decades of reliability progress. *Proceedings Annual Reliability and Maintainability Symposium*, IEEE, New York, pp. 156–159.

169. Kusiak, A., ed. 1993. *Concurrent Engineering : Automation, Tools, and Techniques*. Wiley, New York.

170. Lakner, A. A., and R. T. Anderson. 1985. *Reliability Engineering for Nuclear and Other High Technology Systems*. Elsevier Applied Science, London.

171. Lambert, H. E. 1975. Measures of importance of events and cut sets in fault trees. In *Reliability and Fault Tree Analysis*. R. E. Barlow, J. B. Fussell, and N. D. Singpurwalla, eds. SIAM, Philadelphia.

172. Lambert, M., B. Riera, and G. Martel. 1999. Application of functional analysis techniques to supervisory systems. *Reliability Engineering and System Safety* **64**:209–224.

173. Laprie, J. C., ed. 1992. *Dependability: Basic Concepts and Terminology*. Springer, New York.

174. Lawless, J. F. 1982. *Statistical Models and Methods for Lifetime Data*. Wiley, New York.

175. Lehmann, E. L. 1983. *Theory of Point Estimation*. Wiley, New York.

176. Lesanovský, A. 1988. Multistate Markov models for systems with dependent units. *IEEE Transactions on Reliability* **37**:505–511.

177. Levenbach, G. J. 1957. Accelerated Life Testing of Capacitors. *IRA-Transactions on Reliability and Quality Control, PGRQC* **10**:9–20.

178. Lieblein, J., and M. Zelen. 1956. Statistical investigation of the fatigue life of deep groove ball bearings. *Journal of Research, National Bureau of Standards* **57**:273–316.

179. Lim, T. J. 1998. Estimating system reliability with fully masked data under Brown-Proschan imperfect repair model. *Reliability Engineering and System Safety* **59**:277–289.

180. Limnios, N., and G. Oprisan. 2001. *Semi-Markov Processes and Reliability.* Birkhauser, Basel.

181. Lindqvist, B. H. 1993. The trend renewal process, a useful model for repairable systems. Society of Reliability Engineers, Scandinavian chapter, Annual Conference, Malmö, Sweden.

182. Lindqvist, B. H. 1998. Statistical modeling and analysis of repairable systems. In *Statistical and Probabilistic Methods in Reliability*. V. Ionescu, and N. Limnios, eds. Birkhauser Boston.

183. Lindqvist, B. H. 1999. Repairable systems with general repair. Proceedings from ESREL Conference, 13–17 September, 1999, Munich, Germany.

184. Lindqvist, B. H., and H. Amundrustad. 1998. Markov models for periodically tested components. In *Safety and reliability*, Vol. 1. S. Lydersen, G. K. Hansen, and H. A. Sandtorv, eds. Balkema, Boston

185. Lissandre, M. 1990. *Maîtriser SADT*. Albert Colin, Paris.

186. Lloyd, D. K., and M. Lipow. 1962. *Reliability: Management, Methods and Mathematics.* Prentice-Hall, Englewood Cliffs, NJ.

187. Lorden, G. 1970. On excess over the boundary. *Annals of Mathematical Statistics* **41**:520–527.

188. Luthra, P. 1990. MIL-HDBK-217: What is wrong with it? *IEEE Transactions on Reliability* **39**:518.

189. Lydersen, S., and M. Rausand. 1989. Failure rate estimation based on data from different environments and with varying quality. In *Reliability Data Collection and Use in Risk and Availability Assessment*. V. Colombari, ed. Springer, New York.

190. Malik, M. A. K. 1979. Reliable preventive maintenance policy. *AIIE Transactions* **11**:221–228.

191. Mann, N. K., R. E. Schafer, and N. D. Singpurwalla. 1974. *Methods for Statistical Analysis of Reliability and Life Data.* Wiley, New York.

192. Marshall, A. W., and I. Olkin. 1967. A multivariate exponential distribution. *Journal of the American Statistical Association* **62**:30–44.

193. Martz, H. F., and R. A. Waller. 1982. *Bayesian Reliability Analysis*. Wiley, New York.

194. Meeker, W. Q., and L. A. Escobar. 1998. *Statistical Methods for Reliability Data*. Wiley, New York.

195. Melchers, R. E. 1999. *Structural Reliability: Analysis and Prediction*, 2nd ed. Wiley, Chichester.

196. MIL–HDBK-189. 1981. *Reliability Growth Management*. U.S. Department of Defense, Washington, DC.

197. MIL–HDBK-217F. 1991 *Reliability Prediction of Electronic Equipment*. U.S. Department of Defense, Washington, DC.

198. MIL-STD-756B. 1981. *Reliability Modeling and Prediction*. U.S. Department of Defense, Washington, DC.

199. MIL-STD-882D. 2000. *Standard Practice for System Safety*. U.S. Department of Defense, Washington, DC.

200. MIL-STD 1629A. 1980. *Procedures for Performing a Failure Mode, Effects and Criticality Analysis*. U.S. Department of Defense, Washington, DC.

201. MIL-STD-2155. *Failure Reporting, Analysis and Corrective Action System*. U.S. Department of Defense, Washington, DC.

202. MIL-STD-2173(AS). 1986. *Reliability-Centered Maintenance. Requirements for Naval Aircraft, Weapon Systems and Support Equipment*. U.S. Department of Defense, Washington, DC.

203. Miner, M.A. 1945. Cumulative damage in fatigue. *Journal of Applied Mechanics* **12**:A159–A164.

204. Mitrani, I. 1982. *Simulation Techniques for Discrete Event Systems*. Cambridge University Press, Cambridge, UK.

205. Molnes, E., M. Rausand, and B. H. Lindqvist. 1986. *Reliability of Surface Controlled Subsurface Safety Valves*. SINTEF Report STF75 A86024, SINTEF, Trondheim, Norway.

206. Mosleh, A. 1991. Common cause failures: an analysis methodology and examples. *Reliability Engineering and System Safety* **34**:249–292.

207. Moubray, J. 1997. *Reliability-centered Maintenance II, 2nd ed.* Industrial Press. New York.

208. Nakajima, S. 1988. *Total Productive Maintenance*. Productivity Press, Cambridge, MA.

209. NASA. 2002. *Fault Tree Handbook with Aerospace Applications*. NASA Office of Safety and Mission Assurance, Washington, DC.

210. NASA. 2000. *Reliability Centered Maintenance Guide for Facilities and Collateral Equipment*. NASA Office of Safety and Mission Assurance, Washington, DC.

211. Nelson, W. 1969. Hazard plotting for incomplete failure data. *Journal of Quality Technology* **1**:27–52.

212. Nelson, W. 1971. Analysis of accelerated life test data part. I. The Arrhenius model and graphical methods. *IEEE Transactions on Electrical Insulation* **EI-6**:165–187.

213. Nelson, W. 1972. Theory and application of hazard plotting for censored failure data. *Technometrics* **14**:945–966.

214. Nelson, W. 1975. Analysis of accelerated life test data – least squares methods for the inverse power law model. *IEEE Transactions on Reliability* **24**:103–107.

215. Nelson, W. 1982. *Applied Life Data Analysis*. Wiley, New York.

216. Nelson, W. 1990. *Accelerated Testing: Statistical Models, Test Plans, and Data Analyses*, Wiley, New York.

217. Nelson, W., and T. J. Kielpinski. 1976. Theory for optimum censored accelerated life tests for normal and lognormal life distributions. *Technometrics* **18**:105–114.

218. Neyman, J. 1945. On the problem of estimating the number of schools of fish. *Publications in Statistics*, Vol. 1. University of California, Berkeley.

219. NORSOK Z-013, 2001. *Risk and Emergency Preparedness Analysis*. Norwegian Technology Centre, Oslo.

220. NORSOK Z-016, 1998. *Regularity Management and Reliability Technology*. Norwegian Technology Centre, Oslo.

221. Nowlan, F. S., and H. F. Heap (1978). *Reliability Centered Maintenance*. Dolby Access Press, San Francisco.

222. NUREG-75/014. 1975. *Reactor Safety: An Assessment of Accident Risk in US Commercial Nuclear Power Plants*, WASH-1400. U.S. Nuclear Regulatory Commission, Washington, DC.

223. NUREG-0492 (W. E. Vesely, F. F. Goldberg, N. H. Roberts, and D. F. Haasl). 1981. *Fault Tree Handbook*. U.S. Nuclear Regulatory Commission, Washington, DC.

224. NUREG/CR-1278 (A. D. Swain and H. E. Guttmann).1983.*Handbook of Human Reliability Analysis in Nuclear Power Plant Applications*. U.S. Nuclear Regulatory Commission, Washington, DC.

225. NUREG/CR-4780 (A. Mosleh, K. N. Fleming, G. W. Parry, H. M. Paula, D. H. Worledge, and D. M. Rasmuson). 1988. *Procedures for Treating Common Cause Failures in Safety and Reliability Studies*, Vol. 1: *Procedural Framework and Examples*. U.S. Nuclear Regulatory Commission, Washington, DC.

226. NUREG/CR-4780 (A. Mosleh, K. N. Fleming, G. W. Parry, H. M. Paula, D. H. Worledge, and D. M. Rasmuson). 1989. *Procedures for Treating Common Cause Failures in Safety and Reliability Studies*, Vol. 2: *Analytic Background and Techniques*. U.S. Nuclear Regulatory Commission, Washington, DC.

227. NUREG/CR-5460 (H. M. Paula, and G. W. Parry). 1990. *A Common Defense Approach to the Understanding and Analysis of Common Cause Failures*, Vol. 1. U.S. Nuclear Regulatory Commission, Washington, DC.

228. NUREG/CR-5485. 1998. *Guidelines on Modeling Common-Cause Failures in Probabilistic Risk Assessment*, U.S. Nuclear Regulatory Commission, Washington, DC.

229. NUREG/CR-5801 (A. Mosleh, K. Fleming, G. Parry, H. Paula, D. Worledge, and D. Rasmuson). 1993. *Procedures for Analysis of Common-Cause Failures in Probabilistic Safety Analysis*, Vol. 1. U.S. Nuclear Regulatory Commission, Washington, DC.

230. NUREG/CR-6268. 1998. *Common Cause Failure Data Collection and Analysis System*, Vol. 1. U.S. Nuclear Regulatory Commission, Washington, DC.

231. O'Connor, P. D. T. 2002. *Practical Reliability Engineering*, 4th ed. Wiley, Chichester.

232. OLF. 2001. Guideline on the application of IEC 61508 and IEC 61511 in the petroleum activities on the Norwegian Continental Shelf, The Norwegian Oil Industry Association, OLF report 070 (see http://www.itk.ntnu.no/sil).

233. OREDA. 2002. *Offshore Reliability Data*, 4th ed. OREDA Participants. Available from: Det Norske Veritas, NO-1322 Høvik, Norway.

234. Pagès, A., and M. Gondran (1980). *Fiabilité des Systèmes*. Eyrolles, Paris.

235. Paglia, A., D. Barnard, and D. A. Sonnett. 1991. A case study of the RCM project at V. C. Summer nuclear generating station. *4th International Power Generation Exhibition and Conference*, Tampa, Florida, **5**:1003–1013.

236. Parry, G. W. 1991. Common cause failure analysis: A critique and some suggestions. *Reliability Engineering and System Safety* **34**:309–326.

237. Pearl, J. 2000. *Causality – Models, Reasoning, and Inference*. Cambridge University Press, Cambridge.

238. Péres, F. 1996. Outils d'analyse de performance pour stratégies de maintenance dans les systèmes de production. PhD thesis, Université de Bordeaux I, France.

239. Pham, H., and H. Wang. 1996. Imperfect maintenance. *European Journal of Operational Research* **94**:425–438.

240. Pierskalla, W. P., and J. A. Voelker. 1979. A survey of maintenance models: The control and surveillance of deteriorating systems. *Naval Research Logistics Quarterly* **23**:353–388.

241. Proschan, F. 1963. Theoretical explanation of observed decreasing failure rate. *Technometrics* **5**:375–383.

242. Rausand, M. 1998. Reliability centered maintenance. *Reliability Engineering and System Safety* **60**:121–132

243. Rausand M., and K. Øien. 1996. The basic concepts of failure analysis. *Reliability Engineering and System Safety* **53**:73–83.

244. Ravichandran, N. 1990. *Stochastic Methods in Reliability Theory*. Wiley, New York.

245. Ripley, B. D. 1987. *Stochastic Simulation*. Wiley, New York.

246. Rippon, S. 1975. Browns Ferry fire. *Nuclear Engineering International*, May, p. 461.

247. Ross, S. M. 1970. *Applied Probability Models with Optimization Applications*. Holden-Day, San Francisco.

248. Ross, S. M. 1996. *Stochastic Processes*, 2nd ed. Wiley, New York

249. SAE-ARP 5580. 2001. *Recommended Failure Modes and Effects Analysis (FMEA) Practices for Non-Automobile Applications*. The Society for Advancing Mobility Land, Sea, Air, and Space, 400 Commonwealth Drive, Warrendale, PA 15096-0001, USA

250. SAE JA1012. 2002. *A Guide to the Reliability-Centered Maintenance (RCM) Standard*. Society for Advancing Mobility Land, Sea, Air, and Space, Warrendale, PA.

251. Satyanarayana, A., and A. Prabhakar. 1978. New topological formula and rapid algorithm for reliability analysis. *IEEE Transactions on Reliability* **27**:82–100.

252. Scarf, P. A. 1997. On the modelling of condition based maintenance. In *Advances in Safety and Reliability, Proceedings of ESREL'97, International Conference on Safety and Reliability* **3**:1701–1708.

253. Schmidt, E. R. et al. 1985. Importance measures for use in PRAs and risk assessment. In Proceedings: International Topical Meeting on Probabilistic Safety Methods and Applications. San Francisco, EPRI NP-3912-SR.

254. Schrödinger, E. 1915. Zur Theorie der Fall- und Steigversuche an Teilchen mit Brownscher Bewegung. *Physikalische Zeitschrift* **16**:289–295.

255. Shin, I., T. J. Lim, and C. H. Lie. 1996. Estimating parameters of intensity function and maintenance effect for repairable unit. *Reliability Engineering and System Safety* **54**:1–10.

256. Shooman, M. 1968. *Probabilistic Reliability: An Engineering Approach*. McGraw-Hill, New York.

257. Signoret, J. P. 1986. Availability of petroleum installations using Markov processes and Petri net modelling. In *Risk and Reliability in Marine Technology*. C. G. Soares, ed. Balkema, Rotterdam., pp. 455–472.

258. Singpurwalla, N. D. 1973. Inference from Accelerated Life Tests Using Arrhenius Type Reparametrization. *Technometrics* **15**:289–299.

259. Singpurwalla, N. D., and F. A. Al-Khayyal. 1977. Accelerated life tests using the power law model for the Weibull distribution. In *The Theory and Applications of Reliability with Emphasis on Bayesian and Nonparametric Methods*, Vol. II. C. P. Tsokos and I. N. Shimi, eds. Academic, San Diego, pp. 381–399.

260. Smith, A. M. 1993. *Reliability-Centered Maintenance*. McGraw-Hill, New York.

261. Smith, D. J. 1997. *Reliability, maintainability and Risk, 5th ed.* Butterworth Heinemann, Oxford, UK.

262. Smith, W. L. 1958. Renewal theory and its ramifications. *Journal of the Royal Statistical Society B* **20**:243–302.

263. Smith, W. L., and M. R. Leadbetter. 1963. On the renewal function for the Weibull distribution. *Technometrics* **5**:393–396.

264. Stavrianidis, P., and K. Bhimavarapu. 1998. Safety instrumented functions and safety integrity levels (SIL). *ISA Transactions* **37**:337–351.

265. Strandberg, K. (1992). Elements of a dependability programme. Paper presented to NSDCS'92, The Nordic Seminar on Dependable Computing Systems 1992, 19–21 August, Trondheim, Norway.

266. Summers, A. E., and G. Raney. 1999. Common cause and common sense, designing failure out of your safety instrumented systems (SIS). *ISA Transactions* **38**:291–299

267. Svendsen, A. 2002. Analysis of the reliability of protective relays and local control gear in transformer stations (in Norwegian). MSc-thesis, Department of Electro-Power Technology, Norwegian University of Science and Technology, Trondheim, Norway.

268. Sverdrup, E. 1967. *Law and Chance Variations*, Vol. 1. North Holland, Amsterdam.

269. Sweet, A. L. 1990. On the hazard rate of the lognormal distribution. *IEEE Transactions on Reliability* **39**:325–328.

270. Takács, L. 1956. On a probability problem arising in the theory of counters. *Cambridge Philosophical Society* **32**:488–489.

271. Taylor, H. M., and S. Karlin. 1984. *An Introduction to Stochastic Modeling*. Academic, San Diego.

272. Thompson, W. A., Jr. 1981. On the foundations of reliability. *Technometrics* 23:1–13.

273. Thompson, W. A., Jr. 1988. *Point Process Models with Applications to Safety and Reliability*. Chapman and Hall, London.

274. Trivedi, K. S. 1982. *Probability & Statistics with Reliability, Queuing, and Computer Science Applications*. Prentice-Hall, Englewood Cliffs, NJ.

275. Tweedie, M. C. K. 1946. Inverse statistical variates. *Nature* **155**:453.

276. Tweedie, M. C. K. 1957. Statistical properties of inverse Gaussian distributions I and II. *Annals of Mathematical Statistics* **28**:362–377, 696–705.

277. U.S. Air Force. 1981. Integrated Computer Aided Manufacturing (ICAM) Architecture. Part II, Volume IV, Functional Modeling Manual (IDEF0), Ohio 45433, Air Force Materials Laboratory, Wright-Patterson AFB AFWAL-tr-81-4023.

278. Valdez-Flores, C., and R. M. Feldman. 1989. A survey of preventive maintenance models for stochastically deteriorating single-unit systems. *Naval Research Logistics* **36**:419–446.

279. Vatn, J., P. Hokstad, and L. Bodsberg. 1996. An overall model for maintenance optimization. *Reliability Engineering and System Safety* **51**:241–257.

280. Vatn, J., and H. Svee. 2002. A Risk Based Approach to Determine Ultrasonic Inspection Frequencies in Railway Applications. ESReDA Conference, Madrid, 27–28 May.

281. Vesely, W. E. 1977. Estimating common cause failure probabilities in reliability and risk analysis: Marshall-Olkin specializations. In *Nuclear Systems Reliability Engineering and Risk Assessment*. J. B. Fussell, and G. R. Burdick, eds. SIAM, Philadelphia, pp. 314–341.

282. Vesely, W. E. 1991. Incorporating aging effects into probabilistic risk analysis using a Taylor expansion approach. *Reliability Engineering and System Safety* **32**(3):315–337.

283. Vesely, W. E., and R. E. Narum. 1970. *PREP and KITT, Computer Codes for the Automatic Evaluation of a Fault Tree.* Idaho Nuclear Corp., IN.

284. Viertl, R. 1988. *Statistical Methods in Accelerated Life Testing.* Vandenhoeck and Ruprecht, Göttingen.

285. Villemeur, A. 1988. *Sûreté de Fonctionnement des Systèmes Industriels.* Eyrolles, Paris.

286. Wald, A. 1947. *Sequential Analysis.* Wiley, New York.

287. Wang, H. 2002. A survey of maintenance policies of deteriorating systems. *European Journal of Operational Research* **139**:469–489.

288. Webster, J. T., and Van B. Parr. 1965. A method for discrimination between failure density functions used in reliability predictions. *Technometrics* **7**:1–10.

289. Weibull, W. 1951. A statistical distribution function of wide applicability. *Journal of Applied Mechanics* **18**:293–297.

290. Whitmore, G. A., and V. Seshadre. 1987. A heuristic derivation of the inverse Gaussian distribution. *The American Statistician* **41**(4):280–281.

291. Yañes, M., F. Joglar, and M. Modarres. 2002. Generalized renewal process for analysis of repairable systems with limited failure experience. *Reliability Engineering and System Safety* **77**:167–180.

292. Yun, W. Y., and S. J. Choung. 1999. Estimating maintenance effect on intensity function for improvement maintenance models. 5th ISSAT International Conference on Reliability and Quality in Design, Las Vegas, pp. 164–166.

293. Ødegaard, S. 2002. *Reliability Assessment of a Subsea Production Tree.* Project report. Department of Production and Quality Engineering, Norwegian University of Science and Technology, Trondheim, Norway.

294. Øien, K. 2001. Risk Control of Offshore Installations – A Framework for the Establishment of Risk Indicators. PhD thesis, Department of Production and Quality Engineering, Norwegian University of Science and Technology, Trondheim, Norway.

Author Index

Aalen, O. O., 283
Aarø, R., 453
Akersten, P. A., 240, 287
Al-Khayyal, F. A., 533
Allen, R. A., 177, 332
Amendola, A., 572
Amoia, V., 348
Amundrustad, H., 453, 457–458
Andersen, P., 239, 283
Anderson, R. T., 5
Ansell, J. I., 466, 589
Apostolakis, G., 215, 223
Art, R. J., 184
Ascher, H., 19, 233, 236, 240, 242, 258, 264, 267, 278, 284, 286, 295
Atwood, C. L., 222, 284–285, 299
Aupied, J., 572
Aven, T., 171, 374

Bain, L. J., 286
Barlow, R. E., 61, 103, 108, 123, 223, 245, 247, 260, 325, 380, 387, 481, 487, 489, 491, 515
Barnard, D., 413
Bayes, T., 540
Bedford, T., 108
Bendell, A., 236
Bérenguer, C., 391, 394
Berger, J. O., 540
Bergman, B., 107, 498
Bhattacharayya, G. K., 51

Bhimavarapu, K., 450
Bickel, P. J., 513–514, 529
Billington, R., 177, 332
Birnbaum, Z. W., 48, 134, 185
Blakeley, K. B., 184
Blanche, K. M., 85
Blischke, W. R., 515, 561
Block, H. W., 288–289
Bodsberg, L., 210, 401
Bon, J. L., 260–261, 352
Borgan, Ø., 239, 283
Borges, W. S., 288–289
Bosnak, R. J., 184
Bourne, A. J., 212
Box, G. E. P., 11, 528, 540
Brown, M., 288

Campo, R., 487, 489, 491
Carlin, J. B., 540
Castanier, B., 380, 394
Chan, J. K., 289–290
Chatterjee, P., 139
Cheng Leong, A., 83
Cheok, M. C., 184, 190
Chhikara, R. S., 53, 514, 533
Cho, D. I., 380
Choung, S. J., 293
Christer, A. H., 362, 400
Cocozza-Thivent, C., 35, 231, 241, 243, 248, 262, 267, 285, 302, 355, 477, 502

Cooke, R., 108, 572
Corneliussen, K., 423, 437, 445, 453
Cox, D. R., 51, 248, 264, 267, 285–286, 313, 466, 477, 479–480, 502, 518, 525, 536
Cramér, H., 54
Crow, L. H., 284
Crowder, M. J., 284–286, 297, 466, 471, 519

Dekker, R., 400
Deremer, R. K., 210
Dieulle, L., 391, 394
Dodge, H. F., 1
Dohi, T., 263
Doksum, K. A., 513–514, 529, 535
Dowell III, A. M., 451
Doyen, L., 289–293
Drenick, R. F., 264
Duane, J. T., 284
Dudewicz, E. J., 47–48, 52, 169, 174, 369, 504, 551

Ebeling, C. E., 365, 372
Edwards, G. T., 210–212, 216
Elvebakk, G., 295
Endrenyi, J., 177
Engelhardt, M., 286
Escobar, L. A., 466, 536

Feingold, H., 19, 233, 236, 240, 242, 258, 264, 267, 278, 284, 286, 295
Feldman, R. M., 380
Feller, W., 171, 253
Flamm, J., 572
Fleming, K. N., 210, 213, 215, 217, 221
Folks, J. L., 53, 514, 533
Fox, J., 80
Fries, A., 51
Frøvig, A. T., 30, 456
Fussell, J. B., 193

Gaudoin, O., 286, 289–293
Gay, R. K. L., 83
Gelman, A., 540
Gertsbakh, I., 380
Gill, R., 239
Goldberg, F. F., 96–97
Gondran, M., 332
Graham, A., 582
Grall, A., 391, 394
Gumbel, E. J., 57
Guttmann, H. E., 109

Haasl, D. F, 96–97
Hammer, W., 94
Hansen, G. K., 453
Harris, B., 216, 228

Heap, H. F., 401
Henley, E. J., 85, 178, 184
Hill, W. J., 528
Hoch, R., 413
Hokstad, P., 30, 210, 215, 223, 239, 287, 401, 423, 437, 445, 453, 456
Huang, G. Q., 11
Humphreys, R. A., 221
Hunns, D. M., 212
Hunter, L., 387
Hunter, J. S., 528
Hunter, W. G., 528
Høyland, A., 535

Isham, V., 264
Ishikawa, K., 107

Jelinski, Z., 300
Jensen, F., 525
Jensen, F. V., 108
Joglar, F., 293
Johnson, N. L., 55

Kaio, N., 263
Kalbfleisch, J. D., 466, 475, 477, 479, 525, 537, 589
Kaminskiy, M., 293
Kaplan, E. L., 475, 477, 479
Kaplan, S., 5
Kapur, K. C., 58, 558
Karlin, S., 245
Kawauchi, Y., 372
Keiding, N.|239 Keller, A. Z., 572
Kielpinski, T. J., 533
Kijima, M., 293
Kimber, A. C., 284–286, 297, 466, 471, 519
Klefsjö, B., 107, 284, 498
Knight, C. R., 2
Kotz, S., 55
Krivstov, V. A., 293
Kumamoto, H., 85, 178, 184
Kumar, U., 284
Kusiak, A., 10

Lakner, A. A, 5
Lamberson, L. R., 58, 558
Lambert, H. E., 103, 184
Lambert, M., 80–81
Laprie, J. C., 7–8
Lawless, J. F., 55, 466, 477, 479, 525, 537, 589
Leadbetter, M. R., 256–257
Lehmann, E. L., 560
Lesanovsky, A., 349
Levenbach, G. J., 528
Lewis, P. A., 285–286
Lieblein, J., 496, 522

AUTHOR INDEX

Lim, T. L., 288
Limnios, N., 355
Lindqvist, B. H., 293–295, 453, 457–458, 572
Lipow, M., 58
Lissandre, M., 81
Lloyd, D. K., 58
Luisi, T., 572
Lydersen, S., 572

Malik, M. A. K., 292–293
Mann, N. K., 41, 54–55, 58, 509, 525, 528–529, 532–533, 589
Martel, G., 80–81
Martz, H. F., 540
Meeker, W. Q., 466, 536
Meier, P., 475, 477, 479
Melchers, R. E., 4
Miller, H. D., 51, 313
Miner, M. A., 47
Mishra, S. A., 47–48, 52, 169, 174, 369, 504, 551
Mitrani, I., 378
Modarres, M., 293
Moieni, P., 215, 223
Molnes, E. M., 572
Moranda, P. B., 300
Mosleh, A., 210, 213, 215, 221
Moubray, J., 402
Murthy, D. N. P., 515, 561
Nakajima, S., 414
Narum, R. E., 172
Nelson, W., 55, 283, 472, 480–481, 485, 514, 525, 533, 536
Neyman, J., 11
Nowlan, F. S., 401

Oakes, D., 466, 477, 479–480, 502, 518, 525
Oprisan, G., 355
Osaki, S., 263

Pagès, A., 332
Paglia, A., 413
Parlar, M., 380
Parry, G. W., 184, 190, 210, 212–215, 221
Paula, H. M., 210, 213–215, 221
Pearl, J., 108
Péres, F., 362
Petersen, N. E., 525
Pham, H., 287
Phillips, M. J., 466, 589
Pierskalla, W. P., 380
Poulter, D. R., 212
Prabhakar, A., 170
Prentice, R. L., 466, 475, 477, 479, 525, 537, 589
Proschan, F., 61, 123, 223, 245, 247, 260, 288, 299, 325, 380, 481, 515

Raney, G., 209, 213, 442–443, 446
Rasmuson, D., 210, 213, 215, 221
Rausand, M., 83, 88, 372, 401, 572
Ravichandran, N., 177
Reiner, A. M., 528
Riera, B., 80–81
Ripley, B. D., 377–378
Rippon, S., 210
Roberts, N. H., 96–97
Romig, H. G., 1
Ross, D. T., 81
Ross, S. M., 231, 233, 240, 242–243, 245, 248, 251–252, 258, 260, 267, 272, 279, 302, 305–306, 314–315, 354–355
Roussignol, M., 391, 394
Rubin, D. B., 540

Santomauro, M., 348
Satyanarayana, A., 170
Saunders, S. C., 48
Savits, T. H., 288–289
Scarf, P. A., 362, 400
Schafer, R. E., 41, 54–55, 58, 509, 525, 528–529, 532–533, 589
Schrödinger, E.|512 Sesharde, V., 52
Shaw, L., 289–290
Sherry, R. S., 184, 190
Shewhart, W., 1
Shooman, M., 129
Shrivastava, A. B., 85
Signoret, J. P., 374
Singpurwalla, N. D., 41, 54–55, 58, 509, 525, 528–529, 532–533, 589
Smith, D. J., 5–6, 365
Smith, R. L., 284–286, 297, 466, 519, 471
Smith, W. L., 254, 256–257
Sonnett, D. A., 413
Stern, H. S., 540
Strandberg, K., 10
Straviniadis, P., 450
Sumita, N. A., 293
Summers, A. E., 209, 213, 442–443, 446
Svee, H., 395
Svendsen, A., 349
Sverdrup, E., 501
Swain, A. D., 109
Sweet, A. L., 50
Sweeting, T. J., 284–286, 297, 466, 471, 519

Taylor, H. M., 245
Thompson, W. A., 19, 242, 264, 280
Tiao, G. C., 540
Trivedi, K. S., 177
Tweedie, M. C. K., 52, 72

Valdez-Flores, C., 380

Van Parr, B., 528
Vatn, J., 395, 401
Vesely, W. E., 96–97, 172, 193, 222, 285
Viertl, R., 525, 536
Villemeur, A, 2, 8, 85
Voelker, J. A., 380

Wald, A., 52
Waller, R. A., 540
Walls, L. A., 236
Wang, H., 287, 380

Watson, I. A., 210–212, 216
Webster, J. T., 528
Whitmore, G. A., 52
Worledge, D. H., 210, 213, 215, 221
Wright, F. T., 286

Yañes, M., 293
Yun, W. Y., 293

Zelen, M., 496, 522

Ødegaard, S., 82
Øien, K., 83, 88

Subject Index

Accelerated life test (ALT), 525
   Arrhenius model, 529
   covariates, 525
   experimental design for ALT, 526
   Eyring model, 529
   nonparametric models, 536
   overstress, 525
   parametric models, 528
   PH model, 535–536
   power rule model, 529
   progressive stress (PALT), 526
   step-stress (SALT), 528
   stressor, 525
Accelerated test, 599
Accident, 599
Accidental event, 9
Actuarial approach, 3
Adequacy, 6
Age reduction model, 292
Age replacement, 381, 497
   availability criterion, 385
   replacement period, 381
   time between failures, 383
   Weibull distribution, 383
AIChE, 97, 105, 109, 166, 447, 563, 569
ANSI/ISA S84.01, 447–448, 452
Arrhenius model, 529
As bad as old, 287
As good as new, 27

Associated variables, 223
Availability, 6, 8, 269, 274, 314, 367, 599
   average, 6, 368, 370
   limiting, 271, 369
   mission, 368
   on-stream, 373
   operational, 372
   production, 373
   steady-state, 369

Barlow-Proschan's test, 516
Barrier, 108, 426
Basic event, 599, 98
Bathtub curve, 21
Bayes' theorem, 540
Bayes, Thomas, 540
Bayesian belief network, 74, 107
Bayesian estimator, 544–546
Bayesian inference, 539
   subjective, 557
Bayesian life test sampling plan, 553
Bernoulli trials, 25
β-factor model, 217, 445
Beta function, 574
Binomial coefficient, 25
Binomial failure rate model, 222
Birnbaum's measure of structural importance, 134, 187
Blackwell's theorem, 253

SUBJECT INDEX

Block replacement, 385
    gamma distribution, 388
    limited number of spares, 387
    minimal repair, 387
    Weibull distribution, 386, 388
Boundary conditions, 13
Bridge structure, 130, 136
    reliability, 169
Browns Ferry, 210
BS 5760, 88
Burn-in period, 21

CARA Fault Tree, 164, 171
Cascading failures, 209
Causal analysis, 9
Cause-defense matrix, 214
Cause and effect diagram, 74, 106
Coherent structure, 126, 599
Common cause candidates, 213
Common cause failures, 208, 442, 600
    defensive tactics, 212
    explicitly modeled, 209
    miscalibration, 442
    residual, 209
    taxonomy, 210
Common mode failures, 210
Component importance, 183
Concurrent engineering, 10
Condition-based replacement, 390
Conjugate distribution, 548–549, 551
Conjugate distributions, 547
Consequence analysis, 9
Convolution, 269, 249
Corrosion, 57
Counting process, 231, 233
    forward recurrence time, 238
    independent increments, 236
    nonstationary, 236
    rate, 236
    regular, 237
    ROCOF, 237
    stationary, 236
    stationary increments, 236
Covariates, 466
Cox model, 536
Credibility interval, 546, 549–550
Critical path, 187
Critical path vector, 133
Cumulant-generating function, 52, 72
Cut parallel structure, 132, 169
Cut set, 103, 129
    minimal, 103, 129
        bridge structure, 130
    order, 103

Data set
    censored, 467, 482
        type I, 467, 509
        type II, 468, 508, 511
        type III, 468
        type IV, 468
    complete, 465, 467, 482
    exponentially distributed, 504
    incomplete, 474
    inverse Gaussian, 511
    Weibull, 510
DEF-STD 02-45, 402
Delay-time model, 400
Deliverability, 372
Dependability, 7, 362, 600
Dependence
    cascading failures, 209
    negative, 207, 210
    positive, 207, 215
Design for X, 11
Design review, 600
DFR, 59
DFRA, 62
Diagnostic coverage, 422, 441
Diagnostic self-testing, 422, 440, 443
Distribution function, 17
    empirical, 470
Distribution
    beta, 550
    binomial, 25, 222, 244, 550
        confidence interval, 501
        estimate, 501
    Birnbaum-Saunders, 16, 48, 51
    chi-square, 35, 504, 549
    Erlangian, 35, 61, 174, 393
    exponential, 15, 26, 32, 60
        as good as new, 27
        mixture, 29, 36
        phase-type, 29
    extreme value, 54
        Gumbel - largest extreme, 57
        Gumbel - smallest extreme, 56
        Weibull - smallest extreme, 57
    gamma, 15, 33, 61, 246, 250, 254, 263, 388, 542, 548
    geometric, 26, 383
    Gompertz, 285
    inverse Gaussian, 16, 50, 511, 533
    inverted gamma, 553
    lognormal, 16, 43, 62, 366
    normal, 16, 41, 249, 366
        standard, 41
        truncated, 42
    Poisson, 31, 548
    Rayleigh, 38
    uniform, 60, 245, 377, 552
    Weibull, 16, 37, 61, 154, 250, 510

characteristic lifetime, 38
   three-parameter, 40
   weakest link, 39
Diversity, 213
DNV-RP-A203, 11
Downtime, 364, 600
Downtime distribution, 364
   exponential, 365
      repair rate, 365
   lognormal, 366
   normal, 366
Downtime
   forced outage, 364
   mean system downtime, 366
   planned, 364
   scheduled, 365
   unplanned, 364
   unscheduled, 365

Elementary renewal theorem, 252
EN 50126, 447
EN 50128, 447
EN 50129, 447
Environmental protection, 10
Equipment under control, 419, 450, 600
ERAC, 171
Error, 83
   human, 96, 105
Error factor, 45
ESReDA, 569
Event tree, 9, 74, 109, 117
   barrier, 110
   construction, 111
   initiating event, 109
   quantitative assessment, 112
   safety function, 110
External threats, 76
Eyring model, 529

Fail-safe, 425
Fail safe, 600
Failure, 83, 600
Failure symptom, 601
Failure
   aging, 409, 424
   command fault, 86, 99
   dangerous, 423
      detected, 423
      undetected, 423
   design, 424
   evident, 79, 93, 426
   gradual, 601
   hidden, 79, 93, 601
   incipient, 83
   interaction, 424
   intermittent, 602

   nondangerous, 423
   primary, 85, 99
   secondary, 85, 99, 424
   stress, 424
   sudden, 604
   systematic, 424, 604
   wear-out, 604
Failure cause, 87, 600
   ageing failure, 87
   design failure, 87
   manufacturing failure, 87
   mishandling failure, 87
   misuse failure, 87
   weakness failure, 87
Failure effect, 88, 601
Failure mechanism, 87, 93, 409
Failure mode, 84, 93, 601
Failure mode categories, 84
   catastrophic, 85
   critical, 567
   degraded, 85, 567
   extended, 85
      complete, 85
      partial, 85
   gradual, 85
   incipient, 567
   intermittent, 85
   sudden, 85
Failure rate (function), 15, 18
   cumulative, 480
Failure rate, 8
Failure rate classification, 93
Failure rate reduction model, 289
FAST, 80
Fatal accident rate, 112, 450
Fatigue, 601
Fatigue failure, 46
Fatigue
   failure, 46
   life, 601
Fault, 601
Fault tolerance, 601
Fault tree, 9, 73, 96, 116, 160
   and-gate, 162
   basic event, 97
   construction, 99
   intermediate event, 100
   minimal cut set, 103
   minimal path set, 105
   or-gate, 163
   relation to reliability block diagram, 121
   top event, 96
   top event frequency, 165
   top event probability, 161
      approximation formulas, 163
      upper bound approximation, 164

undeveloped event, 100
Fishbone diagram, 106
Flow network, 374
FMECA, 73, 88, 210, 407
   bottom-up, 91
   detailed, 91
   functional, 90
   interface, 90
   procedure, 92
   process, 95
   product, 95
   top-down, 90
   worksheet, 91
Force of mortality (FOM), 19, 480
Forced outage rate, 369
FRACAS, 562
Function
   auxiliary, 78
   essential, 78
   information, 78
   interface, 78
   off-line, 79
   on-line, 79
   protective, 78
   safety, 78
   superfluous, 79
Function testing, 422
Function tree, 79
Functional analysis, 77
Functional block, 73
Functional failure, 405
Functional failure analysis (FFA), 405
Functional failure analysis, 404
Functional requirement, 78
Functional significant item, 406
Fundamental renewal theorem, 252

Gamma function, 573
Gamma process, 391–392
GIDEP, 562
Greenwood's formula, 479

Happy system, 233, 235
Hardware reliability, 2
Hazard analysis, 450
Hazard plot, 485
Hazard rate, 19
HAZOP, 109, 419, 451
HPP, 31, 240
   as Markov process, 310
   asymptotic properties, 242
   compound, 245
   conditional distribution of failure time, 244
   confidence interval of intensity, 243, 502
   estimator of intensity, 32, 243, 502
   intensity, 31

interoccurrence times, 32
Human reliability, 3

ICDE, 214, 564
IDEF, 82
IEC 50(191), 361
IEC 60812, 88
IEC 61508, 423, 446, 450
   safety life cycle, 447
IEC 61511, 447, 452
IEC 61513, 447
IEC 61882, 419
IEC 62061, 447
IEC 50(191), 83–84, 87
IEC 60300, 8, 10, 402
IEC 60812, 76
IEC 61025, 97
IEEE Std. 352, 88, 92
IEEE Std. 500, 83
IFR, 59
IFRA, 62, 247
Imperfect repair, 387
Imperfect repair process, 287
Imperfect repair
   Brown and Proschan's model, 288
Importance, 183
   Birnbaum's measure, 185, 192, 197, 199, 201
   credible improvement potential, 189
   criticality importance, 192, 198, 200, 202
   Fussell-Vesely's measure, 193, 198, 200, 203
   generalized importance measure, 189
   improvement potential, 189, 198, 200–201
   risk achievement worth, 190, 200, 202
   risk reduction worth, 191, 198, 200, 202
Improvement
   risk achievement worth, 198
Inclusion-exclusion principle, 168, 171
Independent components, 207
Infant mortality, 21, 601
Interarrival time, 232
Interoccurrence time, 232
Irrelevant component, 125
Ishikawa diagram, 106
ISO 10418, 451
ISO 14224, 569
ISO 9000, 10

K-out-of-n structure, 124
Kaplan-Meier estimator, 475, 482
   asymptotic variance, 479
   justification, 477
   nonparametric MLE, 479
   properties, 479
Kaplan-Meier plot, 475
Key renewal theorem, 253
KITT, 172

SUBJECT INDEX 633

Kronecker product, 581
Kronecker sum, 581

Laplace test, 286
Laplace transform, 22, 249, 254, 330, 352, 577
Lattice distribution, 238, 253
Layer of protection analysis (LOPA), 451
Least replaceable item, 77
Life cycle cost (LCC), 10
Life cycle cost, 602
Life cycle profit, 10
Likelihood function, 504, 507–508, 510, 588
Logistic support, 10

Machinery safety directive, 11
Maintainability, 7, 362, 602
Maintainable item, 77, 404, 566
Maintenance, 361, 602
Maintenance cost significant item, 406
Maintenance significant item, 406
Maintenance
  support performance, 602
  action, 409
  age-based, 363
  breakdown, 364
  clock-based, 363
  condition-based, 363, 389
  control limit policy, 390
  corrective, 364, 600
  failure-finding, 364
  interval, 409
  management, 362
  opportunity, 364
  optimization, 400
  predictive, 363, 389
  preventive, 212, 363, 602
  RCM, 603
  repair, 364
  support, 362
Markov chain
  continuous time, 301
  discrete-time, 454–455
  discrete time, 301, 354
  embedded, 305
Markov method, 118, 374
Markov process, 301, 303, 453
  absorbing state, 305, 328
  asymptitic solution, 315
  Chapman-Kolmogorov equations, 310
  common cause failures, 334
  complex systems, 346
  departure rates, 307
  frequency of system failures, 321
  in fault trees, 349
  infinitesimal generator, 307
  instantaneous state, 305

interoccurrence times, 304
irreducible, 315
Kolmogorov equations
  backward, 311
  foreward, 312
Kronecker product, 348
Kronecker sum, 347
mean duration of visit, 320
mean time between system failures, 321
mean time to system failure, 331
skeleton, 305
standby systems, 339
state equations, 313
state transition diagram, 307
stationary transition probabilities, 303
steady-state probabilities, 315, 317
survivor function, 331
system availability, 321
time-dependent solution, 351
time-homogeneous, 303
transition rate matrix, 307
transition rates, 306
visit frequency, 320, 322
Markov property, 303
Markov renewal process, 354
Marlov process
  load-sharing systems, 337
MARS database, 563
Marshall-Olkin's model, 222
Martingale, 238
MDT, 370
Median life, 22
MFDT, 428
MIL-HDBK 217F, 446, 564
MIL-STD 1629A, 76, 88
MIL-STD 2155, 562
MIL-STD 2173 (AS), 402
MIL-STD 882D, 7, 74, 94
Military handbook test, 286
Miner's rule, 47
Minimal repair, 278, 387
  as bad as old, 288
Miscalibration, 446
MLE, 466, 588
MOCUS, 103
Mode, 23
Modular decomposition, 138
Module
  coherent, 138
Modules, 137
Moment generating function, 52
Monte Carlo simulation, 374, 376
  next event simulation, 376, 378
MRL, 15, 24
MTBF, 367
MTBR, 381, 391

MTTF, 8, 15, 22
MUT, 367

NASA, 96, 402
NBU, 62, 260
NBUE, 63
Nelson-Aalen plot, 233, 282, 466
Nelson estimator, 482–483
  justification, 483
Nelson plot, 485
NERC, 6
NHPP, 277
  confidence interval for $W(t)$, 283
  Cox-Lewis model, 285
  linear model, 285
  log-linear model, 285
  mean value function, 279
  minimal repair, 278
  power law model, 284
  relation to HPP, 281
  ROCOF, 278
  time between failures, 280
  time to first failure, 280
  Weibull process, 284
Nonparametric, 469
  estimate, 469
Nonrepairable item, 15
Normal approximation, 501, 503
NORSOK, 372, 451
Nuclear Energy Agency, 564
NUREG, 2, 45, 96, 109
NWU, 62, 260
NWUE, 63

OEE, 415
OLF guideline, 447, 452
Operational mode, 79, 93
Order statistic, 467
OREDA, 28, 36, 83, 222, 297, 429, 446, 563, 566, 572

Part stress analysis prediction technique, 565
Partial stroke testing, 423, 426
Parts count reliability prediction, 565
Path series structure, 131
Path set, 105, 129
  minimal, 105, 129
    bridge structure, 130
  order, 105
PDS, 453
Percentile, 602
Petri net, 374
PF interval, 381, 394, 410
PFD, 427, 448
  2-out-of-3 system, 430
  approximation formulas, 432

Markov approach, 453
  nonnegligible repair time, 436
  parallel system, 429
  series system, 431
  single item, 428
Physical approach, 3
Pivotal decomposition, 136, 152, 167, 185
Posterior distribution, 541, 543
Power rule model, 59, 529
Predictive density, 557
Preliminary hazard analysis, 109
Preventive maintenance policy, 380
Prior distribution, 541–542
  based on empirical data, 556
  interpretation, 555
  noninformative, 551–552
Prism, 565
Probability density function, 17
Probability of failure on demand (PFD), 427, 448
Probability plotting paper, 473
  Weibull, 474
Process demand, 419
Process safety system, 114
Product liability directive, 11
Product limit estimator, 479
Product safety directive, 11
Production regularity, 7, 372
Proportional hazards (PH) model, 535
Protection layer, 114, 419

QRA, 8, 109, 451
Quality, 6

RAC, 563, 565
RAM, 7
RAMS, 8
Random number generator, 377
RCM, 10, 76, 95, 401, 603
  applicability criterion, 412
  cost-effectiveness criterion, 412
  decision logic, 410–411
  implementation, 413
  main steps, 403
  standards, 402
Reactor safety study, 45
Reactor Safety Study, 96
Reactor safety study, 216
Redundancy, 90, 95, 173, 212, 603
  active, 173, 599
  component level, 128
  passive, 173
  switching
    imperfect, 175
    perfect, 173
  system level, 128
Relevant component, 125

SUBJECT INDEX 635

Reliability, 5
Reliability Analysis Center (RAC), 563
Reliability block diagram, 74, 118
   2-out-of-3 structure, 124
   parallel structure, 119
   relation to fault tree, 121
   series structure, 119
Reliability function, 15, 17
Reliability verification, 11
Reliability
   hardware, 2
   human, 3
   software, 2
   structural, 4
Renewal density, 253, 255
Renewal function, 251, 255
   bounds, 262
Renewal process, 247, 386
   age, 258
   alternating, 268
   delayed, 266
   equilibrium, 267
   modified, 266
   remaining lifetime, 258
   superimposed, 263
   synchronous sampling, 267
Renewal reward process, 265
Repair, 603
Repair rate, 365, 370
Repair time distribution, 45
Risk, 449
Risk acceptance criteria, 450
Risk analysis, 8
Risk matrix, 94
ROCOF, 19, 32, 237
   conditional, 289

S-N diagram, 47
Sad system, 233
SADT, 81
SAE-ARP 5580, 88
SAE JA1012, 402
Safety, 7, 603
Safety function, 108, 110
Safety instrumented function, 421
Safety instrumented system, 420
Safety integrity, 448, 603
   safety integrity level, 603
Safety unavailability, 437
Sample coefficient of variation, 470
Sample mean, 470
Sample median, 470
Sample standard deviation, 470
Sample variance, 470
Scheduled function test, 411
Scheduled on-condition task, 410

Scheduled overhaul, 410
Scheduled replacement, 411
Security, 6–7, 603
Semi-Markov process, 353
   skeleton, 354
Severity, 94, 603
Seveso II directive, 563
SIF, 421, 448
SIL, 448
Single failure point, 603
SINTEF, 453, 563
SIS, 420
   actuating items, 420
   logic solver, 420
   sensors, 420
Software reliability, 2
Spurious trip, 438, 452
Square-root method, 215
Staggered testing, 434
Standby
   cold, 173, 175
   partly loaded, 173, 177
State variable, 16
State
   variable, 123, 147
   vector, 123, 147
Step stress test, 604
Stochastic process, 231
Stressor-dependent model, 58
Stressor, 525, 572
Structural importance, 133
   Birnbaum's measure, 134
Structure function, 123
   fault tree, 161
   k-out-of-n structure, 124
   parallel structure, 124
   series structure, 123
Survival probability, 8
Survivor function, 15, 18
   conditional, 24, 27
   empirical, 471
System, 604
System breakdown structure, 91
System reliability, 148
   exact calculation, 166
   k-out-of-n structure, 151, 160
   parallel structure, 150
   series structure, 149
Systematic failure, 604

Tag number, 77
Telcordia, 565
Test
   frequency, 604
Text
   interval, 604

THERP, 109
Tie (in data set), 467, 471
Time to failure, 16
Total probability, 540
Total time on test (TTT), 486
　scaled, 487
TPM, 414
TQM, 10
Trend renewal process, 294
TTT plot, 487
　censored data, 499
TTT transform, 490

　exponential distribution, 492
　scaled, 491
　Weibull, 492

Unavailability, 368
Unbiased estimator, 471
Upper bound approximation, 164
Useful life period, 21

Wöhler curve, 47
Wald's equation, 245, 262, 265
Wear-out period, 21
Wiener process, 51, 533

WILEY SERIES IN PROBABILITY AND STATISTICS

ESTABLISHED BY WALTER A. SHEWHART AND SAMUEL S. WILKS

Editors: *David J. Balding, Noel A. C. Cressie, Nicholas I. Fisher, Iain M. Johnstone, J. B. Kadane, Geert Molenberghs. Louise M. Ryan, David W. Scott, Adrian F. M. Smith, Jozef L. Teugels*
Editors Emeriti: *Vic Barnett, J. Stuart Hunter, David G. Kendall*

The *Wiley Series in Probability and Statistics* is well established and authoritative. It covers many topics of current research interest in both pure and applied statistics and probability theory. Written by leading statisticians and institutions, the titles span both state-of-the-art developments in the field and classical methods.

Reflecting the wide range of current research in statistics, the series encompasses applied, methodological and theoretical statistics, ranging from applications and new techniques made possible by advances in computerized practice to rigorous treatment of theoretical approaches.

This series provides essential and invaluable reading for all statisticians, whether in academia, industry, government, or research.

ABRAHAM and LEDOLTER · Statistical Methods for Forecasting
AGRESTI · Analysis of Ordinal Categorical Data
AGRESTI · An Introduction to Categorical Data Analysis
AGRESTI · Categorical Data Analysis, *Second Edition*
ALTMAN, GILL, and McDONALD · Numerical Issues in Statistical Computing for the Social Scientist
AMARATUNGA and CABRERA · Exploration and Analysis of DNA Microarray and Protein Array Data
ANDĚL · Mathematics of Chance
ANDERSON · An Introduction to Multivariate Statistical Analysis, *Third Edition*
*ANDERSON · The Statistical Analysis of Time Series
ANDERSON, AUQUIER, HAUCK, OAKES, VANDAELE, and WEISBERG · Statistical Methods for Comparative Studies
ANDERSON and LOYNES · The Teaching of Practical Statistics
ARMITAGE and DAVID (editors) · Advances in Biometry
ARNOLD, BALAKRISHNAN, and NAGARAJA · Records
*ARTHANARI and DODGE · Mathematical Programming in Statistics
*BAILEY · The Elements of Stochastic Processes with Applications to the Natural Sciences
BALAKRISHNAN and KOUTRAS · Runs and Scans with Applications
BARNETT · Comparative Statistical Inference, *Third Edition*
BARNETT and LEWIS · Outliers in Statistical Data, *Third Edition*
BARTOSZYNSKI and NIEWIADOMSKA-BUGAJ · Probability and Statistical Inference
BASILEVSKY · Statistical Factor Analysis and Related Methods: Theory and Applications
BASU and RIGDON · Statistical Methods for the Reliability of Repairable Systems
BATES and WATTS · Nonlinear Regression Analysis and Its Applications
BECHHOFER, SANTNER, and GOLDSMAN · Design and Analysis of Experiments for Statistical Selection, Screening, and Multiple Comparisons
BELSLEY · Conditioning Diagnostics: Collinearity and Weak Data in Regression

*Now available in a lower priced paperback edition in the Wiley Classics Library.

BELSLEY, KUH, and WELSCH · Regression Diagnostics: Identifying Influential Data and Sources of Collinearity
BENDAT and PIERSOL · Random Data: Analysis and Measurement Procedures, *Third Edition*
BERRY, CHALONER, and GEWEKE · Bayesian Analysis in Statistics and Econometrics: Essays in Honor of Arnold Zellner
BERNARDO and SMITH · Bayesian Theory
BHAT and MILLER · Elements of Applied Stochastic Processes, *Third Edition*
BHATTACHARYA and JOHNSON · Statistical Concepts and Methods
BHATTACHARYA and WAYMIRE · Stochastic Processes with Applications
BILLINGSLEY · Convergence of Probability Measures, *Second Edition*
BILLINGSLEY · Probability and Measure, *Third Edition*
BIRKES and DODGE · Alternative Methods of Regression
BLISCHKE AND MURTHY (editors) · Case Studies in Reliability and Maintenance
BLISCHKE AND MURTHY · Reliability: Modeling, Prediction, and Optimization
BLOOMFIELD · Fourier Analysis of Time Series: An Introduction, *Second Edition*
BOLLEN · Structural Equations with Latent Variables
BOROVKOV · Ergodicity and Stability of Stochastic Processes
BOULEAU · Numerical Methods for Stochastic Processes
BOX · Bayesian Inference in Statistical Analysis
BOX · R. A. Fisher, the Life of a Scientist
BOX and DRAPER · Empirical Model-Building and Response Surfaces
*BOX and DRAPER · Evolutionary Operation: A Statistical Method for Process Improvement
BOX, HUNTER, and HUNTER · Statistics for Experimenters: An Introduction to Design, Data Analysis, and Model Building
BOX and LUCEÑO · Statistical Control by Monitoring and Feedback Adjustment
BRANDIMARTE · Numerical Methods in Finance: A MATLAB-Based Introduction
BROWN and HOLLANDER · Statistics: A Biomedical Introduction
BRUNNER, DOMHOF, and LANGER · Nonparametric Analysis of Longitudinal Data in Factorial Experiments
BUCKLEW · Large Deviation Techniques in Decision, Simulation, and Estimation
CAIROLI and DALANG · Sequential Stochastic Optimization
CHAN · Time Series: Applications to Finance
CHATTERJEE and HADI · Sensitivity Analysis in Linear Regression
CHATTERJEE and PRICE · Regression Analysis by Example, *Third Edition*
CHERNICK · Bootstrap Methods: A Practitioner's Guide
CHERNICK and FRIIS · Introductory Biostatistics for the Health Sciences
CHILÈS and DELFINER · Geostatistics: Modeling Spatial Uncertainty
CHOW and LIU · Design and Analysis of Clinical Trials: Concepts and Methodologies, *Second Edition*
CLARKE and DISNEY · Probability and Random Processes: A First Course with Applications, *Second Edition*
*COCHRAN and COX · Experimental Designs, *Second Edition*
CONGDON · Applied Bayesian Modelling
CONGDON · Bayesian Statistical Modelling
CONOVER · Practical Nonparametric Statistics, *Second Edition*
COOK · Regression Graphics
COOK and WEISBERG · Applied Regression Including Computing and Graphics
COOK and WEISBERG · An Introduction to Regression Graphics
CORNELL · Experiments with Mixtures, Designs, Models, and the Analysis of Mixture Data, *Third Edition*
COVER and THOMAS · Elements of Information Theory

*Now available in a lower priced paperback edition in the Wiley Classics Library.

COX · A Handbook of Introductory Statistical Methods
*COX · Planning of Experiments
CRESSIE · Statistics for Spatial Data, *Revised Edition*
CSÖRGŐ and HORVÁTH · Limit Theorems in Change Point Analysis
DANIEL · Applications of Statistics to Industrial Experimentation
DANIEL · Biostatistics: A Foundation for Analysis in the Health Sciences, *Sixth Edition*
*DANIEL · Fitting Equations to Data: Computer Analysis of Multifactor Data, *Second Edition*
DASU and JOHNSON · Exploratory Data Mining and Data Cleaning
DAVID and NAGARAJA · Order Statistics, *Third Edition*
*DEGROOT, FIENBERG, and KADANE · Statistics and the Law
DEL CASTILLO · Statistical Process Adjustment for Quality Control
DENISON, HOLMES, MALLICK and SMITH · Bayesian Methods for Nonlinear Classification and Regression
DETTE and STUDDEN · The Theory of Canonical Moments with Applications in Statistics, Probability, and Analysis
DEY and MUKERJEE · Fractional Factorial Plans
DILLON and GOLDSTEIN · Multivariate Analysis: Methods and Applications
DODGE · Alternative Methods of Regression
*DODGE and ROMIG · Sampling Inspection Tables, *Second Edition*
*DOOB · Stochastic Processes
DOWDY, WEARDEN, and CHILKO · Statistics for Research, *Third Edition*
DRAPER and SMITH · Applied Regression Analysis, *Third Edition*
DRYDEN and MARDIA · Statistical Shape Analysis
DUDEWICZ and MISHRA · Modern Mathematical Statistics
DUNN and CLARK · Applied Statistics: Analysis of Variance and Regression, *Second Edition*
DUNN and CLARK · Basic Statistics: A Primer for the Biomedical Sciences, *Third Edition*
DUPUIS and ELLIS · A Weak Convergence Approach to the Theory of Large Deviations
*ELANDT-JOHNSON and JOHNSON · Survival Models and Data Analysis
ENDERS · Applied Econometric Time Series
ETHIER and KURTZ · Markov Processes: Characterization and Convergence
EVANS, HASTINGS, and PEACOCK · Statistical Distributions, *Third Edition*
FELLER · An Introduction to Probability Theory and Its Applications, Volume I, *Third Edition,* Revised; Volume II, *Second Edition*
FISHER and VAN BELLE · Biostatistics: A Methodology for the Health Sciences
*FLEISS · The Design and Analysis of Clinical Experiments
FLEISS · Statistical Methods for Rates and Proportions, *Third Edition*
FLEMING and HARRINGTON · Counting Processes and Survival Analysis
FULLER · Introduction to Statistical Time Series, *Second Edition*
FULLER · Measurement Error Models
GALLANT · Nonlinear Statistical Models
GIESBRECHT and GUMPERTZ · Planning, Construction, and Statistical Analysis of Comparative Experiments
GIFI · Nonlinear Multivariate Analysis
GHOSH, MUKHOPADHYAY, and SEN · Sequential Estimation
GIFI · Nonlinear Multivariate Analysis
GLASSERMAN and YAO · Monotone Structure in Discrete-Event Systems
GNANADESIKAN · Methods for Statistical Data Analysis of Multivariate Observations, *Second Edition*
GOLDSTEIN and LEWIS · Assessment: Problems, Development, and Statistical Issues
GREENWOOD and NIKULIN · A Guide to Chi-Squared Testing

*Now available in a lower priced paperback edition in the Wiley Classics Library.

GROSS and HARRIS · Fundamentals of Queueing Theory, *Third Edition*
*HAHN and SHAPIRO · Statistical Models in Engineering
HAHN and MEEKER · Statistical Intervals: A Guide for Practitioners
HALD · A History of Probability and Statistics and their Applications Before 1750
HALD · A History of Mathematical Statistics from 1750 to 1930
HAMPEL · Robust Statistics: The Approach Based on Influence Functions
HANNAN and DEISTLER · The Statistical Theory of Linear Systems
HEIBERGER · Computation for the Analysis of Designed Experiments
HEDAYAT and SINHA · Design and Inference in Finite Population Sampling
HELLER · MACSYMA for Statisticians
HINKELMAN and KEMPTHORNE: · Design and Analysis of Experiments, Volume 1: Introduction to Experimental Design
HOAGLIN, MOSTELLER, and TUKEY · Exploratory Approach to Analysis of Variance
HOAGLIN, MOSTELLER, and TUKEY · Exploring Data Tables, Trends and Shapes
*HOAGLIN, MOSTELLER, and TUKEY · Understanding Robust and Exploratory Data Analysis
HOCHBERG and TAMHANE · Multiple Comparison Procedures
HOCKING · Methods and Applications of Linear Models: Regression and the Analysis of Variance, *Second Edition*
HOEL · Introduction to Mathematical Statistics, *Fifth Edition*
HOGG and KLUGMAN · Loss Distributions
HOLLANDER and WOLFE · Nonparametric Statistical Methods, *Second Edition*
HOSMER and LEMESHOW · Applied Logistic Regression, *Second Edition*
HOSMER and LEMESHOW · Applied Survival Analysis: Regression Modeling of Time to Event Data
HUBER · Robust Statistics
HUBERTY · Applied Discriminant Analysis
HUNT and KENNEDY · Financial Derivatives in Theory and Practice
HUSKOVA, BERAN, and DUPAC · Collected Works of Jaroslav Hajek— with Commentary
IMAN and CONOVER · A Modern Approach to Statistics
JACKSON · A User's Guide to Principle Components
JOHN · Statistical Methods in Engineering and Quality Assurance
JOHNSON · Multivariate Statistical Simulation
JOHNSON and BALAKRISHNAN · Advances in the Theory and Practice of Statistics: A Volume in Honor of Samuel Kotz
JUDGE, GRIFFITHS, HILL, LÜTKEPOHL, and LEE · The Theory and Practice of Econometrics, *Second Edition*
JOHNSON and KOTZ · Distributions in Statistics
JOHNSON and KOTZ (editors) · Leading Personalities in Statistical Sciences: From the Seventeenth Century to the Present
JOHNSON, KOTZ, and BALAKRISHNAN · Continuous Univariate Distributions, Volume 1, *Second Edition*
JOHNSON, KOTZ, and BALAKRISHNAN · Continuous Univariate Distributions, Volume 2, *Second Edition*
JOHNSON, KOTZ, and BALAKRISHNAN · Discrete Multivariate Distributions
JOHNSON, KOTZ, and KEMP · Univariate Discrete Distributions, *Second Edition*
JUREČKOVÁ and SEN · Robust Statistical Procedures: Asymptotics and Interrelations
JUREK and MASON · Operator-Limit Distributions in Probability Theory
KADANE · Bayesian Methods and Ethics in a Clinical Trial Design
KADANE AND SCHUM · A Probabilistic Analysis of the Sacco and Vanzetti Evidence
KALBFLEISCH and PRENTICE · The Statistical Analysis of Failure Time Data, *Second Edition*

*Now available in a lower priced paperback edition in the Wiley Classics Library.

KASS and VOS · Geometrical Foundations of Asymptotic Inference
KAUFMAN and ROUSSEEUW · Finding Groups in Data: An Introduction to Cluster Analysis
KEDEM and FOKIANOS · Regression Models for Time Series Analysis
KENDALL, BARDEN, CARNE, and LE · Shape and Shape Theory
KHURI · Advanced Calculus with Applications in Statistics, *Second Edition*
KHURI, MATHEW, and SINHA · Statistical Tests for Mixed Linear Models
KLEIBER and KOTZ · Statistical Size Distributions in Economics and Actuarial Sciences
KLUGMAN, PANJER, and WILLMOT · Loss Models: From Data to Decisions
KLUGMAN, PANJER, and WILLMOT · Solutions Manual to Accompany Loss Models: From Data to Decisions
KOTZ, BALAKRISHNAN, and JOHNSON · Continuous Multivariate Distributions, Volume 1, *Second Edition*
KOTZ and JOHNSON (editors) · Encyclopedia of Statistical Sciences: Volumes 1 to 9 with Index
KOTZ and JOHNSON (editors) · Encyclopedia of Statistical Sciences: Supplement Volume
KOTZ, READ, and BANKS (editors) · Encyclopedia of Statistical Sciences: Update Volume 1
KOTZ, READ, and BANKS (editors) · Encyclopedia of Statistical Sciences: Update Volume 2
KOVALENKO, KUZNETZOV, and PEGG · Mathematical Theory of Reliability of Time-Dependent Systems with Practical Applications
LACHIN · Biostatistical Methods: The Assessment of Relative Risks
LAD · Operational Subjective Statistical Methods: A Mathematical, Philosophical, and Historical Introduction
LAMPERTI · Probability: A Survey of the Mathematical Theory, *Second Edition*
LANGE, RYAN, BILLARD, BRILLINGER, CONQUEST, and GREENHOUSE · Case Studies in Biometry
LARSON · Introduction to Probability Theory and Statistical Inference, *Third Edition*
LAWLESS · Statistical Models and Methods for Lifetime Data, *Second Edition*
LAWSON · Statistical Methods in Spatial Epidemiology
LE · Applied Categorical Data Analysis
LE · Applied Survival Analysis
LEE and WANG · Statistical Methods for Survival Data Analysis, *Third Edition*
LePAGE and BILLARD · Exploring the Limits of Bootstrap
LEYLAND and GOLDSTEIN (editors) · Multilevel Modelling of Health Statistics
LIAO · Statistical Group Comparison
LINDVALL · Lectures on the Coupling Method
LINHART and ZUCCHINI · Model Selection
LITTLE and RUBIN · Statistical Analysis with Missing Data, *Second Edition*
LLOYD · The Statistical Analysis of Categorical Data
MAGNUS and NEUDECKER · Matrix Differential Calculus with Applications in Statistics and Econometrics, *Revised Edition*
MALLER and ZHOU · Survival Analysis with Long Term Survivors
MALLOWS · Design, Data, and Analysis by Some Friends of Cuthbert Daniel
MANN, SCHAFER, and SINGPURWALLA · Methods for Statistical Analysis of Reliability and Life Data
MANTON, WOODBURY, and TOLLEY · Statistical Applications Using Fuzzy Sets
MARDIA and JUPP · Directional Statistics
MASON, GUNST, and HESS · Statistical Design and Analysis of Experiments with Applications to Engineering and Science, *Second Edition*
McCULLOCH and SEARLE · Generalized, Linear, and Mixed Models
McFADDEN · Management of Data in Clinical Trials

*Now available in a lower priced paperback edition in the Wiley Classics Library.

McLACHLAN · Discriminant Analysis and Statistical Pattern Recognition
McLACHLAN and KRISHNAN · The EM Algorithm and Extensions
McLACHLAN and PEEL · Finite Mixture Models
McNEIL · Epidemiological Research Methods
MEEKER and ESCOBAR · Statistical Methods for Reliability Data
MEERSCHAERT and SCHEFFLER · Limit Distributions for Sums of Independent Random Vectors: Heavy Tails in Theory and Practice
*MILLER · Survival Analysis, *Second Edition*
MONTGOMERY, PECK, and VINING · Introduction to Linear Regression Analysis, *Third Edition*
MORGENTHALER and TUKEY · Configural Polysampling: A Route to Practical Robustness
MUIRHEAD · Aspects of Multivariate Statistical Theory
MULLER and STOYAN · Comparison Methods for Stochastic Models and Risks
MURRAY · X-STAT 2.0 Statistical Experimentation, Design Data Analysis, and Nonlinear Optimization
MURTHY, XIE, and JIANG · Weibull Models
MYERS and MONTGOMERY · Response Surface Methodology: Process and Product Optimization Using Designed Experiments, *Second Edition*
MYERS, MONTGOMERY, and VINING · Generalized Linear Models. With Applications in Engineering and the Sciences
NELSON · Accelerated Testing, Statistical Models, Test Plans, and Data Analyses
NELSON · Applied Life Data Analysis
NEWMAN · Biostatistical Methods in Epidemiology
OCHI · Applied Probability and Stochastic Processes in Engineering and Physical Sciences
OKABE, BOOTS, SUGIHARA, and CHIU · Spatial Tessellations: Concepts and Applications of Voronoi Diagrams, *Second Edition*
OLIVER and SMITH · Influence Diagrams, Belief Nets and Decision Analysis
PALTA · Quantitative Methods in Population Health: Extensions of Ordinary Regressions
PANKRATZ · Forecasting with Dynamic Regression Models
PANKRATZ · Forecasting with Univariate Box-Jenkins Models: Concepts and Cases
*PARZEN · Modern Probability Theory and Its Applications
PEÑA, TIAO, and TSAY · A Course in Time Series Analysis
PIANTADOSI · Clinical Trials: A Methodologic Perspective
PORT · Theoretical Probability for Applications
POURAHMADI · Foundations of Time Series Analysis and Prediction Theory
PRESS · Bayesian Statistics: Principles, Models, and Applications
PRESS · Subjective and Objective Bayesian Statistics, *Second Edition*
PRESS and TANUR · The Subjectivity of Scientists and the Bayesian Approach
PUKELSHEIM · Optimal Experimental Design
PURI, VILAPLANA, and WERTZ · New Perspectives in Theoretical and Applied Statistics
PUTERMAN · Markov Decision Processes: Discrete Stochastic Dynamic Programming
*RAO · Linear Statistical Inference and Its Applications, *Second Edition*
RAUSAND and HØYLAND · System Reliability Theory: Models, Statistical Methods, and Applications, *Second Edition*
RENCHER · Linear Models in Statistics
RENCHER · Methods of Multivariate Analysis, *Second Edition*
RENCHER · Multivariate Statistical Inference with Applications
RIPLEY · Spatial Statistics
RIPLEY · Stochastic Simulation
ROBINSON · Practical Strategies for Experimenting
ROHATGI and SALEH · An Introduction to Probability and Statistics, *Second Edition*

*Now available in a lower priced paperback edition in the Wiley Classics Library.

ROLSKI, SCHMIDLI, SCHMIDT, and TEUGELS · Stochastic Processes for Insurance and Finance
ROSENBERGER and LACHIN · Randomization in Clinical Trials: Theory and Practice
ROSS · Introduction to Probability and Statistics for Engineers and Scientists
ROUSSEEUW and LEROY · Robust Regression and Outlier Detection
RUBIN · Multiple Imputation for Nonresponse in Surveys
RUBINSTEIN · Simulation and the Monte Carlo Method
RUBINSTEIN and MELAMED · Modern Simulation and Modeling
RYAN · Modern Regression Methods
RYAN · Statistical Methods for Quality Improvement, *Second Edition*
SALTELLI, CHAN, and SCOTT (editors) · Sensitivity Analysis
*SCHEFFE · The Analysis of Variance
SCHIMEK · Smoothing and Regression: Approaches, Computation, and Application
SCHOTT · Matrix Analysis for Statistics
SCHOUTENS · Levy Processes in Finance: Pricing Financial Derivatives
SCHUSS · Theory and Applications of Stochastic Differential Equations
SCOTT · Multivariate Density Estimation: Theory, Practice, and Visualization
*SEARLE · Linear Models
SEARLE · Linear Models for Unbalanced Data
SEARLE · Matrix Algebra Useful for Statistics
SEARLE, CASELLA, and McCULLOCH · Variance Components
SEARLE and WILLETT · Matrix Algebra for Applied Economics
SEBER and LEE · Linear Regression Analysis, *Second Edition*
SEBER · Multivariate Observations
SEBER and WILD · Nonlinear Regression
SENNOTT · Stochastic Dynamic Programming and the Control of Queueing Systems
*SERFLING · Approximation Theorems of Mathematical Statistics
SHAFER and VOVK · Probability and Finance: It's Only a Game!
SMALL and McLEISH · Hilbert Space Methods in Probability and Statistical Inference
SRIVASTAVA · Methods of Multivariate Statistics
STAPLETON · Linear Statistical Models
STAUDTE and SHEATHER · Robust Estimation and Testing
STOYAN, KENDALL, and MECKE · Stochastic Geometry and Its Applications, *Second Edition*
STOYAN and STOYAN · Fractals, Random Shapes and Point Fields: Methods of Geometrical Statistics
STYAN · The Collected Papers of T. W. Anderson: 1943–1985
SUTTON, ABRAMS, JONES, SHELDON, and SONG · Methods for Meta-Analysis in Medical Research
TANAKA · Time Series Analysis: Nonstationary and Noninvertible Distribution Theory
THOMPSON · Empirical Model Building
THOMPSON · Sampling, *Second Edition*
THOMPSON · Simulation: A Modeler's Approach
THOMPSON and SEBER · Adaptive Sampling
THOMPSON, WILLIAMS, and FINDLAY · Models for Investors in Real World Markets
TIAO, BISGAARD, HILL, PEÑA, and STIGLER (editors) · Box on Quality and Discovery: with Design, Control, and Robustness
TIERNEY · LISP-STAT: An Object-Oriented Environment for Statistical Computing and Dynamic Graphics
TSAY · Analysis of Financial Time Series
UPTON and FINGLETON · Spatial Data Analysis by Example, Volume II: Categorical and Directional Data
VAN BELLE · Statistical Rules of Thumb
VESTRUP · The Theory of Measures and Integration
VIDAKOVIC · Statistical Modeling by Wavelets

*Now available in a lower priced paperback edition in the Wiley Classics Library.

WEISBERG · Applied Linear Regression, *Second Edition*
WELSH · Aspects of Statistical Inference
WESTFALL and YOUNG · Resampling-Based Multiple Testing: Examples and Methods for p-Value Adjustment
WHITTAKER · Graphical Models in Applied Multivariate Statistics
WINKER · Optimization Heuristics in Economics: Applications of Threshold Accepting
WONNACOTT and WONNACOTT · Econometrics, *Second Edition*
WOODING · Planning Pharmaceutical Clinical Trials: Basic Statistical Principles
WOOLSON and CLARKE · Statistical Methods for the Analysis of Biomedical Data, *Second Edition*
WU and HAMADA · Experiments: Planning, Analysis, and Parameter Design Optimization
YANG · The Construction Theory of Denumerable Markov Processes
*ZELLNER · An Introduction to Bayesian Inference in Econometrics
ZHOU, OBUCHOWSKI, and McCLISH · Statistical Methods in Diagnostic Medicine

*Now available in a lower priced paperback edition in the Wiley Classics Library.